Applications of
Random Vibrations

N.C. Nigam
S. Narayanan

Springer-Verlag

Narosa Publishing House

Prof. N.C. Nigam
Director
Indian Institute of Technology, Delhi
New Delhi 110 016, INDIA

Prof. S. Narayanan
Department of Applied Mechanics
Indian Institute of Technology, Madras
Madras 600 036, INDIA

D
620.3
NIG

Exclusive distribution in North America (including Mexico), Canada and Europe by
Springer-Verlag Berlin Heidelberg New York.

Exclusive distribution in South East Asia (excluding China and Japan) by
Addison-Wesley Singapore (Pte) Ltd.

For all other countries exclusive distribution by Narosa Publishing House, New Delhi

All export rights for this book vest exclusively with Narosa Publishing House. Unauthorised
export is a violation of Copyright Law and is subject to legal action.

Cover Illustration: Elasto-plastic response of a space frame, in the force space, to the horizontal
components of Taft, 1952 earthquake

ISBN 3-540-19861-X Springer-Verlag Berlin Heidelberg New York
ISBN 0-387-19861-X Springer-Verlag New York Berlin Heidelberg
ISBN 81-85198-58-6 Narosa Publishing House, New Delhi

Printed in India at Rajkamal Electric Press, Delhi 110 033.

Applications of
Random Vibrations

To our students

Preface

This book is a sequel to an earlier book: *Introduction to Random Vibrations* (The MIT Press, 1983) by Nigam. It covers applications of random vibrations to a wide range of problems in civil, mechanical and aerospace engineering. Over the past five decades a vast body of knowledge has accumulated in this area which is scattered in a large number of journals, conference proceedings, research reports and specialised books and monographs. Both authors have compiled and developed the material covered by the book in the course of classroom teaching, theses research, consultancy projects, and series of lectures to practising engineers with interests in aerospace systems, vehicles, earthquakes, wind and offshore engineering. While the physical systems covered by the book differ significantly, the book has pervasive underlying unity. The format adopted to treat each system follows the general sequence: identification of the physical characteristics, stochastic modelling and simulation of the system and its environment; random vibration analysis of system response; and probabilistic design, including optimization. Several examples are included to illustrate the applications of the random vibration theory to the analysis and design of the systems covered.

In the first chapter we illustrate that it is feasible and meaningful to attempt a unified treatment of wide range of diverse systems in one book. Second and third chapters cover respectively the stochastic modelling, simulation, and probabilistic design for random vibration, in a generic format, to provide the basis for a specific treatment in subsequent chapters. Chapters four through nine cover applications to specific systems and environments. These are independent and may be read without reference to each other. Chapter ten on statistical energy analysis stands alone and has been included to cover the special class of problems involving broad band excitations and systems sensitive to high frequency bands.

The book is designed to be used as a companion volume to the earlier book which provides a comprehensive and rigorous treatment of the theory of random vibrations. This has made it possible to cover, in some detail, issues relevant to specific application areas without having to include the underlying theory. The book is intended to be used as text in the classroom, and as a reference for research, and in design offices.

The book is dedicated to our students. Most of the application areas covered by the book formed a part of the thesis research by our students over the past two decades. In dedicating this book to them we acknowledge their contributions, and express our pleasure at the opportunity to work and learn with them.

Several colleagues and former students at IIT Delhi, IIT Madras and University of Roorkee have contributed by reading parts of the manuscript at various stages, and offering valuable comments. We would like to specifically mention: Prof. T.K. Datta and Dr. A.K. Jain, Department of Civil Engineering, IIT Delhi; Prof. K.N. Gupta, Department of Mechanical Engineering, IIT Delhi; Dr. C.V.R. Murty, Department of Civil Engineering, IIT Kanpur; Prof. S. Basu, Department of

Earthquake Engineering, University of Roorkee; Dr. C.S. Manohar, Department of Civil Engineering, IISc Bangalore; Dr. G.V. Raju, Department of Mechanical Engineering, SRKR Engineering College, Bhimavaram. Many of our former and present students at IIT Madras—G.V. Raju, S. Shanmugha Sundaram, A. Elayaperumal, V. Ravi Shankar Rao and S. Senthil—helped at various stages in setting the manuscript on the PC and preparing the index.

The part financial support provided by the Centres for Continuing Education, IIT Madras and IIT Delhi for preparation of the manuscript is gratefully acknowledged. Most of the preliminary typing was done by V.R. Ganesan and M. Kathirvelu at Madras. Some of the art work was initially done by S. Ramanujam and the final drawings were shared by K. Balasundaram, M. Karuppiah and R. Aggarwal.

Our wives, Shashi and Hema; and our children, Tanya and Amiya, and Vidya, provided encouragement during the work on the book through their enthusiastic support and understanding.

N.C. Nigam
S. Narayanan

Contents

1. Introduction

A large class of physical systems operate in random vibration environment. These include: aerospace systems excited by atmospheric and boundary layer turbulence and jet noise; aircraft and vehicles subjected to track induced vibrations; ground based structures excited by earthquakes and wind; and offshore structures excited by wind and hydrodynamic wave-induced loads. In each of these cases, the physical variables—wind velocity, noise pressure field, track profile, ground motion and sea waves—exhibit random fluctuations in both space and time. Further, there is a large measure of uncertainty associated with the occurrence of occasional events, such as, earthquakes, wind and sea storms. Another significant feature of the examples cited above is the strong short- and long-term, time dependent nature of the operating environments. This, coupled with the trend towards large, light and monolithic construction of structural/mechanical systems, makes the consideration of dynamic effects critical to the safety, economics and performance of these systems. A probabilistic approach based on random vibration analysis, therefore, provides a rational and consistent basis for the analysis and design of such systems.

Over the past three decades, applications of random vibration theory for the ·analysis and design of structural/mechanical systems has received considerable attention. In most countries, the design practices and codes now rely explicitly, or implicitly, on such analyses. The research in random vibration theory and applications has grown rapidly, as is evident from the frequent speciality conferences, specialised journals, and the texts: Crandall and Mark (1963); Robson (1963); Lin (1967); Newland (1975, 1985); Bolotin (1961, 1965; 1979); Elishakoff (1983); Vanmarcke (1983); Nigam (1983); Kree and Soize (1983); Yang (1986); Elishakoff and Lyon (1986) Dimentberg (1988); and Spanos and Robson (1990). Several books on vibration theory (Hurty and Rubinstein, 1964; Penzien and Clough, 1975; Thomson, 1975), and some books on special topics (Crandall, 1958, 1963; Fry'ba 1972; Simiu and Scanlan, 1986; Sachs, 1978; Gould and Abu-sitta, 1980; Brebbia and Walker, 1979; Sarpkaya and Isacson, 1981; Patel, 1989; and Lyon, 1975) include a treatment of random vibration theory and applications.

A text on random vibration theory by the first author (Nigam, 1983) covers a brief review of probability theory, and a comprehensive treatment of random processes and random vibration analysis of linear and nonlinear systems. This book is conceived as a companion volume, and it covers the applications of random vibration theory to the analysis and design of a wide range of structural/mechanical systems and operating environments, in a unified manner. Since the basic theory has been covered in the first volume, this book concentrates on the engineering and design aspects, specially engineering approximations to develop practical design procedures. Although the book covers a wide range of physical systems and operating environments, an effort is made to present an integrated approach in the following general format:

   (i) physical characteristics of the environment;
  (ii) stochastic models and simulation of the environment;
 (iii) stochastic models of the excitations induced by system-environment interactions;
  (iv) random vibration analysis of the system; and
   (v) design, including optimization, in a reliability framework.

The stochastic modelling and simulation of random processes provide a powerful tool for the random vibration analysis of systems and have been treated in Chapter 2. The time-series models, which have been applied to a wide range of problems in recent years, are discussed in detail. Fatigue and creep are an integral part of the design of systems operating in sustained vibration and elevated temperature environments. These are treated in Chapter 3. A probabilistic approach to design, including optimization, of structural/mechanical systems operating in random vibration environments is discussed in Chapter 4. The optimization problem is formulated in a reliability framework, and the nonlinear programming techniques and approximations necessary to solve practical problems are discussed. In Chapter 10, we discuss statistical energy analysis, which provides a powerful tool for the analysis of complex systems subjected to high frequency excitation. The remaining Chapters (5–9), cover applications of random vibration theory to aerospace systems (Chapter 5); vehicles (Chapter 6); structures excited by earthquakes (Chapter 7); wind (Chapter 8); and ocean waves (Chapter 9). Each chapter is organised in the general format indicated above. Illustrative examples are included in each chapter. Aspects of modelling, simulation and design specific to a particular system, or environment, are treated in respective chapters.

The book has been conceived and written on the premise, that after the mathematical models of the environment and the system have been formulated, a wide range of physically distinct problems are amenable to a unified treatment, as illustrated in Table 1.1 for earthquakes, wind and ocean waves. We believe that this approach is illuminating for both the students and the practising engineers. In Chapter 1 of the first volume (Nigam, 1983), the underlying unity in the analysis and design of a large class of systems has been discussed in detail, and illustrated with a few examples. The most significant aspect of this approach, which requires deep physical understanding and considerable engineering judgement, is the process of *idealization* in the formulation and validation of physical and mathematical models of the system and the environment. This aspect is emphasized throughout this book.

All engineering systems, except one-shot space systems, are designed to operate over a service life, which may range from 10 to 100 years, or more. Over such time spans (*long-term*), all environments exhibit wide temporal and spatial variations and are, therefore, nonstationary. However, over a short period of time (*short-term*), most physical phenomena may be adequately modelled as a stationary random process in terms of a few fixed parameters called the *conditioning parameters*. A *quasi-stationary* long-term model can, therefore, be constructed by assembling segments of short-term models. Each short-term model is characterised by stochastic properties, such as, probability distributions, mean, variance, level-crossing rate, power spectral density, etc., which are functions of conditioning parameters. The conditioning parameters vary randomly across short-term models, and if their joint

Table 1.1   Earthquakes, wind and ocean waves—Design sequence

Earthquakes	Wind	Ocean waves
Zones (Seismicity—Ground Motion Data) ↓	**Climate** (Wind Speed and Storm Data) ↓	**Climate** (Wave Height and Sea Storm Data) ↓
Geological & Geotechnical Features (Faults, Soil) ↓	**Exposure** (Terrain, Topography) ↓	**Exposure** (Ocean Depth, Shore Profile) ↓
Ground Motion (Stochastic Models) ↓	**Wind Velocity** (Stochastic Models) ↓	**Particle Velocity** (Stochastic Models) ↓
Structure (Dynamic Characteristics, Foundation) ↓	**Structure** (Dynamic Characteristics, Shape, Size, Permeability) ↓	**Structure** (Dynamic Characteristics, Shape, Size, Foundation) ↓
Forces (Base Excitation) ↓	**Forces** (Aerodynamic) ↓	**Forces** (Hydrodynamic) ↓
Structural Response (Random Vibration Analysis- Stress, Deflection, Plastic Deformation) ↓	**Structural Response** (Random Vibration Analysis- Stress, Deflection) ↓	**Structural Response** (Random Vibration Analysis- Stress, Deflection) ↓
Design (Strength, Stiffness, Stability, Ductility)	**Design** (Strength, Stiffness, Stability Fatigue)	**Design** (Strength, Stiffness, Stability, Fatigue)

distributions can be constructed based on the long-term data, the long-term distribution can be determined from the convolution relation:

$$p_L(\bar{x}) = \int \cdots \int_m p_s(\bar{x}\,|\,\bar{a}_s)\, p(\bar{a})\, d\bar{a} \qquad (1.1)$$

in which $\bar{x}$ is the state vector representing the physical quantities of interest; $\bar{a}$ represents the conditioning parameters; $p_L(\bar{x})$ is the long-term distribution; $p_s(\bar{x}|\bar{a})$ is the short-term conditional distribution, and $p(\bar{a})$ is the joint distribution of the vector $\bar{a}$.

The determination of long-term distribution (1.1) is important for practical implementation of reliability based design. A *quasi-nonstationary* model can also be constructed on the same basis, where each short-term segment is nonstationary.

Throughout the book, we rely on the exact and the approximate results of the random vibration theory covered in the first volume (Nigam, 1983). References to equations, figures, examples etc. in the first volume are made in this book, where necessary, through an asterik (*). For example, Eq. (6.51) and Fig. 6.5 are indicated by (6.51*) and Fig. 6.5*, respectively. The material presented in this book has been drawn from a large number of published papers, reports and books. References to these have been provided at the end of each chapter, and cited in the text.

REFERENCES

Bolotin, V.V. 1965. Statistical Methods in Structural Mechanics, Strooiizdat, Moskow (first edition, 1961). (English translation), 1969. San Francisco: Holden-Day.

Bolotin, V.V. 1979. Random Vibration of Elastic Systems (in Russian) Moscow: Nauka.

Brebbia, C.A. and S. Walker, 1979. Dynamics of Offshore Structures. London: Newnes—Butterworth.

Clough, R.W. and J. Penzien, 1975. Dynamics of Structures, New York: McGraw-Hill.

Crandall, S.H. ed. 1958. Random Vibration, Vol. 1. Cambridge; MA: The MIT Press.

Crandall, S.H. ed. 1963. Random Vibration, Vol. 2. Cambridge, MA: The MIT Press.

Crandall, S.H. and W.D. Mark, 1963. Random Vibration in Mechanical Systems. New York: Academic Press.

Dimentberg, M.F. 1988. Statistical Dynamics of Non-linear Time Varying Systems. Somerset: Research Study Press.

Elishakoff, I. 1983. Probabilistic Methods in The Theory of Structures. New York: John Wiley.

Elishakoff, I. and R.H. Lyon, eds. 1986. Random Vibration—Status and Recent Developments. New York: Elsevier.

Fry'ba, L. 1972. Vibration of Solids and Structure Under Moving Loads. Groningen: Noordhoff International.

Gould, P.C. and S.H. Abu-Sitta, 1980. Dynamic Response of Structures to Winds and Earthquakes. Pentech Press.

Hurty, W.C. and M.F. Rubinstien, 1964. Dynamics of Structures. Englewood Cliffs: Prentice-Hall.

Ibrahim, R.A. 1985. Parametric Random Vibration. New York: John Wiley.

Kree, P. and C. Soize, 1983. Mathematics of Random Phenomena-Random Vibration of Mechanical Structures. Dordrecht: Reisel Pub. Co.

Lin, Y.K. 1967. Probabilistic Theory of Structural Dynamics. New York: McGraw-Hill (Second edition, 1976).

Lyon, R.H. 1975. Statistical Energy Analysis of Dynamic Systems: Theory and Applications. Cambridge, MA: The MIT Press.

Newland, D.E. 1984. An Introduction to Random Vibrations and Spectral Analysis. Essex: Longman.

Nigam, N.C. 1983. Introduction to Random Vibrations. Cambridge, MA: The MIT Press.

Patel, M.H. 1989. Dynamics of Offhore Structures. London: Butterworth.

Robson, J.D. 1963. An Introduction to Random Vibrations. Edinburgh: University Press.

Sachs, P. 1978. Wind Forces in Engineering. Oxford: Pergamon Press.

Sarpkaya, T. and M. Isacson, 1981. Mechanics and Wave Forces on Offshore Structures. New York: Van Nostrand Reinhold.

Simiu, E. and R.H. Scanlan, 1986. Wind Effects On Structure. New York: John Wiley.

Spanos, P. and J.B. Robson, 1990. Random Vibration and Statistical Linearization. New York: John Wiley.

Thomson, W.T. 1975. Theory of Vibration with Applications. Englewood Cliffs: Prentice-Hall.

Yang, C.Y. 1986. Random Vibration of Structures. New York: John Wiley.

Zienkewicz, O.C. et al, eds. 1978. Numerical Methods in Offshore Engineering, New York: John Wiley.

2. Modelling and Simulation of Random Processes

2.1 Introduction

The analyses of complex physical phenomena are facilitated by the construction of suitable mathematical models. A mathematical model should be consistent with the quality and quantity of information pertinent to the phenomena under investigation. It should be sufficiently complete so that important physical variables and properties associated with the phenomena are properly reflected in the model. At the same time a model must be simple enough to make the analysis mathematically tractable. An overly complex model which cannot be validated by means of available information and sound scientific methods may tend to complicate the analysis rather than provide a solution to the problem at hand.

There are basically two approaches for constructing models to describe physical phenomena-the deterministic approach and the probabilistic approach. In the deterministic approach, the model leads to a definite prediction, that is, given specific values of the relevant variables of the physical phenomenon, a particular state of the process will be observed. This is accomplished by establishing a functional relationship between different physical variables. In the probabilistic approach, however, such definite predictions are not possible. The different states of the process in the probabilistic model can be predicted only with certain measures of likelihood.

There are many natural phenomena for which all the information necessary for a complete description of the process may not be available. Moreover, in some cases the physical laws governing the progression of events may be so complex that a detailed analysis becomes intractable, or there may be some inherent uncertainties associated with the phenomena. A probabilistic model incorporates these uncertainties in a rational and systematic manner.

A probabilistic model is generally more realistic as compared to a deterministic model. While a deterministic model concerns with the prediction of the definite state of a phenomenon, the probabilistic model helps to investigate the inherent unavoidable and unexplained variations observed in almost all physical phenomena. It quantifies these variations and allows a sensitivity analysis of the idealized phenomena to the uncertainties. It may also provide clues as to the possible causes of the variations leading to a better understanding of the phenomenon. Thus the probabilistic model is dynamic in its approach and provides scope for improving itself as a predictor.

The probabilistic model is generally validated by statistical inference studies on available observations of the random phenomenon. In some cases, the number of observations may be very small, with which no meaningful statistical analysis could be made. The paucity of observed data of the random phenomenon may be due to rarity of its occurrence or due to difficulties in recording the event. It is necessary in such situations to generate artificial records of the phenomenon which could be used as possible observations. In generating the artificial sample records use should be made of the underlying probabilistic model describing the random phenomenon. This process termed as simulation is an important part of the modelling.

In this chapter we discuss the modelling and simulation procedures of random process as applied to random vibration problems. First, the Poisson process models and the filtered white noise model are discussed. The development of time series models is presented next. Subsequently, simulation procedures based on the random process models are discussed.

2.2 Random Process Models

A class of probabilistic models can be used to model random processes. By definition a random process is a parametered family of random variables. The indexing parameters associated with random processes occurring in most engineering applications are either time, or spatial coordinates, or both.

The theory of random processes has been applied to many problems in physics and engineering involving uncertainty. These include the subjects, such as kinetic theory of gases, statistical mechanics, brownian motion, statistical communication theory, turbulent fluid flow, earthquake engineering, aerospace engineering, queuing theory and reliability engineering. In aerospace engineering, the response of aircrafts, missiles and space vehicles to their respective environments is estimated by treating the environmental actions as random processes. For example, the pressure fields generated by boundary layer turbulence, due to jet efflux, the gust velocity component at ambient wind, runway roughness exciting the undercarriage of an aircraft during take-off, landing and taxiing have all been considered as random processes. In a similar manner, the dynamic response of structures to earthquakes and wind loadings have been estimated assuming the excitations to be random processes.

There are three basic approaches to stochastic modelling of random processes. In the first approach, the model takes the form of a known process like the Gaussian, Poisson or Markov and the parameters of the model are estimated by using available data from actual records. In the second method, which is somewhat similar to the first, sample functions of the random process are generated at discrete values of the indexing parameters by passing a sequence of random numbers through appropriately designed filters. The filter weights which are the parameters of the model are estimated by making use of the actual observations of the random phenomena. The third approach is non-parametric in nature and the interest is mainly limited to the estimation of a few 'order-statistics' of the random process. No other information is sought from the model. The spectral decomposition and attendant computations of random processes are examples of this approach. In some cases the stochastic model is simply assumed to be a mathematically convenient process like the Gaussian process with little regard to the information available about the actual process. In such situations, relating the properties of the mathematical model to the actual process must be exercised with extreme caution.

Once a stochastic model has been formulated it has to be validated in terms of its ability to represent the actual process. The rejection or acceptance of a model is generally based on a qualitative and quantitative comparison of the properties predicted by the model to the observed properties of the actual process. Where possible, validation of the models should be based on objective statistical inference tests. In the validation, as well as, in the proper interpretation, and possible refinement of the model, a knowledge of the types of errors that may creep in the formulation of the model is of added significance. The important types of errors that may be introduced are basically due to simplifying assumptions and inaccuracies in the estimation of the model parameters.

We present in the following sections some random process models with particular reference to random vibration applications.

2.3   Poisson Process Models

A large class of random process models of natural phenomena involve counting of events occurring randomly in time and space. The Poisson process, and processes derived from it, are of special interest amongst the general class of counting processes, because of their analytical simplicity and adaptability to represent several natural phenomena.

Natural phenomena like earthquakes, storms, vehicular traffic, gust encounters of an aircraft during flight can be assumed to occur as Poisson type events. The intensity measures of the phenomena during such occurrences, for example the earthquake ground acceleration, or the gust velocity component of wind may also exhibit randomness. The maximum value of the intensity measures is of importance in design applications and consequently the probability of information about the maximum intensity is of interest. If the intensity measures are assumed to be independent and identically distributed random variables, independent of the occurrence of the random event, then, the probability distribution of the maximum intensity measure can be obtained in terms of the parameters of the Poisson process governing the occurrence of the random event and the probability distribution function of the random variables describing the intensity measures.

Consider a random phenomenon E, which occurs as a homogeneous Poisson process with constant mean arrival rate v. Let $t_1, t_2, ..., t_k$ be the times at which the event E occurs. Let Z_i be the random variable representing the intensity measure of E occurring at the time instant t_i. In the sequel, the random variables $Z_1, Z_2, ...$ are assumed to be independently and identically distributed with probability distribution functions $F_{Z_1}(z_1) = F_{Z_2}(z_2) = ... = F_Z(z)$.

Let $Z_{max}(t)$ be the maximum value of Z_i observed in the time interval $(0, t)$. Then, the conditional probability that $Z_{max} \leq z$ given $N(t) = k$ is,

$$P[Z_{max} \leq z \mid N(t) = k] = [F_Z(z)]^k \qquad (2.1)$$

Therefore, the probability distribution function of Z_{max} in the time interval $(0, t)$, can be expressed as,

$$F_{Z_{max}(t)}(z) = \sum_{k=0}^{\infty} P[Z_{max} \leq z \mid N(t) = k] P[N(t) = k]$$

$$= \sum [F_z(z)]^k \frac{(vt)^k}{k!} \exp(-vt)$$

$$= \exp[-vt(1 - F_z(z))] \qquad (2.2)$$

The above stochastic model has been used to model the maximum earthquake ground acceleration in the time interval t, with the intensity measure assumed to be exponentially distributed (Shinozuka, 1972), i.e. $F_z(z) = 1 - \exp[-\alpha(z - z_0)]$,

$\alpha > 0$. Under this assumption, the probability distribution function for the maximum intensity measure is

$$F_{Z_{max}(t)}(z) = \exp\left[- vt \exp\left[- \alpha(z - z_0)\right]\right] \tag{2.3}$$

Equation (2.3) is the first asymptotic distribution of extreme values. The parameters that have to be determined for the model are then the mean arrival rate v and the constants α and z_0, which are specific to a site, earthquake sources, and travel path mechanisms.

The foregoing model can also be employed for the maximum structural response to random excitation due to earthquake and wind loadings.

Example 2.1: Maximum response of the sdf system to earthquake ground acceleration

Consider a structure idealized as a single degree-of-freedom (sdf) system subjected to the earthquake ground acceleration $\ddot{Z}(t)$. The equation of motion of the structure is given by

$$\ddot{X}(t) + 2\zeta\omega_0\dot{X}(t) + \omega_0^2 X(t) = - \ddot{Z}(t) \tag{2.4}$$

where $X(t)$ is the relative displacement between the mass and the base. It can be reasonably assumed that the strong motion portion of the earthquake ground acceleration can be idealized as a segment of a stationary Gaussian process with mean zero and power spectral density (psd) function $\Phi_{\ddot{Z}\ddot{Z}}(\omega)$. If the natural period of the structure T_0 is such that, $NT_0 < T$, where N is approximately the number of cycles required for the response to attain stationary conditions and T is the duration of the segment of the stationary Gaussian process, then the relative displacement can also be regarded as a segment of a stationary Gaussian process of the same duration T. Under these assumptions, the spectral density, and the zero up-crossing rate of the stationary response X are given by,

$$\Phi_{XX}(\omega) = \frac{\Phi_{\ddot{Z}\ddot{Z}}(\omega)}{(\omega_0^2 - \omega^2)^2 + 4\zeta^2\omega^2\omega_0^2} \tag{2.5}$$

$$N_X^+(0) = \left[\int_0^\infty \omega^2\Phi_{XX}(\omega)d\omega / \int_0^\infty \Phi_{XX}(\omega)d\omega\right]^{1/2} = \frac{\omega_0}{2\pi} \tag{2.6}$$

The probability distribution function of the maximum response $X_{max} = \max_{0 \le t \le T} |X(t)|$ can be shown to be (Nigam, 1983)

$$F_{X_{max}}(\alpha) = \exp\left[- \exp\left\{- C_1\left(\frac{\alpha}{\sigma_x} - C_1\right)\right\}\right] \tag{2.7}$$

where

$$C_1 = \{2 \ln[2 N_X^+(0)T]\}^{1/2} \tag{2.8}$$

and σ_x is the standard deviation of X. For those values of X for which $C_1\left(\frac{\alpha}{\alpha_x} - C_1\right) \gg 1$, (2.7) can be approximated by

$$F_{X_{max}}(\alpha) \simeq 1 - \exp\left[-C_1\left(\frac{\alpha}{\sigma_x} - C_1\right)\right] \qquad (2.9)$$

If further we make the assumption that (i) the occurrence of the earthquake is a homogeneous Poisson process with mean arrival rate v, and (ii) each earthquake has the same strong motion duration T and is characterized by the same spectral density $\Phi_{\ddot{z}\ddot{z}}(\omega)$, so that each X_m has the distribution (2.9) the probability distribution function of the maximum relative displacement response during the service life t which is denoted by X_{mm} is

$$F_{X_{mm}(t)}(\alpha) = \exp\left\{-vT\exp\left[-C_1\left(\frac{\alpha}{\sigma_x} - C_1\right)\right]\right\} \qquad (2.10)$$

The parameters of the model that have to be determined in this case are, the mean arrival rate v and the spectral density function $\Phi_{\ddot{z}\ddot{z}}(\omega)$. The parameter C_1 depends on the structural characteristics which are assumed to be known. In the forgoing discussion, we have made simplifying assumptions which are not very realistic. For example, both the duration T, as well as the parameter of psd $\Phi_{\ddot{z}\ddot{z}}(\omega)$ should be treated as random variables conditioned on the occurrence of the earthquakes. In this case the earthquake ground acceleration has to be modelled as a filtered Poisson process or (2.10) has to be treated as conditional distribution and convoluted with the distribution of T and parameters of psd.

2.3.1 Compound Poisson process

Rosenblueth and Bustamante (1962) have used a model for ground acceleration during an earthquake which is essentially a compound Poisson process. Their model assumes that the earthquake ground motion is characterized by a series of acceleration impulses of random magnitude and direction, which occur like the events of a Poisson process. The sum of the integrals of these impulses up to any time, then determines the ground velocity. Thus the ground velocity as a function of time is a compound Poisson process. Cornell (1964) has discussed an example of the compound Poisson process in the fatigue design of a prestressed concrete bridge. In his model, the occurrence of a peak bending moment due to a randomly passing vehicle is assumed to be of the Poisson type and the magnitudes of the peak moments are assumed to be independent and identically distributed random variables.

2.3.2 Filtered Poisson Process Models

By far the most important model based on the Poisson process is the filtered Poisson process model. This is so because, it represents in a very general manner the sum of influences of successive random events occurring as a Poisson process and also because it encompasses in its fold other processes derived from the Poisson process such as the compound Poisson process as special cases. A detailed discussion of the filtered Poisson process and its basic characteristics have been presented in Nigam (1983). The filtered Poisson model has found wide applications in random vibration problems specially in modelling random environment due to earthquakes and wind gusts (Housner, 1955; Cornell, 1964; Merchant, 1964; Lin, 1965; and Racicot and Moses, 1972). The probability structure of a filtered Poisson process can be derived in terms of its characteristic function from which its moment properties can be obtained (Lin, 1967). The response of a linear system to a filtered Poisson process is also a filtered Poisson process, which result can be used to advantage in response statistics calculations.

A filtered Poisson process is defined as

$$X(t) = \sum_{n=1}^{N(t)} w(t, \tau_n, Y_n), t \geq 0 \tag{2.11}$$

where $N(t)$ is a Poisson process with arrival rate v; τ_n is the time of occurrence of the nth event; Y_n are a sequence of independent and identically distributed random variables which are also independent of $N(t)$; and $w(\tau_n, Y_n)$ is an influence function with the property $w(t, \tau_n, Y_n) = 0$ for $\tau_n > t$.

For a large class of problems, the influence function can be expressed as

$$w(t, \tau_n, Y_n) = Y_n w(t, \tau_n) \tag{2.12}$$

where $w(t, \tau_n)$ is a deterministic shape function. For constant arrival rate v, the mean and covariance functions of $X(t)$ are given respectively by (Nigam, 1983)

$$\mu_X(t) = vE[Y] \int_0^t w(t, \tau) d\tau \tag{2.13}$$

$$K_{XX}(t_1, t_2) = vE[Y^2] \int_0^{t_1} w(t_1, \tau) w(t_2, \tau) d\tau; t_1 < t_2 \tag{2.14a}$$

$$\sigma_X^2(t) = vE[Y^2] \int_0^t w^2(t, \tau) d\tau \tag{2.14b}$$

If the arrival rate of the pulse tends to infinity, while amplitudes Y_n tend to zero, $X(t)$ tends to be a normal random process by virtue of the Central Limit Theorem.

For a special case of the shape function of the form

$$w(t, \tau_n) = w(t - \tau_n); t > 0 \tag{2.15}$$

with $w(t - \tau_n)$ tending to zero as $(t - t_n)$ tends to infinity, the random process $X(t)$ is called causal and is independent of absolute time. $X(t)$ becomes a stationary random process whose second order statistics are given by the Campbell's Theorem as

$$\mu_X = v\mu_Y \int_{-\infty}^{\infty} w(u) du \tag{2.16}$$

$$K_{XX}(t_1, t_2) = vE[Y^2] \int_{-\infty}^{\infty} w(u) w(t_2 - t_1 + u) du; t_2 > t_1 \geq 0 \tag{2.17a}$$

and

$$\sigma_X^2(t) = vE[Y^2] \int_{-\infty}^{\infty} w^2(u) du \tag{2.17b}$$

where $u = t - \tau$.

In the following we present an example of filtered Poisson process model for the fluctuating component of wind velocity due to Merchant (1964).

Example 2.2: Wind gust as filtered poisson process

In this model the wind velocity is represented by the randomly arriving gusts superposed on the mean wind velocity. The gusts are assumed to arrive like a Poisson process, and are idealized to have the rectangular shape as shown in Fig. 2.1. The rectangular pulse shape is chosen for its simplicity, and can be justified on physical grounds if gusts are assumed to result from eddies that follow vortex flow within a moving air mass. (Racicot, 1969).

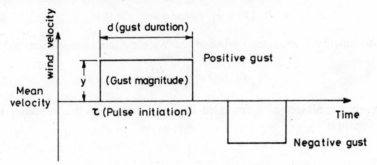

Fig. 2.1   Rectangular pulse shape for random gust.

The fluctuating component of the wind velocity $v(t)$ in the filtered Poisson model is expressed as

$$v(t) = \sum_{n=1}^{N(t)} w(t - \tau_n, Y_n, D_n) \tag{2.18}$$

where τ_n is the time of initiation of the nth gust; Y_n and D_n are independent random variables representing respectively its magnitude and duration; $w(\ )$ is the rectangular impulse shape function, and $N(t)$ is the homogeneous Poisson process with constant mean arrival rate v. Equation (2.18) represents a filtered Poisson process model for the velocity fluctuation. Y_n are assumed to be Gaussian with mean zero and variance σ^2. The duration of the gust D_n is assumed to be exponentially distributed with mean value μ. Thus

$$p_Y(y) = \frac{1}{\sigma\sqrt{2\pi}} \exp(-y^2/2\sigma^2); \quad -\infty < y < \infty \tag{2.19}$$

and

$$p_D(d) = \mu \exp(-\mu d); \quad 0 < d < \infty \tag{2.20}$$

By the use of Campbell's theorem, the autocovariance function of $v(t)$ can be expressed as,

$$R_{vv}(\tau) = v \int_0^\infty E_{YD}[w(t + \tau, y, d) \, w(t, y, d)] \, dt \tag{2.21}$$

In (2.21) E_{YD} represents the expectation operation with respect to the random variables Y and D. As t increases, the autocorrelation becomes a function of τ only and can be shown to be

$$R_{vv}(\tau) = v \int_0^\infty \left\{ \int_0^\infty E_Y[w(s + \tau, y, d) \, w(s, y, d)] ds \right\} p_D(d) \, dd \tag{2.22}$$

For the rectangular pulse the integrand of (2.22) is zero for $d \leq \tau$ and for $s \leq 0$ and $s \geq d - \tau$ when $\tau \geq 0$. Under these conditions (2.22) reduces to

$$R_{vv}(\tau) = v \int_0^\infty \left\{ \int_0^{d-\tau} E_Y[(Y^2)] \, ds \right\} p_D(d) \, dd \tag{2.23}$$

Substituting (2.19) and (2.20) in (2.23) we get

$$R_{vv}(\tau) = \frac{v\sigma^2}{\mu} \exp(-\mu\tau), \mu \geq 0 \tag{2.24}$$

The psd function is obtained by use of the Wiener-Khinchine relations as

$$\Phi_{vv}(\omega) = \frac{1}{\pi} \frac{v\sigma^2}{\mu^2 + \omega^2} \tag{2.25}$$

which when expressed as a one-sided spectral density with the frequency in Hz takes the form

$$G_{vv}(f) = \frac{v\sigma^2}{\pi^2} \left[\frac{1}{(\mu/2\pi)^2 + f^2} \right] \tag{2.26}$$

where $f = \omega/2\pi$.

Thus the parameters of the model that have to be determined are v, μ and σ. These parameters are found to depend on the mean wind velocity at a particular height and the terrain drag coefficient. Merchant (1964) assumed the dependence of the parameters to be in the following form

$$v^* = v/\bar{V}_1, \mu^* = \mu/\bar{V}_1 \quad \text{and} \quad \sigma^* = \sigma/(C_D \bar{V}_1)^{1/2} \tag{2.27}$$

where $\bar{V}_1$ is the mean wind velocity at a reference height of 33 ft and C_D is the terrain drag coefficient. The variables $(v\sigma^2/C_D \bar{V}_1^3)$ and $(\mu/\bar{V}_1)$ are determined by applying a least square fit to the spectral density function given by (2.27) and the Davenport (1961) wind data. The values thus obtained are

$$\mu^* = 0.003375 \quad \text{and} \quad v\sigma^{*2} = \frac{v\sigma^2}{C_D \bar{V}^3} = 0.0187$$

The values given above are only central values and variations of about ± 5% for the parameter $v^*\sigma^{*2}$ and ± 10% for the parameter μ^* are suggested. The comparison between the Merchant and Davenport spectra is shown in Fig. 2.2.

It can be seen from (2.27) that the parameters v and σ appear as the product $v\sigma^2$. Hence it is not possible to obtain v and σ separately from just the spectral density fit. In order to separate v, σ from $v\sigma^2$, Merchant (1964) attempted a comparison with the simulated wind records, in which he varied v and σ. Based on this subjective assessment he suggested a value of $v^*/\mu^* = 22$. Racicot (1969) found that the threshold level crossings of the wind velocity depend significantly on the variations of v and σ for a fixed value of $v\sigma^2$ and for a given value of $C_D \bar{V}_1$. He used the threshold crossing data for the estimation of v^* and σ^* and obtained $v^* = 0.0187$ and $\sigma^* = 1.0$ with variation of about ± 25%. Based on this estimation $v^*/\mu^* \simeq 5.5$, which is significantly different from the value suggested by Merchant. Thus even for the same model the estimates of the parameters are different based on the criteria used to determine them.

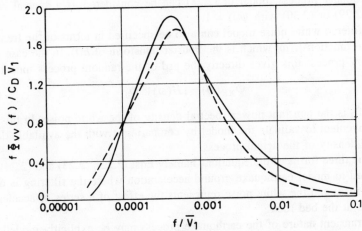

Fig. 2.2 Comparison between Davenport and Merchant psds for wind gust.
———— Davenport, - - - - - Merchant.

2.4 Filtered White Noise Models

A large class of stochastic models can be constructed as the output of a linear system to a white noise input. The properties of the random process modelled as a filtered white noise model can be readily obtained once the parameters signifying the filter transfer function are identified. Mathematically the filtered white noise model is represented by

$$X(t) = \int_{-\infty}^{\infty} h(t - \tau) W(\tau) \, d\tau \qquad (2.28)$$

where $h(t)$ is the impulse response function of a linear time invariant filter and $W(t)$ is the white noise input to the filter. Nonstationarity may be introduced in the model by use of a deterministic modulating function $\psi(t)$ in the form

$$X(t) = \psi(t) \int_{-\infty}^{\infty} h(t - \tau) W(\tau) d\tau \qquad (2.29)$$

Alternatively, a nonstationary random process model can be constructed by passing a nonstationary random process $\psi(t)W(t)$, which is a product of a determinsitic function and white noise, through a linear filter. In such a case the model assumes the form

$$X(t) = \int_{-\infty}^{\infty} h(t - \tau) \psi(t) W(\tau) \, d\tau \qquad (2.30)$$

Random processes $X(t)$ given by (2.29) and (2.30) are in general known as modulated random processes. The two processes are distinguished by the order in which the operations of filtering and modulation are performed. In (2.29) a stationary process is first obtained by passing white noise through an appropriate linear filter and then modulated. Hence it can be called as modulated filtered white noise (MFWN). On the other hand the random process $X(t)$ given by (2.30) is obtained by first modulating the white noise and then filtering the resulting nonstationary random process. It is called as the filtered modulated white noise model (FMWN).

The stationary process model of (2.28) can be considered as a special case of either (2.29) or (2.30) with $\psi(t) = 1$.

The filtered white noise model can also be specified in terms of the frequency response function $H(\omega)$ which is the Fourier transform of $h(t)$. In the case of the stationary process this gives directly the psd of the random process model as

$$\Phi_{XX}(\omega) = |H(\omega)|^2 \, \Phi_0 \qquad (2.31)$$

where Φ_0 is the constant power spectral density of the white noise. In this form it is convenient to validate the model by comparing it with the available data of spectral density of the actual process.

All the three forms of the equations namely Eqs. (2.22), (2.23) and (2.24) have been used to model earthquake ground acceleration where the filtering is due to the soil mass and the white noise represents the effect of a series of acceleration impulses at the bed rock.

The transient nature of the earthquake process may be explicitly modelled by either the MFWN or the FMWN models. If the modulating function is reasonably smooth, it will not significantly affect the frequency content of the random process. The parameters of the model are either the impulse response function $h(t)$ or alternatively the stationary frequency content given by the frequency response function $H(\omega)$ and the modulating function $\psi(t)$. In general the modulating envelope is chosen empirically with only a few parameters to be estimated, such as intensity, duration and build up time.

Though the MFWN model is able to represent the major features of strong motion of an earthquake like average frequency content, intensity and duration it is unable to represent the time varying frequency content of the shift from higher frequencies to lower frequencies toward the end of strong motion records (Smith, 1985). The MFWN model is computationally convenient but no possible physical significance can be attached to the envelope.

The FMWN model on the other hand visualizes the earthquake process as a white noise 'source' which is deterministically modulated and then filtered by the transmission path. The filter characteristics are determined in part by source properties as well as the transmission path. Thus the envelope in this case is physically associated with the source mechanism. The model tacitly assumes determinism of the transmission path and the envelope. Yet it is a reasonable model on physical grounds and exhibits a slight frequency shift with time due to transient behaviour of the filter. If the filter characteristics and modulating functions are the same the difference between MFWN and FMWN models is seen to depend on the smoothness of $\psi(t)$. If $\psi(t)$ is very flat, it may be essentially be assumed constant in (2.30), and taken out of the integral resulting in a MFWN model. Thus, for earthquakes with a long section of quasi-stationary motion, both models will have similar characteristics. The difference will be seen mainly in the transient portions in the buildup and decay of motion. They will show significant differences in modelling shorter duration earthquakes. Boore (1983) has used a FMWN model for a wide range of earthquake durations and successfully reproduced peak velocity and response spectra of strong motion records.

The various forms of the impulse response function $h(t)$ or the frequency response function $H(\omega)$, and the deterministic modulating function $\psi(t)$ assumed by different investigators are presented in Table 2.1.

Table 2.1 Filtered white noise model for earthquake ground acceleration

Author	$h(t)$ or $H(\omega^2)$	$\psi(t)$	Equation Number
1. Tajimi (1960)	$\lvert H(\omega^2)\rvert = \dfrac{\omega_g^4 + 4\zeta_g^2\omega_g^2\omega^2}{(\omega_g^2 - \omega^2)^2 + 4\zeta_g^2\omega_g^2\omega^2}$	1	(2.28)
2. Shinozuka and Sato (1967)	$h(t) = \dfrac{e^{-\zeta_g\omega_g t}}{\omega_{dg}}\sin\omega_{dg}t \quad t>0$ $= 0 \quad t\le 0$	$C(e^{-\alpha t} - e^{-\beta t}), \quad t>0\ (\alpha<\beta)$ $0 \qquad\qquad\qquad\ t\le 0$	(2.29)
3. Shinozuka and Sato (1967)	same as above	same as above	(2.30)
4. Iyengar and Iyengar (1969)	$\lvert H(\omega)^2\rvert = (e^{-C_1^2\omega^2} + A\omega^2 e^{-4C_1^2\omega^2})$	$(a_1 + a_2 t)e^{-nt^n}$ $n = 1,\ \text{or } 2$	(2.29)
5. Amin and Ang (1968)	$h(t) = \dfrac{e^{-\zeta_g\omega_g t}}{\omega_{dg}}\sin\omega_{dg}t \quad t>0$ $= 0 \quad t\le 0$	$\psi^2(t) = 0 \qquad\qquad\ t<0$ $\quad = C_2(t/a)^2 \quad 0\le t\le T_1$ $\quad = C_2 \qquad\qquad T_1\le t\le T_2$ $\quad = C_2 e^{-\alpha_2 t} \qquad T_2\le t$	(2.30)
6. Ruiz and Penzien (1969)	$h(t) = \dfrac{\omega_g^2}{\omega_{dg}}\sqrt{(1-2\zeta_g^2)}\,e^{\zeta_g\omega_g t}\sin\omega_{dg}t$ $\quad + 2\zeta_g\omega_g e^{-\zeta_g\omega_g t}\cos\omega_{dg}t \quad t>0$ $= 0 \qquad\qquad\qquad\qquad\qquad\quad t\le 0$	$\psi^2(t) = 0, \qquad\qquad\ t<0$ $\quad = C_3 \qquad\qquad 0\le t\le t_0$ $\quad = C_3 e^{-\alpha(t-t_0)} \quad t_0<t$	(2.30)

In Table 2.1 ω_g, ζ_g and ω_{dg} are the ground characteristics namely the natural frequency, fraction of critical damping and the damped natural frequency of an equivalent linear single degree of freedom system representing the filtering action of the ground. C, α, β, C_1, C_2, C_3, α_2 and α_3 are other parameters which are estimated by comparing the simulated earthquake records of the model with actual records and also the average velocity spectra with the average velocity spectra of Housner (1947). Another criterion used for testing the validity of the filtered white noise models for earthquake ground acceleration models is to compare the normalized autocovariance function of the strong motion portion of the ground acceleration with the autocovariance function computed from the stationary part of recorded earthquake accelerations by Barstein (1960). In chapter 7 we discuss various approaches to the validation of earthquake ground motion models.

Davenport (1961) has used the filtered white noise model for the fluctuating part of the wind velocity.

Example 2.3:   Earthquake ground acceleration as filtered white noise model

As an example of the filtered white noise model we present, in some detail, the nonstationary form of the filtered white noise model suggested by Shinozuka and Sato (1967) for modelling earthquake ground motion.

Let the action at a particular location to earthquake be characterized by a differential equation of the form

$$L[\eta(t)] = \ddot{\eta}(t) + 2\zeta_g\omega_g\dot{\eta}(t) + \omega_g^2\eta(t) = W(t) \tag{2.32}$$

where $\eta(t)$ is the output of the linear system to the white noise excitation $W(t)$ and $L[\cdot]$ is a linear differential operator. ζ_g and ω_g represent the ground characteristics. The ground velocity $\dot{Z}(t)$ is assumed to be of the form

$$\dot{Z}(t) = \psi(t)\,\eta(t) \tag{2.33}$$

Hence the ground velocity can be expressed as the nonstationary modulated filtered white noise process (MFWN)

$$\dot{Z}(t) = \psi(t)\int_{-\infty}^{\infty} h_\eta(t-\tau)\,W(\tau)d\tau \tag{2.34}$$

where $h_\eta(t)$ is the impulse response function

$$\left.\begin{array}{ll} h_\eta(t) = \dfrac{e^{-\zeta_g\omega_g t}}{\omega_{dj}}\sin(\omega_{dj}t), & t > 0 \\[2mm] \qquad\quad = 0 & \text{otherwise} \end{array}\right\} \tag{2.35}$$

and

$$\omega_{dj} = \omega_g(1 - \zeta_g^2)^{1/2} \tag{2.36}$$

The deterministic modulating function is assumed to be

$$\left.\begin{array}{ll} \psi(t) = (e^{-\alpha t} - e^{-\beta t}), & t \geq 0\ (\beta > \alpha > 0) \\[2mm] \qquad\; = 0 & \text{otherwise} \end{array}\right\} \tag{2.37}$$

From (2.34) the ground displacement $Z(t)$ and ground acceleration $\ddot{Z}(t)$ can be obtained by mean square integration and differentiation respectively, that is,

$$Z(t) = \int_{-\infty}^{t} \psi(\tau)\,\eta(\tau)\,d\tau \qquad (2.38)$$

$$\ddot{Z}(t) = \dot{\psi}(t)\,\eta(t) + \psi(t)\,\dot{\eta}(t) \qquad (2.39)$$

It follows from the properties of the white noise and the function $\psi(t)$ that

$$\left.\begin{array}{l} E[Z(t)] = E[\dot{Z}(t)] = E[\ddot{Z}(t)] = 0 \\[2mm] \lim_{t \to 0} E[Z^2(t)] = \lim_{t \to 0} E[\dot{Z}^2(t)] = 0 \\[2mm] \lim_{t \to \infty} E[\dot{Z}^2(t)] = \lim_{t \to \infty} E[\ddot{Z}^2(t)] = 0 \end{array}\right\} \qquad (2.40)$$

Thus a significant result of the selection of the filter is to ensure the eventual vanishing of the variance of ground velocity and ground acceleration. Further it can be shown that

$$\lim_{t \to 0} E[\ddot{Z}^2(t)] = \frac{\Phi_0\,\omega_{dg}^2}{4\zeta_g\,\omega_g(\zeta_g^2\,\omega_g^2 + \omega_{dg}^2)} \qquad (2.41)$$

and

$$\lim_{t \to \infty} E[Z^2(t)] = \lim_{t \to \infty} \int_0^t \int_0^t \psi(t_1)\,\psi(t_2)\,E[\eta(t_1)\,\eta(t_2)]\,dt_1\,dt_2 \qquad (2.42)$$

with

$$E[\eta(t_1)\,\eta(t_2)] = \frac{\Phi_0 \exp\left[-\zeta_g\,\omega_g |\,t_1 - t_2\,|\right]}{4\zeta_g\,\omega_g\,\omega_{dg}(\zeta_g^2\,\omega_g^2 + \omega_{dg}^2)} \{\omega_{dg} \cos \omega_{dg}\,(t_1 - t_2)$$
$$+ \zeta_g\,\omega_g \sin \omega_{dg}|\,t_1 - t_2\,|\} \qquad (2.43)$$

In (2.41) and (2.43) Φ_0 is the constant power spectral density of the white noise. From (2.39) the autocorrelation function of the ground acceleration becomes

$$E[\ddot{Z}(\tau_1)\,\ddot{Z}(\tau_2)] = \dot{\psi}(\tau_1)\,\dot{\psi}(\tau_2)\,E[\eta(\tau_1)\,\eta(\tau_2)]$$
$$+ \psi(\tau_1)\,\psi(\tau_2)\,E[\dot{\eta}(\tau_1)\,\dot{\eta}(\tau_2)]$$
$$+ \dot{\psi}(\tau_1)\,\psi(\tau_2)\,E[\eta(\tau_1)\,\dot{\eta}(\tau_2)]$$
$$+ \psi(\tau_1)\,\dot{\psi}(\tau_2)\,E[\dot{\eta}(\tau_1)\,\eta(\tau_2)] \qquad (2.44)$$

The acceleration is assumed to attain stationary conditions after a time $t = t^*$ for which the variance of the acceleration becomes stationary, that is when t^* is such that $\dot{\psi}(t^*) = 0$. From (2.39) the acceleration process in the stationary portion can be approximated by,

$$\ddot{Z}(t) \simeq \dot{\eta}(t) \text{ for } t \geq t^* \qquad (2.45)$$

Hence the normalized autocovariance function for the stationary part of the acceleration is approximately the same as that of the normalized autocovariance function of $\dot{\eta}(t)$. This function can be shown to be

$$\rho(\tau) = \exp\left[-\zeta_g \omega_g \,|\, \tau \,|\right]\left\{\cos \omega_{dg}\, \tau - \frac{\zeta_g \omega_g}{\omega_{dg}} \sin \omega_{dg}\, |\, \tau \,|\right\} \qquad (2.46)$$

where $\tau = t_1 - t_2$.

The autocovariance function given by (2.40) is compared with those obtained by computing the time averages of the stationary part of recorded earthquake acceleration from Barstein's work (1960). The comparison is shown to be in good agreement as seen from Fig. 2.3. The values of the ground characteristics assumed in the model are $\omega_{dg} = 12.3$ rad/sec, $\zeta_g \omega_g = 3.86$ per sec, $\alpha = 0.25$ per sec, $\beta = 0.5$ per sec and $\Phi_0 = 2.92 \times 10^6$ cm^2 per sec^5. Further confidence in the validity of filtered white noise model for the ground motion during earthquakes was obtained by simulating the ground motion (displacement, velocity and acceleration) using the model and comparing the relative velocity spectra, computed from the simulated records.

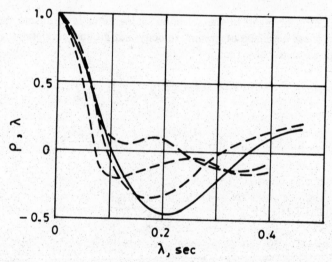

Fig. 2.3   Normalized autocovariance of earthquake acceleration.
———— Shinozuka and Sato, 1967, — — — Barstein, 1960.

2.5   Time Series Models

Another approach to stochastic modelling is to formulate a mechanism by which a sequence of values are generated representing possible observations of the random process at discrete values of the indexing parameter usually time. The model parameters are then estimated on the basis of a comparison of estimated statistics of the generated sequence and the statistics of the actual observations of the random process, which are treated as sample functions drawn out of an ensemble of infinite possibilities. Such a procedure is called by the generic term *time series* modelling. Basically the mechanism is very similar to the filtered white noise model, in that the time series is generated by passing a discrete white noise sequence through a linear filter. The white noise is assumed to be built up by a series of independent shocks drawn from the normal distribution with zero mean and constant variance.

The fundamentals of time series analysis and modelling have been treated in the texts of Box and Jenkins (1971) and Jenkins and Watts (1968), in which the basic characteristics of the various models, their identification, estimation and validation procedures are covered in considerable detail. Of particular importance are the stationary time series models, autoregressive (AR), moving average (MA), mixed autoregressive and moving average (ARMA) models, and their extension to a particular class of nonstationary random process, the autoregressive integrated moving average model (ARIMA).

Time series models have been mostly used in the areas of economic and business planning, production control and industrial engineering. Earlier engineering applications have been mainly confined to seasonal random phenomena like annual rainfall characteristics, and in the areas of meteorological forecasting and hydrology. There are also other important applications in process dynamics, automatic control systems and communication theory.

Applications of time series modelling and simulation technique to random vibration problems are only of relatively recent origin. Gersh (1970) was perhaps the first to systematically extend the concepts of time series modelling to the application of mechanical and structural engineering problems. In a series of papers Gersh and his coworkers (Gersh and Luo, 1972; Gersh and Sharpe, 1973; Gersh and Foutch, 1974; Gersh and Liu, 1976; Gersh and Yonemoto, 1977) have applied time series modelling in the analysis, simulation and identification of random vibration systems involving univariate as well as multivariate stationary time series. They developed a two stage least square procedure for estimating the parameters of ARMA models which reproduce prescribed covariance functions of the random processes to be modelled. The emphasis of their approach was to synthesize the time series, that is, a discrete analogue of the response of a linear multi-degree of freedom system driven by white noise vector. Samaras et al (1985) have proposed an ARMA model for the modelling of multivariate random processes on similar lines.

Polhemus and Cakmak (1981) have used the ARMA models in the modelling and simulation of earthquake ground motions. In their approach, the original nonstationary time series was normalized to yield a stationary time series by dividing the original series by the local estimates of the standard deviation which were estimated by fitting a polynomial to the time dependent variance. An ARMA model is then fitted to the transformed stationary series. Chang, Kwiatkowski and Nau (1982) used least square procedures for the determination of ARMA parameters dealing with the probabilistic characterization and simulation of earthquake ground motions as a scalar random process. The least square procedures primarily involved such time domain quantities as sample and partial autocorrelation functions in addition to power spectra. Gersh and Kitagawa (1985) developed a nonstationary time series for the modelling of earthquake ground motion by means of a smoothness priors-time varying AR coefficient model.

Reed and Scanlan (1981, 1983) made use of scalar ARMA models to characterize and simulate fluctuating wind velocity time histories so that they reproduce the measured turbulence spectra. Wyatt and May (1973) have also used essentially an AR model for the fluctuating wind velocity component.

Spanos and Hansen (1981) and Spanos (1983) dealt with scalar AR, MA, ARMA models to model and simulate ocean wave surface elevation. They proposed a least square procedure wherein the parameters of the model were so determined as to minimize the sum of the square of the difference between the spectrum of the time series derived from the ARMA models and that of the target spectrum at a number of frequencies.

Mignolet and Spanos (1987), and Spanos and Mignolet (1987) in a series of two articles have presented a unified approach for the development of simulation algorithms based on a prior approximation of a target spectral matrix by the response of an AR discrete dynamic system to white noise excitation. They have generalized the approach to multivariate and multi-dimensional random process modelling and simulation. The system of equations leading to the determination of the ARMA coefficients were derived through the minimization of frequency domain errors. The ARMA procedures were exemplified by application to spectral shapes encountered in different technical areas such as earthquake engineering (Kanai-Tajimi spectrum), ocean engineering (Pierson-Muskowitz spectrum) and wind engineering (spectral matrix of turbulent velocities).

Venkatesan and Krishnan (1981) have developed a nonstationary time series model essentially of the ARMA type for the modelling of runway roughness and the associated response of an aircraft for example, pilot location acceleration, loads at some critical points etc., by modelling the corresponding impulse response function.

In this section, we present briefly the important features of the models without going into the details of the procedures of estimation and validation. The models are illustrated with an example of a nonstationary time series pertaining to random vibration problems. The example is due to Polhemus and Cakmak (1981) who fitted an ARMA model for the earthquake ground acceleration. They constructed a stationary time series from a nonstationary ground acceleration after normalizing it with respect to the time varying variance function.

We shall consider in this discussion only discrete univariate time series with observations made at equally spaced discrete times. Such time series can be constructed from a continuous time record by sampling it at appropriate equi-spaced time intervals. It is expedient before considering the different models to define certain basic difference operators as introduced by Box and Jenkins (1971). Let the values of the time series corresponding to the discrete times t_i, $i = 1, 2 \ldots N$ be denoted by $x(t_i) = x_i$. N is called the length of the time series. The *backward shift* operator B is defined by the relation

$$Bx_t = x_{t-1} \qquad (2.47)$$

which implies that $B^m x_t = x_{t-m}$. The *forward difference operator* and the *summation operator* S are defined respectively by

$$\nabla x_t = x_t - x_{t-1} = [1 - B] x_t \qquad (2.48a)$$

$$Sx_t = \sum_{j=0}^{\infty} x_{t-j} \qquad (2.48b)$$

2.5.1   The General Linear Process Model

A time series x_t is said to be a general linear process if each observation x_t may be expressed in the form,

$$x_t = \mu + a_t + \psi_1 a_{t-1} + \psi_2 a_{t-2} + \ldots \qquad (2.49a)$$

where μ and the ψ_i are fixed parameters. The series $(\ldots a_{t-1}, a_t \ldots)$ is a sequence of identically distributed and independent random shocks with mean zero and variance σ_a^2. The time series x_t may be considered as the output obtained by passing the white noise sequence a_t through a linear filter and as such x_t is represented as a weighted sum of the current and past disturbances. A procedure to generate the white noise sequence is given in Appendix 2A. Equation (2.49a) may be expressed in operator notation as,

$$x_t = \mu + \psi(B)\, a_t \tag{2.49b}$$

where

$$\psi(B) = 1 + \psi_1 B + \psi_2 B^2 + \dots \tag{2.50}$$

is the transfer function of the linear filter and B is the backward shift operator defined by (2.47).

The mean, and the autocovariance at lag k, of the process x_t are given by

$$E[x_t] = \mu + E[a_t + \psi_1 a_{t-1} + \dots] \tag{2.51}$$

$$\gamma_k = E[\{x_t - E[x_t]\}\,\{x_{t-k} - E[x_{t-k}]\}]$$

$$= E[(a_t + \psi_1 a_{t-1} + \dots)(a_{t-k} + \psi_1 a_{t-k-1} + \dots)]$$

$$= E[\psi_k a_{t-k}^2 + \psi_1 \psi_{k+1} a_{t-k-1}^2 + \dots] + E[a_t a_{t-k}$$

$$+ \psi_1(a_t a_{t-k-1} + a_{t-1} a_{t-k} + \dots)]$$

$$= \sigma_a^2 \sum_{i=0}^{\infty} \psi_i \psi_{i+k} \tag{2.52}$$

since the random variables $a_t, a_{t-1} \dots$ are mutually independent. In (2.52) $\psi_0 = 1$.

It is evident that for the time series x_t to be stationary the sums $\sum_{i=0}^{\infty} \psi_i$ and $\sum_{i=0}^{\infty} \psi_i \psi_{i+k}$ should converge such that $E(X_t) = \mu$ is a constant, and the autocovariance γ_k depends only on the lag number k. For a general linear process the conditions for stationarity of the time series x_t can be embodied in a single condition that the series $\psi(B)$ in (2.50) which is a polynomial in the operator B should converge for $|B| \le 1$. For a rigorous proof of the above, the readers may refer to Box and Jenkins (1971) and Grenander and Rosenblatt (1957).

Equation (2.49) expresses x_t in terms of the current disturbance a_t and all the past disturbances. An alternative way of expressing x_t is in terms of the current disturbance and all previous observations of the process x_t. Rearranging (2.49) we obtain

$$a_t = x_t - \mu - \psi_1 a_{t-1} - \psi_2 a_{t-2} - \dots \tag{2.53}$$

Since (2.53) holds for any time, for example, for $t - 1$, we can eliminate a_{t-1} from (2.53) and express x_t as

$$x_t = \mu(1 - \psi_1) + \psi_1 x_{t-1} + a_t + (\psi_2 - \psi_1^2)a_{t-2} + (\psi_3 - \psi_1\psi_2)a_{t-3} + \dots \tag{2.54}$$

Similarly, a_{t-2}, a_{t-3}, and so forth may be successively eliminated to yield

$$x_t = \pi_1 x_{t-1} + \pi_2 x_{t-2} + \dots + a_t + \delta \tag{2.55}$$

where the weights π_i are functions of the ψ_i weights and δ is a constant which is a function of μ and the ψ_i weights.

The general linear process defined either by (2.49) or (2.55) has infinite number of ψ_i or π_i parameters respectively. But for practical purposes, we shall be interested in models which use these weights parsimoniously. This is done by way of the autoregressive and moving average process models, which involve only a finite number of these weights.

2.5.2 Autoregressive Processes (AR)

Consider (2.55), where the weights π_i are zero for $i > p$. Then

$$x_t = \varphi_1\, x_{t-1} + \varphi_2\, x_{t-2} + \ldots + \varphi_p\, x_{t-p} + a_t + \delta \qquad (2.56)$$

where the weights are denoted by φ_i instead of π_i to distinguish AR process from the general linear process. The time series model defined by (2.56) is referred to as the *autoregressive process* of order p or just AR(p). In (2.56), x_t is expressed in terms of its own past values and the present shock a_t. It can be shown that for the stationarity of the process x_t it is necessary that the roots of the characteristic equation

$$(1 - \varphi_1 B - \varphi_2 B^2 - \ldots - \varphi_p B^p) = 0 \qquad (2.57)$$

lie outside the unit circle in the complex plane. x_{t-1}, x_{t-2} and so forth can be successively eliminated from (2.56), which can be expressed in the form of the general linear process defined by (2.49), with the infinite order ψ weights being functions of the φ weights. The stationarity condition is then equivalent to requiring the resulting series $\psi(B)$ converge for $|B| \leq 1$. When the time series given by (2.56) is stationary the mean of the AR(p) process is given by

$$E[x_t] = \varphi_1\, E[x_{t-1}] + \varphi_2\, E[x_{t-2}] + \ldots + \varphi_p\, E[x_{t-p}] + E[a_t] + \delta \qquad (2.58)$$

Since for the stationary time series $E[x_t] = E[x_{t-1}] = \ldots = E[x_{t-p}]$ and $E[a_t] = 0$, the mean value of $E[x_t]$ becomes

$$E[x_t] = \frac{\delta}{1 - \varphi_1 - \varphi_2 - \ldots - \varphi_p} \qquad (2.59)$$

The autocovariance of the AR(p) process is obtained in the following manner. Representing by $\tilde{x}_t$ the deviation of x_t from its mean, it can be expressed as an autoregressive process

$$\tilde{x}_t = \varphi_1\, \tilde{x}_{t-1} + \varphi_2\, \tilde{x}_{t-2} + \ldots + \varphi_p\, \tilde{x}_{t-p} + a_t \qquad (2.60)$$

Multiplying (2.60) by $\tilde{x}_{t-k}$ we obtain

$$\tilde{x}_{t-k}\, x_t = \varphi_1\, \tilde{x}_{t-k}\, \tilde{x}_{t-1} + \varphi_2\, \tilde{x}_{t-k}\, \tilde{x}_{t-2} + \ldots + \varphi_p\, \tilde{x}_{t-k}\, \tilde{x}_{t-p} + \tilde{x}_{t-k}\, a_t \qquad (2.61)$$

Taking expectations on both sides of the above equation and noting that $E[\tilde{x}_{t-k}\, a_t]$ vanishes when $k > 0$, since $\tilde{x}_{t-k}$ can only involve disturbances upto time $t - k$, which are uncorrelated with a_t, we get auto-covariance at lag k of the AR(p) model as

$$\gamma_k = \varphi_1\, \gamma_{k-1} + \varphi_2\, \gamma_{k-2} + \ldots + \varphi_p\, \gamma_{k-p},\ k > 0 \qquad (2.62)$$

The variance of the process γ_0 may be directly obtained from (2.60) as

$$\gamma_0 = \varphi_1 \gamma_1 + \varphi_2 \gamma_2 + \ldots + \varphi_p \gamma_p + \sigma_a^2 \qquad (2.63)$$

Dividing (2.63) by γ_0 the normalized autocovariance at lag k is obtained as

$$\rho_k = \varphi_1 \rho_{k-1} + \varphi_2 \rho_{k-2} + \ldots + \varphi_p \rho_{k-p},\ k > 0 \qquad (2.64)$$

Note that the normalized autocovariance is referred to as the autocorrelation in time series parlance. But we shall use the term normalized autocovariance to avoid any possible confusion with the autocorrelation function definition in random process theory.

If we substitute $k = 1, 2, \ldots p$ in (2.64), a set of linear equations are obtained

$$\rho_1 = \varphi_1 + \varphi_2\rho_1 + \ldots + \varphi_p \, \rho_{p-1}$$
$$\rho_2 = \varphi_1\rho_1 + \varphi_2 + \ldots + \varphi_p \, \rho_{p-2}$$
$$\ldots \ldots \ldots \ldots$$
$$\ldots \ldots \ldots \ldots \tag{2.65}$$
$$\rho_p = \varphi_p \, \rho_{p-1} + \varphi_2 \, \rho_{p-2} + \ldots + \varphi_p$$

These are called the Yule-Walker equations and can be used either to evaluate ρ_1 $\ldots$ ρ_p if the φ weights of the model are known or to estimate the φ weights of the model from normalized autocovariance estimated from the actual time histories of the random process.

The simplest examples of the AR process are the AR(1) and AR(2) processes given respectively by

$$x_t = \varphi_1 \, x_{t-1} + a_t + \delta$$
$$x_t = \varphi_1 \, x_{t-1} + \varphi_2 x_{t-2} + a_t + \delta$$

The requirements for the stationarity of the time series are $|\varphi_1| < 1$ for the AR(1) process and that the roots of the characteristic equation $(1 - \varphi_1 B - \varphi_2 B^2) = 0$ lie outside the unit circle for the AR(2) process. The second condition is equivalent to the conditions

$$\varphi_1 + \varphi_2 < 1$$
$$\varphi_2 - \varphi_1 < 1$$
$$|\varphi_2| < 1$$

The mean, autocovariance and the normalized autocovariance at lag k of the AR(1) and AR(2) processes are given by,

AR(1) process:

$$E[x_t] = \frac{\delta}{1 - \varphi_1}$$

$$\gamma_k = \frac{\varphi_1^k \, \sigma_a^2}{1 - \varphi_1^2}$$

and

$$\rho_k = \varphi_1^k$$

AR(2) process:

$$E[x_t] = \frac{\delta}{1 - \varphi_1 - \varphi_2}$$

$$\gamma_k = \varphi_1\gamma_{k-1} + \varphi_2\gamma_{k-2}, \quad k > 0$$

$$\gamma_0 = \varphi_1\gamma_1 + \varphi_2\gamma_2 + \sigma_a^2$$

and

$$\rho_k = \varphi_1 \, \rho_{k-1} + \varphi_2 \, \rho_{k-2}, \quad k > 0$$

where $\rho_0 = 1$.

Substituting $k = 1, 2$ for ρ_k we get

$$\rho_1 = \varphi_1 + \varphi_2 \, \rho_1$$

$$\rho_2 = \varphi_1\rho_1 + \varphi_2$$

which are the Yule-Walker equations for the AR(2) process which can be solved for ρ_1 and ρ_2 and the normalized autocovariance for higher lag numbers can be obtained by using the corresponding recursive relation.

2.5.3 Moving Average Process (MA)

If in the general linear process model defined by (2.49), we introduce parsimony by restricting the ψ_i weights to a finite number that is, $\psi_i = 0$, for $i > q$, then the resulting process is called a *moving average process* of order q or MA(q) and is represented by

$$x_t = \mu + a_t - \theta_1 \, a_{t-1} - \dots \theta_q \, a_{t-q} \tag{2.66}$$

In (2.66) θ_i are introduced instead of ψ_i to distinguish the moving average process from the general linear process. The negative sign is introduced by convention. It is evident that the moving average process is always stationary, since the condition that $\sum\limits_{i=0}^{\infty} \psi_i$ converges is simply that $1 - \sum\limits_{i=1}^{q} \theta_i$ converges. This is always true because there are only finite terms in the sum.

The mean, variance and the autocovariance at lag k for the moving average process are easily obtained as

$$E[x_t] = \mu$$

$$\gamma_0 = \sigma_a^2 \sum\limits_{i=0}^{q} \theta_i^2 \, ; \ \theta_0 = 1$$

and

$$\gamma_k = \begin{cases} \sigma_a^2 \, (-\theta_k + \theta_1\theta_{k+1} + \dots + \theta_{q-k} \, \theta_q), & k = 1, \dots q \\ 0, & k > q \end{cases} \tag{2.67}$$

The normalized autocovariance function is

$$\rho_k = \begin{cases} \dfrac{-\theta_k + \theta_1\theta_{k+1} + \dots + \theta_{q-k} \, \theta_q}{1 + \theta_1^2 + \theta_2^2 + \dots + \theta_q^2}, & k = 1, \dots q \\ 0, & k > q \end{cases} \tag{2.68}$$

Similar to the concept of stationarity of the autoregressive process is the concept of invertibility of the moving average process. Consider the result of expressing a moving average process in the form of (2.55), that is, in terms of the present shock only and the past observations. Any MA process written in this form involves observations back, into the infinite past, which can be considered as AR process of infinite order. For such an inversion of the series to be meaningful, the coefficients

π_i should become small as i gets large and do so fast enough so that the sum $\sum\limits_{i=1}^{\infty} \pi_i$ converges. Thus invertibility of the MA process is the algebraic analog of the stationarity of the AR process and the conditions of invertibility are also analogous, that is, the rocts of the equation $(1 - \theta_1 B - \ldots - \theta_q B^q) = 0$ must lie outside the unit circle in the complex plane.

22.5.4 Mixed Autoregressive Moving Average Processes (ARMA)

We have seen that a finite order MA process can be expressed as an infinite order AR process and vice versa. A natural extension of the AR and MA models would be a class of models having finite number of both autoregressive and moving average terms. Such mixed processes are referred to as *autoregressive-moving average processes,* or ARMA. The autoregressive-moving average process of order p, q, ARMA (p, q) is represented by the time series

$$x_t = \varphi_1 x_{t-1} + \ldots + \varphi_p x_{t-p} + \delta + a_t - \theta_1 a_{t-1} - \ldots - \theta_q a_{t-q} \qquad (2.69)$$

For many series of practical importance, the inclusion of both autoregressive and moving average terms results in a model that has fewer parameters than would be necessary for a satisfactory model of pure AR or pure MA form.

The ARMA (p, q) process will be stationary provided the characteristic equation $\varphi(B) = 0$ has all roots lying outside the unit circle. Similarly the roots of $\theta(B) = 0$ must lie outside the unit circle if the process is to be invertible.

The mean of a stationary process can be expressed as

$$E[x_t] = \varphi_1 E[x_{t-1}] + \ldots + \varphi_p E[x_{t-p}] + \delta + E[a_t] - \ldots - \theta_q E[a_{t-q}]$$

$$= \frac{\delta}{1 - \varphi_1 - \varphi_2 - \ldots - \varphi_p} \qquad (2.70)$$

The covariances are evaluated by a set of equations of the form

$$\gamma_k = E[\tilde{x}_t \tilde{x}_{t-k}] = \varphi_1 E[\tilde{x}_{t-1} \tilde{x}_{t-k}] + \ldots + \varphi_p E[\tilde{x}_{t-p} \tilde{x}_{t-k}] + E[a_t \tilde{x}_{t-k}]$$

$$- \theta_1 E[a_{t-1} \tilde{x}_{t-k}] \ldots - \theta_q E[a_{t-q} \tilde{x}_{t-k}] \qquad (2.71)$$

where $\tilde{x}(t)$, as before, is the deviation of x_t from its mean.
For $k \leq q$, the terms involving $\tilde{x}_{t-k}$ is correlated with all disturbances occurring through period $t - k$. Thus the autocovariances upto lag q will involve the moving average parameters $\theta_1, \ldots \theta_q$. However, for $k > q$, the autocovariance is given by

$$\gamma_k = \varphi_1 \gamma_{k-1} + \ldots + \varphi_p \gamma_{k-p}, \quad k > q \qquad (2.72)$$

because for $k > q$ the terms involving x_{t-k} and the shocks are zero. Consequently, the ρ_k at lags greater than q are given by

$$\rho_k = \varphi_1 \rho_{k-1} + \ldots + \varphi_p \rho_{k-p}, \quad k > q \qquad (2.73)$$

The autoregressive-moving average processes provide a powerful class of models for stationary random processes encountered in practice, because of their flexibility in accounting for a wide range of autocovariance functions.

The simplest autoregressive-moving average model is the ARMA $(1, 1)$ model represented by

$$x_t = \varphi_1 x_{t-1} + \delta + a_t - \theta_1 a_{t-1} \qquad (2.74)$$

The following results for the ARMA (1, 1) model are quite straightforward:

$$E[x_t] = \frac{\delta}{1 - \varphi_1}$$

$$\gamma_0 = \varphi_1 \gamma_1 + \sigma_a^2 + \theta_1(\theta_1 - \varphi_1) \sigma^2$$

$$\gamma_1 = \varphi_1 \gamma_0 - \theta_1 \sigma_a^2$$

$$\gamma_k = \varphi_1 \gamma_{k-1}, \quad k = 2, 3, \ldots \qquad (2.75)$$

Equations (2.75) are solved first for γ_0 and γ_1 to yield

$$\gamma_0 = \frac{1 + \theta_1^2 - 2\varphi_1\theta_1}{1 - \varphi_1^2}$$

and

$$\gamma_1 = \frac{(1 - \theta_1\varphi_1)(\varphi_1 - \theta_1)}{1 - \varphi_1^2} \qquad (2.76)$$

Autocovariances for higher lag numbers are then obtained recursively from the model parameters. The normalized autocovariance at lag 1 is

$$\rho_1 = \frac{(1 - \theta_1\varphi_1)(\varphi_1 - \theta_1)}{1 + \varphi_1^2 - 2\varphi_1\theta_1} \qquad (2.77)$$

For higher lags the following recursive relation holds

$$\rho_k = \varphi_1 \rho_{k-1} \quad k = 2, 3 \qquad (2.78)$$

2.5.5 Autoregressive Integrated Moving Average Process (ARIMA)

Many time series in practice exhibit nonstationary behaviour. There are number of ways in which the time series may depart from stationary behaviour. In some cases, even though the time series is nonstationary, it displays certain uniformity as it evolves in time. This kind of nonstationarity is such that the behaviour of the time series at different segments in time are essentially similar. Such nonstationarity is displayed by series whose successive changes or differences are stationary. Thus the stationary time series models discussed in the previous sections like the AR, MA and ARMA provide the basis for constructing a highly flexible class of models for nonstationary series which exhibit homogeneity if we work with their differences.

We have seen that an ARMA process is stationary if the roots of the characteristic function $\varphi(B) = 0$ lie outside the unit circle. It exhibits nonstationary character if the roots lie inside the unit circle. For the case when the roots of $\varphi(B) = 0$ lie on the unit circle, the process is still nonstationary but the series exhibits certain homogeneity in behaviour as will be seen presently.

Consider the general ARMA model of (2.69) written in operator form

$$\varphi(B)\tilde{x}_t = \theta(B) a_t \qquad (2.79)$$

where

$$\tilde{x}_t = x_t - \frac{\delta}{1 - \varphi_1 - \varphi_2 - \ldots - \varphi_p}$$

$$\tilde{x}_{t-1} = x_{t-1} - \frac{\delta}{1 - \varphi_1 - \ldots - \varphi_p}$$

$$\varphi(B) = 1 - \varphi_1 B - \varphi_2 B^2 - \ldots - \varphi_p B^p$$

and

$$\theta(B) = 1 - \theta_1 B - \theta_2 B^2 - \ldots - \theta_q B^q$$

If (2.79) is to represent a nonstationary time series at least one of the roots of $\varphi(B) = 0$ must either lie on the unit circle or inside it. Let us consider a special case when d of the roots lie on the unit circle and the rest outside it. Then one can represent (2.79) in the form

$$\varphi(B)\, \tilde{x}_t = \Phi(B)\, (1 - B)^d\, \tilde{x}_t = \theta(B)\, a_t \tag{2.80}$$

where $\Phi(B)$ is a stationary autoregressive operator.

Since $\nabla^d \tilde{x}_t = \nabla^d x_t$ for $d \geq 1$, where ∇ is the forward difference operator defined by (2.48a), we can express (2.80) as

$$\Phi(B)\, \nabla^d x_t = \theta(B)\, a_t \tag{2.81}$$

or equivalently

$$\Phi(B)\, w_t = \theta(B)\, a_t \tag{2.82}$$

where

$$w_t = \nabla^d x_t \tag{2.83}$$

Thus we see that the dth difference of the process x_t is stationary and represented by an ARMA process. An alternative way of looking at the process for $d \geq 1$ results from inverting (2.83) to obtain

$$x_t = S^d w_t \tag{2.84}$$

where S is the infinite summation operator defined by (2.48b). This is so because

$$Sw_t = \sum_{i=-\infty}^{t} w_i = (1 + B + B^2 + \ldots) w_t = (1 - B)^{-1} w_t = \nabla^{-1} w_t$$

In expanded notation

$$S^d w_t = \sum_{l=-\infty}^{t} \sum_{k=-\infty}^{l} \sum_{j=-\infty}^{k} \sum_{i=-\infty}^{j} w_i \tag{2.85}$$

Equation (2.85) implies that the process x_t can be obtained by summing up or integrating the stationary process w_t, d times. Hence the model specified by (2.81) is called as the autoregressive-integrated-moving average process (ARIMA). The general ARIMA (p, d, q) process is expressed as

$$(1 - \varphi_1 B - \varphi_2 B^2 - \cdot \varphi_p B^p)\, \nabla^d x_t = (1 - \theta_1 B - \theta_2 B^2 - \ldots \theta_q B^q)\, a_t \tag{2.86}$$

where p is the order of the stationary autoregressive terms, d is the order of integration and q the order of the moving average terms.

We have so far discussed several models which can describe a stationary stochastic process based on the assumption that it is generated by passing a white noise through a linear filter. They take the particular forms of the AR, MA or the ARMA models. We have also seen that for a certain class of nonstationary processes the ARIMA model can be fitted. A successful implementation of such time series modelling requires the identification of a particular model with the actual process, estimation of the model parameters and validation of the model. Procedures for identification, estimation and verification of the model are outside the scope of this text and readers may refer to the book of Box and Jenkins (1971) for these aspects.

Recently the time series models have been extended to describe nonlinear random phenomena (Tong, 1978). Their applications to problems of nonlinear random vibrations can be found in the works of Ozaki (1980, 1981a, 1981o). Such models help to explain typical behaviour exhibited by nonlinear systems like the jump phenomenon and limit cycles which are otherwise obscured in the conventional methods of nonlinear random vibration analysis.

Example 2.4: ARMA model for earthquake ground motion
(Polhemus and Cakmak, 1981)

Polhemus and Cakmak (1981) describe a parametric time series model for earthquake ground motion. For purposes of modelling they considered the observed records of the three components of ground acceleration of the San Fernando Earthquake ($M_L = 6.3$) recorded at 8244 Orion Boulevard (Caltech record No. CO48) which are sampled at uniform intervals of time. Figure 2.4 shows the recorded acceleration in cm/sec^2 for the first 30 secs of the NOOW component of the IC 048 series from the San Fernando earthquake. The record exhibits nonstationary character with the variance changing as a function of time. First a polynomial is fitted to describe the change in variance with time. Each value of the original time series is then normalized by dividing by the corresponding local estimate of the standard deviation to yield a stationary time series. An ARMA model is fitted to the transformed series. A sequence of white noise is generated and passed through the ARMA filter and the resulting series is retransformed to give the time series representing the earthquake ground acceleration.

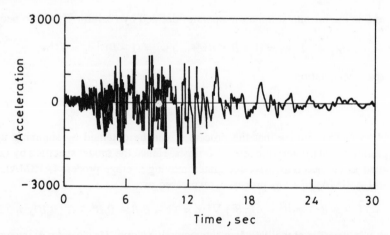

Fig. 2.4 Acceleration record from the San Fernando Earthquake 8244 ORION
BOULEVARD: Component NOOW (Polhemus and Cakmak, 1981).

Let x_t represent the time series corresponding to a particular component of earthquake acceleration. The nonstationary character of the time series may be represented by multiplying the filtered noise by a non-negative time varying function $g(t)$, that is

$$x_t = g(t) \, [\psi(B)a_t] \tag{2.87}$$

where $\psi(B)$ is the transfer function of the ARMA model given by

$$\psi(B) = \frac{1 - \theta_1 B - \theta_2 B^2 - \ldots - \theta_q B^q}{1 - \varphi_1 B - \varphi_2 B^2 - \ldots - \varphi_p B^p} \tag{2.88}$$

In the analysis it is assumed that the mean of x_t is zero and a_t is a sequence of uncorrelated random variables with zero mean and constant variance σ_a^2, that is a white noise process.

$(1 - \sum_{i=1}^{q} \theta_i B_i)$ and $(1 - \sum_{i=1}^{p} \varphi_i B_i)$ are the MA(q) and AR(p) operators respectively. The variance of x_t can be expressed as

$$\sigma_x^2(t) = g^2(t)[1 + \sum_{k=1}^{\infty} \psi_k^2]\sigma_a^2 \tag{2.89}$$

The procedure for the construction of the model defined by (2.81) for the earthquake ground acceleration consists of the following steps.

1. Estimation of the variance function $\sigma_x^2(t)$
2. Construction of a stationary time series by transforming the nonstationary series by dividing it by $\sigma_x(t)$ determined in step 1.
3. Estimation of an ARMA model from the stationary time series.
4. Validation of the model.

Each of these steps are briefly outlined below.

Estimation of Variance Function
The transformation of Box-Cox (1964) is applied to the squared series of x_t to remove any skewness in distribution. The transformation is of the form

$$\left. \begin{array}{ll} Y(\lambda) = \dfrac{Y^\lambda - 1}{\lambda \, m_g^{\lambda-1}} & \lambda \neq 0 \\[3mm] \qquad\quad = m_g \, \log Y & \lambda = 0 \end{array} \right\} \tag{2.90}$$

where Y corresponds to the variance function. m_g is the geometric mean of observations. A polynomial model of the form

$$h(t) = \beta_0 + \beta_1 t + \beta_2 t^2 + \ldots + \beta_k t^k \tag{2.91}$$

is fitted to the expected values of the transformed squared accelerations. A nonlinear least square estimation procedure is adopted to simultaneously determine the transformation parameter λ and the coefficients $\beta_0, \beta_1, \ldots \beta_k$. The value of k is determined by a stepwise procedure and a fourth order polynomial seems appropriate for the purpose and the transformation parameter λ is approximately 0.12. The parameter estimates and approximate standard errors are given in Table 2.2.

Table 2.2   Results of model fitting and variance transformation (Polhemus and Cakmak 1981)

Parameter	Estimate	Standard Error	Sum of Squares
λ_1	0.12	0.01	
β_0	1.05×10^5	0.08×10^5	
β_1	9.59×10^2	0.75×10^2	1.42×10^{12}
β_2	-1.79×10^1	0.20×10^1	1.33×10^{12}
β_3	1.17×10^{-3}	0.20×10^{-3}	0.85×10^{12}
β_4	-2.67×10^{-7}	0.67×10^{-7}	0.06×10^{12}

Construction of Stationary Time Series

The variance $\sigma_x^2(t)$ is estimated by applying the reverse Box-Cox transformation to give

$$\sigma_x^2(t) = [\hat{h}(t)\,\hat{\lambda}\,m_g^{\hat{\lambda}-1} + 1]^{1/\hat{\lambda}}, \quad \hat{\lambda} \neq 0$$

$$= \exp\{\hat{h}(t)/m_g\}, \quad \hat{\lambda} = 0 \tag{2.92}$$

where the top-hats represent estimated values. x_t is scaled now by dividing it by the local estimates of its standard deviation to yield the stationary series

$$x_t^* = x_t/\sigma_x(t) \tag{2.93}$$

Estimation of ARMA Model

An ARMA model is constructed to represent the dynamic behaviour x_t. An ARMA (2, 1) model gives a reasonable fit to the data. The estimated fit for the first horizontal component of the San Fernando series and the approximate standard errors shown in brackets are

$$x_t^* = 1.38\,x_{t-1}^* - 0.502\,x_{t-2}^* + a_t + 0.710\,a_{t-1}$$
$$\quad\;(0.025)\qquad\quad\;(0.025)\qquad\quad\;(0.020) \tag{2.94}$$

with $\sigma_a^2 = 0.109$. ARMA (2, 1) models are also fitted to the second horizontal component and vertical component. ARMA (4, 1) models are also fitted for comparison with ARMA (2, 1) models. Parameter estimates are shown in Tables 2.3 and 2.4. In the case of ARMA (4, 1) models standard errors are larger than ARMA (2, 1) models.

Table 2.3   ARMA (2.1) model estimates (Polhemus and Cakmak 1981)

Component	1	2	q_1	σ_a^2
NOOW	1.380	-0.502	-0.710	0.109
	(0.025)	(0.025)	(0.020)	
S9OW	1.448	-0.556	-0.792	0.078
	(0.023)	(0.023)	(0.017)	
DOWN	1.456	-0.559	-0.789	0.092
	(0.023)	(0.023)	(0.017)	

Validation of Models

The models are validated by comparing the model with the oberved series in terms of the residual autocovariances. Table 2.5 gives the estimated residual normalized

Table 2.4   ARMA (4.1) model estimates (Polhemus and Cakmak, 1981)

Component	1	2	3	4	q_1	σ_a^2
NOOW	1.560	− 0.903	0.293	− 0.050	− 0.594	0.107
	(0.053)	(0.107)	(0.094)	(0.040)	(0.048)	
S9OW	1.648	− 1.036	0.403	− 0.106	− 0.681	0.076
	(0.040)	(0.082)	(0.077)	(0.034)	(0.032)	
DOWN	1.637	− 0.957	0.245	− 0.003	− 0.0687	0.089
	(0.044)	(0.092)	(0.085)	(0.037)	(0.036)	

Table 2.5   Statistics of residuals from fitted ARMA models (Polhemus and Cakmak 1981)

Component	Model	1	2	3	4	5	6
NOOW	(2, 1)	0.06	− 0.06	− 0.08	0.03	0.04	0.10
S9OW	(2, 1)	0.07	− 0.09	− 0.06	0.07	0.00	0.11
DOWN	(2, 1)	− 0.08	− 0.04	− 0.11	0.03	− 0.01	0.10
NOOW	(4, 1)	− 0.00	− 0.01	0.02	− 0.04	0.01	0.07
S9OW	(4, 1)	− 0.00	− 0.01	0.04	− 0.01	− 0.04	0.08
DOWN	(4, 1)	− 0.00	− 0.01	0.01	− 0.00	0.05	0.01

autocovariances for the fitted models. If the residuals are uncorrelated 95% of all estimates should lie within $\pm 2/\sqrt{N}$ about zero, where N is the length of the series. For $N = 1500$ it is about 0.05. It is seen from the table that ARMA (2, 1) model has some correlation in the residuals while ARMA (4, 1) models are better. The model is also validated by simulating a time series using the model and constructing again an ARMA model for the simulated series. This procedure should recover the model parameters with reasonable accuracy. Table 2.6 shows comparison between the models for the original series and the simulated series. The agreement is very close. Polhemus and Cakmak also validated their model by constructing earthquake velocity spectra from the simulated series and comparing them with known spectra.

Table 2.6   ARMA parameter estimates and standard errors from observed and simulated series—NOOW component (Polhemus and Cakmak 1981)

Series	1	2	3	4
Observed	1.374	− 0.496	− 0.710	0.109
	(0.025)	(0.025)	(0.021)	
Simulation No. 1	1.340	− 0.464	0.694	0.123
	(0.026)	(0.026)	(0.021)	
Simulation No. 2	1.362	− 0.491	− 0.729	0.121
	(0.025)	(0.025)	(0.020)	
Simulation No. 3	1.396	− 0.528	− 0.708	0.117
	(0.024)	(0.024)	(0.020)	

2.6   Simulation of Random Processes

For a meaningful application of the random process models, it is essential that the assumed probability structure and the assumption of stationarity or nonstationarity should in some measure fit the actual process reasonably well. This is usually done by statistical inference studies on the model in conjunction with available sampling functions of the random process. Sometimes it becomes necessary to artificially generate sample functions of the processes for the analysis and design of systems operating in the random environment. The stochastic model being a valid representation of the random process must be capable of generating the required sample functions. Thus the procedure of modelling and simulation of random processes go together.

The need to simulate the random processes can be appreciated by considering the earthquake phenomenon. Severe earthquakes are rare in occurrence. In earthquake resistant design of structural systems one has to contend with only a few strong-motion records of actual earthquakes. This lack of information about the random process can be overcome by numerically generating artificial earthquake records of required severity on the digital computer through stochastic models which reflect the basic characteristics of the earthquake process.

There are many problems in random vibration theory for which conventional time and frequency domain analyses are not applicable. For example, the response of highly nonlinear structures to random excitation cannot be estimated by the usual procedures based on Markov-vector approach, the equivalent linearization or perturbation methods. Again for nonstationary excitations of a general nature the frequency domain calculations of response statistics are not always possible. The time domain integrals are also cumbersome to evaluate, if not impossible. The estimation of failure statistics in such cases is an even more difficult problem. By generating sample functions of an excitation process in these cases, the governing differential equations can be numerically integrated to build sample functions of the response which can be processed to give the required statistics of the response, its extreme values and other failure statistics, such as first-passage time, level crossings, etc. Simulation techniques are also useful in the solution of problems associated with random material properties and random eigen values.

In the following sections we shall discuss the various procedures by which random processes may be simulated on the digital computer with special reference to random vibration applications.

2.7 Simulation Using Poisson and Filtered Poisson Process Models

In section 2.3.1 we had seen that the probability structure of the maximum intensity measure of a random process which exhibits randomness both in its time of occurrence and in the measure of its intensity can be obtained in terms of the parameters of the Poisson process governing the occurrence of the random event and the probability distribution function governing its magnitude. For example, (2.3) may be a representation for the probability distribution of the maximum intensity measure of an earthquake at a particular location. The model is determined by the parameters ν, α and z_0. Once these parameters are estimated, the sample values of the random variable $Z_{max}(t)$ representing the maximum intensity measure of the earthquake can be generated by use of random numbers to fit the appropriate distribution given in (2.3).

The Monte Carlo technique can similarly be applied to simulate the filtered Poisson process models. Artificial sample function can be generated by use of random numbers fitting the probability distributions of τ_n and Y_n and with the knowledge of the impulse shape function $w(\cdot)$ in (2.11).

For example, Racicot (1969) simulated the fluctuating component of the wind velocity by generating sample values of the random variables τ, Y and D in (2.18) using random numbers. For the computer simulation he adopted an average wind velocity of 45 mph at a reference height of 33 ft, with parameter $v = 1.0$ gust/sec, $\mu = 0.25$/sec and $\sigma = 5$ ft/sec. The random variables τ and D follow the exponential distribution and Y the Gaussian distribution. Sample values of these random variables are generated in a straight forward manner using uniform random variates between 0 and 1.

First, a table in a vector form of pulse starting times, pulse duration for a given number N_p of pulses is made. The total amount of simulated time is estimated as the product of the pulse rate v and N_p. For ease of computation, the termination of a gust pulse of magnitude Y was considered as the initiation of a new pulse with magnitude $- y$. The sum of the two resulting modified pulses gives the required rectangular pulse shape which is shown in Fig. 2.5. The time of a new pulse initiation is then the sum of the original pulse starting time τ and pulse duration

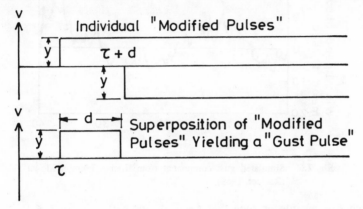

Fig. 2.5.   Superposition of modified wind pulses (Racicot, 1969).

time d. As in the simulation process it is possible to have a new pulse initiated before the termination of an old pulse, the table of starting times which includes both initiation and termination times with associated magnitudes is sorted so that the modified pulse starting times occurred consecutively. Starting at $t = 0$, the time is incremented at small steps of Δt, checking at each increment whether a modified pulse has been initiated. If so, the magnitude of that pulse is added to an accumulative sum of all previous pulses. The simulated wind velocity record is compared with an as actual wind velocity record and the comparison is good as seen from Fig. 2.6.

2.8   Simulation Using Filtered White Noise Model

Consider in general the nonstationary random process modelled by the filtered white noise process as given in (2.29) and (2.30). Simulation of $X(t)$ specified by (2.29) can be effected on a digital computer in which independent Gaussian random numbers are generated representing the white noise process $W(t)$. A procedure for generating the white noise sequence is given Appendix 2A. Sample values $X(t)$ of the random process $X(t)$ at discrete time points can be generated by approximating the integral by the summation

$$x(t) = x(k\,\Delta t) \simeq \ \psi(k\,\Delta t) \sum_{j=-\infty}^{k} h\left\{k\,\Delta t - \left(j - \tfrac{1}{2}\right)\Delta t\right\} W\left\{\left(j - \tfrac{1}{2}\right)\Delta t\right\} \Delta t \qquad (2.95)$$

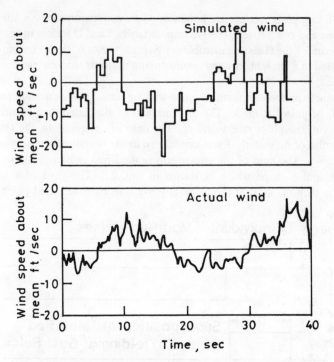

Fig. 2.6. Simulated gust component from filtered Poisson Model (Racicot, 1969).

in which k is an integer such that $t = k\,\Delta t$ and $x(t) = 0$, for $k \le 0$ $(t \le 0)$ as $\psi(t) = 0$ for $t \le 0$. Similarly the process specified by (2.30) can be simulated as

$$x(t) = x(k\,\Delta t) \simeq \sum_{j=1}^{k} h\left\{k\,\Delta t - \left(j - \tfrac{1}{2}\right)\Delta t\right\} \psi\left\{\left(j - \tfrac{1}{2}\right)\Delta t\right\} W\left\{\left(j - \tfrac{1}{2}\right)\Delta t\right\} \Delta t \quad (2.96)$$

At every time instant $(j - 1/2)\Delta t$, an independent Gaussian variate is generated on the computer and the white noise $W(t)$ is considered to remain constant during the time interval $[(j - 1)\Delta t, j\,\Delta t]$ and equal to $W\{(j - 1/2)\Delta t\}$ which represents an approximation to an ideal white noise with substantially uniform spectrum over a wide band compared with the band width of the system. Though the lower bound in the index j in (2.95) extends to $-\infty$, it is sufficient in actual summations to include only a finite number of terms as the system is assumed to have a decaying memory of preceding excitations. The presence of convolution terms in the summation makes this procedure expensive in terms of computer time. Substantial reduction in simulation time is possible if the impulse response function $h(t - \tau)$ can be written as the product of separate functions of t and τ only or as the sum of such products.

As an example let us consider the simulation of earthquake ground motion by use of the filtered white noise model described in section (2.4), with the ground velocity $\dot{z}(t)$ given by (2.34) with $h(t)$ and $\psi(t)$ given respectively by (2.35) and (2.37). Then, $\dot{z}(t)$ can be expressed as

$$\dot{z}(t) = \frac{\psi(t)\,e^{-\zeta_g \omega_g t} \cdot H(t)}{\omega_{dg}} \left\{\sin(\omega_{dg} t)\,I_c(t; -\infty, 0) - \cos(\omega_{dg} t)\,I_s(t; -\infty, 0)\right\} \quad (2.97)$$

where $H(t)$ is the Heaviside unit step function and the function I_c and I_s are defined by the integrals

$$\left.\begin{array}{c} I_c(t; t_0, \zeta) \\[2mm] I_s(t; t_0, \zeta) \end{array}\right\} = \int_{t_0}^{t} e^{(\zeta_g \omega_g - \zeta)u} \left\{\begin{array}{c} \cos \omega_{dg} u \\[2mm] \sin \omega_{dg} u \end{array}\right\} W(u)\, du \tag{2.98}$$

Hence $\dot{z}(t)$ can be directly simulated by use of (2.95).

Once $\dot{z}(t)$ is simulated the ground displacement $z(t)$ and the ground acceleration $\ddot{z}(t)$ can be obtained by integration and differentiation respectively. They can also be simulated directly as shown by Shinozuka and Sato (1967) by use of the following relations. As $z(t) = \displaystyle\int_{-\infty}^{t} \psi(\tau)\, \eta(\tau) d\tau$ it can be expressed as

$$\begin{aligned} z(t) = [&-\{L_1(\alpha)e^{-\alpha t} - L_1(\beta)e^{-\beta t}\} \{\sin (\omega_{dg} t) I_c(t, -\infty, 0) \\ &- \cos (\omega_{dg} t) I_s(t, -\infty, 0)\} - \{L_2(\alpha)e^{-\alpha t} \\ &- L_2(\beta)e^{-\beta t}\} \{\cos (\omega_{dg} t) I_c(t, -\infty, 0) \\ &+ \sin (\omega_{dg} t) I_s(t, -\infty, 0)\} e^{-\zeta_g \omega_g t} - \{L_1(\alpha) \\ &- L_1(\beta)\} I_s(0; -\infty, 0) + \{L_2(\alpha) - L_2(\beta)\} I_c(0; -\infty, 0) \\ &+ L_2(\alpha) \cdot I(t; \alpha) - L_2(\beta)\, I(t; \beta)] \frac{H(t)}{\omega_{dg}} \end{aligned} \tag{2.99}$$

Since $\ddot{z}(t) = \dot{\psi}(t)\, \eta(t) + \psi(t)\, \dot{\eta}(t)$ it can be expressed as

$$\begin{aligned} \ddot{z}(t) = [&\{-(\alpha + \zeta_g \omega_g)e^{-\alpha t} \\ &+ (\beta + \zeta_g \omega_g)e^{-\beta t}\} \{\sin (\omega_{dg} t) I_c(t; -\infty, 0) \\ &- \cos (\omega_{dg} t) I_s(t; -\infty, 0)\} e^{-\zeta_g \omega_g t} \\ &+ \omega_{dg} \psi(t) e^{-\zeta_g \omega_g t} \{\cos (\omega_{dg} t) I_c(t; -\infty, 0) \\ &+ \sin (\omega_{dg} t) I_s(t; -\infty, 0)\}] \frac{H(t)}{\omega_{dg}} \end{aligned} \tag{2.100}$$

In (2.99) and (2.100)

$$I(t; \zeta) = \int_0^t e^{-\zeta u} W(u)\, du$$

$$L_1(\zeta) = (\zeta + \zeta_g \omega_g)/[(\zeta + \zeta_g \omega_g)^2 + \omega_{dg}^2]$$

and

$$L_2(\zeta) = \omega_{dg}/[(\zeta + \zeta_g \omega_g)^2 + \omega_{dg}^2] \tag{2.101}$$

Hence $z(t)$ and $\ddot{z}(t)$ can be simulated by use of formula (2.95). A typical sample function of ground acceleration, velocity and displacement records generated by Shinozuka and Sato (1967) is shown in Fig. 2.7.

Essentially a similar method has been employed in the works of Housner and

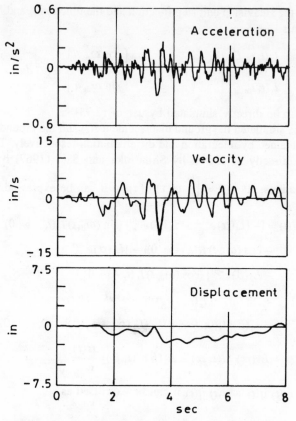

Fig. 2.7. Simulated earthquake ground motion from filtered white noise model (Shinozuka and Sato, 1967).

Jennings (1964), Levy, Kozin and Moorman (1971), Amin and Ang (1968), Saragoni and Hart (1974) and numerous others for the simulation of earthquake ground motions with slight modifications.

2.9 Simulation by Sum of Cosine Functions with Specified Spectral Density

In the simulation methods of the previous sections, the random process was generated either by determining random numbers corresponding to specified distribution of the random process or as an output of an appropriate filter driven by a simulated white noise sequence, and confined mainly to a single variate single dimensional random process. Though conceptually an extension of the above methods to multi-dimensional and multi-variate processes is possible, the computational efforts involved in such procedures are prohibitive even with the present generation computers.

Based on the idea of Rice (1944, 1945) that a stationary random process can be represented as a sum of cosine functions with random phase angles Shinozuka (1971) has developed a powerful method of simulating the sample functions of a general multi-variate and multi-dimensional homogeneous Gaussian random processes with specified spectral and cross spectral densities. The idea has been initially made use of by Goto and Toki (1969) for the simulation of earthquake ground motion and later by Borgman (1969) for the simulation of ocean waves. In both the cases the simulations are confined to a single variate and single dimensional random process. Borgman (1969) also gives a method of simulating a multi-variate random process which is essentially similar to passing a white noise through linear filters and on the lines of Franklin (1965). Shinozuka (1972) later improved upon his original method by simulating a random process as a series of cosine functions with weighted amplitudes and almost evenly spaced frequencies resulting in an increased rate of convergence and reduction in computer time. This method also avoided the periodicity of the generated sample function which was present in his original method. An extension of the approach to a class of multi-dimensional nonstationary random processes called the oscillatory processes is also presented.

Shinozuka and his coworkers have applied the simulation techniques to a number of practical problems in random vibration. Some of them include the response of a nonlinear panel to boundary layer turbulence (Vaicaitis and Jan, 1971; Shinuzoka and Wen, 1972) nonlinear dynamics of off shore structures (Shinozuka and Wen, 1971) simulation of wind velocities and response analysis of structures to wind loads (Wen and Shinozuka, 1971; and Shinozuka and Levy, 1977), fatigue under random loading (Itagaki and Shinozuka, 1972) and random eigen value problems (Shinozuka and Astill, 1972).

From the view point of failure of structures operating in a random vibration environment, the statistics regarding the peaks, envelope and level crossings of a random process are of utmost importance. Though these statistics can be obtained from the simulated sample functions of the random process, such a procedure is cumbersome, as it involves cataloguing of peaks, troughs and crossings at a large number of sample functions. To circumvent this difficulty Yang (1972a) has proposed an efficient and direct way of simulating the envelope of a stationary and evolutionary nonstationary random process by extending Shinozuka's method.

In this section, we shall discuss the approach due to Shinozuka (1972) for the simulation of the random processes as a sum of cosine functions and the method due to Yang (1972a) for the simulation of random envelope processes.

2.9.1 Simulation of a Single Variate One Dimensional Random Process

A one dimensional single variate stationary random process with mean zero can be expressed in spectral representation as

$$X(t) = \int_{-\infty}^{\infty} e^{i\omega t} \, dZ(\omega) \tag{2.102}$$

where $Z(\omega)$, called the spectral process is orthogonal, i.e. $E[dZ(\omega)dZ^*(\omega)] = 0$ for $\omega_1 \neq \omega_2$. The auto-correlation function of $X(t)$ can then be expressed as

$$R_{xx}(\tau) = E[X(t)X(t+\tau)] = \int_{-\infty}^{\infty} e^{i\omega \tau} E[|dZ(\omega)|^2] \tag{2.103}$$

Equating $E[|dZ(\omega)^2|]$ to $\Phi_{xx}(\omega)d\omega$, where $\Phi_{xx}(\omega)$ is the spectral density of $X(t)$, (2.103) reduces to the well-known Weiner-Khinchine relation. When $X(t)$ is a real process (2.102) becomes

$$X(t) = \int_{-\infty}^{\infty} [\cos \omega t \, dU(\omega) + \sin \omega t \, dV(\omega)] \tag{2.104}$$

where $U(\omega)$ and $V(\omega)$ for any $\omega \geq 0$ are two mutually orthogonal processes both real with orthogonal increments,

$$E[dU(\omega)]^2 = E[dV(\omega)]^2 = G_{xx}(\omega)d\omega \qquad (2.105)$$

where $G_{xx}(\omega) = 2\Phi_{xx}(\omega)$ is the one sided spectral density function. The above conditions are satisfied if we define

$$dU(\omega_k) = [2G_{xx}(\omega_k)\Delta\omega]^{1/2} \cos\varphi_k$$

$$dV(\omega_k) = -[2G_{xx}(\omega_k)\Delta\omega]^{1/2} \sin\varphi_k \qquad (2.106)$$

where $\varphi_k(k = 1, 2, ...)$ are independent and identically distributed random phase angles with a uniform density function $1/2\pi$ in the interval $[0, 2\pi]$. The stationary random process can be simulated by subsitution of (2.106) in (2.104) and approximating the integral in (2.104) in the following form:

$$x(t) = \sum_{k=1}^{N} [2G_{xx}(\omega_k)\Delta\omega]^{1/2} \cos(\omega_k' t + \varphi_k) \qquad (2.107)$$

This is the essence of Shinozuka's method of simulation. In his notation the simulation formula is expressed as

$$x(t) = \sqrt{2} \sum_{k=1}^{N} A_k \cos(\omega_k' t + \varphi_k) \qquad (2.108)$$

where

$$A_k = [\Phi_{xx}(\omega_k)\Delta\omega]^{1/2}, \quad \omega_k' = \omega_k + \delta\omega_k$$

$$\omega_k = \omega_l + \left(k - \frac{1}{2}\right)\Delta\omega, \quad \Delta\omega = (\omega_u - \omega_l)/N$$

is the frequency increment, $\delta\omega_k$ is a small random frequency varying uniformly between $-\Delta\omega'/2$ and $\Delta\omega'/2$ with $\Delta\omega' \ll \Delta\omega$ and φ_k are uniform random numbers distributed in the region $[0, 2\pi]$. ω_l and ω_u are frequency bounds such that $\Phi_{xx}(\omega)$ does not have significant values in the region $\omega_l < \omega < \omega_u$. The small random frequency $\delta\omega_k$ is introduced to avoid the periodicity in the generated sample function.

It can be shown the process simulated by (2.108) is consistent with its spectral representation. The autocorrelation function and consequently the power spectral density respectively converge to the target autocorrelation and target spectral density of the random process when $N \to \infty$ as $1/N^2$. Moreover, the simulated process is ergodic upto second moment regardless of the size N, and tends to a Gaussian random process as $N \to \infty$ by virtue of the central limit theorem.

2.9.2 Simulation of Multi-dimensional and Multi-variate Random Processes
The simulation procedure represented by (2.108) can be extended directly to a multi-dimensional homogeneous random process $X(\bar{t})$ in the following form:

$$x(\bar{t}) = \sqrt{2} \sum_{k=1}^{N} A_k(\bar{\omega}_k) \cos(\bar{\omega}_k' \cdot \bar{t} + \varphi_k) \qquad (2.109)$$

where $A(\bar{\omega}_k) = [\Phi_{xx}(\bar{\omega}_k)\Delta\omega_1\Delta\omega_2 ... \Delta\omega_n]^{1/2} = [\Phi_{xx}(\bar{\omega}_k)\Delta\bar{\omega}]^{1/2}$ and $N = N_1 N_2 ... N_n$, $\bar{t}$ is the n dimensional vector of indexing parameters and $\bar{\omega}_k$ is the corresponding frequency (wave number) vector $\bar{\omega}'_k = \bar{\omega}_k + \delta\bar{\omega}_k$, is a small random frequency vector introduced to avoid the periodicity of the simulated process and $\bar{\omega}_k \cdot \bar{t}$ represents the vector dot product. In expanded notation (2.109) can be expressed as

$$x(\bar{t}) = \sqrt{2} \sum_{k_1=1}^{N_1} \cdots \sum_{k_n=1}^{N_n} [\Phi_{xx}(\omega_{1k_1}, \ldots \omega_{nk_n})\Delta\omega_1 \ldots \Delta\omega_n]^{1/2}$$

$$\cos(\omega'_{1k_1}t_1 + \ldots + \omega'_{nk_n}t_n + \varphi_{k_1 \ldots k_n}) \tag{2.110}$$

where $\varphi_{k_1 \ldots k_n} = \varphi_k$ is the independent random phase angle uniformly distributed between 0 and 2π, $(\Delta\omega_1, \ldots \Delta\omega_n) = \left(\dfrac{\omega_{1u} - \omega_{1l}}{N_n}, \ldots \dfrac{\omega_{nu} - \omega_{nl}}{N_n}\right)$, t_i is the ith index-

ing parameter, ω_i the corresponding frequency or wave number, ω_{il}, ω_{iu} respectively are lower and upper bounds on ω_i beyond which the spectral density function $\Phi_{xx}(\omega_1, \ldots \omega_n)$ has insignificant values, $\omega_{ik_i} = \omega_{il} + (k_i - 1/2)\Delta\omega_i$ and $\omega'_{ik_i} = \omega'_{ik_i} + \delta\omega_i$, in which the index k_i varies from 1 to N_i and i from 1 to n. $\delta\omega_i$ is the small random frequency introduced to avoid the periodicity of the simulated process, uniformly distributed between $-\Delta\omega'_i/2$ and $\Delta\omega'_i/2$ with $\Delta\omega'_i \ll \Delta\omega_i$. As for the one dimen-sional process, the n dimensional process represented by (2.109) possesses the autocorrelation function and mean square spectral density function identical to that of the parent process, as $N_1, N_2 \ldots$ tend to ∞. The ergodicity up to second moment can also be established.

Simulation of multivariate and multidimensional homogeneous random process by sum of cosine functions is given by the following procedure. Consider a set of m homogeneous processes of n dimensions $X_j(\bar{t}), j = 1, 2, \ldots, m$ with zero mean vector and cross spectral density matrix $\Phi_{\bar{X}\bar{X}}(\bar{\omega})$, defined by

$$\Phi_{\bar{X}\bar{X}}(\bar{\omega}) = \begin{bmatrix} \Phi_{11}(\bar{\omega}) & \Phi_{12}(\bar{\omega}) & \ldots & \Phi_{1n}(\bar{\omega}) \\ \Phi_{21}(\bar{\omega}) & \Phi_{22}(\bar{\omega}) & \ldots & \Phi_{2n}(\bar{\omega}) \\ \ldots & \ldots & \ldots & \ldots \\ \ldots & \ldots & \ldots & \ldots \\ \Phi_{m1}(\bar{\omega}) & & & \Phi_{mn}(\bar{\omega}) \end{bmatrix} \tag{2.111}$$

where $\Phi_{jk}(\bar{\omega})$ is the n fold Fourier transform of the cross correlation function $R_{jk}(\bar{\tau})$, $\bar{\omega}$ and $\bar{\tau}$ respectively denoting the frequency (wave number) and parametric lag vectors. As $R_{jk}(\bar{\tau}) = R_{jk}(-\bar{\tau})$, $\Phi_{\bar{X}\bar{X}}(\bar{\omega})$ is Hermitian and non-negative definite.

When the cross spectral density matrix $\Phi_{\bar{X}\bar{X}}(\bar{\omega})$ is positive definite it can be decomposed in the following form

$$\Phi_{\bar{X}\bar{X}}(\bar{\omega}) = H_{\bar{X}\bar{X}}(\bar{\omega}) H^*_{\bar{X}\bar{X}}(\bar{\omega})^T \tag{2.112}$$

where $H_{\bar{X}\bar{X}}(\bar{\omega})$ is a lower triangular matrix and $H^*_{\bar{X}\bar{X}}(\bar{\omega})$ is an upper triangular matrix, the asterisk denoting complex conjugation and T the transpose. The elements of the matrix $H_{\bar{X}\bar{X}}(\bar{\omega})$ can be obtained as

$$H_{kk}(\bar{\omega}) = [D_k(\bar{\omega})/D_{k-1}(\bar{\omega})]^{1/2}, k = 1, 2, ..., n \qquad (2.113)$$

where $D_k(\bar{\omega})$ is the kth principal minor of $\Phi_{\bar{X}\bar{X}}(\bar{\omega})$ with $D_0 = 1$ and

$$H_{jk}(\bar{\omega}) = H_{kk}(\bar{\omega}) \left| \Phi_{\bar{X}\bar{X}} \begin{pmatrix} 1, 2 \ldots k-1, j \\ 1, 2 \ldots k-1, k \end{pmatrix} \right| \Big/ D_k(\bar{\omega}) \qquad \begin{array}{l} k = 1, 2, \ldots m \\ j = k+1, \ldots m \end{array} \qquad (2.114)$$

where $\left| \Phi_{XX} \begin{pmatrix} 1, 2 \ldots k-1, j \\ 1, 2 \ldots k-1, k \end{pmatrix} \right|$ is the determinant of a matrix obtained by deleting all elements except those belonging to $(1, 2, ..., k-1, j)$th rows and $(1, 2, ..., k-1, k)$th columns of matrix $\Phi_{\bar{X}\bar{X}}(\bar{\omega})$.

As the real and imaginary parts of the cross spectral density can be expressed in polar form as

$$H_{jk}(\bar{\omega}) = |H_{jk}(\bar{\omega})| \, e^{i\theta_{jk}(\bar{\omega})} \qquad (2.115)$$

where

$$\theta_{jk}(\bar{\omega}) = -\theta_{jk}(-\bar{\omega}), \quad \text{with } \theta_{jj}(\bar{\omega}) = 0$$

The multivariate multi-dimensional random process $\bar{X}(t)$ can now be simulated by the sum

$$x_j(\bar{t}) = \sum_{r=1}^{j} \sum_{k=1}^{N} |H_{jr}(\bar{\omega}_k)| \sqrt{(2\Delta\bar{\omega})} \cos[\omega'_k \cdot \bar{t} + \theta_{jr}(\bar{\omega}_k) + \varphi_{rk}] \qquad (2.116)$$

where $\bar{\omega}_k, \bar{\omega}'_k, \Delta\bar{\omega}, N$ and φ_{rk} have the same connotation as defined for the n dimensional process in (2.109) and $\theta_{jr}(\bar{\omega}_k) = \tan^{-1}[\text{Im } H_{jr}(\bar{\omega}_k)/\text{Re } H_{jr}(\bar{\omega}_k)]$, Im and Re referring respectively to the imaginary and real parts. It can be shown that the processes $x_j(\bar{t}), j = 1, ... m$, as simulated by (2.115) possess the target cross correlation functions and hence the target cross-spectral density with respect to ensemble averages. For digital simulation of sample functions of $x_j(\bar{t})$, in (2.116) φ_{rk} are replaced by their realized values using Monte-Carlo techniques.

Since the cross-spectral density matrix $\Phi_{\bar{X}\bar{X}}(\bar{\omega})$ is known to be only non-negative definite, special consideration and slight modification are needed in the simulation of $\bar{X}(t)$, when $\Phi_{\bar{X}\bar{X}}(\bar{\omega})$ has a zero principal minor. The vanishing of the principal minor is due to two reasons. Whenever the mean square spectral density of a component process has a zero value corresponding to the frequency vector $\bar{\omega}$ or when the coherence between two component processes of the multivariate processes at a frequency $\bar{\omega}$ is unity the principal minor of $\Phi_{\bar{X}\bar{X}}(\bar{\omega})$ will be zero. To simulate $\bar{X}(t)$ in such situations the following procedure can be adopted.

Let $X_j(\bar{t})$ be the jth component of the multi-variate random vector $\bar{X}(t)$ for which the mean square spectral density $\Phi_{X_j X_j}(\bar{\omega}) = 0$ at frequency $\bar{\omega}$. Then the elements of the cross spectral densities associated with this component are also zero corresponding to this frequency, that is, $\Phi_{X_j X_k}(\bar{\omega}) = 0$, $k = 1, ..., m$. It can be shown that a sufficient condition for such a case is, $H_{jk}(\bar{\omega}) = 0, k = 1, 2, ... j$ and $H_{kj}(\bar{\omega}) = 0, k = j+1, 2, ... m$, where $H_{jk}(\bar{\omega})$ is the jkth element of the lower triangular matrix $H_{\bar{X}\bar{X}}(\bar{\omega})$ in (...112). With this information the remaining elements of the matrix $H_{\bar{X}\bar{X}}(\bar{\omega})$ can be determined by factorizing the submatrix obtained by deleting the jth row and the jth column of $\Phi_{\bar{X}\bar{X}}(\bar{\omega})$ in the form given in (2.112), and the simulation procedure following the same sequence.

A coherence of unity between two component processes of $\bar{X}(t)$ say between $X_j(t)$ and $X_{j+1}(t)$ at a frequency $\bar{\omega}$ implies complete linear dependence between them at this frequency and the existence of a transfer function $g(\bar{\omega})$ such that

$$\Phi_{X_{j+1}X_{j+1}}(\bar{\omega}) = |g(\bar{\omega})|^2 \, \Phi_{X_j X_j}(\bar{\omega}) \tag{2.117}$$

$$\Phi_{X_j X_{j+1}}(\bar{\omega}) = g(\bar{\omega}) \, \Phi_{X_j X_j}(\bar{\omega}) \tag{2.118}$$

In this case, $H_{\bar{X}\bar{X}}(\bar{\omega})$ can be solved for this particular wave number $\bar{\omega}$ in the following manner. First, the size of both matrices $\Phi_{\bar{X}\bar{X}}(\bar{\omega})$ and $H_{\bar{X}\bar{X}}(\bar{\omega})$ are reduced by deleting their $j + 1$th row and column. The reduced $H_{\bar{X}\bar{X}}(\bar{\omega})$ is solved from the reduced $\Phi_{\bar{X}\bar{X}}(\bar{\omega})$ by use of (2.113) and (2.114). The deleted elements of the matrix $H_{\bar{X}\bar{X}}(\bar{\omega})$ can be obtained from the relations

$$H_{j+1,k} = gH_{jk} \qquad \text{for } k = 1, 2, \dots j$$

$$H_{j+1,j+1} = 0$$

$$H_{k,j+1} = 0 \qquad \text{for } k = j + 2, \dots, m \tag{2.119}$$

If two component processes, again say the jth and the $j + 1$th are completely correlated then $g(\bar{\omega})$ as defined above will reduce to a real constant g_0 independent of frequencies. This will lead to $D_k(\bar{\omega}) = 0, k = j + 1, j + 2, \dots m$, for all frequencies. For this case it is necessary to simulate only $m - 1$ component processes without the $j + 1$th component, for it can be simply obtained in the form

$$x_{j+1}(t) = g_0 X_j(t) \tag{2.120}$$

With the proper combination of the above procedures, it is possible to simulate even when more than one component of $\Phi_{\bar{X}\bar{X}}(\bar{\omega})$ vanishes and when more than two components are completely coherent or completely correlated and a combination thereof.

As an example of digital simulation of a multi-dimensional random process consider a two dimensional homogeneous Gaussian process $v(t, s)$ with mean zero and spectral density given by

$$\Phi_{vv}(\omega, k) = \frac{C_D L^2}{2\pi^2} \frac{|\omega|}{(1 + C^2\omega^2)^{4/3}} \frac{\alpha |\omega|}{\pi(\alpha^2\omega^2 + k^2)}$$

where t and s represent time and distance respectively and correspondingly ω and k the frequency and wave number. Essentially the foregoing equation is the model for the fluctuating part of the wind velocity along a straight line direction s adopted by Davenport (1961), at a reference altitude of 33 ft. C_D is the drag coefficient, α is a constant and $C = L/2\pi U$, where $L = 4000$ ft and U is the mean wind velocity at the reference altitude. For $U = 40$ mph, $\alpha = 0.02$ ft sec and $C_D = 0.03$, the simulated wind velocity records for $s = 0.50$ and 200 ft are shown in Fig. 2.8.

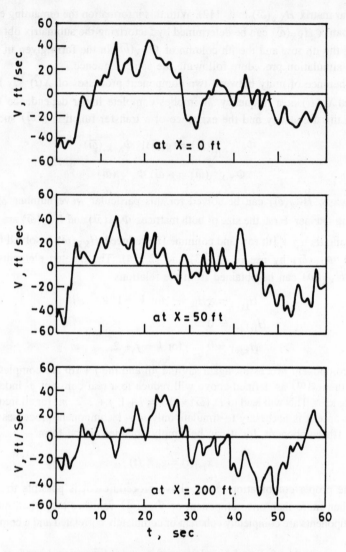

Fig. 2.8.   Simulation of wind velocities by sum of cosine functions
(Shinozuka and Jan, 1972).

2.9.3 Simulation of Nonhomogeneous (Nonstationary) Processes

The simulation procedure of the previous section can be directly extended to multi-dimensional nonhomogeneous processes having evolutionary power. For example, consider the evolutionary nonstationary random process $X(t)$ having the spectral representation

$$X(t) = \int_0^\infty B(t, \omega) \, [\cos \omega t \, dU(\omega) + \sin \omega t \, dV(\omega)] \qquad (2.121)$$

where $B(t, \omega)$ is the deterministic modulating function characterizing the nonstationarity of the process. The mean square value of the random process is,

$$E[X^2(t)] = \int_0^\infty B^2(t, \omega) \, G_{XX}(\omega) \, d\omega \qquad (2.122)$$

where $\{B^2(t, \omega)G_{XX}(\omega)d\omega\}$ is the evolutionary spectral density function.

Thus a real nonstationary random process with evolutionary power can be simulated by the sum

$$x(t) = \sqrt{2} \sum_{k=1}^{N} [B^2(t, \omega) \, \Phi_{XX}(\omega_k) \Delta \omega]^{1/2} \cos{(\omega_k' t + \varphi_k)} \qquad (2.123)$$

where the notations are the same as in (2.100). In particular if $B(t, \omega) = B(t)$, $X(t)$ can be simulated by simply multiplying the stationary process simulated with spectral density $\Phi_{XX}(\omega)$ by $B^2(t)$. The extension to multi-dimensional nonhomogeneous processes with evolutionary power is straight forward. It can be simulated as

$$\bar{x}(t) = \sqrt{2} \sum_{k=1}^{N} A(\bar{t}, \bar{\omega}_k) \cos{(\bar{\omega}_k' \bar{t} + \varphi_k)} \qquad (2.124)$$

where $A(\bar{t}, \bar{\omega}_k) = [B^2(\bar{t}, \bar{\omega}) \, \Phi_{XX}(\bar{\omega}_k) \, \nabla \bar{\omega}]^{1/2}$, $B(\bar{t}, \bar{\omega})$ is the multidimensional modulating function and the remaining notations are as in (2.109).

2.9.4 Simulation of Random Envelope Processes

The central idea of simulating a random process by a sum of cosine functions is extended by Yang (1972a) to the envelope of the random process which is of importance in the failure analysis of structures subjected to random excitation. Yang used the envelope definition of Cramer and Leadbetter (1967) with the envelope process $X_e(t)$ given by

$$X_e(t) = [X^2(t) + \hat{X}^2(t)]^{1/2} \qquad (2.125)$$

where $\hat{X}(t)$ is the Hilbert transform of the random process $X(t)$ which is obtained by passing it through a filter with gain $g(\omega)$ given by

$$g(\omega) = \begin{cases} i & \omega < 0 \\ 0 & \omega = 0 \\ -i & \omega > 0 \end{cases} \qquad (2.126)$$

Hence $X(t)$ is real and has the spectral representation

$$\hat{X}(t) = \int_0^\infty \sin{\omega t} \, dU(\omega) - \cos{\omega t} \, dV(\omega) \qquad (2.127)$$

where $dU(\omega)$ and $dV(\omega)$ are as defined in (2.104) and (2.105). $\hat{X}(t)$ can now be simulated by the series

$$\hat{X}(t) = \sum_{k=1}^{N} [2G_{XX}(\omega_k) \, \Delta \omega_k]^{1/2} \sin{(\omega_k t + \varphi_k)} \qquad (2.128)$$

where φ_k are uniform random variates between 0 and 2π. It can be shown that the above series possesses autocorrelation function identical with the target autocorrelation function as $N \to \infty$ and the central limit theorem ensures the normality of $\hat{X}(t)$. It can further be shown that the autocorrelation function of the series $\hat{X}(t)$ converges to the autocorrelation function of $X(t)$ in (2.107) and hence the random process $X(t)$ can be alternatively simulated by the series $\hat{X}(t)$.

The envelope process $X_e(t)$ can therefore be simulated as,

$$X_e(t) = [X^2(t) + \hat{X}^2(t)]^{1/2} \tag{2.129}$$

Using equally spaced frequency increments $\Delta\omega_1 = \Delta\omega_2 = \ldots = \Delta\omega$, an improvement in the simulation efficiency can be achieved by noting that $X(t)$ and $\hat{X}(t)$ can be expressed as

$$X(t) = \sqrt{(\Delta\omega)}\, \text{Re}\, z(t); \hat{X}(t) = \sqrt{(\Delta\omega)}\, \text{Im}\, z(t) \tag{2.130}$$

where $z(t) = \sum_{k=1}^{N} \{[2G_{XX}(\omega_k)]^{1/2}\, e^{i\varphi k}\} e^{i\omega_k t}$ is the finite complex Fourier transform

of $[2G_{XX}(\omega)]^{1/2}\, e^{i\varphi}$. The envelope $X_e(t)$ can then be simulated as

$$X_e(t) = \sqrt{\Delta\omega}\, |z(t)| \tag{2.131}$$

Therefore the sample functions of the random process $X(t), \hat{X}(t)$ and as well as the envelope $X_e(t)$ are obtained by performing a Fourier transform on $[2G_{XX}(\omega)]^{1/2}\, e^{i\varphi}$ which can be efficiently done by the fast Fourier transform (FFT) technique.

The simulation of the envelope process by use of (2.129) requires the generation of N random phases φ_k uniformly distributed in the interval 0 to 2π. The number N should be taken in such a way that the one-sided power spectral density $G_{XX}(\omega)$ has insignificant values for $\omega \leq \omega_1$ and $\omega \geq \omega_N$. The series in (2.131) exhibits periodicity but the period can be made as large as possible by choosing a small $\Delta\omega$.

The simulation method can be extended to the envelope of an evolutionary nonstationary random process for which the envelope is defined by Yang (1972b)

$$X_e(t) = [X^2(t) + \hat{X}^2(t)]^{1/2} \tag{2.132}$$

where $X(t)$ has the spectral representation given in (2.121) and $\hat{X}(t)$ the spectral representation

$$\hat{X}(t) = \int_0^\infty B(t, \omega)\, [\sin \omega t\, dU(\omega) - \cos \omega t\, dV(\omega)] \tag{2.133}$$

Hence $\hat{X}(t)$ can be simulated by the series

$$\hat{X}(t) = \sqrt{2} \sum_{k=1}^{N} [B^2(t, \omega_k)\, \Phi_{XX}(\omega_k)\, \Delta\omega]^{1/2} \sin(\omega_k t + \varphi_k) \tag{2.134}$$

and $X_e(t)$ by

$$X_e(t) = [X^2(t) + \hat{X}^2(t)]^{1/2} \tag{2.135}$$

where $X(t)$ and $\hat{X}(t)$ are respectively given by (2.123) and (2.134).

An an example the envelope process of a wide band process representing the road surface roughness is simulated by Yang (1972a). Typical sample functions of the process and its envelope having one-sided power spectral density $G_{XX}(\omega)$ equal to $2\Phi_0$ for $\omega \leq \omega_0$ and to $2\Phi_0(\omega_0/\omega)C_2$ for $\omega > \omega_0$, with $\omega_0 = 2\pi$, $\Phi_0 = 0.5/\omega_0$ and $C_2 = 2.0$ are shown in Fig. 2.9. In this example the process is normalized such that the standard deviation σ is 1 inch and the vehicle is travelling at a speed V of 40 mph. It is observed from the figure that the envelope does not necessarily pass through all the peaks which is typical of wide band processes.

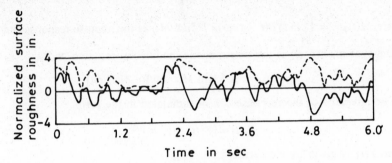

Fig. 2.9. Simulation of envelope of road roughness (Yang. 1972 a)

A simulation technique described here for the envelope process of a one dimensional single variate random process can be extended without much difficulty to the simulation of the envelope processes of multi-variate and multi-dimensional homogeneous and nonhomogeneous random processes with evolutionary power.

2.9.5   Simulation of Random Response Processes
The simulation of random processes by the sum of cosine functions with the random phase provides a basis for directly simulating the response of linear time invariant systems to random excitation. For example consider $X(t)$ as simulated by (2.107) to be the input to a linear time invariant single degree of freedom vibrating system with frequency response function $H(\omega)$. The random response $Y(t)$ can be directly simulated by the series

$$Y(t) = \sum_{k=1}^{N} [2G_{XX}(\omega_k)\Delta\omega_k]^{1/2} |H(\omega_k)| \cos(\omega_k t + \varphi_k) \qquad (2.136)$$

Similarly one can simulate the response envelope process by first simulating the series

$$\hat{Y}(t) = \sum_{k=1}^{N} [2G_{xx}(\omega_k)\Delta\omega_k]^{1/2} |H(\omega_k)| \sin(\omega_k t + \varphi_k) \qquad (2.137)$$

and then the response envelope by

$$Y_e(t) = [Y^2(t) + \hat{Y}^2(t)]^{1/2} \qquad (2.138)$$

The FFT technique can also be efficiently employed to simulate the response processes by expressing $Y(t)$ and $\hat{Y}(t)$ in the form

$$Y(t) = \sqrt{(\Delta\omega)}\,\mathrm{Re}\,z'(t);\ \hat{Y}(t) = \sqrt{(\Delta\omega)}\,\mathrm{Im}\,z'(t) \quad \text{and} \quad Y_e(t) = \sqrt{(\Delta\omega)}\,|z'(t)|$$

where

$$z'(t) = \sum_{k=1}^{N} \{[2G_{XX}(\omega_k)\Delta\omega_k]^{1/2} \,|\, H(\omega)\,|\, e^{i\varphi_k}\} e^{i\omega_k t} \qquad (2.139)$$

If the excitation process $X(t)$ is nonstationary with evolutionary power, the response of the linear system to such an excitation can be simulated by

$$Y(t) = \sum_{k=1}^{N} [2G_{XX}(\omega_k)\Delta\omega_k]^{1/2} \,|\, M(t, \omega_k)\,|\cos[\omega_k t + \delta(t, \omega_k) + \varphi_k] \qquad (2.140)$$

where $M(t, \omega) = \int_0^t h(\tau) B(t - \tau, \omega) e^{-i\omega\tau} d\tau$, $h(t)$ is the impulse response function of the system, $B(t, \omega)$ is the deterministic modulating function and

$$\delta(t, \omega) = \tan^{-1}[\operatorname{Im} M(t, \omega)/\operatorname{Re} M(t, \omega)]$$

The envelope of the response can be simulated in the form

$$Y_e(t) = [Y^2(t) + \hat{Y}^2(t)]^{1/2} \qquad (2.141)$$

where $\hat{Y}(t)$ is given by the series

$$\hat{Y}(t) = \sum_{k=1}^{N} [2G_{XX}(\omega_k)\Delta\omega_k]^{1/2} \,|\, M(t, \omega_k)\,|\sin[\omega_k t + \delta(t, \omega_k) + \varphi_k] \quad (2.142)$$

The extension to the simulation of the random response of linear multi-degree of freedom systems and continuous systems is straightforward through the use of the normal mode approach in conjunction with the simulation procedure for multivariate and multi-dimensional random processes.

The theory of random response of linear systems under stochastic loading is well developed and therefore the simulation of random response in such situations is not of great practical advantage. However, in the case of highly nonlinear systems excited by random loading, the simulation techniques provide an useful tool in the solution of the random response, where other methods like the Markov-vector formulation, perturbation and equivalent linearization methods cannot be used to advantage. In the simulation method, sample functions of the excitation process are generated as the sum of cosine functions and the nonlinear differential equations characterizing the system are numerically integrated to yield corresponding sample functions of the random response from which the response statistics can be estimated.

Example 2.5:   Response of nonlinear panel to boundary layer turbulence
(Vaicaitis and Jan, 1971)

As an application of the simulation method to multi-dimensional random response processes let us take the example of the response of a nonlinear panel to turbulent boundary layer excitation. This problem was considered in detail by Shinozuka and Jan (1972), Vaicaitis and Jan (1971), Shinozuka and Wen (1972) and Vaicaitis et al (1972).

Consider the nonlinear random vibration of a simply supported rectangular panel (Fig. 2.10) with turbulent flow on one side and cavity flow on the other and having a geometric nonlinearity. In the analysis the fluid-solid interaction effects have also been considered. The panel motion is governed by the following differential equations in non-dimensional form with appropriate boundary conditions

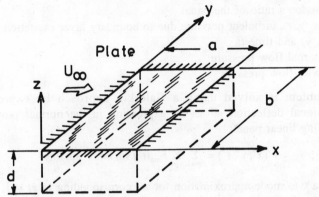

Fig. 2.10. Geometry of nonlinear panel with cavity (Vaicaitis and Jan, 1971).

$$\nabla'^4 w' = \frac{\partial^2 \psi'}{\partial y'^2} \frac{\partial^2 w'}{\partial x'^2} + \frac{\partial^2 \psi'}{\partial x'^2} \frac{\partial^2 w'}{\partial y'^2} - 2 \frac{\partial^2 \psi'}{\partial x' \partial y'} \frac{\partial^2 w'}{\partial x' \partial y'}$$

$$- \beta' \frac{\partial w'}{\partial t'} - \frac{\partial^2 w'}{\partial t'^2} + p' - p'_e + p'_c$$

$$\nabla'^4 \psi' = 12(1 - v^2) \left\{ \left(\frac{\partial^2 w'}{\partial x' \partial y'} \right)^2 - \frac{\partial^2 w'}{\partial x'^2} \frac{\partial^2 w'}{\partial y'^2} \right\}$$

with
$$w'(0, y', t') = w'(1, y', t') = w'(x', 0, t') = w'(x', 1, t') = 0$$

$$\frac{\partial^2 w'}{\partial x'^2}(x', 0, t') = \frac{\partial^2 w'}{\partial x'^2}(x', 1, t') = \frac{\partial^2 w'}{\partial y'^2}(0, y', t') = \frac{\partial^2 w'}{\partial y'^2}(1, y', t') = 0$$

In the above equations the nondimensional quantities are defined as

$$w' = w/h, x' = x/a, y' = y/b, \psi' = \psi/D, t' = \frac{t}{ab}(D/\rho)^{1/2}$$

$$\beta' = \beta a b/(\rho D)^{1/2}, p' = a^2 b^2 p/(hD), p'_e = a^2 b^2 p_e/(hD)$$

$$p'_c = a^2 b^2 p_c/(hD), \text{ and } \nabla'^4 = \left[(b/a)\frac{\partial}{\partial x'^2} + (a/b)\frac{\partial^2}{\partial y'^2} \right]^2$$

where the different notations are
  $w(x, y, t)$ — lateral deflection of panel
  ψ — Airy stress function of panel membrane stress
  ρ — surface mass density of the panel
  β — coefficient of linear viscous damping of the panel
  a — panel length in x-direction
  b — panel length in the y-direction
  h — thickness of the panel
  D — $= Eh^3/12(1 - v^2)$, the flexural rigidity of the panel
  E — Young's modulus of the panel

v — Poisson's ratio of the panel

p — $p(x, y, t)$, turbulent pressure due to boundary layer excitation at location (x, y) and time t

p_e— external flow pressure

p_c— cavity flow pressure

The problem is solved using a Galerkin approach by expressing the nondimensional deflection w' as an expression in the normal modes of the corresponding linear panel:

$$w'(x', y', t') = \sum_{m=1} \sum_{n=1} b_{mn}(t') \sin m\pi x' \sin n\pi y'$$

Assuming a two mode approximation for w' corresponding to $m = 1, 2$ and $n = 1$, the solution for the stress function ψ' with average inplane boundary conditions namely

$$\int_0^a \int_0^b \frac{\partial u}{\partial x} dx\, dy = \int_0^a \int_0^b \frac{\partial v}{\partial y} dx\, dy = 0$$

is given by

$$\psi' = \frac{\pi^3 E h^3}{16 D(1 - v^2)\alpha^2} \{[(1 + v\alpha^2)b_1^2 + (4 + v\alpha^2)b_2^2]y'^2 + [(\alpha^2 + v)b_1^2$$

$$+ (\alpha^2 + 4v)b_2^2]\alpha^2 x'^2\} + \frac{E h^3 \alpha^2}{4D} \{(b_1^2 \cos 2\pi x'/8 - b_1 b_2 \cos \pi x'$$

$$+ b_1 b_2 \cos 3\pi x'/9 + b_2^2 \cos 4\pi x'/32) + [(b_1^2 + 4b_2^2)/8\alpha^4$$

$$+ 9b_1 b_2 \cos \pi x'/(1 + 4\alpha^2)^2 - b_1 b_2 \cos 3\pi x'/(9 + 4\alpha^2)^2] \cos 2\pi y'\}$$

where u and v are the inplane displacements and α is the aspect ratio of the plate $\alpha = a/b$, and for simplicity the notations $b_1 = b_{11}$ and $b_2 = b_{21}$ have been adopted.

Substituting the normal mode expansion in the equation of motion along with the solution for ψ' the following set of nonlinear ordinary differential equations in the coefficients b_1 and b_2 are obtained

$$\ddot{b}_1 + \beta_1 \dot{b}_1 + (C_{10} + C_{11}b_1^2 + C_{12}b_2^2)b_1 = 4(F_1' + F_{1e}' + F_{1c}')$$

$$\ddot{b}_2 + \beta_2 \dot{b}_2 + (C_{20} + C_{21}b_1^2 + C_{22}b_2^2)b_2 = 4(F_2' + F_{2e}' + F_{2c}')$$

where

$$C_{10} = (\alpha + 1/\alpha)^2 \pi^4 \qquad C_{20} = (\alpha + 4/\alpha)^2 \pi^4$$

$$C_{11} = \frac{3}{4} \pi^4 [(1 - v^2)(\alpha^2 + 1/\alpha^2) + 2(2v + \alpha^2 + 1/\alpha^2)]$$

$$C_{12} = \frac{3}{4} \pi^4 \{(1 - v^2)[4(\alpha^2 + 1/\alpha^2) + 81\alpha^2/(1 + 4\alpha^2)^2$$

$$+ \alpha^2/(9 + 4\alpha^2)^2] + 2(5v + \alpha^2 + 4/\alpha^2)\}$$

$$C_{21} = C_{12}$$

$$C_{22} = \frac{3}{4} \pi^4 [(1 - v^2)(\alpha^2 + 16/\alpha^2) + 2(8v + \alpha^2 + 16/\alpha^2)]$$

$$\beta_{11} = \beta_{21} = \beta'$$

and F_1', F_{1e}' ... etc. are the nondimensional generalized forces given by expressions of the form

$$F_r'(t') = \int_0^1 \int_0^1 p'(x', y', t') \sin m\pi x' \sin n\pi y' \, dx' \, dy'$$

$$F_{re}'(t') = \int_0^1 \int_0^1 p_e'(x', y', t') \sin m\pi x' \sin n\pi y' \, dx' \, dy', r = 1, 2$$

Now consider the simulation of the generalized forces $F_1'(t')$ and $F_2'(t')$ corresponding to the turbulent boundary layer excitation. The semi-empirical multi-dimensional one-sided power spectral density function of subsonic turbulent boundary layer pressure according to Bull (1967) is

$$
\begin{aligned}
G_{pp}(\omega, k_1, k_2) = \ & \{0.715 \times 10^{-6} q^2 \delta^*/(\pi^2 U_\infty) \, [3.7 \exp(-2\omega\delta^*/U_\infty) \\
& + 0.8 \exp(-0.47 \, \omega\delta^*/U_\infty) - 3.4 \exp(-8\omega\delta^*/U_\infty)]\} \\
& - \left(\frac{\omega}{U_c}\right)^2 \{(0.1\omega/U_c)^2 + (\omega/U_c + k_1)^2 \, [(0.715\omega/U_c)^2 + k_2^2]\}^{-1}
\end{aligned}
$$

where U_∞ is the free stream velocity, $q = 1/2 \, \rho_\infty U_\infty^2$ is the dynamic pressure, δ^* is the boundary layer thickness and U_c is the convection speed. k_1 and k_2 are the wave numbers corresponding to x and y directions and ω is the frequency.

The boundary layer pressure $p(x, y, t)$ and hence the nondimensional prsesure $p'(x', y', t')$ can be simulated as a three dimensional random process according to (2.109). Sample functions of the generalized forces $F_r'(t')$ corresponding to the boundary layer pressure are digitally generated by replacing $p'(x', y', t')$ in the integral by the simulated series. Advantage is taken of the form of the normal modes which the sinusoidal functions and the simulated series for $p(x', y', t')$ which is again a sum of cosine functions in the integrals. Hence the indicated integration in the space domain can be effected in closed form and the generalized forces are expressed as a sum of cosine functions in the parameter t only. For details of the procedure readers may refer to Vaicaitis and Jan (1971).

The fluid motion in the cavity is described by the velocity potential ψ_c satisfying the wave equation

$$\nabla^2 \psi_c - \frac{1}{a_c^2} \frac{\partial^2 \psi_c}{\partial t^2} = 0$$

with the boundary conditions

$$\frac{\partial \psi_c}{\partial x}(0, y, z, t) = \frac{\partial \psi_c}{\partial x}(a, y, z, t) = 0$$

$$\frac{\partial \psi_c}{\partial y}(x, 0, z, t) = \frac{\partial \psi_c}{\partial y}(x, b, z, t) = 0$$

$$\frac{\partial \psi_c}{\partial z}(x, y, -d, t) = 0$$

$$\frac{\partial \psi_c}{\partial z}(x, y, 0, t) = w(x, y, 0, t)$$

The pressure in the cavity is obtained as

$$p_c = -\rho_c \frac{\partial \psi_c}{\partial t}$$

where a_c is the local speed of sound in the cavity and ρ_c is the fluid density inside the cavity. For a shallow cavity Dowell (1969) has shown that the generalized force F'_{rc} corresponding to the cavity pressure p'_c can be expressed as

$$F'_{rc} = -\frac{a^2 b^2 \rho_c a_c^2}{hD \, \pi^4 d} \frac{r[1 - (-1)^r]}{r^2} \sum_{n=1} \frac{n(1 - (-1)^n) b_n}{n^2}$$

The effect of the external fluid pressure on the panel response due to its motion is neglected in the example.

Substituting the expression for the generalized force corresponding to the cavity pressure in the equations for the generalized coordinates b_1 and b_2 and making use of the simulated generalized force corresponding to the boundary layer pressure excitation, the set of non-linear equations are integrated numerically. b_1 and b_2 thus generated at discrete times are used in the normal mode expansion to give sample functions of the random response of the panel at particular locations.

A typical example of the time history of the nonlinear panel response for the parameters $\rho_\infty = \rho_c = 0.00089$ slugs/ft^3, $a_\infty = a_c = 995$ ft/sec, $U_\infty = 800$ ft/sec, $\delta^* = 0.157$ in, $U_c = 0.65 \, U_\infty$ and a rms pressure $\sigma_p = 0.0028 \, \rho_\infty U_\infty^2$, $a = 10$ in., $b = 20$ in., $d = 5$ in., $v = 0.3$, $\rho = 0.1$ h lbs/in^2, $E = 10^7$ psi, $\beta_1 = \beta_2 = 0.901$ simulated by Vaicaitis and Jan (1971) is shown in Fig. 2.11. The response exhibits narrow band characteristics. The number of the temporal and spatial sample points used for the simulation of the generalized forces are $N_1 = 100$, $N_2 = 40$ and $N_3 = 40$.

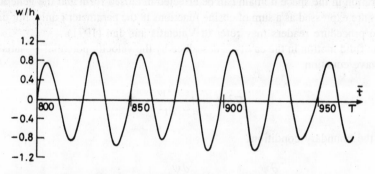

Fig. 2.11. Time history of panel response with cavity (Vaicaitis and Jan. 1971).

2.10   Simulation of Random Processes Using FFT

The Fast Fourier Transform (FFT) technique developed by Cooley and Tukey (1965), and subsequent improvements on the basic algorithm, have been used in a large number of engineering applications involving spectral analysis. Vaicaitis et al (1974), Borgman (1969), Shinozuka (1974) and Yang (1972a) have applied the FFT technique in the simulation of random processes with specified spectral characteristics. But in their works the FFT algorithm is used only during certain stages of the simulation process rather than as a basic philosophy of the simulation procedure. For example Borgman used the FFT algorithm to determine the linear filter weights through which was passed a white noise sequence for the simulation of multi-variate random processes. Vaicaitis used the FFT technique to compute the spectral density of the random response of a nonlinear panel from the simulation series for the response. Shinozuka and Yang showed that the simulation of the random process by a sum of cosine functions could be effected through a discrete Fourier transform for which the FFT algorithm could be efficiently used.

Wittig and Sinha (1975) have used the FFT algorithm in a more direct manner to simulate sample functions of multi-variate Gaussian random processes. They generate discrete frequency functions which correspond to the Fourier transform of the time series. Effecting an inverse Fourier transform on the discrete frequency functions by use of the FFT technique the required sample functions of the random process are obtained. In the sequel we shall discuss their method of simulation.

Consider a set of m homogeneous random processes $X_j(t)$, $j = 1, 2, \ldots, m$ with zero mean vector and one-sided cross spectral density matrix given by

$$\mathbf{G}_{\overline{X}\overline{X}}(\omega) = \begin{bmatrix} G_{11}(\omega) & \ldots & G_{1m}(\omega) \\ \ldots & \ldots & \ldots \\ G_{m1}(\omega) & & G_{mm}(\omega) \end{bmatrix} \qquad (2.143)$$

Let Δt be the time interval between discrete points in the time series to be simulated and N the total number of sample points. Hence the frequency increment for the FFT analysis is $\Delta\omega = 1/(N\Delta t)$. Assume that $G_{\overline{X}\overline{X}}(\omega)$ can be factored into a lower triangular matrix and upper triangular matrix in the form

$$\mathbf{G}_{\overline{X}\overline{X}}(\omega = \mathbf{H}(\omega)\,\mathbf{H}^{*T}(\omega) \qquad (2.144)$$

A typical element of the matrix in (2.144) in expanded notation can be expressed as

$$G_{ij}(\omega) = \sum_{l=1}^{i} H_{il}(\omega)\,H_{jl}^{*}(\omega) \qquad (2.145)$$

Elements of the **H** matrix can be evaluated by the procedure explained in section (2.9.2). Sampling at discrete frequencies $k\Delta\omega$ (2.145) becomes

$$G_{ij}(k\Delta\omega) = \sum_{l=1}^{i} H_{il}(k\Delta\omega)\,H_{jl}^{*}(k\Delta\omega) \qquad (2.146)$$

Now the random process $X_j(t)$ can be simulated at discrete time intervals as,

$$X_j(n\Delta t) = \frac{1}{N} \sum_{k=0}^{N-1} F_j(k\Delta\omega) \exp\left(\frac{i\,2\pi kn}{n}\right) \qquad (2.147)$$

where $F_j(k\Delta\omega)$ are generated in the following manner. First a set of independent Gaussian random numbers β_{ik} and η_{ik} are generated with mean zero and mean square value equal to $0.5 \cdot F_j(k\Delta\omega)$ are obtained from the random numbers thus generated as

$$F_j(k\Delta\omega) = \left[\frac{N}{2\Delta t}\right]^{1/2} \sum_{i=1}^{j} H_{ji}(k\Delta\omega)\alpha_{ik} \tag{2.148}$$

where $\alpha_{ik} = \beta_{ik} + i\,\eta_{ik}$, or in matrix notation

$$\begin{bmatrix} F_1 \\ F_2 \\ \dots \\ F_m \end{bmatrix} = \left(\frac{N}{2\Delta t}\right)^{1/2} \begin{bmatrix} H_{11} & 0 & \dots & 0 \\ H_{21} & H_{22} & 0 & 0 \\ \dots & \cdot & \dots & \dots \\ H_{m1} & H_{m2} & \dots & H_{mm} \end{bmatrix} \begin{bmatrix} \alpha_{1k} \\ \alpha_{2k} \\ \dots \\ \alpha_{mk} \end{bmatrix} \tag{2.149}$$

Thus the required sample function $x_j(t)$ in (2.147) is the discrete Fourier transform of $F_j(\omega)$. It has been shown that the simulated series has cross spectral density matrix identical with that of the target spectral density matrix and that the simulated process is Gaussian. The simulation is effected by use of FFT technique on the series $F_j(k\Delta\omega)$. There is a substantial reduction in computer time by using this method in comparison with Shinozuka's method of sum of cosine functions essentially because of the use of the FFT algorithm. The speed ratio between evaluating $m(m + 1)/2$ series of N cosine terms in Shinozuka's method and taking the FFT of m series each with N terms corresponding to Wittig and Sinha's method is

$$\text{Speed ratio} = \frac{N}{4P}\frac{(m+1)}{2} \tag{2.150}$$

where $P = \log_2 N$.

Further improvement in this method has been suggested by Hudspeth and Borgman (1979) who use a stacked FFT algorithm of the complex Fourier coefficients F_j instead of the unstacked algorithm of Witting and Sinha. The method has advantages over that of Witting and Sinha in terms of computer memory. A possible extension of the FFT method of simulation to multi-variate and multi-dimensional random processes can be found in the work of Rajasekhara (1976).

2.11 Simulation Using Time Series Models

In sections (2.9) and (2.10) we have discussed the aspects of digital simulation of spectrum consistent sample functions of a Gaussian random process by means of the sum of cosine functions with random phase angles. For a stationary process, the coefficients of the cosine functions are dependent only on the frequency, while for an evolutionary random process they are functions of both time and frequency. The procedure for such simulation is very general and applicable to multivariate and multidimensional random processes.

The sample functions generated by this procedure are especially helpful when numerical solutions are needed in Monte Carlo studies involving responses of nonlinear structures to random excitation, structures with random material properties and structures subjected to random parametric excitation.

When strong nonlinearities are present, when the randomness in material properties is substantial and when the parametric excitation is severe, the Monte Carlo procedures are often the only reliable methods for the solution of such problems. However, these methods can become computationally time consuming. The computational cost can considerably be reduced by use of FFT techniques in the superposition of the harmonic components. This operation however, requires a large computer memory which may become prohibitive depending on the random process whose sample functions have to be generated. This difficulty which exists even when a simple one dimensional univariate random process is simulated over a long period of time is compounded further in the case of simulation of multi-variate and multi-dimensional random processes as the number of component processes and their dimensionality increase.

In this context, the time series models discussed in section (2.5) provide the most direct and efficient method of generating sample functions of a random process consistent with certain properties of the random process. The digital generation of the sample functions of the random process is accomplished by recursively obtaining its sample values at discrete times in the AR, MA or ARMA models, once the model coefficients φ_i and θ_i are estimated. The models require only the generation of appropriate order of sequences of independent Gaussian random variates a_ts. Nonstationary processes can be simulated by the ARIMA model. The simulation using the time series models requires only the storage of the appropriate model coefficients in the computer memory, the generation of the white noise sequence and the recursive computation of the time series. Such tasks can be easily done even with small computers. Another advantage of the method is that analog version of the time series can be generated in real time which can be used in random vibration experimentation.

A typical simulated earthquake accelerogram by the use of ARMA model of example 2.4 is shown in Fig. 2.12. The simulated record exhibits the different features of the actual earthquake record (Polhemus and Cakmak, 1981).

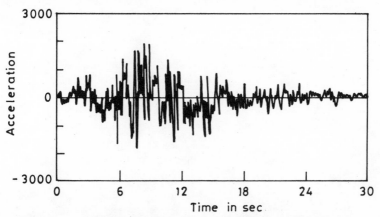

Fig. 2.12.   Simulated earthquake ground acceleration from ARMA model (Polhemus and Cakmak, 1981).

The time series representation of random processes and their simulation have been extended to multivariate random processes based on the development of multiple input-multiple output AR and ARMA discrete models by Gersh and Yonemoto (1977), Spanos and Hansen (1981), Samii and Vandiver (1984), Samaras et al (1985), Spanos and Shultz (1985, 1986) and to multi-dimensional random processes by Mignolet (1987). Some of the intrinsic features of the applicability of these procedures to random vibration problems are brought out in these papers.

Mignolet and Spanos (1987) and Spanos and Mignolet (1987) in a series of two recent articles have developed systematically ARMA simulation algorithms for multi-variate random processes. Their procedure is based on first obtaining an approximation of the target spectral matrix by the response of an AR discrete dynamic system to white noise excitation. Then a number of procedures of determining lower order ARMA approximations by relying on the initial higher order AR approximation of a given spectral matrix are presented in their investigation. The system of equations leading to the determination of the ARMA coefficients are derived on the basis of minimization of frequency domain errors. The different procedures are analyzed and compared by matching of the auto and cross-correlations of the target and simulated processes. Some of the properties like stability and invertibility of the procedures are discussed. A few examples demonstrating the reliability and efficiency of the time series simulation procedures are presented by estimating the spectra of the Tajimi-Kanai earthquake spectrum, Pierson-Muscowitz ocean wave spectrum and the trivariate spectral matrix of fluctuating wind velocity and comparing them with corresponding target spectra.

The simulation of a random process by the various models, like the filtered white noise model or the time series models involve the generation of a stationary Gaussian white noise sequence. This is generally accomplished by Monte Carlo procedures.

2A.1 Random Number Generation

The most important step in the application of Monte Carlo simulation is the generation of random numbers corresponding to prescribed probability distribution functions. Automatic generation of random numbers is most effectively and efficiently done on a digital computer. The first step in the generation of random numbers associated with a specific probability distribution function is to generate random numbers corresponding to the uniformly distributed random variable over the interval 0 to 1.

Let X be the random variable with the given probability distribution function $F_X(x)$. We are required to generate random numbers which can serve as sample values of the random variable X. It can be easily verified that the transformed random variable given by the functional relation $U = F_X(x)$ is uniformly distributed over the interval 0 to 1. If $\{u_n\} = \{u_1, u_2, ..., u_n\}$ is a set of sample values of U obtained by some procedure, the corresponding set of values $\{x_n\} = \{x_1, x_2, ..., x_n\}$ obtained by the inverse transformation $x_i = F_X^{-1}(u_i)$ will be random numbers associated with the random variable X. Thus the generation of uniformly distributed random numbers between 0 and 1 is most basic to the generation of random numbers with general probability distribution.

Although any independent trial like tossing a coin or throwing a dice can be used in the generation of random numbers, as mentioned earlier from practical considerations and general applications, they are most effectively generated by recursive algorithms. The procedure can be fully automated in the digital computer. Uniformly distributed random numbers are generated recursively by calculating the residues of modulus m from a linear transformation. The recursive relation which is often used for this purpose is

$$x_{i+1} = (ax_i + b) \text{ (modulo } m) \tag{2A.1}$$

where a, b and m are non-negative integers. If k_1 is the integer part of the ratio $(ax_i + b)/m$, then the corresponding residue of the modulus m is

$$x_{i+1} = ax_i + b - mk_i \tag{2A.2}$$

Dividing the values obtained in (2A.2) by the modulus m, we obtain

$$u_{i+1} = x_{i+1}/m \tag{2A.3}$$

which form a set of uniformly distributed random numbers in the interval $(0, 1)$.

In the computerized procedure mentioned above, the random numbers are generated in a purely deterministic way and hence are not strictly random. They can be duplicated exactly. For this reason they are called "pseudo random" numbers. The generated pseudo random number sequence is periodic with the period being less than the modulus m. In a Monte Carlo simulation method the period of the pseudo random sequence should be greater than the number of random numbers that will be used in the simulation procedure. This will insure the randomness of the sequence in the particular application. Thus a large value of m is needed for a sufficiently large period. It can be shown that the numbers generated with large m appear to be uniformly distributed and statistically independent (Knuth, 1969). Satisfactory results have been observed (Rubinstein, 1981) with $m = 2^{35}$, $a = 2^7 + 1$ and $b = 1$ for binary computers.

Another common recursive relation for generating uniform random numbers is the congruential generator defined by

$$x_{i+1} = ax_i \text{ (modulo } m) \tag{2A.4}$$

and

$$u_i = x_i/m \tag{2A.5}$$

Equations (2A.4) and (2A.5) are used in the IBM system 360 as a uniform random number generator with $a = 16808$ and $m = 2^{31} - 1$.

Random numbers thus generated can be tested for statistical independence and for uniform distribution by run tests and goodness of fit tests. The procedures with the parameter values suggested above have been tested statistically and shown to give satisfactory results (Rubinstein, 1981).

2A.2 White Noise Simulation

To generate sample functions which approach a stationary Gaussian white noise process, first a sequence of pairs of statistically independent random numbers $\{u_{2n}\} = \{u_1, u_2, ..., u_{2n-1}, u_{2n}\}$ all of which have a uniform probability distribution over the range $0 \le u \le 1$ is generated. Given this sequence $\{u_{2n}\}$ of uniform random variates one can construct a new sequence of pairs of statistically independent random numbers $\{x_{2n}\} = \{x_1, x_2, ..., x_{2n-1}, x_{2n}\}$ using the relations

$$x_{2i-1} = (-2 \ln u_{2i-1})^{1/2} \cos (2\pi u_{2i}) \qquad i = 1, 2, ... \tag{2A.6}$$

$$x_{2i} = (-2 \ln u_{2i-1})^{1/2} \sin (2\pi u_{2i}) \qquad i = 1, 2, ... \tag{2A.7}$$

It can be shown that the sequence $\{x_{2n}\}$ form independent samples of the Gaussian distribution with mean 0 and variance 1 (Box and Muller, 1958; Franklin, 1965).

A sample function $a_t(t)$ can now be constructed by assigning the values x_1, x_2, ... x_n to n successive ordinates spaced at equal intervals Δt along a time abscissa and by joining them by straight lines implying linear variation of the ordinates over each time interval. The time scale is chosen such that the initial time is a uniform random variate in the time interval $0 < t < \Delta t$.

A complete ensemble of p such sample functions $a_r(t)$, $r = 1, 2, ..., p$ can be obtained by repeating the procedure p times. If the intensity of the process is changed by muliplying each ordinate x_i by the normalization factor $(2\pi\Phi_0/\delta t)^{1/2}$, where Φ_0 is a constant, the resulting new ensemble $w_r(t) = (2\pi\Phi_0/\delta t)^{1/2}a_r(t)$ $r = 1, 2 ..., p$ constitutes a stationary Gaussian random process with autocorrelation function (Clough and Penzien 1975)

$$R_{WW}(\tau) = \begin{cases} \dfrac{2\pi\Phi_0}{\Delta t}\left[\dfrac{2}{3} - \left(\dfrac{\tau}{\Delta t}\right)^2 + \dfrac{1}{2}\left(\dfrac{|\tau|}{\Delta t}\right)^3\right] - \Delta t \le \tau \le \Delta t \\[4mm] \dfrac{2\pi\Phi_0}{\Delta t}\left[\dfrac{4}{3} - \dfrac{2|\tau|}{\Delta t} + \left(\dfrac{\tau}{\Delta t}\right)^2 - \dfrac{1}{6}\left(\dfrac{|\tau|}{\Delta t}\right)^3\right] \\[2mm] \qquad\qquad -2\Delta t \le \tau \le -\Delta t \\[2mm] \qquad\qquad \Delta t \le \tau \le 2\Delta t \\[4mm] 0 \qquad\qquad \tau \le -2\Delta t;\ \tau \ge 2\Delta t \end{cases} \tag{2A.8}$$

Taking Fourier transform of (2A.8) gives the power spectral density function of the random process

$$\Phi_{ww}(\omega) = \Phi_0 \frac{6 - 8 \cos(\omega \Delta t) + 2 \cos(2\omega \Delta t)}{(\omega \Delta t)^4}, \quad -\omega < \omega < \infty \qquad (2A.9)$$

As Δt approaches zero $R_{ww}(\tau)$ in (2A.8) approaches

$$R_{ww}(\tau) = 2\pi \Phi_0 \delta(\tau) \qquad (2A.10)$$

and $\Phi_{ww}(\omega)$ in (2A.9) $\to \Phi_0$ in the limit as $\Delta t \to 0$, and the process $w(t)$ becomes Gaussian white noise. The expression for $\Phi_{ww}(\omega)$ remains within 5% of Φ_0 for $\omega \Delta t < 0.57$ and within 10% for $\omega \Delta t < 0.76$. Thus the time interval may be chosen sufficiently small to approximately generate the white noise process to within a given tolerance to any desired frequency.

REFERENCES

Amin, M. and A.H.S. Ang, 1968. Nonstationary stochastic model of earthquake motion. J. Eng. Mech. Div. Proc. ASCE. **94(EM2):** 559–583.

Barstein, M.F., 1960. Application of probability methods for designing the effect of seismic forces on engineering structures. Proc. Second World Conference on Earthquake Engineering, Tokyo, Japan: 1467–1482.

Boore, D.M., 1983. Stochastic simulation of high frequency ground motions based on seismological models of the radiated spectra. Bull. Seis. Soc. Amer. **73(6):** 1865–1894.

Borgman, L.E., 1969. Ocean wave simulation of engineering design. J. Water Ways. Proc. ASCE. **95(WW4):** 557–583.

Box, G.E.P. and D.R. Cox, 1964. An analysis of transformations. J. Royal Statistical Society. B26: 211–252.

Box, G.E.P. and G.M. Jenkins, 1971. Time series analysis, Forecasting and Control. Holden Day, San Francisco.

Box, G.E.P. and M.E. Muller, 1958. A note on the generation of random normal deviates. Annals of Math. Stat. **29:** 610–611.

Bull, M.K., 1967. Wall pressure fluctuations associated with subsonic turbulent boundary layer flow. J. Fluid. Mech. **28:** 719–754.

Chang, M.K., Kwiatkowski, J.W. and R.F. Nau, 1982. ARMA models for earthquake ground motions. Earthquake Eng. Struct. Dyn. **10:** 651–662.

Clough, R.W. and J. Penzien, 1975. Dynamics of Structures. McGraw-Hill, New York.

Cooley, J.W. and J.W. Tukey, 1965. An algorithm for the machine calculation of complex Fourier series. Math. of Computation. **19:** 297–301.

Cornell, C.A., 1964. Stochastic process models in structural engineering. Tech. Rep. 34. Dept. of Civil Eng. Stanford University.

Cramer, H. and M.F. Leadbetter, 1967. Stationary and Related Stochastic Processes. John Wiley and Sons, New York.

Davenport, A.G., 1961. The application of statistical concepts to the loading of structures. Proc. Inst. Civil Engineers. **19:** 449–473.

Dowell, E.H., 1969. Transmission of noise from a turbulent boundary layer through a flexible plate with a closed cavity J. Acous. Soc. Amer. **46(1):** 238–252.

Franklin, J.N., 1965. Numerical simulation of stationary and nonstationary Gaussian random processes. SIAM Rev. **7(1):** 68–80.

Gersh, W., 1970. Estimation of the autoregressive parameters of a fixed autoregressive moving-average time series. IEEE. Trans. Automatic Control AC-15: 583–588.

Gersh, W. and D. Foutch, 1974. Least-squares estimates of structural system parameters using covariance function data. IEEE. Trans. Automatic Control AC-19: 898–903.

Gersh, W. and G. Kitagawa, 1985. A time varying AR coefficient model for modelling and simulating earthquake ground motion. Earthquake Eng. Struct. Dyn. **13:** 243–254.

Gersh, W. and R.S. Liu, 1976. Time series models for the synthesis of random vibration systems. Trans. ASME. J. Appl. Mech. **98:** 159–168.

Gersh, W. and S. Luo, 1972. Discrete time series synthesis of randomly excited structural system response. J. Acous. Soc. Amer. **51**: 402–408.

Gersh, W. and D.R. Sharpe, 1973. Estimation of power spectra with finite order autoregressive models. IEEE Trans. Automatic Control. **AC-18**: 367–369.

Gersh, W. and J. Yonemoto, 1977. Synthesis of multivariate random vibration systems: A two stage least squares ARMA model approach. J. Sound Vib. **52(4)**: 553–565.

Goto, H. and K. Toki, 1969. Structural response to nonstationary random excitation. Proc. Fourth World Conference on Earthquake Engineering, Santiago, Chile: 130–144.

Grenander, U., and M. Rosenblatt, 1957. Statistical Analysis of Stationary Time Series. John Wiley and Sons, New York.

Housner, G.W., 1947. Characteristics of strong motion earthquakes. Bull. Seis. Soc. Amer. **37(1)**: 19–31.

Housner, G.W., 1955. Properties of strong ground motion earthquakes. Bull. Seis. Soc. Amer. **45(3)**: 197–218.

Housner, G.W. and P.C. Jennings, 1964. Generation of artificial earthquakes. J. Eng. Mech. Div. Proc. ASCE. **90(EM1)**: 113–150.

Hudspeth, R.T. and L.E. Borgman, 1979. Efficient FFT simulation of digital time sequences. J. Eng. Mech. Div. Proc. ASCE. **105(EM2)**: 223–235.

Itagaki, H. and M. Shinozuka, 1972. Application of Monte Carlo technique to fatigue failure analysis under random loading. ASTM, STP 511, 168–184.

Iyengar, R.N. and K.T.S. Iyengar, 1969. A nonstationary random process model for earthquake accelerations. Bull Seis. Soc. Amer. **59(3)**: 1163–1188.

Jenkins, G.M. and D.G. Watts, 1968. Spectral Analysis and its Applications. Holden Day. San Fransisco.

Knuth, D.E., 1969. The Art of Computer Programming: Seminumerical Algorithms. Vol. 2. Addison-Wesley. Massachusetts.

Levy, R., Kozin, F. and B.B. Moorman, 1971. Random processes for earthquake simulation. J. Eng. Mech. Div. Proc. ASCE. **97(EM2)**: 495–517.

Lin, Y.K., 1965. Nonstationary excitation and response in linear systems treated as sequences of random pulses. J. Acous. Soc. Amer. **38(3)**: 453–460.

Lin, Y.K., 1967. Probabilistic Theory of Stgructural Dynamics. McGraw Hill, New York.

Merchant, D.H., 1964. A stochastic model for wind Gusts. Tech. Rep. 48. Dept. of Civil Eng., Stanford University.

Mignolet, M.P., 1987. ARMA Simulation of Multivariate and Multi-dimensional Random Processes. Ph.D. Thesis. Rice University, Houston, Texas.

Mignolet, M.P. and P.D. Spanos, 1987. Recursive simulation of stationary multivariate random processes. Part I. Trans. ASME J. App. Mech. **54**: 674–680.

Nigam, N.C., 1983. Introduction to Random Vibrations. Cambridge: The MIT Press.

Ozaki, T., 1980. Nonlinear time series models for nonlinear vibrations. J. App. Prob. **17**: 84–93.

Ozaki, T., 1981a. Nonlinear phenomena and time series models Proc. 43rd Session of Int. Statistical Institute. Buenos Aires, Argentina.

Ozaki, T., 1981b. Nonlinear threshold autoregressive models for nonlinear random vibrations. J. App. Prob. **18**: 443–450.

Polhemus, N.W. and A.S. Cakmak, 1981. Simulation of earthquake ground motions using ARMA models. Earth. Eng. Struct. Dyn., **9(4)**: 343–354.

Racicot, R.L., 1969. Random vibration analysis—Application to wind loaded structures. Ph.D. Thesis. Case Western Reserve Univ.

Racicot, R.L. and F. Moses., 1972. Filtered Poisson process for random vibration problems. J. Eng. Mech. Div. Proc. ASCE. **98(EM1)**: 159–175.

Rajasekhara, H.S., 1976. Digital simulation applications in random vibrations. M. Tech. Thesis., Dept. App. Mech., IIT, Madras.

Reed, D.A. and R.H. Scanlan, 1981. ARIMA representation of turbulence spectra and longitudinal integral scales. Proc. Joint Conference of the U.S.–Japan cooperative program in natural resources. Tokyo, Japan.

Reed, D.A., and R.H. Scanlan, 1983. Time series analysis of cooling tower wind loading. J. Str. Div. Proc. ASCE 109 (STR2): 538–554.

Rice, S.O., 1944. Mathematical analysis of random noise. Bell Sys. Tech. J. **23**: 282–332. Reprinted in Selected Papers on Noise and Stochastic Processes. N. Wax, ed. 1954, New York: Dover.

Rice, S.O., 1945. Mathematical analysis of random noise. Bell Sys. Tech. J. **24**: 46–156. Reprinted in Selected Papers on Noise and Stochastic Processes. N. Wax, ed. 1954, New York: Dover.

Rosenblueth, E. and J.K. Bustamante, 1962. Distribuion of structural response to earthquakes. J. Eng. Mech. Div. Proc. ASCE. **88**(EM3): 75–106.

Rubinstein, R.Y., 1981. Simulation and Monte Carlo Method. John Wiley and sons, New York.

Ruiz, P. and J. Penzien, 1969. Probabilistic study of behaviour of structures during earthquake. Rep. EERC 69–3. University of California, Berkeley.

Samaras, E., Shinozuka, M. and A. Tsurui, 1985. ARMA representation of random processes. J. Eng. Mech. Div. Proc. ASCE. **111**(EM3): 449–461.

Samii, K. and J.K. Vandiver, 1984. A numerically efficient technique of the simulation of random wave forces on offshore structures. Proc. Sixteenth Annual Offshore Technology Conference, Houston, Texas, OTC **4811**: 357–367.

Saragoni, G.R. and G.C. Hart, 1974. Simulation of artificial earthquakes. Earthquake. Eng. Struct. Dyn. **2**: 249–267.

Shinozuka, M., 1971. Simulation of multivariate and multi-dimensional random processes. J. Acous. Soc. Amer. **49**(1): 357–368.

Shinozuka, M., 1972. Monte Carlo solution of structural dynamics. Tech. Rep. 19. Dept. of Civil Eng. and Eng. Mech., Columbia University.

Shinozuka, M., 1974. Digital simulation of random processes in engineering mechanics with the aid of FFT techniques in Stochastic Problems in Mechanics Ed. Ariaratnam, S.T., and H.H.E. Leipholz. Study 10. Solid Mech. Div., University of Waterloo Press: 277–286.

Shinozuka, M. and J.C. Astill, 1972. Random eigenvalue problems in structural mechanics. AIAA Journal. **10**(4): 456–462.

Shinozuka, M. and C.M. Jan., 1972. Digital simulation of random processes and its applications. J. Sound and Vib. **25**(1): 111–128.

Shinozuka, M. and R. Levy, 1977. Digital generation of along wind velocity field. J. Eng. Mech. Div. Proc. ASCE. 103(EM4): 689–700.

Shinozuka, M. and Y. Sato, 1967. Simulation of nonstationary random process. J. Eng. Mech. Div. Proc. ACE. **93**(EM1): 11–40.

Shinozuka, M. and Y.K. Wen, 1971. Nonlinear dynamic analysis of offshore structure. A Monte Carlo approach. Proc. First Int. Symp. on Stochastic Hydraulics. University of Pittsburgh Press: 507–521.

Shinozuka, M. and Y.K. Wen, 1972. Monte Carlo solution of nonlinear vibrations. AIAA Journal. **10**(1): 37–40.

Smith, K.C. 1985. Stochastic analysis of the seismic response of secondary systems. Resp. EERL 85–01. Earthquake Engineering Research Laboratory. California Institute of Technology, Pasadena, California.

Spanos, P.T.D., 1983. ARMA algorithms for ocean wave modelling. Trans. ASME J. Energy. Res. Tech. **105**: 300–309.

Spanos, P.T.D. and J.E. Hansen, 1981. Linear prediction theory for digital simulation of sea waves. Trans. ASME J. Energy. Res. Tech. **103**: 243–249.

Spanos, P.D. and M.P. Mignolet, 1987. Recursive simulation of stationary multivariate random processes. Part II. Trans. ASME. J. App. Mech. **54**: 681–687.

Spanos, P.T.D. and K.P. Schultz, 1985. Two stage order of magnitude matching for the Von-Karman turbulence spectrum. Proc. Fourth Int. Conf. on Struct. Safety and Reliability ICOSSAR, **1**: 211–218.

Spanos, P.T.D. and K.P. Schultz, 1986. Numerical synthesis of trivariate velocity realizations of turbulence. Int. J. Nonlin. Mech. **21**(4): 269–277.

Tajimi, H., 1960. A statistical method for determining the maximum response of a building structure during an earthquake. Proc. Second World Conf. Earthquake Engg. **2**: 781–797.

Tong. H., 1978. An approach to nonlinear time series modelling, Tech. Rep. 82. Dept. of Mathematics. (Statistics) UMIST, Tokyo, Japan.

Vaicaitis, R. and C.M. Jan., 1971. Nonlinear panel response and noise transmission from a turbulent boundary layer by a Monte Carlo approach. Tech. Rep. 13. Dept. of Civil Eng. and Eng. Mech. Columbia University.

Vaicaitis, R., Jan. C.M. and M. Shinozuka, 1972. Nonlinear panel response from a turbulent boundary layer. AIAA Journal. **10**(7): 895–899.

Vaicaitis, R., Dowell, E.H. and C.S. Ventres, 1974. Nonlinear panel response by a Monte Carlo approach. AIAA Journal. **12**(5): 685–692.

Venkatesan, C. and V. Krishnan, 1981. Stochastic modelling of an aircraft traversing a runway using time series analysis. J. Aircraft. **18**(2): 115–120.

Wen, Y.K. and M. Shinozuka, 1971. Monte Carlo solution of structural response to wind load. Proc. Third. Int. Conference on Wind effects on buildings and structures. Tokyo, Japan.

Wittig, L.E. and A.K. Sinha, 1975. Simulation of multi-correlated random processes using FFT algorithm. J. Acous. Soc. Amer. **58**(3): 630–634.

Wyatt, T.A. and H.I. May, 1973. The generation of stochastic load functions to simulate wind loading on structures. Earthquake Eng. and Struct. Dyn. **1**: 217–224.

Yang, J.N. 1972a. Simulation of random envelope processes. J. Sound and Vib. **21**(1): 73–85.

Yang, J.N., 1972b. Nonstationary envelope processes and first excursion probability. Quart. Tech. Rev. Jet Propulsion Laboratory **1**: 1–12.

3. Fatigue and Creep Under Random Vibration

3.1 Introduction

Materials when subjected to cyclic, repeated or fluctuating loads may fail in fatigue at a stress level much lower than that is required to cause failure under static conditions. According to the American Society of Testing Materials (ASTM), fatigue failure is described as the 'process of progressive localized permanent structural damage occurring in a material subjected to conditions which produce fluctuating stresses and strains at some point or points and which may culminate in cracks or complete fracture after a sufficient number of fluctuations'.

The fatigue life of a structural component is found to depend in a complex manner on a large number of factors, such as, the material properties, the fluctuating load sequence; the geometry; size and surface finish of the component; environmental factors like temperature and pressure; and the environment, for example, gaseous or aqueous.

The fact that fatigue failure of materials is critical for the reliability of machine elements and structural components, in general, and for aircraft structures in particular, has spurred intense research activity in the field of fatigue-fracture phenomenon. In view of the increasing trend towards high performance aerospace vehicles and high speed machinery, fatigue has become one of the most important operational and design consideration for these structural systems.

The earliest theories of fatigue were based on the macroscopic observations of the relationship between the applied stress and the number of cycles to failure. Because of the relative ease of obtaining data in this form and its easy application to fatigue design, the bulk of fatigue test data is presented even today in the form of S-N curves or the Wöhler (1860) diagrams, after the German engineer who while investigating the fatigue failure of railway axles, both by testing full scale models and laboratory specimens of the materials, obtained such data for the first time. The S-N curves exhibit a large scatter which is primarily due to the inherent uncertainties underlying the fatigue phenomenon as a result of the variability of the myriad of microscopic influences—dislocations, lattice defects, grain boundaries, strain hardening properties, impurities and alloy phases of the material.

It is now generally accepted that the fatigue failure process involves three phases: (i) the fatigue crack initiation phase in which microscopic slip bands within the individual grain of the material form minute cracks; (ii) the crack propagation phase in which the micro cracks grow in size and (iii) the final phase of unstable rapid crack growth to fracture when the crack attains a critical size. The prediction of fatigue life based on the S-N curves rests on the use of a suitable cumulative damage criterion representing the progressive deterioration of the material during the cyclic loading. The damage criterion may be considered to signify, in a gross way, many physical changes that occur in the material during the process of crack initiation, its propagation and final fracture.

The simplest and perhaps the most widely applied theory of fatigue damage is the linear cumulative damage law of Palmgren (1924) and Miner (1945), in which the material is assumed to suffer cumulative damage proportional to the number of cycles at constant stress amplitude. In the case of multi level stress cycles, fatigue failure is assumed to occur when the sum of the damage at each stress level calculated on the above basis attains the value of unity.

Due to the simplicity of the Palmgren-Miner hypothesis, the cumulative damage analysis is usually performed on the basis of constant stress amplitude tests even for complex stress histories. The main criticism of this approach is that the loading sequence and stress interaction effects are disregarded resulting in inaccurate and sometimes nonconservative estimates of fatigue life. Several modifications of the linear cumulative damage law to account for the load sequencing and stress interaction effects have been proposed. The notable among them are due to Corten and Dolan (1956), Freudenthal-Heller (1959), Shanley (1952) and Marco and Starkey (1954). In spite of these modifications the Palmgren-Miner linear damage rule is frequently used because of its simplicity, and experimental fact that other more complex cumulative damage theories do not always yield a significant improvement in failure prediction. This is essentially so, because of the complex nature of the fatigue process itself and because of the fact that most of the damage theories rely heavily on empiricism rather than a basic understanding of the physical process. The problem is compounded by the fact that often a material is subjected to random fluctuating loads and not to pure sinusoidal loads or combinations thereof.

The limitations of the cumulative damage theories have led to an increasing use of fracture mechanics concepts and crack propagation theories, which treat the fatigue phenomenon in its entirety from the stage of crack nucleation and its growth up to failure. The crack propagation theories usually assume that micro cracks are already present in the material which grow under the action of stress amplifications from the crack tips at a rate proportional to a power of the local stress intensity factor. Failure criteria are established in terms of a critical size of the crack. Paris (1962, 1963) and his co-workers pioneered this approach, based on the Griffith-Irwin's (1920, 1957) fracture mechanics theory.

Although crack propagation approach is a significant improvement over the cumulative damage theories fatigue life estimates based on this approach are often at variance with actual data. It is due to the fact that different crack propagation theories are based on one or more of the many phenomena such as the piling up of dislocations, accumulation of plastic deformations, fracture of molecular bonds, condensation of voids etc. on which the fatigue process depends.

The basic uncertainties characterizing the fatigue process have also led to a statistical approach to the problem. Freudenthal and Gumbel (1962) and Yokobori (1965) have done pioneering work in this field. It must however be emphasized that just as the micro structural theories or the cumulative damage theories by themselves cannot offer an accurate basis for fatigue strength calculations, a statistical analysis bereft of any relation to the actual, physical, metallurgical and structural characteristics will be totally inadequate for fatigue life prediction.

Apart from the inherent uncertainty underlying the fatigue phenomenon, the fluctuating load cycles during the service life of most mechanical, civil and aerospace structural systems are random in nature. For example, the load environment of an aircraft due to gusts, jet noise, runway roughness, etc. is random in nature. Likewise, the wind and ocean wave loads on structural systems cause random fluctuating loads that may cause fatigue failure.

Traditionally fatigue life prediction under random loads has been based on cumulative damage theories and constant amplitude fatigue test data. The absence of a well defined stress history and the difficulty of identifying a cycle compatible with the sinusoidal history presents lot of problems in such an approach. Fatigue tests under programmed and simulated random loads have therefore been performed providing a wealth of data. Swanson (1968) has carried out extensive tests to obtain such data. Fracture mechanics theories with variable rates of crack growth have also been proposed (Rice et al, 1965) and the relevant statistics of peak loads and rise and fall of the stress history have been investigated. Special counting methods to identify the number of cycles of stress reversals, in the absence of well defined cyclic process have also been devised mainly based on the closing of the stress strain hysteresis loop. These enable the application of cumulative damage theories in a more direct manner to random load fatigue estimation.

Creep is the phenomenon of gradual and permanent deformation of materials subjected to mechanical and thermal stresses over extended periods of time. This is especially true of ductile materials which accumulate large plastic strains under loading and thermal environments leading to failure. Creep rupture is an extension of the creep process to the limiting condition where the stressed member actually separates into two parts. The interaction of creep and stress rupture with cyclic stressing and the fatigue process has not been clearly understood but is of great importance in many modern engineering systems. The excessive deformation due to creep is not the only important consideration, but other consequences of the creep process may also be important. These include creep rupture, thermal relaxation, creep and rupture under multi-axial states of stress, cumulative creep effects and effects of combined creep and fatigue.

Different types of creep are exhibited in materials such as logarithmic creep at low temperatures, high temperature creep and diffusional creep. At high temperatures and under constant or slowly varying loads, the strain accumulation is found to be a time dependent phenomenon. At low temperatures and cyclic loads superimposed on constant loads, cycle dependent strain accumulation is observed. Considerable work has been done in the theoretical and experimental study of creep failure under constant or deterministically varying sustained loading. Randomness in creep behaviour of material arises mainly because of two reasons: (i) due to random thermal and loading environment, and (ii) due to random material properties. While some work has been done on the statistical nature of creep due to random material behaviour, very little attention has been paid to creep behaviour under random loading. Cumulative damage theories similar to fatigue theories have been proposed for creep under random loading.

This chapter presents different aspects of fatigue under random vibration, namely, the cumulative damage theories and fracture mechanics concepts for prediction of fatigue lives of components under random loading, and the statistical nature of fatigue phenomenon. It also presents an introduction to creep under random vibration and the evaluation of failure criteria.

3.2   The Mechanisms of Fatigue

Many theories of fatigue crack initiation have been proposed based on the formations of extrusions and intrusions in the material. The first observations concerning these formations were made by Cottrell (1956) on precipitation hardened aluminum alloys. It has been assumed that overaging softens the material resulting in extrusions in the active slip bands. The slip bands under cyclic loading give rise to surface grooves and ridges as a result of the reversed slip on adjacent slip planes caused by load reversal. This is shown in Fig. 3.1. These grooves and ridges may be either sharp or smoothly rounded corrugations. If a number of planes slip, the striations are shallow and undulating, while if only a few closely spaced planes slip, sharply defined crevices are formed. The extrusions thus formed are accelerated by the to and fro motion of dislocations during the cyclic loading. Many times, the fatigue cracks are detected at these extrusions lending credence to this mechanism of crack initiation.

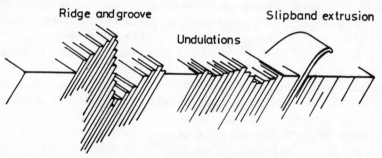

Fig. 3.1.   Grooves and ridges formed by reversed loading and regions of slip.

The presence of intrusions and extrusions observed in a pure metal like copper by Cottrell and Hull (1957) showed that the process of diffusion of dislocations could not be considered the only reason for crack initiation. Another possible mechanism of the fatigue crack initiation is the formation of cavities by agglomeration of edge dislocations. These cavities then constitute fatigue nuclei from which the cracks grow.

The growth of fatigue crack takes place under the action of the cyclic loading along the slip planes oriented in the direction of the maximum shearing stress. There is virtually no change observed in the mechanism of the crack growth as long as the crack continues to grow along the active slip planes. This is usually observed in case of low stress cycles and slow crack growth rates and is termed as Stage-I crack growth. If periodic stress cycles, notches and stress risers exist the Stage-II crack growth occurs which is governed by the maximum principal normal stress in the neighbourhood of the crack tip. The crack grows roughly in a direction perpendicular to the direction of the maximum principal normal stress. The fracture surface in this case is characterized by smooth striations and beach marks whose width and density can be related to the stress level. It is by now fairly well understood that the fatigue crack propagation is the consequence of plastic deformation at the crack tip.

Many micro structural theories have been proposed for the physical phenomena controlling the process of crack propagation under cyclic loading. These theories are mainly based on the strain hardening mechanism, the motion of dislocations, plastic deformations, grain boundaries and inter molecular bonds and are sometimes combined with statistics and probability theory. Notable among the theories are those due to Orawan (1939), Afanasiev (1940), Dehlinger (1940), Head (1953), Freudenthal (1946), Mott (1952) and Yokobori (1956). The mechanism of crack propagation after the crack has been initiated is explained on a bigger scale by theories based on linear elastic fracture mechanics and elastic plastic fracture mechanics concepts as alluded to briefly earlier. Finally the crack length attains a critical size leading to failure. The final fracture region always shows evidence of plastic deformation produced just prior to the final rupture.

Although the micro structural theories of fatigue are by themselves important in understanding the fatigue process, the phenomenological aspects of fatigue failure and the macroscopic behaviour of the material are of considerable interest to the designer in avoiding fatigue failures during the life of a component or a structure. Some of the macroscopic aspects to be considered in the fatigue design methodology include: the nature of the applied stress cycles—whether a simple completely reversed alternating stress pattern or stress cycles with a non zero mean, random or sinusoidal stress cycles, multi-axial states of stress; size, shape and surface finish of the material, the effects of environmental factors like humidity, corrosive media, temperature, the effects of variation of material properties; the effects of damage accumulation at various levels of stress cycles and so on. In this chapter we shall concentrate only on the fatigue behaviour of materials under random loading conditions.

3.3   Fatigue Stress Histories

In designing a fatigue sensitive component in a structure or a machine, the fatigue behaviour of engineering materials to different types of loadings that might occur during the service life of the structure is of critical interest. The simplest fatigue stress spectrum to which a component may be subjected to is a zero-mean sinusoidal stress-time history of constant amplitude and constant frequency applied for a specified number of cycles. Such a stress history is called a completely reversed cyclic stress and is shown in Fig. 3.2a. A second type of stress history is very similar to the first one except that it has a non-zero mean and is shown in Fig. 3.2b. A particular case of the non-zero mean stress history with zero minimum shown

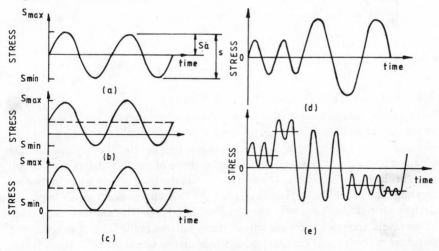

Fig. 3.2.   Cyclic stress histories.

in Fig. 3.2c, and known as released tension, is also of practical interest. Some parameters of interest in fatigue design procedures and life predictions can be defined with reference to these figures. For example, $s_{\max}, s_{\min}, s_{\text{mean}}$ are respectively the maximum, minimum and the arithmetical mean of the stress history, $\Delta s = s_{\max} - s_{\min}$ is the stress range; and $s_a = \Delta s /2$ is the alternating stress amplitude. Other complicated stress patterns which retain some form of the sinusoidal nature are shown in Figs. 3.2d and 3.2e. In the former the mean stress is zero, but there is combination of two or more amplitudes and frequencies. In the latter, not only the stress amplitudes and the frequencies vary, but the mean stress also varies from cycles to cycles.

Figure 3.3 shows a typical stress history that might be encountered by an aircraft structural component during a mission that includes, taxi, takeoff, gusts, maneouvers and landing. In this case, the stress pattern is random and distinct cycles of stress cannot be identified. Therefore design methodologies and methods of fatigue life predictions under such loadings are very important.

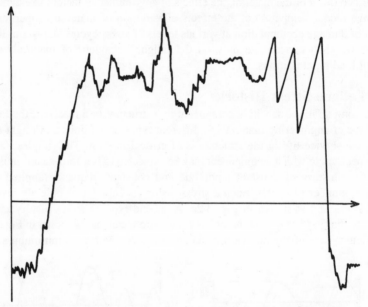

Fig. 3.3.   Typical random stress history encountered by an aircraft.

3.4   Fatigue Test Data—The S-N Curves

Basic fatigue data of a material is usually displayed on a plot of cyclic stress level versus the logarithm of number of cycles of stress applications to failure or alternatively, on a log-log plot of stress versus fatigue life. Usually the data are obtained by testing standard specimens in the shape of small cylinders in push-pull tests, or flat bars in cantilever bending tests, or in the form of thin wires in rotating bending machines. These plots are called the S-N curves and furnish important design information. Generally the test data is obtained for constant amplitude stress levels, though some variations in stress patterns in the form of programmed sequence of loadings or random loadings to stimulate actual realistic situations are also possible (Collins, 1981; Swanson, 1968). Full scale fatigue tests of proto-type models, especially aircrafts, are also carried out. To translate the laboratory test data into actual design practice requires the consideration of the influence of non-zero mean stress, varying stress amplitude, environment, size and surface finish on the fatigue characteristics of the material.

The S-N curves obtained even under a stress history of constant amplitude exhibit a large amount of scatter as shown in Fig. 3.4. The scatter in the fatigue data can be attributed to many factors as explained earlier and reflects the inherent uncertainties underlying the fatigue phenomenon. It must be recognized that there is not a single S-N curve for a given material, but a family of S-N curves. This representation treats the number of cycles to failure, N, in a constant amplitude stress test, as a random variable. It is possible to represent the families of S-N curves in a diagram with the probability of failure P of the test specimen, in other words the probability that the specimen will fail in fatigue in less number of cycles than obtained from the diagram. Such a family of S-N curves called the S-N-P curves can be obtained by estimating the relevant probabilities by constructing appropriate histograms based on fatigue test conducted on a large number of samples of statistical significance. It has been observed in practice that the random variable N follow either the log-normal or the Weibull distribution for most of the materials.

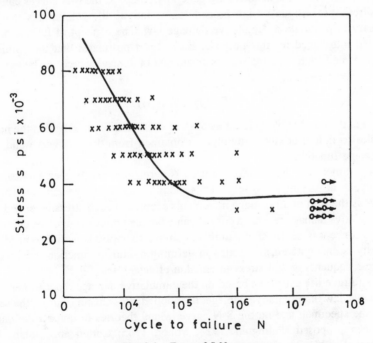

Fig. 3.4. Form of S-N curve.

Another approach for the treatment of scatter is to fit a mean curve to the fatigue test data. Generally, the S-N curves in the literature refers to the mean S-N curve. We shall refer only to the mean S-N curve in the discussion to follow.

The mean S-N curves generally exhibit two types of behaviour depending on the material. For ferrous alloys and titanium the S-N curves are characterized by a steep branch in the relatively short life range but approach a stress asymptote called the fatigue limit or endurance limit at longer lives. The endurance limit is the stress below which the specimen is theoretically supposed to sustain infinite number of cycles of stress reversals without fatigue failure. The non-ferrous alloys do not exhibit an asymptote and the S-N curve continues to drop off indefinitely. There is no fatigue limit for such materials though they exhibit a relatively flat curve in long life range.

Based on the fatigue test data it has been generally observed that for many engineering materials, the S-N curve can be approximated by a relation of the form

$$Ns^b = c \tag{3.1}$$

where b and c are constants that depend mainly on the material and shape of the specimen. The value of b for engineering materials such as steel and aluminum ranges from 4 to 20. The relation (3.1) plotted on a log-log paper with the log (s) as ordinate and log (N) as abscissa represents a straight line with slope equal to $-1/b$.

3.5   Cumulative Damage Theories

Since the real life stress histories are generally complex, the S-N curves cannot be used directly in fatigue design procedures or fatigue life estimations making it necessary to postulate a cumulative damage law. The concept of fatigue damage has been introduced for this purpose. Basically it postulates that the component under repeated cyclic loading suffers an amount of damage which can be expressed as

$$dD = f(n,\ s,\ N(s)) \tag{3.2}$$

where n is the number of cycles at a constant stress amplitude s, $N(s)$ is the number of cycles to failure at stress amplitude s obtained from the S-N curve and dD is the damage function.

It is assumed that the damage incurred is permanent and further application of stress cycles at varying stress levels cause additional cumulative damage. The total damage is the sum of all damage increments accrued at each stress level and when the cumulative damage reaches a critical value fatigue failure occurs. The cumulative damage concept is heuristic in nature and simple to apply. However, considerable difficulty is encountered in practice in defining a damage function incorporating the stress sequencing and stress interaction effects.

Thus fatigue life estimates based on the cumulative damage theories depend on a particular law of cumulative damage, the applied stress history during the service life of the specimen and on the S-N curve. Many theories of cumulative damage have been proposed (Palmgren, 1924; Miner, 1945; Corten and Dolan, 1956; Freudenthal and Heller, 1959; Marco and Starkey, 1954; Hashin and Rotem, 1978; Bogadanoff and Kozin 1980). We shall discuss the Palmgren-Miner cumulative damage theory.

3.5.1   Palmgren-Miner Cumulative Damage Theory

The first cumulative damage law was proposed by Plamgren (1924) and later developed by Miner (1945). The linear damage law is still widely used in spite of many limitations, because of its simplicity in application and because of the fact that many other theories developed to account for the limitations do not necessarily give better and accurate estimates of fatigue life. The basic assumptions of the Palmgren-Miner damage rule are:

1. Each alternate stress cycle inflicts a certain damage in the material equal to the reciprocal of the number of cycles of failure at the stress level given by the S-N curve. This damage is irrespective of the relative position of the stress cycle in the overall stress history and independent of the magnitude of stress levels before or after it.
2. The total damage due to the overall stress history is the sum of all damages of each individual cycle.
3. The fatigue failure of the material is deemed to occur when the cumulative damage attains a value unity.

Thus if D_i is the fraction of damage suffered by the material due to n_i number of cycles at stress level s_i it is given by

$$D_i = \frac{n_i}{N_i} \qquad (3.3)$$

where N_i is the number of cycles to failure at s_i given by the S-N curve. The total cumulative damage D is then given by

$$D = \sum_i D_i = \sum_i \frac{n_i}{N_i} \qquad (3.4)$$

and failure occurs when $D = 1$.

The main limitations of the Palmgren-Miner linear damage theory are that no consideration is given to the order of various stress applications and the damage is assumed to accumulate at the same rate at a given stress level without regard to the past history. Experiments reveal that both effects may have a profound influence on the fatigue characteristics of the structure. For example, for a sequence of stress cycles starting from a high stress amplitude followed by a stress cycles of lower amplitudes (high-low), the Miner's sum D at failure is generally less than one. On the other hand for a low-high sequence of stress loading, it is generally greater than 1. Experimental values of the Miner's sum at the time of failure often range from to 0.25 to 4, depending on the nature of stress history. For a random stress history, there is no well defined sequencing of high and low stress levels and the experimental Miner's sum nearly approaches the value of unity with the values ranging from about 0.6 to 1.5. Since in many applications the fluctuating stress is random, the Palmgren-Miner linear damage rule is often satisfactory for failure prediction.

3.6 Fatigue Under Random Loading

As mentioned earlier the stress history is generally a random process. It is important, therefore to develop methods for fatigue life estimates for random stress histories. We shall assume that the S-N curve for the material is a well defined mean S-N curve, which implies that the fatigue properties of the material are pseudo deterministic and the randomness in the problem arises only due to the random nature of the applied load.

3.6.1 Mean Fatigue Damage Under Stationary Narrow Band Random Vibration

For a random stress history shown in Fig. 3.3, the application of the Palmgren-Miner cumulative damage law is not straight forward, because the stress cycles cannot be clearly identified. In such cases, special methods of identification of the stress cycles must be devised which we shall discuss subsequently. However, when the stress history is a narrow band random process, the Palmgren-Miner hypothesis can be directly applied. A typical sample function of narrow band random stress history is shown in Fig. 3.5. For a narrow band process the expected number of cycles is nearly equal to the expected number of maxima as the probability of encountering positive troughs and negative peaks in the stress history is negligibly small.

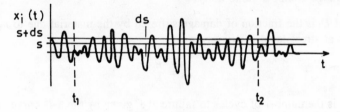

Fig. 3.5.   Typical narrow band stress history.

Consider the narrow band stress history $X(t)$ shown in Fig. 3.5. The expected number of peaks occurring between the levels s and $s + ds$ in the time interval $[t, t + T]$ is

$$N^+(s; T) - N^+(s + ds; T) \qquad s > 0 \tag{3.5}$$

where $N^+(s; T)$ represents the expected number of crossings of the level s with positive slope in the time interval $(t, t + T)$. This is so, because for a narrow band random process, there will be atmost a single crossing or no crossing at all of the level $s + ds$ with positive slope for every crossing of the level s with positive slope.

According to the Palmgren-Miner damage rule the expected incremental damage at the stress level s therefore is

$$E[D_s(t, t + T)] = \frac{N^+(s; T) - N^+(s + ds; T)}{N(s)} \tag{3.6}$$

where $N(s)$ is the number of cycles required to cause fatigue failure at the stress level s. Equation 3.5. can be expressed in terms of differentials in the following way

$$E[D_s(t, t + T)] = -\frac{1}{N(s)} \frac{\partial N^+(s; T)}{\partial s} ds \tag{3.7}$$

If $N^+(s, t)$ represents the expected rate of crossing of the level s with positive slope at time t

$$E[D_s(t, t + T)] = -\frac{1}{N(s)} \frac{\partial}{\partial s} \left[\int_t^{t+T} N^+(s, t) \, dt \right] ds \tag{3.8}$$

Hence, the total expected damage at all stress levels in the interval $(t, t + T)$ is obtained by integration as,

$$E[D(t, t + T)] = -\int_0^\infty \frac{1}{N(s)} \left\{ \int_t^{t+T} \frac{\partial}{\partial s} N^+(s, t) \, dt \right\} ds \qquad (3.9)$$

For a narrow band stress history noting that the probability density of the peaks is given by (Nigam, 1983)

$$p_s(s, t) = \frac{-\int_t^{t+T} \dfrac{\partial N^+}{\partial s} (s, t) \, dt}{\int_t^{t+T} N^+(0, t) \, dt} \qquad (3.10)$$

Equation (3.9) can be alternatively expressed as

$$E[D(t, t + T)] = \int_0^\infty \frac{1}{N(s)} \left\{ \int_t^{t+T} N^+(0, t) \, p_s(s, t) \, dt \right\} ds \qquad (3.11)$$

For a stationary random process $N^+(s; t)$ is independent of t and (3.9) reduces to

$$E[D(T)] = -T \int_0^\infty \frac{1}{N(s)} \frac{\partial N^+(s)}{\partial s} \, ds = N^+(0)T \int_0^\infty \frac{p_s(s) \, ds}{N(s)} \qquad (3.12)$$

Substituting the relation of the S-N curve from (3.1) into (3.12), we obtain

$$E[D(T)] = -\frac{T}{c} \int_0^\infty s^b \frac{\partial N^+(s)}{\partial s} \, ds = \frac{N^+(0)T}{c} \int_0^\infty s^b p_s(s) \, ds \qquad (3.13)$$

The rate of expected fatigue damage per unit time is therefore,

$$E[DR] = -\frac{1}{c} \int_0^\infty s^b \frac{\partial N^+(s)}{\partial s} \, ds = \frac{N^+(0)}{c} \int_0^\infty s^b p_s(s) \, ds \qquad (3.14)$$

For a narrow band stationary stress history the peak density can be expressed as

$$p_S(s) = -\frac{1}{N^+(0)} \frac{\partial N^+(s)}{\partial s} \qquad (3.15)$$

If the parent process is Gaussian, using the standard expressions for $N^+(0)$ and $N^+(s)$ the peak distribution (3.15) reduces to the Rayleigh distribution

$$p_s(s) = \frac{s}{\sigma_X^2} \exp \left(-\frac{s}{2\sigma_X^2} \right) \qquad 0 \leq s < \infty \qquad (3.16)$$

where σ_X is the standard deviation of the stress.

Substituting (3.16) into (3.13) the total expected damage in an interval of duration T for a narrow band stationary Gaussian random process is obtained as

$$E[D(T)] = \frac{N^+(0)T}{c\sigma_X^2} \int_0^\infty s^{b+1} \exp \left(-\frac{s^2}{2\sigma_X^2} \right) ds$$

$$= \frac{N^+(0)T}{c} (\sqrt{2} \sigma_X)^b \, \Gamma \left(1 + \frac{b}{2} \right) \qquad (3.17)$$

where $N^+(0) = \dfrac{1}{2\pi} \dfrac{\sigma_{\dot{x}}}{\sigma_X}$ is the expected rate of zero crossings, and $\Gamma(\cdot)$ is the Gamma function. The above formula was first derived by Miles (1954).

A modified fatigue law to give greater weight to stress cycles of higher amplitudes has been proposed by Shanley (1956). The Shanley's hypothesis assumes a S-N curve of the form

$$Ns^{q^b} = c \tag{3.18}$$

The total cumulative damage function remains the same as before given by the Palmgren-Miner damage rule. Usually a value of $q = 2$ is assumed for most of the materials. The modification can easily be incorporated in calculating the expected fatigue damage under stationary narrow band random vibration. The result is

$$E[D(T)] = \frac{N^+(0)\,T}{c} \, (\sqrt{2}\,\sigma_X)^{q^b} \, \Gamma\left(1 + \frac{q^b}{2}\right) \tag{3.19}$$

The average time to failure T_f in fatigue according to Palmgren-Miner rule can be obtained by equating the expected fatigue damage level given by (3.17) to unity; that is,

$$E[T_f] = \frac{c}{N^+(0)\,(\sqrt{2}\,\sigma_X)^b \, \Gamma\left(1 + \dfrac{b}{2}\right)} \tag{3.20}$$

3.6.2 Mean Fatigue Damage Under Non-stationary Narrow Band Random Vibration

Equation (3.8) was derived without any restriction on the stationarity or nonstationarity of the random process and is generally applicable to a nonstationary random stress history provided it is narrow band. In the case of nonstationary stress history the related average crossing rates, expected number of peaks per unit time at a stress level all become functions of time.

Assuming $X(t)$ to be a nonstationary Gaussian process with mean zero the expected rate of crossings of the level s with positive slope can be shown to be (Shinozuka and Yang, 1971)

$$N^+(s, t) = \frac{1}{2\pi} \frac{\sigma_{\dot{X}(t)}}{\sigma_{X(t)}} \sqrt{(1 - \rho^2)} \, \exp\left(-\frac{s^2}{2\sigma_X^2}\right) \{\exp(-\lambda^2)$$
$$+ \sqrt{\pi}\lambda \, (1 + \operatorname{erf} \lambda \,)\} \tag{3.21}$$

where $\sigma_{X(t)}$, $\sigma_{\dot{X}(t)}$ are respectively the standard deviations of the stress rate and the stress, which are functions of time and

$$\rho = E[X(t)\dot{X}(t)]/(\sigma_{X_{(t)}}\sigma_{\dot{X}_{(t)}}) \quad \text{and} \quad \lambda = (s\rho/\sigma_X\sqrt{1-\rho^2}) \tag{3.22}$$

For a stationary random process $\rho = 0$ and $\sigma_{\dot{x}}$ and σ_X are independent of time and (3.21) reduces to the stationary value. Using (3.21) to obtain $\dfrac{\partial N^+(s, t)}{\partial s}$ and substituting in (3.8) with the S-N law the expected fatigue damage during the interval $(t, \, t + T)$ has been shown to be (Shinozuka and Yang, 1971) as

$$E[D(t, t + T)] = \frac{1}{c} 2^{b/2} \int_{t}^{t+T} \sigma_X^b N^+(0, t) \left\{ (1 - \rho^2)^{b/2} \Gamma\left(1 + \frac{b}{2}\right) \right.$$

$$+ \rho \frac{b}{2} \sqrt{\pi/(1 - \rho^2)} \; \Gamma\left(\frac{b + 1}{2}\right)$$

$$\left. + \frac{\rho b}{(1 - \rho^2)} \Gamma\left(1 + \frac{b}{2}\right) {}_2F_1\left[\frac{1}{2}, 1 + b/2, \frac{3}{2}, -\frac{\rho}{\sqrt{1 - \rho^2}}\right] \right\} dt \quad (3.23)$$

where $_2F_1$ is the hyper geometric function and $N^+(0, t)$ is the nonstationary zero crossing rate.

Roberts (1968) has also obtained a similar expression for the mean fatigue damage under nonstationary narrow band random vibration for odd values of b.

Crandall and Mark (1963) in their book have presented an example for the calculation of the expected fatigue life of a cantilever beam with tip mass modelled as a single degree of freedom vibrating system and subjected to mean zero Gaussian white noise excitation at the base. The calculation is based on the Palmgren-Miner hypothesis. The cross section of the beam assumed square with side $a = 0.25$ in. and length $L = 4.0$ inches. The material of the beam is assumed to have an Young's modulus $E = 10.3 \times 10^6$ psi and the tip mass $m = 7.28 \times 10^{-4}$ lb sec²/ in. and a damping factor $\zeta = 0.01$. The white noise acceleration at the base is assumed to have a constant psd $\Phi_0 = 5930$ (in./sec²)²/(rad/sec). The constants in the fatigue law have been assumed as $b = 6.0$ and $c = 6.4 \times 10^{31}$. For these data the expected time to failure based on (3.20) is 115.54 seconds.

The transient response of a sdf system to stationary white noise excitation is nonstationary in character. Caughey and Stumpf (1961) have obtained closed form expressions for the transient response. The expected time to failure in fatigue in the transient response can be estimated using (3.23) with time dependent transient statistics. However, the indicated integration can be performed only numerically.

For many applications the peak distribution can be approximated by the Weibull distribution (Shinozuka and Yang, 1971)

$$F_s(s, T) = 1 - \exp\left\{ -\frac{1}{\alpha} \left(\frac{s}{\sigma}\right)^\alpha \right\} \quad (3.24)$$

where σ and α depend on the characteristics of the non-stationary random process and the duration T. Using (3.24) in (3.13) the expected fatigue damage level can be computed as

$$E[D(t, T)] = \frac{\bar{N}}{c} \int_0^\infty x^b p_s(s, T) \, ds = \frac{\bar{N}}{c} \sigma^b \alpha^{b/\alpha} \Gamma(1 + b/\alpha) \quad (3.25)$$

where $\bar{N}$ is the expected number of stress peaks in $(t, t + T)$, that is,

$$\bar{N} = E[N] = \int_t^{t+T} N^+(0, t) \, dt \quad (3.26)$$

Equation (3.26) can be approximated by

$$\bar{N} \cong T/T_0 \cong (\omega_0 T/2\pi)$$

where T_0 is the natural period of the system and ω_0 is the natural frequency.

Once the parameters α and σ are determined (3.25) can be used in a straight forward manner to estimate the fatigue damage in the case of nonstationary narrow band random loading. Note that the Rayleigh distribution can be obtained as a special case of the Weibull distribution by setting $\alpha = 2.0$.

3.6.3 Variance of Fatigue Damage Under Stationary Narrow Band Random Vibration

The evaluation of the variance of the fatigue damage according to the Miner's rule is a cumbersome task. Crandall, Mark and Khabbaz (1962) derived an expression for the variance of the fatigue damage and estimated the variance by a numerical integration procedure under the following assumptions:

(i) the exponent b in the fatigue law is odd.

(ii) the stress history $X(t)$ is a stationary Gaussian narrow band random process being the response of a lightly damped strongly resonant system to a broad band stationary excitation.

Considering the response of a sdf system to white noise excitation and assuming that the stress response is linearly related to the displacement response the expression for the variance of the fatigue damage during an interval of duration T can be expressed as

$$\frac{\sigma_D(T)}{E[D(T)]} = \frac{1}{\sqrt{N^+(0)T}} \left\{ \frac{f_1(b)}{\zeta} - \frac{f_2(b)}{\zeta^2 N^+(0)T} + \frac{f_3(b)}{N^+(0)T} \right\}^{1/2} \qquad (3.27)$$

where $\sigma_D(T)$ is the standard deviation of the damage and ζ is the damping ratio. The functions, $f_1(b)$, $f_2(b)$ and $f_3(b)$ are given in Table 3.1 for odd integral values of b (Crandall and Mark, 1963).

Table 3.1 Quantities in (3.27) for b odd (Crandall and Mark, 1963)

b	$f_1(b)$	$f_2(b)$	$f_3(b)$
1	0.0414	0.00323	0.0796
3	0.369	0.0290	0.212
5	1.28	0.0904	0.679
7	3.72	0.223	2.33
9	10.7	0.518	8.28
11	31.5	1.23	30.0
13	96.7	3.06	111.2
15	308.0	8.11	415.0

In most applications the expected number of cycles $N^+(0)T$ will be sufficiently large so that the third term in (3.27) can be neglected. Again in many cases $N^+(0)T$ will be large enough to make $\zeta N^+(0)T \gg 1$ so that the second term will also be negligible compared to the first. Under these assumptions the expression for the variance of the fatigue damage becomes.

$$\frac{\sigma_D(T)}{E[D(T)]} \cong \left\{ \frac{f_1(b)}{N^+(0)T} \right\}^{1/2} \qquad (3.28)$$

It should be noted that while the expected value of the fatigue damage varies as T the standard deviation of the damage increases only in proportion to $\sqrt{T}$. For (3.28) to be valid the damping ratio ζ should be less than about 0.05.

The variance of the fatigue damage according to the Palmgren-Miner law can be expressed as

$$\sigma_D^2 = E[D^2(T)] - E^2[D(T)] \tag{3.29}$$

For a stationary stress history using conditional probability concepts and relative frequency interpretation of probabilities the mean square value of the damage can be expressed as

$$E[D^2(T)] = \frac{1}{c^2} \int_0^\infty \int_0^\infty \int_0^T \int_0^T p_{s_1 s_2}(s_1, t_1; s_2, t_2) \, s_1^b s_2^b R_{MM}(t_1, t_2) \, dt_1 dt_2 ds_1 ds_2$$

$$\tag{3.30}$$

where $p_{s_1 s_2}(s_1, t_1; s_2, t_2)$ represents the joint probability density function of stress peaks at two times and $R_{MM}(t_1, t_2)$ represents the auto correlation function between the number of peaks per unit time at times t_1 and t_2 respectively. For a stationary random process both $p_{s_1 s_2}(s_1, t_1; s_2, t_2)$ and $R_{MM}(t_1, t_2)$ become functions only of the time difference $\tau = t_1 - t_2$ and not on the individual times t_1 and t_2. Even when the stress response $X(t)$ is Gaussian it is difficult to get the expressions for the joint peak density function and the auto correlation of the expected number of peaks per unit time, and to evaluate the integrals in (3.30).

Shinozuka [1966] has derived an expression for the variance of the fatigue damage in a different manner which agrees with the results of Crandall et al (1962). In the sequel we shall briefly present his approach of estimating the variance of the fatigue damage.

Let the specimen be subjected to a stationary narrow band random stress history containing n cycles at different stress levels. Without regard to the sequencing of the stress levels which is a basic assumption of the Palmgren-Miner cumulative damage law, the incremental damage suffered by the specimen during the ith stress cycle

$$\delta_i = \frac{1}{N_i} \tag{3.31}$$

where N_i is the number of cycles to failure corresponding to the stress level of the ith stress reversal. Hence the total damage suffered by the material during the n cycles will be,

$$D_n = \sum_{i=1}^n \delta_i = \sum_{i=1}^n \frac{1}{N_i} \tag{3.32}$$

Obviously D_n, δ_i are random variables. Since the stress history is assumed to be stationary the sequence δ_i is also stationary and hence its mean is a constant. Using the fatigue law representing the S-N curve

$$E[\delta_i] = E[\delta] = \frac{1}{c} \int_0^\infty s^b p_s(s) \, ds \tag{3.33}$$

and

$$E[D_n] = E[\sum_{i=0}^{n} \delta_i] = \frac{n}{c} \int_0^{\infty} s^b p_s(s) \, ds \tag{3.34}$$

During a time interval T the number of cycles of stress reversals n is also a random variable and hence the expected fatigue damage during the interval T can be obtained by using conditional expectation concepts

$$E[D(T)] = \sum_{n=0}^{\infty} E[D_n] P_N(n) = \sum_{n=0}^{\infty} E[\sum_{i=0}^{n} \delta_i] P_N(n) \tag{3.35}$$

$$= E[\delta] \sum_{n=0}^{\infty} n P_N(n) = E[\delta] E[n] \tag{3.36}$$

where $P_N(n)$ is the probability that the number of cycles in the time interval T is equal to n.

As $E[n] = N^+(0)T$, combining (3.33) and (3.36), we get

$$E[D(T)] = \frac{N^+(0)T}{c} \int_0^{\infty} s^b \, p_s(s) \, ds$$

which is the same as (3.13). Now consider the variance of D_n,

$$\sigma_{D_n}^2 = E[D_n^2] - E^2[D_n] = \sum_{i}^{n-1} \sum_{j=0}^{n-1} E[\delta_i \delta_j] - E^2[D_n] \tag{3.37}$$

we can write

$$\sum_{i}^{n-1} \sum_{j=0}^{n-1} E[\delta_i \delta_j] = \sum_{i=0}^{n-1} E[\delta_i^2] + \sum_{i=j=0}^{n-1} \sum_{i \neq j}^{n-1} E[\delta_i \delta_j] \tag{3.38}$$

which can be simplified for stationary sequence of δ_i to (Shinozuka, 1966)

$$\sum_{i}^{n-1} \sum_{j=0}^{n-1} E[\delta_i \delta_j] = n\sigma_\delta^2 + n^2 E^2[\delta] + 2 \sum_{k=1}^{n-1} (n-k)\{E[\delta_0 \delta_k] - E^2[\delta]\} \tag{3.39}$$

where

$$\sigma_\delta^2 = E[\delta^2] - E^2[\delta] \tag{3.40}$$

Substitute (3.39) in (3.37),

$$\sigma_{D_n}^2 = n\sigma_\delta^2 + 2 \sum_{k=1}^{n-1} (n-k)\,\psi_\delta(k) \tag{3.41}$$

where

$$\psi_\delta(k) = E[\delta_0 \delta_k] - E^2[\delta_0] \tag{3.42}$$

is the auto-covariance function of the sequence δ_i corresponding to the lag number k.

Similar to (3.36) the mean square fatigue damage during the time interval T can be shown to be

$$E[D^2(T)] = \sum_{n=0}^{\infty} E[D_n^2] P_N(n)$$

$$= \sum_{n=0}^{\infty} \left\{ n\sigma_\delta^2 + n^2 E^2[\delta] + 2 \sum_{k=1}^{n-1} (n-k)\,\psi_\delta(k) \right\} P_N(n) \tag{3.43}$$

If $\psi_\delta(k)$ can be estimated then the variance of the fatigue damage can be evaluated as

$$\sigma_D^2(T) = E[D^2(T)] - E^2[D(T)] \tag{3.44}$$

Let us consider the example of the cantilever. The various statistics regarding the fatigue damage are easily seen to be

$$E[\delta] = \frac{(\sqrt{2}\,\sigma_x)^b}{c}\, \Gamma\left(1 + \frac{b}{2}\right)$$

$$E[D_n] = \frac{n}{c}\, (\sqrt{2}\,\sigma_x)^b\, \Gamma\left(1 + \frac{b}{2}\right)$$

$$E[D(T)] = \frac{N^+(0)}{c}\, T(\sqrt{2}\,\sigma_x)^b\, \Gamma\left(1 + \frac{b}{2}\right)$$

$$E[\delta^2] = \frac{(\sqrt{2}\,\sigma_x)^{2b}}{c^2}\, \Gamma(1 + b)$$

$$\sigma_\delta^2 = \frac{(\sqrt{2}\,\sigma_x)^{2b}}{c^2}\, \{\Gamma(1 + b) - \Gamma^2(1 + b/2)\}$$

and

$$\psi_\delta(k) = \frac{1}{c^2}\, \{E[s_0^b s_k^b] - (\sqrt{2}\,\sigma_x)^b\, \Gamma^2(1 + b/2)\} \tag{3.45}$$

Shinozuka (1966) obtained an approximate expression for $E[s_0^b s_k^b]$ assuming the stress peaks to be jointly Gaussian and for even integral values of b. The expression is,

$$E[s^b(t)s^b(t + \tau)] = \sum_{i=0}^{b/2} A_i E^{b-2i}\, [s^b(t)s^b(t + \tau)]\, E^{2i}\, [s^2(t)] \tag{3.46}$$

where

$$A_i = \left(\frac{b!}{i!\, 2^i}\right)^2 \frac{1}{(b - 2i)!} \tag{3.47}$$

After some lengthy calculations (Shinozuka, 1966) the variance of the fatigue damage during a time interval T can be shown to be

$$\sigma_{D(T)}^2 = E[D^2(T)] - E^2[D(T)] = \frac{N^+(0)T}{2}\, \sigma_\delta^2[1 + 2\Psi_0^*(b)/\zeta] + \sigma_n^2\, E[\delta^2] \tag{3.48}$$

where σ_n^2 is the variance of the number of cycles in the duration T and

$$\Psi_0^*(b) = \frac{1}{\pi} \sum_{i=0}^{b/2-1} \frac{a_i}{(b - 2i)} \tag{3.49}$$

where

$$a_i = \frac{A_i}{\sum\limits_{i=0}^{b/2-1} A_i} \tag{3.50}$$

σ_n^2 can be evaluated by numerical integration from a result due to Leadbetter and Cryer (1965a). For $N^+(0)T \gg 1$ the second term in (3.48) may become negligible in comparison with the first term and (3.48) can be expressed as

$$\frac{\sigma_{D(T)}}{E[D(T)]} = \frac{1}{\sqrt{2N^+(0)T}} \frac{[\Gamma(1+b) - \Gamma(1+b/2)]^{1/2} [1 + 2\Psi_0^*(b)/\zeta]^{1/2}}{\Gamma(1+b/2)} \quad (3.51)$$

Comparing (3.51) which is valid for even integer values of b with (3.27) which is valid for odd values of b, the equivalence between the function $f_1(b)$ (Crandall, Mark and Khabhaz, 1962) and $\Psi_0^*(b)$ (Shinozuka, 1966) can be shown to be

$$f_1(b) = \frac{[\Gamma(1+b) - \Gamma^2(1+b/2)][\zeta + 2\Psi_0^*(b)]^{1/2}}{2\Gamma^2(1+b/2)} \quad (3.52)$$

$f_1(b)$ for different even values of b are evaluated from (3.52) using (3.47), (3.50) and (3.49) and given in Table 3.2. The results agree well with the trend of results given in Table 3.1.

Table 3.2   Quantity in (3.52) for b even
(Shinozuka, 1966)

b	$f_1(b)$
2	0.1841
4	0.8213
6	2.5517
8	7.6093
10	23.8701

3.6.4   Fatigue Damage Under Wide Band Random Vibration

For a wide band random stress history as shown in Fig. 3.6 it is not immediately obvious exactly what constitutes a cycle or how cycles should be counted. Dowling (1972) has provided a summary of cycle counting methods for obtaining statistics of stress ranges for fatigue prediction when the stress histories are wide band random processes. Two methods of cycle counting, the range pair method and the rain flow method are widely adopted in practice and the salient features of both the methods are described by Dowling (1972) in the following manner.

Range Pair Counting Method

'The range pair counting method counts a strain range as a cycle if it can be paired with a subsequent straining of equal magnitude in the opposite direction. For a complicated history, some of the ranges counted as cycles will be simple ranges during which the strain does not change direction, but others will be interrupted by smaller ranges which will also be counted as cycles'. Cycle counting by the range pair method is illustrated in Fig. 3.6. The counted ranges are marked with solid lines and the paired ranges with dashed lines.

'Each peak is taken in order as the initial peak of a range, except that a peak is skipped if the part of the history immediately following it has already been paired with a previously counted range. If the initial peak of a range is a minimum, a cycle is counted between this minimum and the most positive minimum which occurs before the strain becomes more negative than the initial peak of the range'.

STRAIN

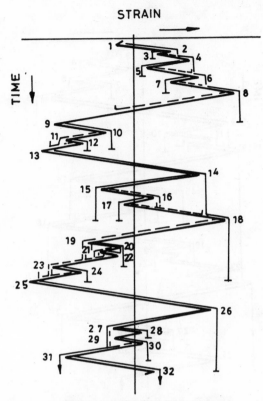

Fig. 3.6. Range pair counting method.

For example in Fig. 3.6 a cycle is counted between peak 1 and peak 8, peak 8 being the most positive maximum before the strain becomes more negative than peak 1. 'If the initial peak of a range is a maximum, a cycle is counted between this maximum and the most negative minimum which occurs before the strain becomes more positive than the initial peak of the range'. For example in Fig. 3.6 a cycle is counted between peak 2 and peak 3, peak 3 being the most negative minimum before the strain becomes more positive than peak 2. 'Each range that is counted is paired with the next straining of equal magnitude in opposite direction'. Hence in the range pair method complete cycles are counted rather than half cycles. For example, the part of the range between peaks 8 and 9 is paired with the range counted between peaks 1 and 8.

Rain-flow Method

The rain flow counting method is illustrated in Fig. 3.7. The strain-time history is plotted so that the axis is vertically downward, and the lines connecting the strain peaks are imagined to be a series of roofs. Several rules are imposed on rain dripping down these roofs so that the cycles and half cycles are defined. Rain flow begins sequentially at the inside of a strain peak. The rain flow initiating at each peak is allowed to drip down and continue, except that if it initiates at a minimum, it must stop when it comes opposite a minimum more negative than the minimum from which it initiated. For example in Fig. 3.7, begin at peak 1 and stop opposite peak 9, peak 9 being more negative than peak 1. A half cycle is thus counted

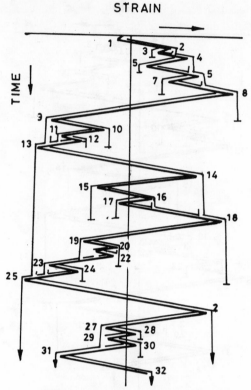

Fig. 3.7. Rain flow counting method.

between peaks 1 and 8. Similarly if the rain flow initiates at a maximum, it must stop when it comes oppposite a maximum more positive than the maximum, from which it was initiated. Thus we may begin at peak 2 and stop opposite peak 4, counting a half cycle between peaks 2 and 3. A rain flow must also stop if it meets the rain from a roof above. For example, the half cycle beginning at 3 ends beneath peak 2. Note that every path of the strain time history is counted and counted only once.

When this procedure is applied to a strain history, a half cycle is counted between the most positive maximum and the most negative minimum. Assume that of these two the positive maximum occurs first. Half cycles are also counted between the most positive maximum and the most negative minimum that occurs before it in the stress history, between this minimum and most positive maximum occurring previous to it and so on to the beginning of the history. After the most negative minimum in the history, half cycles are counted which terminate at the most positive maximum occurring subsequently in the history, the most negative minimum occurring after this maximum and so on to the end of the history. The strain ranges counted as half cycles therefore increase in magnitude to the maximum and then decrease.

All other strainings are counted as interruptions of these half cycles or as interruptions of these interruptions, etc., and will always occur in pairs of equal magnitude to form full cycles. The rain flow counting method corresponds to the stable cyclic stress-strain behaviour of a metal in that all strain ranges counted as cycles will form closed stress-strain hysteresis loops and those counted as half cycles will not. This is illustrated in Fig. 3.8.

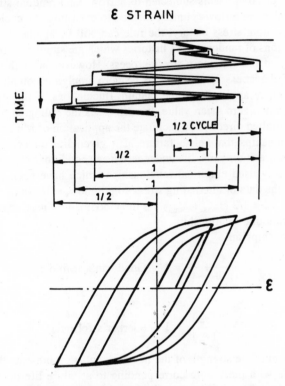

Fig. 3.8.   Cycles and half cycles in rain flow counting method and hysteresis loops.

Once the cycles are counted by any one of these methods, a variable amplitude history can be reduced to a series of reversals whose maximum and minimum values are defined. Mean stress is then known for each reversal and a completely reversed stress cycle can be obtained using the modified Goodman's formula about which we will discuss subsequently. The Palmgren-Miner damage rule can then be applied for each individual stress cycle to estimate the fatigue damage, which can be added cumulatively to yield the total fatigue damage.

In the case of a stationary wide band random stress history, generally a large number of sample functions are generated corresponding to the given power spectral density of the random process by any one of the simulation methods discussed in Chapter 2. The counting algorithm is programmed on the computer to estimate the number of cycles corresponding to each sample function. The statistics of the damage such as the mean damage and the variance of the damage can be obtained by statistical inference studies on the generated sample functions. Such a method is adopted by Wirsching and Shehata (1977) to obtain the mean fatigue damage of components subjected to wide band random stresses using the rain flow counting method. Simple modifications suitable for computer implementation of the rainflow counting algorithm have been proposed in the literature (Wetzel, 1977; Downing and Socie 1982).

3.6.5   Mean Stress Effects

We have seen that the Miner's rule can be used to advantage in the estimation of fatigue damage of components subjected to narrow band random stress histories. With the help of the rain flow counting method to determine the cycles in the case of wide band random stress history, the rule can still be applied to an ensemble of sample functions of random stress histories which are simulated from the power spectral density of the stationary stress process. However, in both the cases, the mean value of the stress is assumed to be zero implying completely reversed cycles. But in many applications, the random fluctuations of the stress response oscillate about a non zero mean value. In such cases the Palmgren-Miner damage rule cannot be applied directly. To facilitate the application of Miner's rule certain empirical relationships that relate failure at a given life under non-zero mean conditions to failure at the same life under zero mean conditions have been evolved. Two of the widely used empirical relationships which have been obtained after curve fitting to the data of alternating stress amplitude $s_a = 1/2\,(s_{max} - s_{min})$ versus the mean stress $s_m = 1/2\,(s_{max} + s_{min})$ are the Goodman's linear relationships and the Gerber's parabolic relationship.

These relationships are

$$\frac{s_a}{s_N} + \frac{s_m}{s_u} = 1 \quad \text{(Goodman's formula)} \tag{3.53}$$

and

$$\frac{s_a}{s_N} + \left\{\frac{s_m}{s_u}\right\}^2 = 1 \quad \text{(Gerber's formula)} \tag{3.54}$$

where s_m is the ultimate strength of the material, and s_N represents the equivalent stress amplitude with mean zero corresponding to a failure life of N cycles. But usually a modified form of the Goodman relationship is used in practice. The equivalent stress with mean zero, that would cause the same amount of fatigue damage as the stress history under non zero mean conditions depends on the range of values of the mean stress and the yield point of the material. Further details of applying the modified Goodman formula are given in Collins (1981). In the case of random stress history, the application of the modified Goodman's formula becomes still more complicated, because of the fact that the alternating stress amplitude s becomes a random variable.

3.7   Renewal Theory Model for Fatigue Life Prediction

A probabilistic model for fatigue life has been presented by Parzen (1959) based on renewal theory. In the model it is assumed that the damage suffered by the specimen during a load application at stress level s is a random variable. If $d_i(s)$ is the damage incurred during the ith cycle the total damage is the sum of the damages corresponding to all stress cycles, that is

$$D_n(s) = \sum_{i=1}^{n} d_i(s) \tag{3.55}$$

The random variables $d_i(s)$ are assumed to be independent and identically distributed. It is considered that failure occurs when $D_n(s)$ exceeds a critical value D^*. The number of cycles to failure in terms of the number of load applications is the smallest integer such that

$$D_n(s) \geq D^* \tag{3.56}$$

As $d_i(s)$ and hence $D_n(s)$ are random variables the fatigue life $N(s)$ is also a random variable in this model even under constant stress amplitude. It can be shown from renewal theory concepts (Smith, 1958) that when the expected value $E[d(s)]$ of the individual damage is small compared to D^*, the expected value and the variance of $d(s)$ are

$$E[d(s)] = \frac{D^*}{E[N(s)]} \tag{3.57}$$

and

$$\sigma_{d_s}^2 = \frac{\text{var}\,[N(s)]}{E^3[N(s)]}\,D^{*2} \tag{3.58}$$

Consider a specimen subjected to stress levels s_1, s_2, ... s_r respectively n_1, n_2, ... n_r times, where n_1, n_2, ... n_r are random variables. Then the damage done by n_k applications of s_k is

$$D_k = \sum_{i=1}^{n_k} d_i(s_k) \tag{3.59}$$

Taking conditional expectations, it can be shown that

$$E[D_k] = E[n_k]\,E[d(s_k)] \tag{3.60}$$

$$\sigma_{D_k}^2 = E[n_k]\,\sigma_{d_k}^2 + \sigma_{n_k}^2\,E^2[d_k] \tag{3.61}$$

$$E[\{D_k - E[D_k]\}\{D_l - E[D_l]\}] = K_{n_k n_l}\,E[d_k]\,E[d_l] \tag{3.62}$$

where $K_{n_k n_l}$ is the covariance of n_k and n_l.

The total damage is given by

$$D = \sum_i D_i \tag{3.63}$$

The expected value and variance of D can be shown to be

$$E[D] = D^* \sum_{k=1}^{r} \frac{E[n_k]}{E[N_{sk}]} \tag{3.64}$$

and

$$\sigma_D^2 = (D^*)^2 \sum_{k=1}^{r} \left\{ E[n_k] \frac{\text{var}\,[N_{sk}]}{E^3[N_{sk}]} \right\} + (D^*)^2 \left\{ \text{var}\left[\sum_{k=1}^{r} \frac{E_k}{E[N_{sk}]} \right] \right\} \tag{3.65}$$

where var $[\cdot]$ represents the variance of the quantity inside the brackets.

For a stationary narrow band Gaussian random history the expected number of maxima between s and $s + ds$ is given by $N^+(0)\,p_s(s)\,ds$. From this the expected damage level during an interval of duration T can be expressed as

$$E[D(T)] = N^+(0)T \int_0^\infty \frac{p_s(s)\,ds}{E[N_s]} \tag{3.66}$$

where $E[N_s]$ can be obtained from fatigue test data under constant stress amplitude. This expression is very similar to the expression for the expected fatigue damage calculated on the basis of Palmgren-Miner law.

There have been many attempts to improve on the idea of Parzen to treat the incremental damage as a random variable. Birnbaum and Saunders (1965) also treat the fatigue damage as a random variable and give a probabilistic interpretation for the Miner's rule. Other notable works along this line include those of Sweet and Kozin (1968), Hashin and Rotem (1978), Shimkova and Tanaka (1980), Tsuri (1981), and Birnbaum and Saunders (1969). Bogdanoff (1978a) has proposed a new cumulative damage model of a very general nature that includes the major sources of variability in the fatigue phenomenon. The model is based on a Markov process assumption of the damage accumulated during a cycle of stress reversal. The model holds promise to treat a wide variety of cumulative damage phenomena like fatigue crack growth, creep and wear in a simple manner as evidenced by the follow up work by Bogdanoff and his coworkers (Bogdanoff and Kreiger, 1978; Bogdanoff, 1978b; Bogdanoff and Kozin, 1980, 1982).

3.8 Fracture Mechanics Approach to Fatigue Under Random Loading

3.8.1 Some Fundamental Aspects of Linear Fracture Mechanics Theory

The problem of crack propagation under cyclic loading has been approached in different ways. The Griffith (1920)—Irwin (1968) fracture mechanics theory of crack propagation is a well established procedure for determining crack growth behaviour. It is observed in fracture behaviour that the magnitude of the nominal applied stress which causes fracture is related to the crack size. Materials like aluminum are characterized by two distinct stages of crack growth rate when subjected to tensile loads. Initially the crack growth rate is very slow which abruptly changes to a rapid rate of crack growth. This sudden change establishes an important property of the material called the fracture toughness which can be used as a design criterion in fracture prevention.

The simplest model for estimating the stress distribution in the neighbourhood of a crack is through linear elastic fracture mechanics theory. It should be noted however, that in any real engineering material there is a plastic zone formation at the tip of the crack. As long as this zone remains small compared to the dimensions of the crack the linear elastic model gives reasonable results which can be improved by the inclusion of a correction factor to account for the crack tip plasticity.

There are three distinct modes of crack deformation namely the crack opening mode: (Mode I) in which the crack surfaces move directly apart; the edge sliding mode (Mode II) in which the crack surfaces slide over each other in a direction perpendicular to the leading edge of the crack; and the tearing or parallel shear mode (Mode III) in which the crack surfaces slide with respect to each other in a direction parallel to the leading edge of the crack. In the sequel we shall consider only mode I crack propagation. The analyses for other modes are essentially similar.

Based on the methods developed by Westergard (1939), Irwin (1957) derived the expressions for the stress field near the crack tip using a linear elastic stress analysis. The expressions for the state of stress for an infinite plane under plane stress conditions are

$$\tau_{xx} = \frac{K}{\sqrt{2\pi r}} \cos \frac{\theta}{2} \left[1 - \sin \frac{\theta}{2} \sin \frac{3\theta}{2} \right] \tag{3.67}$$

$$\tau_{yy} = \frac{K}{\sqrt{2\pi r}} \cos \frac{\theta}{2} \left[1 + \sin \frac{\theta}{2} \sin \frac{3\theta}{2} \right] \tag{3.68}$$

$$\tau_{xy} = \frac{K}{\sqrt{2\pi r}} \sin \frac{\theta}{2} \cos \frac{\theta}{2} \cos \frac{3\theta}{2} \tag{3.69}$$

For plane strain conditions

$$\tau_{zz} = v(\tau_{xx} + \tau_{yy}) \quad \text{and} \quad \varepsilon_{zz} = \gamma_{xz} = \gamma_{yz} = 0 \tag{3.70}$$

where τ_{xx}, τ_{yy} etc. and ε_{zz}, γ_{xz} etc. connotate to the usual notations of stress and strain.

In (3.67) to (3.70) K is a stress intensity factor which represents the strength of the stress field surrounding the tip of the crack and is a function of the applied stress level and crack length, v is the Poisson's ratio of the material and r and θ are polar coordinates of the material element under consideration (Fig. 3.9).

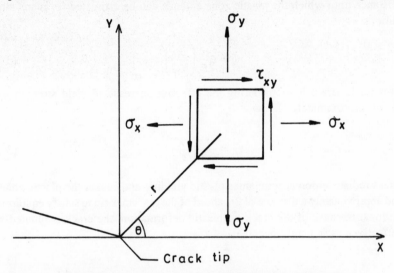

Fig. 3.9.   Stress distribution in the neighbourhood of the crack.

The concepts of linear elastic fracture mechanics may also be independently developed in terms of the strain energy release rate G. A simple linear relation connects the stress intensity factor and the strain energy release rate

$$G = K^2/E \quad \text{(for plane stress)} \tag{3.71}$$

and

$$G = \frac{K^2}{E}(1 - v^2) \quad \text{(for plane strain)} \tag{3.72}$$

Values of the stress intensity factor K for various loadings and configurations can be calculated using elasticity theory involving analytical, numerical and experimental methods. The common reference value of K is for a centre crack of length $2a$ in an infinite sheet subjected to a uniform tensile stress S which is given by

$$K = S\sqrt{\pi a} \tag{3.73}$$

The stress intensity factors for other crack geometries, configurations and loadings are usually modifications of the above equation.

Another parameter which is useful as a design criterion is the critical stress intensity factor K_c which is the value of K associated with rapid crack extension. For a given material K_c has a lower limiting value as the state of strain approaches the conditions of plane strain. This lower bound is a characteristic of the material and is called as the plane strain fracture toughness K_{Ic}. The subscript I refers to the fact that these fracture occur almost by mode I crack opening. K_c can be used as a failure criterion in fracture, in that fracture is predicted to occur if

$$S\sqrt{\pi a} \geq K_c \tag{3.74}$$

Often a simple correction can be applied to account for crack tip plasticity. The correction factor can be obtained by estimating the extent of the plastic zone near the crack tip. Equating τ_{yy} obtained from (3.68) to the yield stress of the material, the distance upto which the plastic zone extends can be estimated for plane stress conditions as

$$r_{y\tau} = \frac{1}{2\pi}\left(\frac{K}{\tau_{yp}}\right)^2 \tag{3.75}$$

and for plane strain by considering the effective increase in yield strength by a factor of approximately $\sqrt{3}$ as

$$r_{y\varepsilon} = \frac{1}{6\pi}\left(\frac{K}{\tau_{yp}}\right)^2 \tag{3.76}$$

Stress redistribution accompanies plastic yielding and causes the plastic zone to extend approximately a distance of $2r_y$ ahead of the real crack tip to satisfy equilibrium conditions. Because of the crack tip plastic deformation, the crack is blunted and the effective crack length becomes

$$a' = a + r_y \tag{3.77}$$

3.8.2 Fatigue Life Prediction Using Fracture Mechanics Theory

As already mentioned, it is well recognized that the fatigue failure process involves three stages. A crack initiation phase occurs first, followed by a crack propagation phase. Finally when the crack reaches a critical size, the final phase of unstable rapid crack growth to fracture completes the failure process. Realistic modeling of the crack initiation phase should precede the fracture mechanics approach, which can then be used to estimate the life from the crack propagation phase.

The most promising model of the crack initiation phase is based on the local stress-strain concept. It rests on the premise that the site of crack initiation is analogous to the fatigue response of a small smooth specimen subjected to the same cyclic strains and stresses. The cyclic stress-strain response of the smooth specimen can be determined by carefully simulated experimental procedures or by computer simulation. Time to crack initiation is usually estimated by application of any one of the cumulative damage theories, most often the Palmgren-Miner damage rule. For complex and random stress histories the special cycle counting methods can be employed in the use of the Palmgren-Miner hypothesis.

A fatigue crack that has been initiated by cyclic loading or any other pre-existing flaw in the material will grow under sustained cyclic loading until it reaches a critical size. From that stage, it will propagate rapidly and cause catastrophic failure. A significant portion of the fatigue life of the specimen will be spent in the fatigue initiated crack growing to a critical size. Thus the rate of crack propagation is an important parameter in the estimation of fatigue life. Many models have been proposed for predicting the crack growth rate in terms of the material properties and the external applied load. In a comparative analysis of the different crack growth models, Paris and Erdogan (1963) determined that the crack growth rate could be approximated by an expression of the form

$$\frac{da}{dN} = f(\Delta s, a, c) \tag{3.78}$$

where Δs, is the alternating nominal stress range, a is the crack length and c is a parameter that depends upon mean load and material properties. The stress intensity factor K depends upon the applied stress and crack size and so it may be concluded that the fatigue crack growth rate should be related to the range of the stress intensity factor K. Fatigue crack growth data for many engineering materials reveal that the crack growth rate can be expressed as

$$\frac{da}{dN} = c(\Delta K)^b \tag{3.79}$$

where b is the slope of the log (da/dN) versus log (K) plot. With this background, it is possible to consider the fatigue behaviour of a material subjected to random vibration using fracture mechanics concepts.

3.8.3   Fatigue Crack Propagation Under Random Vibration

CRACK INITIATION: When the specimen is subjected to cyclic loading for sometime a fatigue crack will be initiated. The time for crack initiation T_i is in general influenced by the size, shape, and material properties of the component. T_i is a random variable even under constant stress amplitude and it is more so in the case of random loading. It has been shown that the probability density function for the crack initiation time may be represented by a two parameter Weibull distribution (Eggwertz, 1974; Freudenthal, 1946)

$$P_{T_i}(t) = \alpha\beta^{-1}(t/\beta)^{\alpha-1} \exp\left(-\frac{t}{\beta}\right)^\alpha \tag{3.80}$$

where α is a shape parameter and β is a scale parameter. α and β are usually estimated from fatigue test results. One could also estimate the parameter β by use of S-N diagram data and cumulative damage hypothesis.

CRACK PROPAGATION: Once the crack is initiated further application of the stress cycles causes the crack to grow in size. For most materials the crack size is large in comparison with the crack tip plastic zone and hence the fracture mechanics theory can be applied to predict the crack propagation rate under random loading. Equation (3.79) gives the crack growth rate. As the structure is subjected to a random stress history the stress field near the crack tip will be random and consequently the stress intensity factor will also be a random process. For Gaussian random loading it has been verified experimentally (Rice, Beer and Paris, 1965; Paris and Erdogan, 1963) that the crack growth rate follows a similar law as for deterministic loading, that is

$$\frac{da}{dN} = cE[\Delta K^b] \tag{3.81}$$

where ΔK is the range of stress intensity factor and b and c are material constants and $E[\cdot]$ represents the expectation operation.

From (3.73) it can be shown that (3.81) can be expressed as

$$\frac{da}{dN} = Ca^{b/2} E[s^b] \tag{3.82}$$

Note that in (3.82) the constant $C = c\pi^{b/2}$. Thus the crack growth rate per cycle under random loading can be estimated if the bth power of the rise and fall statistics of the stress range is available. Rice, Beer and Paris (1965) have given a method of estimating the statistics in terms of joint probability density functions of the stress and its derivatives and by numerical integration schemes. Yang (1974) has proposed a closed form expression for the relevant statistics by use of extreme point process theory.

For many of the engineering materials the value of b is approximately equal to 4 in which case (3.82) reduces to

$$\frac{da}{dN} = Ca^2 E[s^4] \tag{3.83}$$

Assume now the stress history $X(t)$ is a zero mean stationary Gaussian random process. In such a case an approximate expression for the average fourth power of the stress range is (Rice, Beer and Paris, 1965)

$$E[s^4] = 3\pi \frac{\sigma_{\dot{X}}^5}{\sigma_{\ddot{X}}} \tag{3.84}$$

where $\sigma_{\dot{X}}$ and $\sigma_{\ddot{X}}$ are the standard deviations of the first and second derivatives of the process $X(t)$ with respect to t. Integrating (3.83) between crack initiation and time t, we obtain

$$\frac{1}{a_0} - \frac{1}{a(t)} = C\int_0^t E[s^4] \frac{dN}{dt} dt \tag{3.85}$$

where a_0 is the initial crack size and $a(t)$ is the crack size at time t. The rate of stress cycles per unit time dN/dt can be approximated by the expected rate of zero crossings of the process $X(t)$

$$\frac{dN}{dt} = N^+(0) = \frac{1}{2\pi} \frac{\sigma_{\dot{X}}}{\sigma_X} \tag{3.86}$$

Substituting (3.86) into (3.85) and integrating, the crack size $a(t)$ at time t is determined as

$$a(t) = \frac{a_0}{1 - \frac{3}{2} Ca_0 \dfrac{\sigma_{\dot{X}}^6}{\sigma_X \sigma_{\ddot{X}}} t} \tag{3.87}$$

One way to define fracture failure is in terms of the residual strength of the material. After fatigue crack initiation, the ultimate strength decreases because of the presence of the crack. Based on fracture mechanics, the relationship between the residual strength R of a structure containing a crack and the crack size is given by the Griffith-Irwin equation

$$K_c = R\left(\frac{\pi a}{2}\right)^{1/2}$$
(3.88)

in which K_c is the critical stress intensity factor. The relation (3.88) holds upto the point where R is equal to the ultimate strength R_0. As a result there is a critical crack size a_0 beyond which the strength R_0 starts decreasing following (3.88).

Let t_c be the time required to reach the crack size a_c. Then from (3.87).

$$t_c = \frac{2}{3}\frac{\sigma_X}{C}\frac{\sigma_{\dot{X}}}{\sigma_{\dot{X}}^6 a_0}\left[1 - \frac{a_0}{a_c}\right]$$
(3.89)

The residual strength at time t_m after the critical crack size has been reached can be obtained using (3.87) and (3.88) in conjunction, that is

$$R(t_m) = R_0\left[1 - \frac{2}{3}\frac{a_c C \sigma_{\dot{X}}^6}{\sigma_X \sigma_{\dot{X}}} t_m\right]$$
(3.90)

In fail safe structures the crack is not allowed to grow beyond a particular size. In such a case the residual strength at time t after crack initiation can be assumed to be of the form

$$R(t) = R_0\left\{1 - (1 - \xi)\left[\frac{a(t) - a_0}{a_s - a_0}\right]^{1/2}\right\}$$
(3.91)

where a_s is the fail safe crack size and $R_0\xi$ is the residual strength corresponding to that size.

It can now be assumed that the fatigue fracture would occur if the residual strength in the material is exceeded by the cyclic stress process.

FAILURE PROBABILITY: The probability of failure in fracture of the specimen can be posed as a first passage problem of the exceedance of the stress of the level corresponding to the diminishing residual strength of the material.

Consider the average failure rate $h_0(R_0, t)$ per cycle for the threshold R_0. $h_0(R, t)$ is the probability of exceedance by $X(t)$ of the level R_0 at time t given that it has not exceeded the level upto that instant. The average failure rate can be expressed as (Lin, 1967)

$$h_0(R_0) = \frac{N^+(R_0)}{M_C}$$
(3.92)

where $M_C \geq 1$ is the average clump size of the process $X(t)$. Usually the residual strength R_0 will be very high compared to the standard deviation of the stress response in which case M_C can be conservatively assumed to be unity.

The ultimate strength R_0 for most structures is a random variable. Hence the failure rate h_0 can be expressed as

$$h_0 = \int_0^\infty h_0(s, t)\, p_{R_0}(r)\; dr \tag{3.93}$$

where $p_{R_0}(r)$ is the probability density function of the ultimate strength of the material. For a stationary Gaussian random stress history $X(t)$ with mean zero

$$N^+(r, t) = \frac{1}{2\pi} \frac{\sigma_{\dot{X}}}{\sigma_X} \exp\left(-\frac{r^2}{2\sigma_X^2}\right) \tag{3.93}$$

If the material strength is also assumed to be a Gaussian random variable with parameters μ_{R_0} and σ_{R_0} (3.93) can be integrated to yield

$$h_0 = \frac{1}{4\pi} \frac{\sigma_X}{\sqrt{\sigma_X^2 + \sigma_{R_0}^2}} \exp\left[-\frac{\mu_{R_0}^2}{2(\sigma_{R_0}^2 + \sigma_X^2)}\right]\left[1 + \mathrm{erf}\left\{\frac{\mu_{R_0}\sigma_X^2}{\sigma_{R_0}\sigma_X\sqrt{\sigma_X^2 + \sigma_{R_0}^2}}\right\}\right] \tag{3.94}$$

If it is assumed that the mean residual strength of the material varies after crack initiation in the same manner given by (3.91), while the variance remains the same, the failure rate at different times when the material strength starts deteriorating can be obtained by replacing μ_{R_0} in (3.94) by $\lambda\mu_{R_0}$, where

$$\lambda = \left[1 - (1 - \xi)\left\{\frac{a(t) - a_0}{a_s - a_0}\right\}^{1/2}\right] \tag{3.94}$$

Since the time to crack initiation is a random variable using conditional probability functions the probability of failure within the intended service life T of the structure can be obtained as

$$P_f = \int_0^T \left[1 - \exp\left(-th_0 - \int_0^{T-t} h(t)\, dt\right)\right] p_T(t)\, dt$$

$$+ \int_T^\infty \left[1 - e^{-th_0}\right] p_T(t)\, dt \tag{3.96}$$

where $p_T(t)$ represents the probability density function of the time of crack initiation. The first term of (3.96) is the probability of failure with crack initiation within the duration $(0, T]$ and the second term represents the probability of failure without crack initiation in $(0, T]$.

The above analysis is on the lines of Yang and Trapp (1974) who used the above method for the reliability analysis of aircraft structures during its mission of ground, ground-air, gust load and landing cycles. Included in their analysis is also the consideration of replacement of defective components after inspection. Yadav (1976) has also used essentially the same approach in determining the probability of failure of an aircraft under carriage system both under stationary and nonstationary random loadings.

Recently Oh (1980) has proposed a diffusion model based on Markov process assumption for the crack size at different times in its growth to estimate the mean fatigue life of components under random loading. He used the Fokker-Planck equation approach in combination with a perturbation method for the analysis. Rau (1970) has also used the Markov process assumption for the crack length. He has obtained expressions for the probability of failure under some restrictive assumptions.

3.9   Creep Under Random Vibration

Ductile materials when subjected to certain loading and temperature environment, accumulate large strain leading to failure. If a structure or a specimen is subjected to uniaxial tension, which produces a nominal stress above certain critical value, it undergoes in addition to a possible plastic response, a gradual elongation at the rate dependent on the stress level. The specimen breaks after a certain time of sustained stress, the time t_c to fracture being dependent on the stress level. Figure 3.10 shows schematically the creep rupture process indicating that the elongation has a faster rate and consequently the time to fracture is shorter for a greater sustained stress level. If cyclic stresses are superimposed over the tensile stress, cycle dependent creep of the material is observed. The cycle dependent creep behaviour is similar to the time dependent creep behaviour of material.

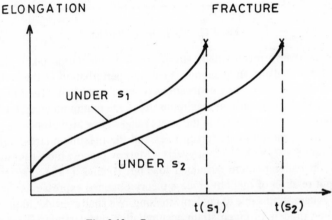

Fig. 3.10.   Creep rupture process.

Though much work has been done on creep behaviour of materials under constant or deterministically varying loading, only very little work has been done on creep under random loading. Two possible types of random loading that can cause creep behaviour in materials are (1) a sustained tensile loading whose amplitudes change in a random manner and (2) random fluctuating loads over a mean tensile load. The two types of loadings are schematically shown in Fig. 3.11. Experimental studies conducted by Tarleja (1974) on carbon steel specimen subjected to random loading superimposed on a tensile mean stress showed accumulation of strains with the number of cycles. He observed the familiar features of time dependent creep with the initial instantaneous increase in strain consisting of the plastic strain corresponding to the largest peak in the random stress history and the transient creep increasing at a decreasing rate until steady state was reached. In the steady state, strains increased gradually due to increase in stress, until instability of deformation occurred leading to rupture.

Feltner and Sinclair (1963) suggest a model of cycle dependent creep at low temperatures on the basis of a point defect-dislocation model. Under cyclic stresses

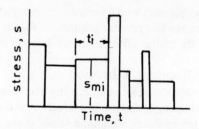

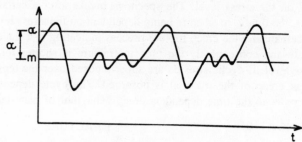

Fig. 3.11.   Cyclic tensile loading.

superimposed over a mean tensile stress, strain accumulation takes place due to the oscillations of dislocation segments causing perturbation of the concentration of the point defects, permitting dislocations to move over them. Their experimental results show that, for a constant maximum stress, the minimum creep rates increase exponentially with the increase in the stress range. They also propose a power law relation between the minimum creep rates and the maximum stress value.

Kameswara Rao (1977, 1978) suggests a Markov process model for creep strain accumulation under random sustained loading. Bieniek (1963, 1965) using the input-output relation of random process theory obtained expressions for the mean and mean square strains due to random stressing. We shall consider in the following section some models of creep strain accumulation under random loading.

3.9.1   Creep Damage Under Random Loading

Consider the material to be subjected to a random stress history as shown in Fig. 3.11b. Let $t_c(s)$ denote the time to failure due to creep strain accumulation. It can be assumed that the relative damage Δ_i due to a sustained stress s_i for a period of t_i is given by (Freudenthal, 1966; Shinozuka, 1977).

$$\Delta_i = \frac{t_i}{t_c(s_i)} \tag{3.97}$$

Hence, the relative damage done by sustained stresses at levels $s_1, s_2, \ldots s_n$ for periods of $t_1, t_2, \ldots, t_n$ respectively is

$$\Delta = \Sigma \Delta_i = \sum_{i=1}^{n} \frac{t_i}{t_c(s_i)} \tag{3.98}$$

Failure is assumed to occur when the total relative damage Δ attains a value of unity. The above hypothesis is very much similar to the Palmgren-Miner fatigue damage rule. It is also assumed that the variance of the stress is so small compared with the mean of the process, that fatigue and effect of negative stress may be disregarded. Let s_c be a threshold stress level below which no creep damage occurs.

Let us define

$$d\Delta = \frac{dt}{t[s(t)]} = g_c\{s(t)\} \quad \text{if } s(t) \geq s_c \tag{3.99}$$

$$= 0 \quad \text{otherwise}$$

Then, the damage Δ_t done by the stress process $s(t)$ in the time interval $(0, t)$ according to the damage rule given by (3.98) is

$$\Delta_t = \int_0^t \frac{d\tau}{t_c[s(\tau)]} = \int_0^t g_c\{s(\tau)\} \, d\tau \tag{3.100}$$

Assuming $s(t)$ is a stationary process with mean μ_s and variance σ_s^2,

$$E[\Delta t] = tE[g_c(s)] = \int_{s_c^*}^{\infty} g_c(x + \mu_s)p_{s*}(x) \, dx \tag{3.101}$$

where $s^*(t)$ is the random process obtained by subtracting the mean from $s(t)$, that is

$$s^*(t) = s(t) - \mu_s(t) \tag{3.102}$$

and $p_{s*}(x)$ is the probability density function of $s^*(t)$. Furthermore,

$$E[\Delta_t^2] = 2\int_0^t (t - \tau) R_{gg}(\tau) \, d\tau \tag{3.103}$$

where

$$R_{gg}(\tau) = E[g_c\{s(\tau_1)\} \, g_c\{s(\tau_1 + \tau)\}]$$

$$= \int_{s_c^*}^{\infty} \int_{s_c^*}^{\infty} g_c(x_1 + \mu_s) \, g_c(x_2 + \mu_s)p_{s_1^* s_2^*}(x_1, x_2) \, dx_1 dx_2 \tag{3.104}$$

is the auto correlation function of $g\{s(t)\}$ and $P_{s_1^* s_2^*}(x_1, x_2)$ is the joint probability density function of $s_1^*(\tau_1)$ and $s_2^*(\tau_1 + \tau)$. According to experimental results of Taira and Koterazawa (1961) $g_c(s)$ may be assumed in the form of a power law

$$g_c(s) = as^\alpha = a[s^*(t) + \mu_s]^\alpha \tag{3.105}$$

where a and α are material constants which depend on the shape of the specimen and the environment. Typical values of α for aluminum alloy at 500°C and for 0.15% carbon steel at 450°C are 8 and 32 respectively.

If $s(t)$ is a stationary Gaussian process with mean μ_s and standard deviation σ_s it can be shown that (Leadbetter and Cryer, 1965b) that

$$E[s_t] = \frac{at \, \sigma_s^\alpha}{\sqrt{2\pi}} \int_{\mu_s/\sigma}^{\infty} (x + \mu_s/\sigma_s)^\alpha \exp\left(-\frac{x^2}{2}\right) dx \tag{3.106}$$

The variance of the damage could also be expected in an integral form containing the standardized Gaussian density function.

Another formulation due to Tarleja (1974) considers the time rate of creep strain increase under stationary random loading about a mean essentially in the same manner as fatigue crack growth rate. A power law of the form

$$\frac{d\bar{\varepsilon}}{dN} = L\sigma_{s_a}^m \sigma_{s_p}^n \tag{3.107}$$

is assumed where $\bar{\varepsilon}$ is the mean strain and σ_{s_a} and σ_{s_p} are the root mean square values of the stress amplitude and the difference of the peak stress and a critical stress required to initiate persistent strain accumulation. The critical stress is taken to be the lower yield stress of the material. Failure is defined in terms of the Hoff's rupture (1953) theory which estimates the time of the cross-sectional area of a conceptual specimen reducing to zero under the action of creep strain accumulation. The statistics of σ_{s_a} and σ_{s_p} can be obtained from the original stress history $X(t)$ and its peak distribution.

Bieniek (1963, 1965) using the input-output relations of random process theory obtained the expressions for the mean and auto correlation function of the strain under creep assuming it to be a random process. His analysis included two models one for linear viscoelastic materials and the other for nonlinear viscoelastic materials. For a linear viscoelastic material the strain $\varepsilon(t)$ under a uniaxial tensile stress $X(t)$ which is taken to be a random process can be expressed as

$$\varepsilon(t) = \frac{1}{E} X(t) + \frac{1}{E}\int_0^t \varphi(t-\tau) X(\tau) \, d\tau \tag{3.108}$$

where E and φ are material constant and material function respectively. The mean and auto-correlation function of the strain can be expressed in the following manner:

$$\mu_\varepsilon(t) = \frac{1}{E}\mu_{X(t)} + \int_0^t \varphi(t-\tau)\mu_{X(\tau)}\, d\tau \tag{3.109}$$

and

$$
\begin{aligned}
R_{\varepsilon\varepsilon}(t_1, t_2) = \frac{1}{E^2}\Bigg\{ & R_X(t_1, t_2) + \int_0^{t_1}\varphi(t_1-\tau)R_{XX}(t_2, \tau)\, d\tau \\
& + \int_0^{t_2}\varphi(t_2-\tau)R_{XX}(t_1, \tau)\, d\tau \\
& + \int_0^{t_1}\int_0^{t_2}\varphi(t_1-\tau_1)\varphi(t_2-\tau_2)R_{XX}(\tau_1, \tau_2)\, d\tau_1\, d\tau_2
\end{aligned} \tag{3.110}
$$

For nonlinear viscoelastic material the creep strain can be expressed as

$$\varepsilon(t) = \frac{X(t)}{E_M} + \frac{1}{\eta_k}\int_0^t \varepsilon^{-(t-\theta)E_k/\eta_k X(\theta)d\theta} + \int_0^t \frac{X(\theta)}{\eta_M[X(\theta)]}\, d\theta \tag{3.111}$$

Equation (3.111) represents the stress-strain relation of the Burger's body that couples a Kelvin body with parameters E_k and η_k, and a Maxwell body with parameters E_M and η_M in series as shown in Fig. 3.12. η_M is assumed to be given by a nonlinear function of the form

$$\eta_M\{X(t)\} = [B \mid X(t)\mid^{n-1}]^{-1} \tag{3.112}$$

In (3.111) the second and third terms on the right hand side might represent phenomenologically the primary and secondary creep. For the nonlinear viscoelastic

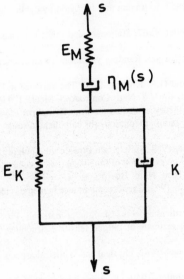

Fig. 3.12.   Nonlinear viscoelastic model.

material the mean and mean square values of the strains can again be obtained from the statistics of the random process $X(t)$.

REFERENCES

Afanasiev, N.N., 1940. J. Tech. Phy., USSR 12, 10, 1553.

ASTM, Standard definitions of terms relating to fatigue testing and statistical analysis of data, ASTM designation E 206–72.

Bienick, M.P., 1963. The effect of random loading on creep deformation in a nonlinear material. US. Airforce Aero. Sys. Div. ASD-TDL-63-646.

Bieniek, M.P., 1965. Creep under random loading. AIAA Journal. Vol. 3, No. 8 1959–1961.

Birnbaum, Z.W. and S.C. Saunders, 1965. A probabilistic interpretation of Miner's rule. SIAM J. App. Math. 16(3) 637–52.

Birnbaum, Z.W. and S.C. Saunders, 1969. A new family of life distributions. J. App. Prob. 6(2) 219–327.

Birnbaum, Z.W. and S.C. Saunders, 1969. Estimation for a family of life distributions with applications to fatigue. J. App. Prob. 6(2), 328–337.

Bogdonoff, J.L., 1978. A new cumulative damage model-part 1. ASME J. App. Mech. 45(2) 246–250.

Bogdonoff, J.L. and W. Kreiger, 1978. A new cumulative damage model—part 2 ASME J. App. Mech. 45(2) 251–257.

Bogdonoff, J.L., 1978. A new cumulative damage theory-part 3 ASME J. App. Mech. 45(4), 733–739.

Bogdonoff, J.L. and F. Kozin, 1980. A new cumulative damage model-part 4 ASME J. App. Mech. 47(1), 40–44.

Bogdonoff, J.L. and F. Kozin, 1982. On nonstationary cumulative damage models. ASME J. App. Mech. 49, 37–42.

Caughey, T.K. and H.J. Stumpf, 1961. Transient response of a dynamic system under random excitation. ASME J. App. Mech. 28(4), 563–566.

Collins, J.A., 1981. Failure of materials in mechanical design. Analysis, prediction and prevention. John Wiley & Sons, New York.

Corten, H.T. and T.J. Dolan, 1956. Cumulative fatigue damage. Int. Conf. Fatigue Metals. Inst. Mech. Engg., 235–246.

Cottrell, A.H. 1956. Point defects and the mechanical properties of metals at low temperatures. Symposium on point defects in metals and alloys. Inst. Metals. London Monograph 23.

Cottrell, A.H. and D. Hull, 1957. Extrusion and Intrusion by cyclic slip in copper. Proc. Roy. Soc. London A242, 211–213.

Cramer, H. and M.R. Leadbetter, 1967. Stationary and related stochastic processes, John Wiley & Sons. New York.

Crandall S.H. and W.D. Mark, 1963. Random vibrations in mechanical systems. Academic Press, New York, London.

Crandall, S.H., W.D. Mark and G.R. Khabaz, 1962. The variance in Palmgren-Miner damage due to random vibration. Proc. 4th U.S. Natl. Confr. App. Mech. 1, 119–126.

Dehlinger, U., 1940. On the theory of endurance limit. Zeiitchrift Physics 4, 115, 625–638.

Dowling, N.E., 1972. Fatigue failure predictions for complicated stress-stain histories. J. Mat., 7(1), 71–87.

Downing, S.D. and D.F. Socie, 1982. Simple rain flow counting algorithm Int. J. Fatigue, 4, 31–40.

Eggwertz, E., 1974. Statistical investigation of time until first, second and third cracks in wing panel. Aero. Res. Inst. Sweden. Tech. Note. Hu-150.

Feltner, C.E. and G.M. Sinclair, 1963. Cyclic stress induced creep of closed packed metals. Joint Int. Conf. Creep. 2.

Freudenthal, A.M., 1946. Statistical aspects of fatigue of materials. Proc. Roy. Soc. A187, 416–429.

Freudenthal, A.M., 1966. Life estimate of fatigue sensitive structures. Proc. 7th Int. Con. IABSE. Rio de Janeiro. 511–517.

Freudenthal, A.M. and E. Gumbel, 1962. Physical and statistical aspects of fatigue. Adv. App. Mech. 4, 117–159.

Freudenthal, A.M. and R.A. Heller, 1959. On stress interaction in fatigue and cumulative damage rule. J. Aero. Space Sci. 26(7), 431–442.

Freudenthal, A.M. 1966. Introduction to the mechanics of solids. John Wiley & Sons, New York.

Griffith, A.A. 1920. The phenomena of rupture and flow in solids. Phil. Tans. Roy. Soc. London. Series A 221, 163.

Hashin, Z. and A. Rotem, 1978. A cumulative damage theory of fatigue failure. Mat. Sci. Engg., 34, 147–160.

Head, A.K. 1953. The growth of fatigue cracks. Phil. Mag. 7, 44, 925–938.

Hoff, N.J., 1953. The necking and ruptures of rods subjected to constant tensile loads. ASME. J. App. Mech. 20, 105–108.

Irwin, G.R., 1957. Analysis of stress and strains near the end of a crack traversing a plate, ASME J. App. Mech. 24, 361–364.

Irwin, G.R., 1968. Linear fracture Mechanics, Fracture transition and fracture control, Engg. Frac. Mech, 1, 241–257.

Kameswara Rao, C.V.S., 1977. Creep damage under stationary random loading. ASME J. Appl. Mech. 44(4): 763–764.

Kameswara Rao, C.V.S., Creep strain accumulation under random loading. ASCE. J. Engg. Mech. Div. 104: 1457–1461.

Leadbetter, M.R., and J.D. Cryer, 1965a. The variance of the number of zeroes of a stationary normal process. Bull. Am. Math. Soc. 71: 561–563.

Leadbetter, M.R., and J.D. Cryer, 1965b. Curve crossings by normal processes and reliability implications. SIAM Rev. 7: 241–250.

Lin, Y.K., 1967. Probabilistic theory of structural dynamics, McGraw-Hill. New York.

Marco, S.M. and W.L. Starkey, 1954. A concept of fatigue damage. Trans. ASME. 76, 627–632.

Miles, J.W., 1954. On structural fatigue under random loading. J. Aero. Sci. 21: 753–762.

Miner, M.A., 1945. Cumulative damage in fatigue. ASME J. App. Mech. 12(3): 159–164.

Mott, N.F., 1952. A theory of work-hardening of metal crystals. Phil. Mag. 7, 43: 1151–1178.

Oh, K.P., 1980. The prediction of fatigue life under random loading, a diffusion model. Int. J. Fat. 2: 99–104.

Orawan, E., 1939. Theory of the fatigue of metals. Proc. Roy. Soc. Lond. A171: 79–106.

Palmgren, A. 1924. Die Lebensdaner Von Kugellagren. Zeitschrift des Vereines Deutscher Ingenieure, 58, 339–341.

Paris, P.C., 1962. The growth of cracks due to variations in load. Ph.D. Thesis, Leehigh University.

Paris, P.C., and F. Erdogan, 1963. A critical analysis of crack propagation loss. ASME. J. Basic Engg. 85(4): 528–543.

Parzen, E., 1959. On models for the probability of fatigue failure of a structure. NATO 244.

Rau, I.S., 1970. On fatigue crack propagation under stationary random loading. Int. J. Engg. Sci. 8: 175–189.

Roberts, J.B., 1968. Structural fatigue under nonstationary random loading. J. Mech. Engg. Sci. 8: 392–405.

Rice, J.R., F.P. Beer and P.C. Paris, 1965. On the prediction of some random loading characteristics relevant to fatigue. Acoustical Fatigue in Aerospace Structures, (ed.) Trapp, W.J. and D.M. Forney, Syracuse University Press.

Shanley, F.R., 1952. A theory of fatigue based on unbonding during reverse slip. Rand Corp. p. 350.

Shanley, F.R., 1956. A proposed mechanism of fatigue failure. Colloquim on Fatigue, Stockholm, Sweden, IUTAM Symposium, Springer-Verlag, Berlin, 251–259.

Shimkova, T. and S. Tanaka, 1980. A statistical consideration of Miner's rule. Int. J. Fatigue. : 165.

Shinozuka, M., 1966. Application of stochastic processes to fatigue, creep and catastrophic failures. Seminar on Application of Statistics in Structural Mechanics, Purdue University.

Shinozuka, M., 1972. Response to earthquakes creep and fatigue failure, Session IV G. Lecture Notes, ed. N.C. Nigam. Probabilistic Methods in Engineering. IIT Kanpur.

Shinozuka, M. and J.N. Yang, 1971. Peak structural response to nonstationary random excitation. J. Sound. Vib. 16: 501–517.

Smith, W.L. 1958. Renewal theory and its ramifications. J. Roy. Stat. Soc. SerB 20: 243–302.

Swanson, S.R., 1968. Random load fatigue testing. A state of the art survey. Mat. Res. Std. 8(4): 10–44.

Sweet, A.L. and F. Kozin, 1968. Investigation of a random cumulative damage theory. J. Mat, 3(4): 802–823.

Taira, S. and R. Koterazawa, 1961. Investigation on dynamic creep and rupture of a low carbon steel. Bull. Jap. Soc. Mech. Engs. 4: 238–246.

Talreja, R., 1974. Fatigue and cycle dependent creep under stationary Gaussian random loads. The Danish Center for Applied Mathematics and Mechanics. Rept. No. 5.5.

Tsuri, A., 1981. Probability of a fatigue failure as a statistic. Int. J. Fatigue, 3: 125–127.

Westergaard, H.M., 1939. Bearing pressures and cracks. ASME J. App. Mech, 66: 49.

Wetzel, R.M. ed., 1977. Fatigue under complex loading, SAE Publications.

Wirsching, P.H. and A.M. Shehata, 1977. Fatigue under wide band random stresses using rain flow method. ASME J. Engg. Mat. Tech. 99: 205–211.

Wöhler, A., 1860. Verschue uber die Festigkeit der Eisenbahnwagen-Achsen. Zeitschrift fur Bauwesen.

Yadav, D., 1976. Response of moving vehicles to ground induced excitation. Ph.D. thesis, IIT Kanpur.

Yang, J.N., 1974. Statistics of random loading to fatigue. ASCE J. Engg. Mech, Div. 100: 469–475.

Yang, J.N. and W.J. Trapp, 1974. Reliability analysis of aircraft structures under random loading and periodic inspection. AIAA J, 12: 1663–1630.

Yokobori, T., 1956. On the frequency dependence of the fatigue of metals. J. Phy. Soc. of Japan. 11: 715–716.

Yokobori, T., 1965. The strength, fracture and fatigue of materials. Noordhoff, Groningen.

4. Design of Structural/Mechanical Systems in Random Vibration Environment

4.1 Introduction

Many mechanical and structural systems operate in random vibration environments. These environments include the acoustic pressure field due to jet noise or boundary layer noise incident on aerospace structures; the gust loading on aircrafts; the ground motion during earthquakes; the wind loading on structures; the ground induced vibration in vehicles; the wave loading on off-shore structures etc. The uncertainties inherent in the underlying phenomena necessitate a probabilistic approach as a realistic and rational basis for analysis and design of such systems.

The deterministic and probabilistic approaches to design differ in basic philosophy and principles. In a deterministic framework system design completely discounts the possibility of failure. In this approach, the environment, the system and their interaction are represented as though they were fully determined. The designer is trained to believe that by a proper choice of design variables and safe operating limits, the levels of system performance would never be exceeded. It is assumed, as it were, that the system is immune to failure and will survive indefinitely. Overload capabilities are built into the design through safety factors which are often selected arbitrarily. The safety factor is an artifice by which the uncertainties involved in the design problem are recognized and taken into consideration, albeit in an indirect and arbitrary manner.

The elements of uncertainty inherent in almost all engineering design problems, implies that no matter how much is known about the phenomenon system behaviour is incapable of being predicted precisely. Under these circumstances there would always exist some likelihood of system failure, or system malfunction. However large may be the margins of safety allowed for in the design, the system cannot be deemed to be absolutely safe. There can in fact be no 'never fail' system.

The inadequacy of deterministic design in completely discounting possibilities of failure and the large uncertainties exhibited by natural phenomena and material behaviour have led to the extensive study and introduction of probabilistic design concepts or reliability based design. Reliability, which is the probability of system survival may be defined as the probability that a component part, equipment or system will satisfactorily perform its intended function under given circumstances such as environmental conditions, limitations as to operating time and frequency and thoroughness of maintenance for a specified period of time.

In the probabilistic approach the system is designed for a specified probability of failure depending on the consequences of failure. The uncertainties in the design situations are taken into account through appropriate probabilistic descriptions of the environment, system characteristics and the interaction between them. It should not however be construed at this stage that reliability based design is the 'end all' of all design problems. In problems, where the level of uncertainty is relatively small a deterministic approach to system design may be adequate. Even though the contingency of failure is not explicitly admitted, an adequate factor of safety would ensure very low failure probability levels. Moreover, a systematic application of reliability concepts in design requires realistic and meaningful probability information regarding the physical phenomena and system parameter variations. Pertinent and sufficient data, collection of which can become expensive and cumbersome, may not be available in all cases for the statistical inference studies. Assumptions regarding the probability distributions in the absence of sufficient and reliable data in the reliability based design may introduce more uncertainties into the problem instead of treating the inherent uncertainties.

Concepts that belong in the realm of engineering reliability such as reliability function, failure rate etc., and the selection of these with the recurrence period are quite useful in reliability based design of engineering systems. We shall consider the applications of these concepts with particular reference to the design of structural/ mechanical systems in a reliability framework.

The significant developments that have occurred in the area of structural reliability can be found in the comprehensive review on the subject by the task committee on structural safety of the American Society of Civil Engineers (Ang, 1972). Freudenthal (1956, 1962), Freudenthal, Garrets and Shinozuka (1966), Cornell (1967), Ang and Amin (1968, 1969), Ang (1973), Moses and Stevenson (1970). Stevenson and Moses (1970) consider such problems as estimation of failure probabilities with assumed probability distributions for the applied loads and resisting strengths; sensitivity of failure probability with the probability distributions especially near the tail portions of the distributions; and estimation of the overall probability of failure in terms of failure probabilities of individual members and the reliability of statically determinate and indeterminate structures. Haugen (1968, 1980) in his books treats probabilistic design of mechanical systems. Most of these works have been concentrated on structures subjected to static load conditions.

The application of reliability concepts to the design of structural/mechanical systems resisting dynamic loads is not very extensive. Shinozuka (1974) presents the basic approach and the difficulties that may be encountered in reliability based design for dynamic loads. Much of the work in this direction has been limited to estimation of failure probabilities corresponding to certain failure modes (Shinozuka, 1964; Yang and Heer, 1971).

In random vibration problems, specially if the level of uncertainties is high it becomes important and essential that the design is made within a probabilistic framework. In such a formulation the complexities involved in computing the random response for the evaluation of reliability estimates in different failure modes of the structural/mechanical system are carried through to the design problem. The most important of these difficulties is the treatment of the time parameter that exists either directly or indirectly in the design problem due to the dynamic nature of the response.

The general problem of mechanical design in random vibration environment has been discussed in detail by Mains (1958). Design methodologies of structural systems to resist earthquakes, wind loads, ocean wave forces are outlined in the works of Rosenblueth (1964), Davenport (1961a, 1961b, 1967), Newmark and Rosenblueth (1971), Racicot (1969), Malhotra and Penzien (1970), Sarpkaya and Issacson (1981) and numerous others. Clarkson (1962), Maestrello (1968) present basic design procedures to be adopted for aircraft structural systems to jet noise, boundary layer noise and other intense acoustic environments.

Optimization provides a formal framework for design in which one selects a particular design out of a set of feasible designs which is optimal in a specified sense. This process involves the selection of the design parameters in a systematic manner to satisfy the constraints in the problem and to optimize an objective function. Nigam (1983) has outlined the basic iterative cycle involving a sequence of steps for the implementation of the design of a structural/mechanical system against uncertainties, including optimization.

Design of optimum vehicle suspensions and optimum vibration isolators and absorbers to random excitation has been the subject of research by Bender (1967), Karnopp (1968), Karnopp and Trikha (1969), Trikha and Karnopp (1969), Sekiguchi and Iida (1970, 1973), Fujiwara and Murotsu (1974), Wirsching and Campbell (1974), Ayorinde and Warburton (1980), Warburton (1981, 1982), Jacquot and Hoppe (1973), Dahlberg (1977, 1979), Kadam (1983).

A general framework of optimum design of structures in random vibration environment with probability of failure constraints has been outlined by Narayanan and Nigam (1978). The application of such a formulation to different practical problems, such as earthquake and wind loading, jet noise excitation can be found in detail in the works of Narayanan (1975) and Narayanan and Nigam (1976). Other works in this area include those of Davidson, Felton and Hart (1977, 1980), Rao (1981) and Cederkvist (1982).

In this chapter we present various aspects involved in the design of structural/ mechanical systems operating in random vibration environments, including optimization. We discuss the formulation of the constraints in probabilistic terms on the dynamic response parameters such as displacements, stresses and acceleration representing the safe functioning of the system; the importance of constraints on natural frequencies and fatigue life of the structure, the design of optimum isolators and absorbers for random vibration. The reduction of the optimization problem to a mathematical programming problem and the method of solution are outlined. Practical applications with illustrative examples are presented.

4.2 Design Problem Formulation

The problem of reliability based design of structural/mechanical systems subjected to random vibration can be stated in the following general manner.

Find $\bar{d}$, the design vector, whose elements are the design variables, such that they satisfy the following constrains.

$$P[\bigcup_{h=1}^{n} \{ g_{hk} (\bar{X}(\bar{d}, \bar{s}, t)) \geq r_{hk}\}] \leq p_k, \quad k = 1, 2, \ldots m \tag{4.1}$$

$$g_j(\bar{d}) \leq \alpha_j, \quad j = 1, \ldots r \tag{4.2}$$

$$\omega_j^L \leq \omega_j(\bar{d}) \leq \omega_j^U, \quad j = 1, 2, \ldots l \tag{4.3}$$

In the foregoing inequalities $\bar{X}(\bar{d}, \bar{s}, t)$ is the random dynamic response vector with arguments $\bar{s} \in S$ and $t \in T$ representing the space domain and time parameters respectively. Double subscripts are adopted in the failure probability constraints (4.1) to treat different failure modes, each of which can be considered to be a combination of different events. For example g_{hk} may represent the stress in the hth member of a multimember structure, the subscript k referring to the failure mode based on the stress response of the structure or g_{hk} may be the acceleration corresponding to the hth generalized coordinate of a multidegree of freedom vibrating system, with the subscript k referring to the failure mode based on the acceleration response, or g_{hk} may be some function of the deflection at a critical location of a continuous structure, the subscript h denoting the particular location and k referring to the failure mode based on the displacement response of the structure. The quantity r_{hk} which represents the constraint on the function g_{hk} may be either deterministic or a random variable. p_k represents the specified probability of failure in the kth failure mode. $P[\cdot]$ denotes the probability of the event inside the brackets and $\cup$ the union operation. The constraints g_j of (4.2) are deterministic in nature and may include such restrictions as may be imposed by limits on the design variables (side constraints). ω_j represents the jth natural frequency of the system, and ω_j^L and ω_j^U represent respectively the lower and upper bounds. The constraints (4.3) could have been included in (4.2), but because of the importance of natural frequency constraints in design for vibration, they have been specified separately.

Optimization can be included in the formulation of the design problem by requiring that some performance criterion or utility factor represented by the objective function $W(\bar{d})$ is optimized. Thus the general problem of optimum design in random vibration environment can be stated as:

Find $\bar{d}^*$ to optimize:

$$W(\bar{d}) \tag{4.4}$$

subject to the constraints (4.1) to (4.3).

In the constraints expressed by (4.1), the probability of failure is defined over a period of time $[0, T]$. This adds a dimension of difficulty to the design problem. To obtain the optimum solution, the probability of the union of events specifying failure must be evaluated. This can be done if acceptable upper bounds for the probabilities expressed by the left hand side of inequalities (4.1) are established. Let E_{hk} be the event

$$E_{hk} = [\,g_{hk}\,(\bar{X}(\bar{d}, \bar{s}, t)) \geq r_{hk}] \qquad_{0 \leq t \leq T} \tag{4.5}$$

and

$$P[E_{hk}] = q_{hk}(\bar{d}, \bar{s}, t, T) \tag{4.6}$$

The constraints (4.1) can now be expressed as

$$P[\bigcup_{h=1}^{n} E_{hk}] \leq p_k, \quad k = 1, 2, \ldots m \tag{4.7}$$

Upper bounds on the probability $P[\bigcup_{h=1}^{n} E_{hk}]$ can be established by considering the nautre of the individual events. If E_{hk} are independent and the probabilities q_{hk} are small a close upper bound for $P[\bigcup_{h=1}^{n} E_{hk}]$ can be expressed as

$$P[\bigcup_{h=1}^{n} E_{hk}] \leq \sum_{h=1}^{n} q_{hk} \tag{4.8}$$

If the events E_{hk}s are fully nested,

$$P[\bigcup_{h=1}^{n} E_{hk}] \le \max_{h} q_{hk} \tag{4.9}$$

In general (Vanmarcke, 1972),

$$P[\bigcup_{h=1}^{n} E_{hk}] \le q_{1k} \sum_{h=2}^{n} a_h q_{hk} \tag{4.10}$$

where a_h is a conditional probability defined in terms of the survival events $\bar{E}_{hk}$ as

$$a_h = P[\bar{E}_{1k} \cap \bar{E}_{2k} \cap \ldots \cap \bar{E}_{h-1k} | \bar{E}_{hk}], \quad h = 2, 3, \ldots n \tag{4.11}$$

and

$$\bar{E}_{1k} = [\underset{0 \le t \le T}{g_{hk}} (\bar{X}(\bar{d}, \bar{s}, t)) < r_{hk}] \tag{4.12}$$

The values of a_h depend on the ordering of the component events in the particular failure mode. Evaluation of (4.11) would involve multiple integrations even if the needed joint distributions were available. A useful set of upper bonds for a_h can be obtained when all except one of the survival events are eliminated from (4.11), that is,

$$a_h \le P[E_{lk} | E_{hk}] = a_{hl}, \quad l = 1, \ldots h - 1 \tag{4.13}$$

The closest upper bound to a_h is obtained by selecting l which minimizes the set of upper bounds

$$a_h^* = \min_{l=1}^{h-1} (a_{hl}) \tag{4.14}$$

Vanmarcke (1972) has given an approximate result for the probabilities a_{hl}

$$a_{hl} = \frac{1 - P[g_{hk} \ge \max \{r_{hk}, r_{hk} - \mu_{hk}^M + \sigma_{hk}^M \mu_{lk}^M / (\sigma_{lk}^M | \rho_{hl}^M |)\}]}{P[g_{hk} \ge r_{hk}]} \tag{4.15}$$

where μ_{ik}^M and σ_{ik}^M are the mean and standard deviation of the safety margin $M_{ik} = r_{ik} - g_{ik}$ and ρ_{ij}^M is the correlation coefficient between M_{ik} and M_{jk}.

Substitute (4.11), (4.13) and (4.15) in (4.10):

$$P[\bigcup_{h=1}^{n} E_{hk}] \le \sum_{h=1}^{n} a_h^* q_{hk} \tag{4.16}$$

where $a_1^* = 1, a_2^* = a_2 = P[\bar{E}_{1k} | E_{2k}],$

$a_3^* = \min \{P[\bar{E}_{1k} | E_{3k}], P[\bar{E}_{2k} | E_{3k}]\}, \ldots$

Thus the probability in (4.1) satisfies the following inequality:

$$\max_{h} (q_{hk}) \le P[\bigcup_{h=1}^{n} E_{hk}] \le \sum_{h=1}^{n} a_h^* q_{hk} \le \sum_{h=1}^{n} q_{hk} \tag{4.17}$$

The dependence of the probabilities in the above inequality on the design variable vector $\bar{d}$, time t, space coordinates $\bar{s}$ and time interval T is implicit.

In a large class of random vibration problems, the dynamic response $\bar{X}$, can be treated as a stationary random vector process which ensures the time independence of the mean and the variance of the response. When reliability estimates can be made in terms of these statistics, the time parameter t can be directly eliminated from the constraints (4.1). It can be shown that,

$$P[\,g_{hk}\,(\bar{X}(\bar{d},\,\bar{s},\,t)) \geq r_{hk}] = P[\max_{t \in [0,T]}\,\{g_{hk}(\bar{X}(\bar{d},\,\bar{s},\,t)) \geq r_{hk}\}] \qquad (4.18)$$

The probability expressed in (4.18) is the first excursion failure probability. Useful upper bounds have been established and can be used conservatively in the optimal design problem. If we express by $U_{hk}(\bar{d},\,T)$, the upper bound on the probability in (4.18) then in the sequence of inequalities in (4.17) we can replace q_{hk} by U_{hk}.

Generally, the failure of mechanical/structural systems in random vibration can be specified in the following ways:

(i) First excursion failure.
(ii) Failure expressed by the fractional occupation time exceeding a given value.
(iii) Failure due to accumulation of damage as in fatigue or creep.

The failure mechanisms in the first two failure modes and the estimation of failure probabilities have been discussed by Nigam (1983). Fatigue failure mechanism under random vibration is discussed in Chapter 3.

Frequency is an important parameter in all vibration problems. Constraints on frequency are typical of optimal design problems in the dynamic response regime. The large dynamic responses that may result because of resonance conditions can be avoided by requiring the natural frequencies of the system to be away from excitation frequencies. Frequency is also an important factor for passenger comfort in the design of vehicle suspensions. Phenomena such as flutter, vortex induced vibrations, coincidence excitation of panels to acoustic excitation also depend on frequency. Further in the design of systems subjected to random vibration, the constraints on frequency can be specified from a knowledge of the power spectral density function. If the natural frequencies of the system lie in regions where the power spectral density of the excitation has small values, the response levels are reduced *per se* leading to a better design. In addition, if the natural frequencies are in the region where the power spectral density of the excitation is more or less uniform, response calculations pertinent to design can be considerably simplified using the local white noise assumption.

Apart from the probabilistic constraints there may also be deterministic constraints to be satisfied by a feasible design. For example, the stresses induced at a critical location of a structure due to wind loads consists of a mean stress and a fluctuating random stress. It is meaningful in the design problem to impose the condition that the mean stress is less than a safe limit which may be specified as a deterministic constraint, while the constraint on the random fluctuating component of the stress can be expressed in probabilistic terms. Sometimes the constraint representing system failure instead of being expressed probabilistically, may be expressed directly in terms of the response statistics such as the mean square value. Side constraints, which impose restrictions on the values of the design variables and which may arise because of the need to satisfy space requirements and to conform to the practical limits on available sections may also be part of the deterministic constraints which are expressed by the general inequality (4.2). Replacing the inequality (4.4) by one of the inequalities (4.17) the optimization problem in random vibration environment as expressed by (4.1) to (4.4) can be reduced to a standard nonlinear programming problem. The available methods of solving nonlinear programming problems can then be used to obtain the solution of the optimum design problem.

4.3   Optimum Vibration Isolators for Random Excitation

The conventional problem of vibration isolation of a single degree of freedom system subjected to a harmonic exciting force is to determine suitable values of spring constant and damping ratio such that the force transmitted to the base is minimized. Alternatively, if a single degree of freedom system is subjected to a base excitation, the isolation problem is to minimize the dynamic response of the mass. Under certain conditions both the problems are identical.

Consider the problem of finding the optimum parameters of a vibration isolation system for the single degree of freedom system subjected to a random support motion, $Z(t)$ as shown in Fig. 4.1. Let the criterion to find the optimum values of the spring stiffness k and the damping coefficient c be such that some measure of the absolute acceleration of the mass is minimized. In the absence of any constraint on the relative motion $Y(t)$ between the mass and the base, the optimum value of the stiffness reduces to a degenerate case since the maximum absolute acceleration can be made arbitrarily small for a very flexible isolator. In the design of vibration isolation systems, the clearance, or the sway space, between the mass and the base must be controlled. A stiff isolator is required to keep the relative motion between

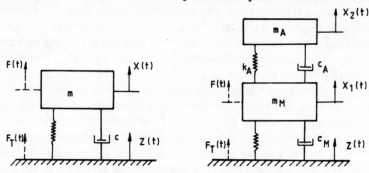

Fig. 4.1   (a) Vibration isolator. Single degree of freedom system.
(b) single degree of freedom vibration absorber attached to main system.

the mass and the base small. For harmonic excitation these conflicting requirements lead to the problem of finding an optimum system which minimizes the objective function of the form

$$W_m = \alpha_m \ddot{X}_{max} + Y_{max} \qquad (4.19)$$

where α_m is a weight factor to be selected by the system designer to indicate a range of preferences between small deflections (α_m small) and small vibration levels (α_m large). However, when the input $Z(t)$ is random, and only its second order statistics are known, an equivalent criterion can be expressed in terms of the mean square values of the response. Such a criterion takes the form

$$W_a = E[Y^2] + \alpha_a E[\ddot{X}^2] \qquad (4.20)$$

and has been employed by Bender (1967). Trikha and Karnopp (1969) have developed an alternate criterion by considering the minimization of an objective function of the form

$$W = y^2 + \alpha(m\ddot{x})^2 \qquad (4.21)$$

where y and $m\ddot{x}$ are sample values of the relative displacement and the force transmitted to the mass, such that

$$P[\mid Y(t) \mid \leq y] = 1 - p_1 \qquad (422)$$
$$\underset{0 \leq t \leq T}{}$$

and

$$P[\mid m\ddot{X} \mid \leq m\ddot{x}] = 1 - p_2 \qquad (4.23)$$
$$\underset{0 \leq t \leq T}{}$$

where α is again a weight factor, T is the total operation time, and p_1 and p_2 are specified probability of failures in the corresponding modes which are usually taken to be small.

Assuming $Z(t)$ to be a mean zero Gaussian random process, the response $\ddot{X}(t)$ and $Y(t)$ are also Gaussian due to the linearity of the system. Assuming the crossings of the levels y and $m\ddot{x}$ to be independent (Poisson events) and for low values of probabilities p_1 and p_2, approximate expressions for y^2 and $f^2 = (m\ddot{x})^2$ can be obtained in terms of mean square values and p_1 and p_2. (Trikha and Karnopp, 1969). These are

$$Y^2 \simeq 2\sigma_Y^2 \left[\ln \frac{T\sigma_{\dot{Y}}}{\pi\sigma_Y p_1} \right] \qquad (4.24)$$

$$f^2 \simeq 2\sigma_F^2 \left[\ln \frac{T\sigma_{\dot{F}}}{\pi\sigma_F p_2} \right] \qquad (4.25)$$

In (4.24) and (4.25) σ_Y and σ_F represent respectively the standard deviations of $Y(t)$ and $F(t) = m\ddot{X}(t)$. Trikha and Karnopp (1969) have given expressions for σ_Y, $\sigma_{\dot{Y}}$, σ_F and $\sigma_{\dot{F}}$ in terms of the impulse response function of the system. When substituted into the objective function the formulation leads to an integral equation in the impulse response function, the solution of which yields the optimum transfer function for the isolator that minimizes the criterion function W. A variational technique is used to obtain the optimal solution. A similar formulation for optimum vehicle suspension design which leads to a mathematical programming problem is considered later in this chapter and in Chapter 6.

We consider now an alternative fomulation of optimum isolation design for stationary random vibration. The problem is stated as follows (Fujiwara and Murotsu, 1974).

Find optimum values of the parameters k and c to minimize

$$W(k, c) = E[\ddot{X}^2(t, k, c)] \qquad (4.26)$$

such that

$$g_1(k, c) = E[Y^2(t, k, c)] \leq M \qquad (4.27)$$

Thus in the above formulation the optimization is performed with respect to the mean square absolute acceleration of the mass, with a constraint on mean square value of the relative displacement. Even though, the problem is not explicitly formulated in a probabilistic format, the objective function and the constraint are functions of the response statistics which depend on the probability structure of the base excitation. If the base acceleration $\ddot{Z}(t)$ is stationary, the power spectral density functions of $\ddot{X}$ and Y are given by

$$\Phi_{\ddot{X}\ddot{X}}(\omega) = |H_{\ddot{X}}(\omega)|^2 \, \Phi_{\ddot{Z}\ddot{Z}}(\omega) \tag{4.28}$$

and

$$\Phi_{YY}(\omega) = |H_Y(\omega)|^2 \, \Phi_{\ddot{Z}\ddot{Z}}(\omega) \tag{4.29}$$

where the frequency response functions are,

$$H_{\ddot{X}}(\omega) = \frac{\omega_n^2 + 2i\zeta\omega\,\omega_n}{\omega_n^2 - \omega^2 + 2i\zeta\omega\,\omega_n} \tag{4.30}$$

and

$$H_Y(\omega) = -\frac{1}{\omega_n^2 - \omega^2 + 2i\zeta\omega\,\omega_n} \tag{4.31}$$

The mean square values are obtained by integration of the corresponding power spectral densities over the frequency range.

Example 4.1: (Fujiwara and Murotsu, 1974)
Given

$$\Phi_{\ddot{Z}\ddot{Z}}(\omega) = \Phi_0\omega_r^2/(\omega^2 + \omega_r^2) \tag{4.32}$$

find optimum values of the parameters k^* and c^* for the isolation problem described by (4.26) and (4.27). In (4.32) Φ_0 is a constant representing the intensity of support acceleration and ω_r is a reference frequency.

From random vibration theory, the stationary mean square responses are obtained as (Nigam, 1983)

$$\frac{E[\ddot{X}^2(t)]}{\Phi_0\omega_r} = \frac{\omega_n(1 + 2\zeta\omega_n/\omega_r + 4\zeta^2)}{4\zeta\omega_r(1 + 2\zeta\omega_n/\omega_r + \omega_n^2/\omega_r^2)} \tag{4.33}$$

and

$$\frac{E[Y^2(t)]}{\Phi_0\omega_r^3} = \frac{1 + 2\zeta\omega_n/\omega_r}{4\zeta(\omega_n/\omega_r)^3\,(1 + 2\zeta\omega_n/\omega_r + \omega_n^2/\omega_r^2)} \tag{4.34}$$

The optimization problem becomes a nonlinear programming problem with two variables and a single constraint. Fujiwara and Murotsu (1974) have solved the problem using the Sequential Unconstrained Minimization Technique (SUMT) of Fiacco and McCormick (1968) in which an unconstrained penalty function of the form

$$P(\zeta, \omega_n/\omega_r, r) = W(\zeta, \omega_n/\omega_r) + \frac{r}{M - g_1(\zeta, \omega_n/\omega_r)} \tag{4.35}$$

$$= E[\ddot{X}_1^2(t)] + \frac{r}{M - E[Y^2(t)]} \tag{4.36}$$

is sequentially minimized for decreasing values of r, a positive constant. Note that the design variables are represented in terms of the damping ratio ζ and the frequency ratio ω_n/ω_r. A design chart in terms of the non-dimensional parameter $N = M\omega_r^3/\Phi_0$

is presented in Fig. 4.2 for determining the optimum values ζ_{opt} and $(\omega_n/\omega_r)_{opt}$. The mean square acceleration of the mass corresponding to the optimum value

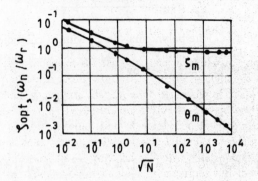

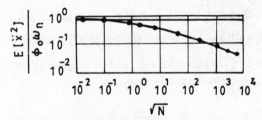

Fig. 4.2.   Optimum damping and natural frequency for vibration isolator (Fujiwara and Murotsu [1974]) $(\theta_m = \omega_n/\omega_r)$.

is also plotted alongside. It is found that the minimum value of $E[\ddot{X}^2(t)]$ occurs corresponding to the maximum value of clearance available, that is,

$$E[Y^2(t)] = M \quad \text{(at optimum)} \tag{4.37}$$

Thus the abscissa in Fig. 4.2 may be considered as a relative displacement for the optimum system. It is also observed that the optimum value of mean square absolute acceleration of the mass becomes smaller for increased values of M, indicating the conflicting behaviour of $\ddot{X}(t)$ and $Y(t)$.

4.4   Optimum Vibration Absorbers for Random Excitation

Vibration absorbers can be effectively used to reduce the random response of structural/mechanical systems just as they are used in reducing the harmonic response. Simple expressions for optimum absorber parameters can be obtained for a single degree of freedom absorber attached to a main single degree of freedom system whose mass is subjected to white noise excitation (Fig. 4.1b). The absorber parameters and the main mass parameters are denoted respectively by the subscripts A and M. Using the expressions given in Crandall and Mark (1963) for a two degree of freedom system, when the main mass is subjected to white noise excitation with constant power spectral density Φ_0, Jacquot and Hoppe (1973) obtain optimal values of uncoupled absorber damping ratio $\zeta_A = C_A/(2m_A\omega_A)$ for specified values of the mass ratio $\mu = m_A/m_M$ and the frequency ratio ω_A/ω_M, where $\omega_A = (k_A/m_A)^{1/2}$ and $\omega_M = (k_M/m_M)^{1/2}$ are the uncoupled natural frequencies of the absorber and the main mass.

The variance of the response $X_M(t)$ can be expressed as

$$\sigma^2_{X_M} = \frac{\pi \Phi_o \omega_M}{2k_M^2} \frac{\varphi_1(\zeta_A)}{\varphi_2(\zeta_A)} \tag{4.38}$$

where $\varphi_1(\zeta_A)$ and $\varphi_2(\zeta_A)$ are polynomials in ζ_A with its coefficients dependent on the mass ratio μ, the tuning ratio $\gamma = \omega_A/\omega_M$ and the main mass system damping ratio ζ_M. The polynomials are given by

$$\varphi_1(\zeta_A) = B_1 \zeta_A^3 + B_2 \zeta_A^2 + B_3 \zeta_A + B_4$$

$$\varphi_2(\zeta_A) = A_1 \zeta_A^3 + A_2 \zeta_A^2 + A_3 \zeta_A + A_4 \tag{4.39}$$

with

$$A_1 = 4\zeta_M \gamma^2(1 + \mu), \, A_2 = [\gamma\mu + 4\gamma\zeta_M^2 + \gamma^3\zeta_M^2 (1 + \mu)]$$

$$A_3 = [\zeta_M + \zeta_M\gamma^4(1 + \mu)^2 + 4\gamma^2\zeta_M^3 - 2\gamma^2\zeta_M] \text{ and } A_4 = \gamma^3\zeta_M\mu$$

and

$$B_1 = 4\gamma^2(1 + \mu), \, B_2 = 4[\gamma\zeta_M + \gamma^3\zeta_M(1 + \mu)]$$

$$B_3 = [1 - \gamma^2(2 + \mu) + \gamma^4(1 + \mu)^2 + 4\zeta_M^2\gamma^2] \text{ and } B_4 = \gamma^3\zeta_M\mu$$

An optimum value of ζ_A to minimize $\sigma^2_{X_M}$ can be obtained by differentiating (4.38) with respect to ζ_A and equating it to zero.

$$\frac{\partial \sigma^2_{X_M}}{\partial \zeta_A} = 0 \tag{4.40}$$

Equation (4.40) can be written as a fourth order polynomial of the form

$$\zeta_A^4 + C_1 \zeta_A^3 + C_2 \zeta_A^2 + C_3 \zeta_A + C_4 = 0 \tag{4.41}$$

with

$$C_1 = (\zeta_M/\gamma\mu)[1 + \gamma^4(1 + \mu)^2 + 2\zeta_M^2\gamma^2 - 2\gamma^2 + \gamma^2(1 + \mu)^2] \tag{4.42}$$

$$C_2 = \gamma^2\zeta_M^2 + \{\gamma^2\mu(2 + \mu) + 4\zeta_M(1 - \zeta_M) - \mu[1 + \gamma^4(1 + \mu)^2]\} \tag{4.43}$$

$$C_3 = - \gamma \mu\zeta_M/\{2(1 + \mu)\} \tag{4.44}$$

and

$$C_4 = - \zeta_M^2 \gamma^2\mu/\{4(1 + \mu)\} \tag{4.45}$$

Equation (4.41) can be solved numerically and the optimum damping ratio ζ_A can be obtained for specific values of ζ_M, γ and μ. The results are given in graphical form by Jacquot and Hoppe (1973).

Optimum Absorber Parameters without Damping in the Main System
In absorber designs it is not only the absorber damping ratio which is of interest, but also the frequency ratio γ. Ayorinde and Warburton (1980) following the same approach minimize the mean square displacement of the main mass with respect to both the absorber damping ratio ζ_A and the tuning ratio γ. Using relations

$$\frac{\partial \sigma^2_{X_M}}{\partial \zeta_A} = 0; \frac{\partial \sigma^2_{X_M}}{\partial \gamma} = 0 \tag{4.46}$$

optimum absorber parameters can be obtained for a given mass ratio μ and for white noise excitation of the mass. If for simplicity the main system damping is neglected ($\zeta_M = 0$), the minimizing conditions lead to closed form expressions for optimal parameters ζ_A and γ

$$\gamma_{opt} = \left(1 + \frac{1}{2}\mu\right)^{1/2} \Big/ (1 + \mu) \tag{4.47}$$

$$(\zeta_A)_{opt} = \left[\frac{\mu\left(1 + \frac{3}{4}\mu\right)}{4(1 + \mu)(1 + \mu/2)}\right]^{1/2} \tag{4.48}$$

The corresponding value of the mean square response is

$$(\sigma^2_{X_M})_{opt} = \frac{2\pi\Phi_o\omega_M}{k^2_M}\left[\frac{1 + \frac{3}{4}\mu}{\mu(1 + \mu)}\right]^{1/2} \tag{4.49}$$

Equations (4.47) and (4.48) are of similar form to the classical expressions for optimum absorber parameters for harmonic excitation of an ideal system (Den Hartog, 1956). For white noise excitation of the main mass γ_{opt} is greater for a given mass ratio μ than for harmonic excitation and $(\zeta_A)_{opt}$ is smaller for random excitation than the corresponding value for harmonic excitation.

Instead of minimizing the mean square displacement of the main mass if the mean square velocity of the main mass is minimized, the optimum absorber parameters for an undamped main mass system are given by

$$\gamma_{opt} = \frac{1}{(1 + \mu)^{1/2}}; \ (\zeta_A)_{opt} = \mu/4 \tag{4.50}$$

and

$$(\sigma^2_{\dot{X}_M})_{opt} = \frac{2\pi\Phi_o\omega^3_M}{k^2_M(\mu(1 + \mu))^{1/2}} \tag{4.51}$$

Crandall and Mark (1963) consider a two degree of freedom system of which the base is subjected to a white noise acceleration and obtain closed form expressions for the variances of the relative responses of each mass and their accelerations. Their numerical results show that with proper choice of absorber parameters considerable attenuation of the main system response is possible. As a part of a study of optimum absorber parameters to minimize the response of shear buildings which are subjected to white noise base acceleration Wirsching and Campbell (1974) present graphs of γ_{opt} and $(\zeta_A)_{opt}$ with respect to μ for an absorber which minimizes the mean square relative response of the main mass for this type of excitation and for various values of the main system damping ratio ζ_M. As the base excitation to a structure is often assumed to be a white noise in earthquake engineering problems, simple expressions for the optimum absorber parameters for an undamped main system are useful. If $Y = X_M - Z$ is the relative displacement between the main mass and the base, the variance of the relative displacement Y can be obtained by setting $\zeta_A = 0$ in the expression given by Crandall and Mark (1963), as

$$\sigma_Y^2 = \frac{2\pi\Phi_o}{\omega_M^3} \left\{ \frac{1}{\mu\gamma\zeta_A} \left[1 - (1 + \mu^2)(2 - \mu)\gamma^2 \right. \right.$$

$$\left. \left. + (1 + \mu)^4\gamma^4 \right] + 4\zeta_A\gamma(1 + \mu)^3/\mu \right\} \tag{4.52}$$

Applying the minimizing conditions

$$\frac{\partial \sigma_Y^2}{\partial \zeta_A} = 0 \quad \text{and} \quad \frac{\partial \sigma_Y^2}{\partial \gamma} = 0 \tag{4.53}$$

$$\gamma_{opt} = \left[\frac{1 - \mu/2}{1 + \mu} \right]^{1/2}; \quad (\zeta_A)_{opt} = \left[\frac{\mu(1 - \mu/4)}{4(1 - \mu/2)(1 + \mu)} \right]^{1/2} \tag{4.54}$$

and

$$(\sigma_Y^2)_{opt} = \frac{2\pi\Phi_o}{\omega_M^3} \left[\frac{(1 + \mu)^3(1 - \mu/4)}{\mu} \right]^{1/2} \tag{4.55}$$

Optimum absorber parameters that minimize the mean square acceleration of the main mass $E[\ddot{X}_M^2]$ for a white noise base excitation are the same as those that minimize $E[X_M^2]$ for a white noise force excitation of the main mass given by (4.47) and (4.48). This is so because the corresponding frequency response functions differ only by a multiplicative constant. The optimal value of the mean square acceleration is

$$(\sigma_{\ddot{X}_M}^2)_{opt} = 2\pi\Phi_o/\omega_M \left[\frac{1 + \frac{3}{4}\mu}{\mu(1 + \mu)} \right]^{1/2} \tag{4.56}$$

Likewise if the main mass is excited by white noise and if the objective is to minimize the mean square force transmitted to the foundation the optimum values of the absorber parameters are

$$\gamma_{opt} = \frac{(1 + \mu/2)^{1/2}}{1 + \mu}; \quad (\zeta_A)_{opt} = \frac{\mu(1 + 3\mu/4)}{4(1 + \mu/2)(1 + \mu)} \tag{4.57}$$

$$(\sigma_{F_T}^2)_{opt} = 2\pi\Phi_o\omega_M(1/\mu)^{1/2} \left[\frac{1 + \frac{3}{4}\mu}{(1 + \mu)} \right]^{1/2} \tag{4.58}$$

Optimum Absorber Parameters with Damping in the Main System
If damping is included in the main system in the form of a linear viscous dashpot, the optimum absorber parameters can still be obtained in closed form for the case of a white noise force excitation of the main mass, or a white noise base excitation as the case may be that minimize the mean square values of different response quantities of the main mass. In both the cases, the spectral densities of the response parameters can be obtained by considering the appropriate complex frequency response function. For the various cases, the complex frequency response function can be represented by

$$H_j(r) = (A_j + iB_j)/(C + iD) \qquad (4.59)$$

where A_j, B_j, C and D are functions of the tuning ratio $\gamma = \omega_A/\omega_M$, mass ratio $\mu = m_A/m_M$, the excitation frequency ratio $r = \omega/\omega_M$ and the uncoupled damping ratios ζ_A and ζ_M. The functions C and D are given by

$$C = (\gamma^2 - r^2)(1 - r^2) - \mu\gamma^2 r^2 - 4\zeta_A\zeta_M\gamma r^2 \qquad (4.60)$$

$$D = 2\zeta_A r\gamma(1 - r^2 - \mu r^2) + 2\zeta_M r(\gamma^2 - r^2) \qquad (4.61)$$

The functions A_j and B_j are tabulated by Warburton (1982) for various excitations and response parameters for a damped single degree of freedom system with an attached damped vibration absorber and are reproduced here in Table 4.1.

Table 4.1. Complex frequency response functions $H_j(r)$ for various excitations and response parameters of damped single degree of freedom system with attached absorber

$$H_j(r) = A_j + iB_j/(C + iD) \text{ and } R_j = |H_j(r)|$$

Case j	Excitation	Response parameters	R_j	A_j	B_j
1	$Fe^{i\omega t}$	X_M	$k_M X_M/F$	$\gamma^2 - r^2$	$2\zeta_A r\gamma$
2	$Fe^{i\omega t}$	$\dot{X}_M$	$k_M \dot{X}_M/(F\omega_M)$	$-2\zeta_A r^2\gamma$	$r(\gamma^2 - r^2)$
3	$Fe^{i\omega t}$	$\ddot{X}_M$	$m_M \ddot{X}_M/F$	$-r^2(\gamma^2 - r^2)$	$-2\zeta_A r^3\gamma$
4	$Fe^{i\omega t}$	F_T	F_T/F	$\gamma^2 - r^2 - 4\zeta_A\zeta_M\gamma r^2$	$2\zeta_A r\gamma + 2\zeta_M r(\gamma^2 - r^2)$
5	$\ddot{Z}e^{i\omega t}$	$Y_M = X_M - Z$	$\omega_M^2 Y_M/\ddot{Z}$	$\gamma^2(1 + \mu) - r^2$	$2\zeta_A r\gamma(1 + \mu)$
6	$\ddot{Z}e^{i\omega t}$	$\ddot{X}_M$	$\ddot{X}_M/\ddot{Z}$	as in case 4	as in case 4
7	$\ddot{Z}e^{i\omega t}$	X_M	$\omega_M^2 X_M/\ddot{Z}$	$-A_4/r^2$	$-B_4/r^2$
8	Force $b\omega^2 e^{i\omega t}$	X_M	$m_M X_M/b$	$-A_3$	$-B_3$
9	Force $b\omega^2 e^{i\omega t}$	F_T	$F_T/b\omega^2$	$r^2 A_4$	$r^2 B_4$

F_T—force transmitted to the base.

The nondimensional variance of the appropriate response can be expressed by N_j,

$$N_j = \frac{1}{2\pi} \int_{-\infty}^{\infty} |H_j(r)|^2 \, dr \qquad (4.62)$$

The relation between N_j and the corresponding variance σ_j^2 are given in Table 4.2 (Warburton, 1982).

For cases 3, 8 and 9 of Table 4.1, the transfer function $H(r) \to 1$ as $r \to \infty$. For case 7, $H(0) \to \infty$. Thus for these cases an infinite mean square response is predicted for white noise excitation and therefore not considered henceforth. Using integrals given by Crandall and Mark (1963), the nondimensional response ratios can be represented as

$$N_j = I_j(\mu, \gamma, \zeta_A, \zeta_M)/L(\mu, \gamma, \zeta_A, \zeta_M) \qquad (4.63)$$

Table 4.2.   Nondimensional mean square response to be minimized

Case	Excitation		Optimized nondimensional response parameter	
j	Type	Applied to	σ_j^2	N_j
1	Force	Main mass	$\sigma_{X_M}^2$	$\dfrac{\sigma_{X_M}^2 k_M^2}{2\pi\Phi_o\omega_M}$
2	Force	Main mass	$\sigma_{X_M}^2$	$\dfrac{\sigma_{X_M}^2 k_M^2}{2\pi\Phi_o\omega_M}$
3	Force	Main mass	$\sigma_{F_T}^2$	$\dfrac{\sigma_{F_T}^2}{2\pi\Phi_o\omega_M}$
4	Acceleration	Base	$\sigma_{Y_M}^2$	$\dfrac{\sigma_{Y_M}^2\omega_M^3}{2\pi\Phi_o}$
5	Acceleration	Base	$\sigma_{\dot{X}_M}^2$	$\dfrac{\sigma_{\dot{X}_M}^2}{2\pi\Phi_o\omega_M}$

where

$$L(\mu, \gamma, \zeta_A, \zeta_M) = 4[\mu\gamma\zeta_A^2 + \zeta_A\zeta_M\{1 - 2\gamma^2 + \gamma^4(1 + \mu)^2 + 4\gamma^2\zeta_A^2(1 + \mu)$$
$$+ 4\gamma\zeta_A\zeta_M[1 + \gamma^2(1 + \mu)] + 4\gamma^2\zeta_M^2\} + \mu\gamma^3\zeta_M^2] \tag{4.64}$$

The optimum parameters to be determined are the tuning ratio γ and the damping ratio of the absorber ζ_A and are obtained as before from the equations $\partial N_j/\partial\gamma = 0$ and $\partial N/\partial\zeta_A = 0$. For the different cases, the functions I_j are given by

$$I_1 = \zeta_A[1 - \gamma^2(2 + \mu) + \gamma^4(1 + \mu)^2] + \mu\gamma^3\zeta_M$$
$$+ 4\gamma^2\zeta_A^2\zeta_M[1 + \mu^2(1 + \mu)] + 4\gamma^2\zeta_A\zeta_M^2 \tag{4.65a}$$

$$I_2 = \zeta_A[1 - 2\gamma^2 + \gamma^4(1 + \mu)] + \mu\gamma^3\zeta_M + 4\gamma^2\zeta_A^3$$
$$+ 4\gamma\zeta_A^2\zeta_M(1 + \gamma^2) + 4\gamma^2\zeta_A\zeta_M^2 \tag{4.65b}$$

$$I_4 = \zeta_A[1 - \gamma^2(2 + \mu) + \gamma^4(1 + \mu)^2] + \mu\gamma^3\zeta_M$$
$$+ 4\gamma^2\zeta_A^3(1 + \mu) + 4\gamma\zeta_A^2\zeta_M[1 + \gamma^2(1 + \mu)]$$
$$+ 4\zeta_A\zeta_M^2[1 - \gamma^2 + \gamma^4(1 + \mu)] + 4\mu\gamma^3\zeta_M^3$$
$$+ 16\zeta_A\zeta_M^2[\gamma^2\zeta_A^2 + \zeta_A\zeta_M(\gamma + \gamma^3) + \zeta_M^2\gamma^2] \tag{4.65c}$$

$$I_5 = \zeta_A[1 - (2 - \mu)(1 + \mu)^2\gamma^2 + (1 + \mu)^4\gamma^4] + \zeta_M[\mu^2\gamma$$
$$+ \mu\gamma^2(1 + \mu)^2] + 4\gamma^2(1 + \mu)^3\zeta_A^3$$
$$+ 4\gamma(1 + \mu)^2\zeta_A^2\zeta_M[1 + (1 + \mu)\gamma^2] + 4\gamma^2(1 + \mu)^2\zeta_A\zeta_M^2 \tag{4.65d}$$

For chosen values of μ and ζ_M numerical values of optimum parameters can be obtained by an efficient iterative procedure or an unconstrained minimization technique using derivatives of the functions N_j with respect to γ and ζ_A, which can be obtained in closed form Warburton (1982). The optimum values of γ_{opt} and $(\zeta_A)_{opt}$ for different combinations of μ and ζ_M are given in Tables 4.3 to 4.6.

Table 4.3   Optimum parameters for absorbers attached to damped 1-DOF systems

Case 1. Random force with white noise spectral density Φ_o applied to main mass.

Response: Variance of displacement of main mass $\sigma^2_{X_M}$.

For 1-DOF system $N = \sigma^2_{X_M} k^2_M / 2\pi \Phi_o \omega^3_M$.

μ	Main System Damping ζ_M	Main System		
		γ_{opt}	$(\zeta_A)_{opt}$	N_{opt}
0.01	0	0.9926	0.04981	9.988
	0.01	0.9921	0.04981	7.632
	0.02	0.9916	0.04981	6.118
	0.05	0.9901	0.04981	3.748
	0.1	0.9877	0.04981	2.222
0.03	0	0.9781	0.08565	5.752
	0.01	0.9773	0.08565	4.893
	0.02	0.9765	0.08565	4.239
	0.05	0.9741	0.08565	2.985
	0.1	0.9700	0.08565	1.959
0.1	0	0.9315	0.1525	3.126
	0.01	0.9302	0.1525	2.856
	0.02	0.9289	0.1525	2.624
	0.05	0.9250	0.1525	2.098
	0.1	0.9186	0.1525	1.554
0.2	0	0.8740	0.2087	2.189
	0.01	0.8724	0.2087	2.053
	0.02	0.8708	0.2087	1.932
	0.05	0.8662	0.2087	1.635
	0.1	0.8586	0.2087	1.291

We observe from the tables that γ_{opt} decreases as ζ_M increases except for case 2. The dependence of γ_{opt} on ζ_M is greatest for case 5. The optimum absorber damping ratio $(\zeta_A)_{opt}$ is insensitive to changes in the main system damping ζ_M except for case 2, in which it gradually increases as ζ_M increases. For small mass ratios the optimum tuning ratio γ_{opt} is very nearly equal to unity, but for higher mass ratios the optimum tuning ratio gradually decreases. The mean square response of the main mass even without an absorber decreases rapidly with increases in its damping and hence for small mass ratios and relatively large values of ζ_M, the addition of optimum absorber produces only a small decrease in the mean square response.

The concepts of optimum absorber designs can be extended to multi degree of freedom and continuous systems subjected to random excitation. The indicated expressions for absorber parameters of a single degree of freedom system are of practical significance if the real systems are replaced by equivalent single degree of freedom systems. Using the normal mode analysis such equivalence is established by Warburton (1981) for minimization of the mean square value of the absolute response.

4.5   Optimum Vehicle Suspension Design

Single Degree of Freedom Suspension

The problem of optimum vibration isolator design of a single degree of freedom system to random base excitation considered in Section 4.3 is of practical significance

Table 4.4   Optimum parameters for absorbers attached to damped 1-DOF systems

Case 2. Random force with white noise spectral density Φ_o applied to main mass.

Response: Variance of velocity of main mass $\sigma_{\dot{X}_M}^2$.

For 1-DOF system $N = \sigma_{\dot{X}_M}^2 k_M^2 / 2\pi\Phi_o\omega_M^3$.

μ	Main System Damping ζ_M	Main System		
		γ_{opt}	$(\zeta_A)_{opt}$	N_{opt}
0.01	0	0.9950	0.05000	9.950
	0.01	0.9956	0.05008	7.604
	0.02	0.9964	0.05018	6.096
	0.05	1.0000	0.05056	3.736
	0.1	1.0102	0.05153	2.215
0.03	0	0.9853	0.08660	5.689
	0.01	0.9863	0.08684	4.840
	0.02	0.9874	0.08710	4.194
	0.05	0.9921	0.08805	2.954
	0.1	1.0040	0.09022	1.940
0.1	0	0.9535	0.1581	3.015
	0.01	0.9550	0.1588	2.755
	0.02	0.9568	0.1596	2.532
	0.05	0.9633	0.1623	2.026
	0.1	0.9783	0.1680	1.501
0.2	0	0.9129	0.2236	2.041
	0.01	0.9148	0.2250	2.915
	0.02	0.9171	0.2265	1.802
	0.05	0.9250	0.2313	1.526
	0.1	0.9424	0.2412	1.206

in the suspension design of vehicles excited by random road roughness. Essentially two formulations of the problem were considered in that section. In one formulation due to Trikha and Karnopp (1969) a composite objective function to satisfy the conflicting requirements of minimal response of the acceleration of the mass and the relative displacement during a time of operation was developed. The acceleration and the relative displacement levels were chosen such that the probabilities of excedence of such levels during the time period T by the corresponding random processes were equal to specified probabilities. Assuming the random processes to be Gaussian, and its crossings over a level to be independent Poisson events and for low probabilities of excedence, approximate expressions were derived in terms of the mean square response and its derivative and the specified probabilities for the objective function. Parameters of optimum isolator design in terms of its impulse response function were obtained as solution of an integral equation through a variational approach. In another formulation optimum absorber natural frequency and damping ratio were determined minimizing the mean square acceleration response of the mass with a constraint on the mean square relative displacement.

Dahlberg (1977) formulates the problem of optimal vehicle suspension design when considered as a single degree of freedom system within a reliability framework. The formulation is somewhat similar to that of Trikha and Karnopp (1969), except that it is formulated as a mathematical programming problem with a number of constraints. The design variables are the natural frequency ω_n and the damping

Table 4.5   Optimum parameters for absorbers attached to damped 1-DOF systems

Case 1. Random force with white noise spectral density Φ_0 applied to main mass.

Response: Variance of force transmitted to frame, σ_{FT}^2.

For 1-DOF system $N = \sigma_{FT}^2./2\pi\Phi_0\omega_M$.

μ	Main System Damping ζ_M	Main System		
		γ_{opt}	$(\zeta_A)_{opt}$	N_{opt}
0.01	0	0.9926	0.04981	9.988
	0.01	0.9921	0.04981	7.635
	0.02	0.9916	0.04981	6.128
	0.05	0.9902	0.04982	3.786
	0.1	0.9885	0.04990	2.310
0.03	0	0.9781	0.08565	5.752
	0.01	0.9773	0.08565	4.895
	0.02	0.9765	0.08565	4.246
	0.05	0.9742	0.08568	3.015
	0.1	0.9713	0.08587	2.038
0.1	0	0.9315	0.1525	3.126
	0.01	0.9302	0.1525	2.857
	0.02	0.9289	0.1525	2.628
	0.05	0.9253	0.1527	2.119
	0.1	0.9206	0.1532	1.615
0.2	0	0.8740	0.2087	2.189
	0.01	0.8724	0.2087	2.054
	0.02	0.8709	0.2087	1.935
	0.05	0.8667	0.2090	1.651
	0.1	0.8611	0.2101	1.341

ratio ζ of the suspension. In the sequel we present the design problem formulation and the optimal solutions.

The optimization problem of the vehicle suspension can be stated as, find optimum values of ω_n, the natural frequency and ζ the damping ratio to minimize

$$E[\ddot{X}_{max}] \tag{4.66}$$

subject to

$$\underset{0 \leq t \leq T}{P[|\,Y(t)\,| \geq y_0]} \leq p_1 \tag{4.67}$$

$$\zeta_L \leq \zeta \leq \zeta_U \tag{4.68}$$

$$\omega_{nL} \leq \omega \leq \omega_{nU} \tag{4.69}$$

$$\sigma_Y \leq 0.3\, y_0 \tag{4.70}$$

where as before $Y(t) = X(t) - Z(t)$ is the relative displacement, $Z(t)$ is the random excitation representing the road roughness, ζ_L, ζ_U, ω_{nL} and ω_{nU} are lower and upper bounds on the corresponding variables to avoid unrealizable systems and y_0 is a safe threshold level and p_1 represents a specified probability of failure.

OBJECTIVE FUNCTION: The expected value of the largest maximum of a stationary random process over a sampling period T can be approximately given by (Cartwright, 1958)

Table 4.6   Optimum parameters for absorbers attached to damped 1-DOF systems

Case 1. Random accelatation with white noise spectral density Φ_o applied to frame main mass. Response: Variance of relative displacement of main mass σ_Y^2.

For 1-DOF system $N = \sigma_Y^2 \omega_M^3 / 2\pi\Phi_o$.

μ	Main System Damping ζ_M	Main System		
		γ_{opt}	$(\zeta_A)_{opt}$	N_{opt}
0.01	0	0.9876	0.04981	10.138
	0.01	0.9850	0.04981	7.743
	0.02	0.9819	0.04981	6.205
	0.05	0.9704	0.04982	3.798
	0.1	0.9436	0.04982	2.249
0.03	0	0.9636	0.08566	6.058
	0.01	0.9592	0.08566	5.110
	0.02	0.9545	0.08566	4.424
	0.05	0.9380	0.08567	3.109
	0.1	0.9032	0.08569	2.036
0.1	0	0.8861	0.1527	3.602
	0.01	0.8789	0.1527	3.285
	0.02	0.8714	0.1528	3.014
	0.05	0.8468	0.1529	2.399
	0.1	0.7991	0.1531	1.765
0.2	0	0.7906	0.2097	2.865
	0.01	0.7815	0.2098	2.680
	0.02	0.7721	0.2099	2.516
	0.05	0.7421	0.2103	2.113
	0.1	0.6862	0.2112	1.649

$$E[\ddot{X}_{max}] \simeq \sqrt{\pi/2} \left[N^+(0)T - \frac{N^+(0)T(N^+(0)T-1)}{2! \, 2} \right.$$

$$\left. + \frac{N^+(0)T(N^+(0)T-1)(N^+(0)T-3)}{3! \, 3} - \frac{(-1)^{N^+(0)T}}{\sqrt{N^+(0)T}} \right] \sigma_{\ddot{x}} \quad (4.71)$$

An asymptotic expression for $E[\ddot{X}_{max}]$ is (Davenport, 1964)

$$E[\ddot{X}_{max}] \simeq [(2 \ln N^+(0)T)^{1/2} + 0.5772/(2 \ln N^+(0)T)^{1/2}]\sigma_{\ddot{x}} \quad (4.72)$$

with the expected rate of zero crossings with positive slope given by $N^+(0) = \frac{1}{2\pi}\left(\frac{\sigma_{\dddot{x}}}{\sigma_{\ddot{x}}}\right)$. Equation (4.71) is used for $N^+(0)T < 20$ and (4.72) for $N^+(0)T \geq 20$.

An alternate formulation for the objective functions is as follows. Instead of minimizing the expected maximum of the acceleration an extreme value of $\ddot{X}(t)$, $\ddot{x}_p$ during the time period T may be minimized such that, $\ddot{x}_p$ will be exceeded only with a small probability $P << 1$ during the sampling period T. $\ddot{x}_p$ can be estimated by the mode of $X(t)$ during a sampling period of length T/P.

$$\ddot{x}_p \simeq (2 \ln N^+(0)T/P)^{1/2}\sigma_{\ddot{x}} \quad (4.73)$$

CONSTRAINTS: The probability expression on the left hand side of (4.67) can be expressed as

$$P[|Y(t)| \geq y_0] \leq P_2 + (1 - P_2)P_1 \qquad (4.74)$$
$$\scriptstyle 0 \leq t \leq T$$

where P_2 is the probability that $|Y(t)| \geq y_0$ at $t = 0$ and P_1 is the probability that $|Y(t)|$ will cross y_0 with positive slope at least once in a sampling period T. With the assumption that the level crossings are Poisson events

$$P_1 = 1 - \exp (1 - N^+(y_0) T) \qquad (4.75)$$

and

$$P_2 = \frac{2}{\sqrt{2\pi}\,\sigma_Y} \int_{y_0}^{\infty} \exp (- y^2/2\sigma_y^2)\, dy \qquad (4.76)$$

It has been assumed that the random process $Y(t)$ is Gaussian. With the assumption that $\sigma_Y \ll y_0$, (4.76) can be approximated by

$$P_2 \cong \frac{2\sigma_y}{\sqrt{2\pi}\,y_0} (1 - \sigma_Y^2/y_0^2) \exp (- y^2/2\sigma_y^2) \qquad (4.77)$$

The assumption that $\sigma_Y \ll y_0$ implies that $P_2 \ll 1$. In addition if $N^+(y_0)T \ll 1$,

$$P_1 = 1 - \exp (- N^+(0)T) \cong N^+(0)T \qquad (4.78)$$

And finally if T is large $P_2 \ll P_1$. Thus the expression in (4.74) can in some cases be simplified to

$$P[|Y(t)| \geq y_0] \cong N^+(0)T \qquad (4.79)$$
$$\scriptstyle 0 \leq t \leq T$$

The constraint (4.70) is included so that the approximations underlying Eqs. (4.75) and (4.76) will be valid, at the same time restricting the standard deviation of the suspension working space.

RESPONSE STATISTICS EVALUATION: Assume the input spectral density of the road undulations, a mean zero stationary Gaussian random process to be

$$\Phi_{ZZ}(\omega) = \Phi_r \omega_r^2/\omega^2, \quad \omega_L < |\omega| < \omega_U \qquad (4.80)$$

where Φ_r is a scaling factor and ω_r is a reference frequency for a particular road. The standard deviation of the input displacement and acceleration after integration are

$$\sigma_Z = [2\Phi_r\omega_r^2(1/\omega_L - 1/\omega_U)]^{1/2}$$

$$\sigma_{\ddot{Z}} = \left[\frac{2}{3} \Phi_r\omega_r^2(\omega_U^3 - \omega_L^3)\right]^{1/2} \qquad (4.81)$$

Response statistics $\sigma_{\ddot{x}}^2$ and σ_Y^2 needed in the objective function and the constraint functions are obtained in the usual way of integrating the corresponding spectral density functions over the frequency range ω_L to ω_U after multiplying (4.80) by the square of the modulus of the respective frequency response functions.

Example 4.2: Optimum suspension design (Dahlberg, 1977)

Numerical results are presented by Dahlberg (1977) for the following data: $\omega_L = 0.1\omega_r$, $\omega_U = 1000\omega_r$, $T = 600/\omega_r$ and $p_1 = 0.001$. The optimization problem is solved by the sequential unconstrained minimization technique (SUMT) with a mixed interior-exterior penalty function for the unconstrained minimization. The results of the optimization problem are shown in Fig. 4.3 which give optimum combinations of ζ and ω_n for various values of y_0. The numbers in the figure give relative mean values of the largest acceleration maxima $E[\ddot{X}_{max}]/\sigma_{\ddot{z}}$. The numbers in the parenthesis are relative standard deviations $\sigma_{\ddot{x}}/\sigma_{\ddot{z}}$. For the level of $y_0 = 0.9\sigma_y$, we observe that the relative displacement $Y(t)$ will exceed y_0 during a period $T = 600/\omega_r$ with a probability $P = 0.001$ for the optimum parameters of the suspension $\omega_n = 10$ rad/s and $\zeta = 0.31$. The corresponding value of $E[\ddot{X}_{max}] = 0.05433\sigma_{\ddot{z}}$ and $\sigma_{\ddot{x}} = 0.01135\sigma_{\ddot{z}}$. For the same level of y_0 small reductions in $E[\ddot{X}_{max}]$ and $\sigma_{\ddot{x}}$ are possible with a slight decrease in the optimum value of ω_n and adjusting the value of ζ correspondingly. For example from the figure for the optimum combinations of $\omega_n = 8$ rad/s and $\zeta = 0.4$, $E[\ddot{X}_{max}] = 0.05314\sigma_{\ddot{z}}$ and $\sigma_{\ddot{x}} = 0.01109\sigma_{\ddot{z}}$. If the level on the relative displacement y_0 is changed to σ_Z, for $\omega_n = 10$ rad/s. $E[\ddot{X}_{max}] = 0.04509\sigma_{\ddot{z}}$ and $\sigma_{\ddot{x}} = 0.00943\sigma_{\ddot{z}}$ for an optimum damping ratio of $\zeta = 0.25$.

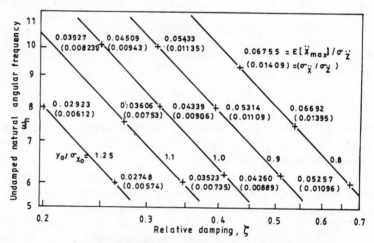

Fig. 4.3. Optimum combinations of ζ and ω_n for single degree of freedom vehicle suspension. (Dahlberg [1977]).

Two Degree of Freedom Suspension

Consider a two degree of freedom model for the vehicle suspension as in Fig. 4.4. m_1 and m_2 represent respectively the unsprung and sprung masses. The optimal parameters to be determined are the uncoupled suspension natural frequency $\omega_2 = \sqrt{k_2/m_2}$ and damping ratio $\zeta_2 = c_2/2\sqrt{k_2m_2}$ for a constant value of $\omega_1 = \sqrt{k_1/m_1}$ and $\zeta_1 = c_1/2\sqrt{k_1m_1}$. The road excitation $Z(t)$ is assumed to be a mean zero stationary Gaussian random process. For a constant speed of the vehicle the spectral density of the response quantities can be obtained in the usual manner for the linear system.

OBJECTIVE FUNCTION: The objective function to be minimized is formulated as a composite function in terms of a ride comfort criterion and road holding criterion.

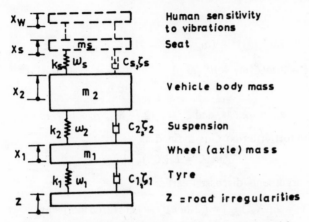

Fig. 4.4.   Two degree of freedom vehicle suspension model.

The ride comfort is quantified by means of a comfort (discomfort) number C which is defined in terms of the expected value of the maximum weighted vertical acceleration of the driver. This is obtained by scaling the acceleration of the passenger-seat combine by suitable weight factors in the different frequency ranges. For example, if we denote the seat passenger acceleration by $\ddot{X}_s$, the weighted acceleration for ride comfort criterion is,

$$\ddot{X}_{ws} = f(\omega)\ddot{X}_s \tag{4.82}$$

According to ISO standard 2631, the weighting function $f(\omega)$ in the different frequency ranges are (Dahlberg, 1978)

$$f(\omega) = \begin{cases} (\omega/8\pi)^{1/2} & 2\pi \leq \omega \leq 8\pi \\ 1 & 8\pi \leq \omega \leq 16\pi \\ 16\pi/\omega & 16\pi \leq \omega \leq 160\pi \end{cases} \tag{4.83}$$

The response below 1 Hz is also of importance in terms of human comfort as motion sickness is caused mainly in the frequency range of 0.1–0.6 Hz. The weight factor $f(\omega)$ for frequencies below 1 Hz can be taken as (Allen, 1975)

$$f(\omega) = 1.75 \, \omega^{-0.69}, \qquad 0.2\pi \leq \omega \leq 2\pi \tag{4.84}$$

The power spectral density of the weighted acceleration $\ddot{X}_{ws}$ is.

$$\Phi_{\ddot{X}_{ws}\ddot{X}_{ws}}(\omega) = f^2(\omega)\,|H_s(\omega)|^2\,|H_{\ddot{X}_2}(\omega)|^2\,\Phi_{ZZ}(\omega) \tag{4.85}$$

where $H_s(\omega)$ represents the amplitude of harmonic acceleration of the seat $\ddot{X}_s$ due to a unit harmonic acceleration of the mass m_2 and $H_{\ddot{X}_2}(\omega)$ represents the amplitude of harmonic acceleration of mass m_2 due to unit harmonic excitation at the base Z. For the model shown in Fig. 4.4 where the seat is considered as a single degree of freedom system attached to the sprung mass m_2,

$$H_{\ddot{X}_2}(\omega) = \omega^2(\omega_2^2 - 2i\zeta_2\omega\omega_2)(\omega_1^2 - 2i\zeta_1\omega\omega_1)/\Delta \tag{4.86}$$

and

$$H_s(\omega) = (\omega_s^2 + 2i\zeta_s\omega\omega_s)/(\omega_s^2 - \omega^2 + 2i\zeta_s\omega\omega_s) \tag{4.87}$$

where

$$\Delta = \mu(\omega_2^2 + 2i\omega\zeta_s\omega_2)^2 - (\omega^2 - \omega_2^2 - 2i\omega\zeta_2\omega_2)[\omega^2 - \omega_1^2$$
$$- \mu\omega_2^2 - 2i\omega(\zeta_1\omega_1 + \mu\zeta_2\omega_2)] \tag{4.88}$$

and

$$\omega_s = \sqrt{k_s/M_s} \text{ and } \zeta_s = c_s/2\sqrt{k_sM_s}, \ \mu = m_2/m_1$$

The ride comfort criterion C is defined as

$$C = E[(\ddot{X}_{ws})_{max}] \tag{4.89}$$

Following the arguments discussed earlier

$$C = E[(\ddot{X}_{ws})_{max}] = [(2 \ln n)^{1/2} + 0.5772/(2 \ln n)^{1/2}]\sigma_{\ddot{X}_{ws}} \tag{4.90}$$

where $n = \sigma_{\dddot{X}_{ws}} T/(2\pi\sigma_{\ddot{X}_{ws}})$ and T is the time period.

The other criterion in the composite objective function is the road holding. Let N represent the wheel-road contact force which is also a stationary Gaussian random process. Choose a deviation N_l from the static mean $E[N]$ of the wheel-road contact force. Let P_{Nl} be the probability that the contact force N will exceed $E[N] + N_l$ at least once in a time period T. Optimal road holding is defined as when P_{Nl} is minimum.

$$P_{Nl} = P_{N_1} + (1 - P_{N_1})P_{N_2} \tag{4.91}$$

with

$$P_{N_1} = \{\sigma_N/(\sqrt{2\pi}\,N_l)\}\,(1 - \sigma_N^2/N_l^2)\exp(-N_l^2/2\sigma_N^2)$$
$$P_{N_2} = 1 - \exp[-(\sigma_N T/2\pi\,\sigma_N)\exp(-N_l^2/2\sigma_N^2)] \tag{4.92}$$

where P_{N_1} is the probability that $N \geq E[N] + N_l$ at a fixed time $t = t_0$ and P_{N_2} is the first excursion probability that N will cross the level $E[N] + N_l$ from below at least in the time period $t_0 \leq t \leq t_0 + T$.

The usual assumptions of independent crossings and $N_l \gg \sigma_N$ have been made. As the randomly varying part of N is symmetric around its mean, P_N also gives the probability of having a contact force smaller than $E[N] - N_l$. Thus if $N_l = E[N]$, P_{Nl} gives the probability of losing the road contact. This formulation does not give any information about how often and when it happens, or for how long a time the road contact is lost.

Now the composite objective function is given by

$$W(\omega_2, \zeta_2) = \alpha C/C_{opt} + (1 - \alpha)P_{Nl}/(P_{Nl})_{opt} \tag{4.93}$$

where α is a weight factor $(0 \leq \alpha \leq 1)$ and C_{opt} and $(P_{Nl})_{opt}$ are known levels of optimal ride comfort and road holding when only one of them is considered for optimization.

CONSTRAINTS: As before a probabilistic constraint on the suspension working space is imposed. It requires that the probability P_{y_0} that the relative displacement $|Y(t)| = |X_2 - X_1|$ will exceed a given level y_0 in some time period $t_0 \leq t \leq t_0 + T$ is less than a specified probability p_1. That is,

$$P_{Y_0} = p_1 + (1 - p_1)p_2 \leq p_1 \tag{4.94}$$

where P_1 is the probability that $| Y(t) | \geq y_0$ at a fixed time t_0 and P_2 is the probability that $| Y(t) |$ will cross the level y_0 with a positive slope at least once during the sampling period T. Approximate formulae for P_1 and P_2 are

$$P_1 = (2\sigma_Y/y_0\sqrt{2\pi})(1 - \sigma_Y^2/y_0^2) \exp(-y_0^2/2\sigma_Y^2) \tag{4.95}$$

and

$$P_{N_2} = 1 - \exp[(-\sigma_Y T/\pi\sigma_Y) \exp(-y_0^2/2\sigma_Y^2)] \tag{4.96}$$

DESCRIPTION OF ROAD UNEVENNESS: The power spectral density of the road profile is assumed to be given by

$$\Phi_{ZZ}(\omega) = \Phi(\omega_r) (\omega_r/\omega)^2, \qquad |\omega| \leq \omega_r$$
$$= \Phi(\omega_r) (\omega_r/\omega)^{1.5}, \qquad |\omega| \geq \omega_r \tag{4.97}$$

where ω_r is a reference frequency. The power spectral density $\Phi(\omega_r)$ can be represented by

$$\Phi(\omega_r) = \Phi(n_r)/(4\pi v) \tag{4.98}$$

where $n_r = \omega_r/2\pi v$ is a reference spatial frequency and v is the speed of the vehicle. For an avearge road $\Phi(n_r)$ may vary in the range 32×10^{-6} m^3/cycle to 128×10^{-6} m^3/cycle. From the form of the spectral density function, the spectral density functions of the relevent response quantities are obtained and the response statistics required in the objective function and the constraints are determined by integration.

Example 4.3:   (Dahlberg, 1978)
The optimization problem of the vehicle suspension design with the objective functin (4.93) and the constraint (4.94) is solved by SUMT for the following parameters. $m_1 = 50$ kg, $m_2 = 500$ kg, $v = 20$ m/sec, $n_r = 1/2\pi$ cycles/m, $\omega_L = 0.2\pi$, rad/sec, $\omega_U = 160\pi$ rad/sec, $\omega_1 = 60$ rad/sec, $\zeta_1 = 0.06$, $\Phi(n_r) = 30 \times 10^{-6}$ m^3/cycle, $\zeta_s = 0.3$, $\omega_s = 5$ rad/sec, $T = 600$ sec and $N_l = (1.74 \times 10^6$ N/m$) \sqrt{\omega_r \Phi_{ZZ}(\omega_r)}$.

The results of the optimization program are in the form of graphs which are given in Figs. 4.5 and 4.6, for different values of α and for different levels y_0. A constraint $\omega_2 \geq 0.1$ rad/sec was introduced without which the optimization for the ride comfort alone would have yielded the degenerate case $\omega_2 = 0$. From Fig. 4.6a it is observed that the optimal ride comfort can be improved with improved levels of suspension working space. From Fig. 4.5, this corresponds to an optimal weaker spring dashpot system. From Fig. 4.6b, we find that the optimum road hodling improves with increase in y_0 up to a certain limit only.

For a level $y_0/\sigma_Z = 0.15$, the optimal comfort number C and the optimal road holding number P_N for $\alpha = 1$ and $\alpha = 0$ respectively are $C_{opt} = 8.3 \times 10^{-3} \sigma_{\ddot{Z}}$ and $(P_{nl})_{opt} = 0.058$. Choosing $\alpha = 0.5$ and for $y_0/\sigma_Z = 0.15$, the optimal suspension parameters from Fig. 4.5 are $(\omega_2)_{opt} = 4.9$ rad/sec and $(\zeta_2)_{opt} = 0.5$. For these values the corresponding optimal values of C and P_{Nl} can be read off from Figs. 4.6a and 4.6b as $(C_{opt})_{\alpha=0.5} = 14.3 \times 10^{-3} \sigma_{\ddot{Z}}$ and $(P_{Nl})_{opt(\alpha=0.5)} = 0.066$.

If the parameters ω_2 and ζ_2 are too far away from those giving optimum road holding, $P_{nl} \to 1$ as the second term in (4.93) is no longer a function of ω_2 and ζ_2. The minimization of $W(\omega_2, \zeta_2)$ then yields the same ω_2 and ζ_2 as for optimal comfort. This is the reason why the points d_6 and e_6 coincide in Fig. 4.5.

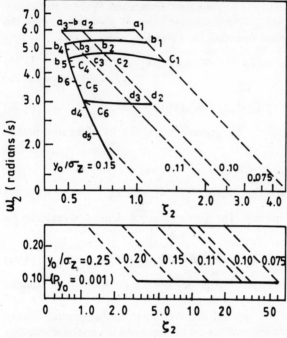

Letter	α
a	0.0
b	0.5
c	0.7
d	0.9
e	1.0

Index	y_0/σ_z
1	1.7
2	2.2
3	2.4
4	3.2
5	4.2
6	5.2

Fig. 4.5. Optimum combination of τ and ω_n for two degree of suspension model (Dahlberg 1978).

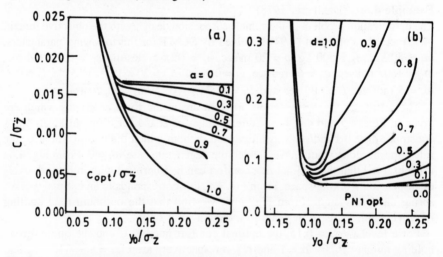

Fig. 4.6 (a) Relative comfort number $C/\sigma_{\ddot{x}}$ as a function of suspension working space (Dahlberg, 1978). (b) Road holding number P_n for small α values as a function of suspension working space (Dahlberg [1978]).

The optimal values of the suspension parameters have been determined for one particular speed. The optimal values will be different for different speeds. Then the passive suspension system considered in the above example will not be adequate. A speed controlled optimal active suspension system have to be adopted in such cases. Dahlberg, (1979a) has discussed the important considerations in such designs.

Higher degrees of freedom models including front and rear wheel suspensions, the pitching and bouncing degrees of freedom, with more design variables and additional constraints can similarly be formulated for vehicle suspension optimization problems. These can be found in the works of Dahlberg (1979b, 1980) and Kadam (1983) and are also discussed in Chapter 6. An introduction to optimal active suspension design using stochastic optimal control theory is also presented in Chapter 6.

4.6   Structural Optimization in Random Vibration Environment

In the examples so far considered, the emphasis was on the direct minimization of the dynamic response, such as mean square acceleration or the expected maximum response over a specific period of time, or the probability of the response exceeding a safe level during a time interval. Also, they were concerned with discrete vibrating systems and for a simple analysis of the results and understanding of the concepts were confined only to single degree and two degree of freedom vibrating systems. In the case of structural systems subjected to random excitation the optimal design criterion may be based on the minimization of other objectives, such as minimum weight or minimum cost which do not directly depend on the response, but on the structural dimensions and properties which may be considered as the design variables. The safe operation of the structural system can be expressed in terms of behavioural constraints, which in a reliability framework can be expressed as functions of the design variables and represent the different modes in which the structure can fail.

The probability of failure of the structure can be specified in two ways. In the first each failure mode is treated separately and the probability of failure in each mode is required to be less than a prescribed low value. Another way is to treat all the failure modes together and specify a probability limit for the overall probability of failure. It is more often advantageous to adopt the first method as it is more realistic and meaningful to conceive failure in each individual failure mode. Moreover estimation of the overall failure probability of the structure in terms of the failure probabilities in individual modes may prove to be very difficult.

The structural optimization problem can now be formulated in the general format discussed in Section 4.2. The probability of failure constraints can be in terms of the first excursion failure, fractional occupation time or fatigue failure. While some of the constraints can be expressed explicitly in terms of probabilities, other constraints may have to be expressed in terms of response statistics. We illustrate the formulation and solution of such problems through a number of examples that follow.

4.6.1   Cantilever Beam with Tip Mass

Consider a beam of thin walled rectangular section supporting a tip mass as shown in Fig. 4.7a. The optimization problem is to determine optimum dimensions of the cross-section such that the beam is of minimum weight satisfying certin constraints. We assume the beam to be of negligible mass compared to the tip mass. Under this assumption it can be idealized as a single degree of freedom system. The beam is subjected to a mean zero stationary Gaussian white noise excitation of constant power spectral density Φ_0 at the support.

OBJECTIVE FUNCTION: The design is completely defined by the design variable

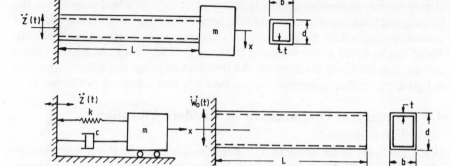

Fig. 4.7 (a) Cantilever beam with tip mass; (b) Uniform cantilever beam.

vector $\bar{d}$, given by $\bar{d} = \{b, d, t\}^{\mathrm{T}}$, that is $d_1 = b$, $d_2 = d$ and $d_3 = t$. Since the beam is of uniform cross-section and fixed length L minimizing the weight is equivalent to minimizing the area of cross-section of the beam. Thus the objective funciton is

$$W(\bar{d}) = 2d_3(d_1 + d_2) \tag{4.99}$$

assuming the thickness of the beam is very much smaller than the other two dimensions.

FREQUENCY CONSTRAINTS: The natural frequencies of the cantilever beam is required to be greater than a lower bound. So the frequency constraint is expressed in the form

$$\omega_n(\bar{d}) \geq \omega_L \tag{4.100}$$

where ω_L is the specified lower limit on the natural frequency. In a normalized form the frequency constraint becomes

$$g_1(\bar{d}) = 1 - \omega_L/\omega_n(\bar{d}) \geq 0 \tag{4.101}$$

For the single degree of freedom system considered $\omega_n(\bar{d}) = (3EI/mL^3)^{1/2}$, where I is the moment of inertia of the cross-section, m the tip mass and E the Young's modulus of the material. The moment of inertia in terms of the design variables is given by

$$I(\bar{d}) = \frac{1}{12}(6d_1 d_2^2 d_3 - 12\,d_1 d_2 d_3^2 + 2d_2^3 d_3 - 12d_2^2 d_3^2) \tag{4.102}$$

FATIGUE DAMAGE CONSTRAINT: The constraint on the fatigue life of the structure is expressed in terms of the expected rate of fatigue damage based on the linear cumulative damage law of Palmgren and Miner. It is required that the expected rate of fatigue damage is less than a specified value. That is,

$$\frac{1}{T}E[D(\bar{d})] \leq D_1 \tag{4.103}$$

or in normalized form

$$g_2(\bar{d}) = 1 - \frac{N^+(0)}{\beta D_1}(\sqrt{2\sigma_s})^{\alpha}\Gamma(1 + \alpha/2) \geq 0 \tag{4.104}$$

where $N^+(0)$ is the expected rate of zero crossings, σ_s is the standard deviation of the maximum stress at the root of the cantilever, $\Gamma(\cdot)$ is the gamma function, α and β are constants in the fatigue law $NS^\alpha = \beta$ and $E[D(\bar{d})]$ is the expected rate of fatigue damage and σ_s is given by

$$\sigma_s = \frac{3Ed}{2L^2} \left(\frac{\pi \Phi_0}{2\zeta \omega_n^3} \right)^{1/2} \tag{4.105}$$

where ζ is the damping factor.

STRESS CONSTRAINT: The stress constraint in this problem is expressed probabilistically in terms of the first excursion failure probability. The first excursion failure probability is evaluated in two ways. In the first case, the crossings of the stress response over the specified level are assumed to be independent (Poisson assumption). Under this assumption the first excursion failure probability is given by

$$p_f(\bar{d}) = 1.0 - \exp\left[-\omega_n T/\pi \exp\left(-S_y^2/2\sigma_s^2 \right) \right] \tag{4.106}$$

where S_y is the yield stress of the material which is assumed deterministic in this analysis and T is the time of operation of the random process. In the other case the two state zero-one process formed by the stress response crossings is assumed Markovian. Under this assumption the first excursion failure is obtained as (Vanmarcke, 1969)

$$p_f(\bar{d}) = 1.0 - \exp\left[-\frac{\omega_n T}{\pi} \left\{ \frac{1 - \exp\left(-S_y q\sqrt{(\pi/2)}/\sigma_s \right)}{\exp\left(\frac{1}{2} S_y^2/\sigma_s^2 \right) - 1} \right\} \right] \tag{4.107}$$

where q is a shape factor of the spectral density of the stress response given in terms of the spectral moments as

$$q = (1 - \lambda_1^2/\lambda_0\lambda_2)^{1/2} \tag{4.108}$$

where the spectral moments λ_js are defined by

$$\lambda_j = \int_0^\infty \omega^j G_s(\omega)\, d\omega \tag{4.109}$$

where $G_s(\cdot)$ is the one sided power spectral density of the stress response. The stress constraint can now be stated as

$$P[\underset{0 \leq t \leq T}{|S(t)|} \geq S_y] \leq p_1 \tag{4.110}$$

In the normalized form it is written as

$$g_3(\bar{d}) = 1.0 - p_f(\bar{d})/p_1 \geq 0 \tag{4.111}$$

ACCELERATION CONSTRAINT: The acceleration constraint is expressed in terms of the fractional occupation time of the acceleration response. For the stationary Gaussian random process $A(t)$ we require that the expected ratio of the time of the acceleration response spending above a safe level a_1 to the total duration of the response is less than a specified quantity p_2. The acceleration constraint can then be expressed as

$$g_4(\bar{d}) = 1 - \text{erfc} \, (a_1/\sqrt{2}\,\sigma_A)/p_2 \geq 0 \qquad (4.112)$$

where erfc$(\cdot)$ is the complementary error function and σ_A is the standard deviation of the absolute acceleration of the mass. It may be noted that erfc $(a_1/\sqrt{2}\,\sigma_A)$ represents a lower bound on the first excursion failure probability of excedence of the process $A(t)$ of the level a_1 (Shinozuka, 1964).

BUCKLING CONSTRAINTS AND SIDE CONSTRAINTS: The local buckling constraints on the web and flange of the box beam are expressed as

$$g_5(\bar{d}) = 1 - S_y/[k_1 E(d_3/d_2)^2] \geq 0$$

$$g_6(\bar{d}) = 1 - S_y/[k_2 E(d_3/d_1)^2] \geq 0 \qquad (4.113)$$

For the validity of the Euler beam assumption to hold in the optimization problem side constraints on the design variables are expressed in the form

$$g_7(\bar{d}) = 1 - 10d_1/L \geq 0$$

$$g_8(\bar{d}) = 1 - 10d_2/L \geq 0 \qquad (4.114)$$

In addition to the above, the non-negative restrictions on the design variable vector give rise to additional constraints of the form

$$g_9(\bar{d}) = d_1 \geq 0$$

$$g_{10}(\bar{d}) = d_2 \geq 0 \qquad (4.115)$$

$$g_{11}(\bar{d}) = d_3 \geq 0$$

OPTIMIZATION PROBLEM STATEMENT: The structural optimization problem of the cantilever beam with the tip mass can now be stated as:

Find $\bar{d}*$ to minimize the objective function (4.99) subject to the constraints

$$g_i(\bar{d}) \geq 0 \quad i = 1, 11$$

given by the appropriate inequalities from (4.101) to (4.115).

In the above form the problem is in the format of a typical nonlinear programming problem. The problem is solved (Narayanan and Nigam, 1978) using SUMT with interior penalty funciton of the P function form, namely,

$$P(\bar{d}, r) = W(\bar{d}) + r \sum_{i=1} \frac{1}{g_i(\bar{d})} \qquad (4.116)$$

in conjunction with the Fletcher and Powell (1963) gradient method for the unconstrained minimization and a golden section rule for the one dimensional search. The problem is solved for the following data:

$\Phi_0 = 0.02$ (ft/sec^2)^2/Hz, $M = 1.0$ lb-sec^2/in., $\zeta = 0.02$, $L = 200$ in., $\omega_L = 10\pi$ rad/sec, $p_2 = 10^{-3}$, $a_1 = 3864$ in./sec^2, $S_y = 20000$ lbs/in^2., $E = 10.6 \times 10^6$ psi, $\alpha = 6.0$, $\beta = 6.4 \times 10^{31}$, $k_1 = 21.72$, $k_2 = 3.62$, $T = 3600$ sec.

Two values of p_1 at constant fatigue damage level D_1 and two values of D_1 at constant value of p_1 are considered. The results are tabulated in Tables 4.7 to 4.9 both for the assumption of independent crossings and the assumption that the crossings constitute a Markov process. The tables also show convergence to almost the same optimum values starting from two different initial designs. From the results presented we can draw the following conclusions.

(1) For the low failure probability levels in the problem the Poisson and Markov assumptions regarding the first passage failure do not have any significant change in the optimal design with the Poisson assumption being slightly conservative.

Table 4.7   Cantilever beam with tip mass

$p_1 = 10^{-3}$ $D_1 = 5 \times 10^{-8}$	r	d_1	d_2	d_3	$W(\bar{d})$	$W(\bar{d}) + r \sum_i \frac{1}{g_i(\bar{d})}$
Initial design	0.8451	10.0	18.0	0.3	16.8	37.826
Optimum design (P)	0.8451×10^{-9}	8.43	19.992	0.192	10.944	10.944
Optimum design (V)	0.8451×10^{-8}	8.349	19.997	0.191	10.808	10.810
Initial design	0.9019	14.0	14.0	0.35	19.6	40.122
Optimum design (P)	0.9019×10^{-8}	8.465	19.874	0.193	10.959	10.960
Optimum design (V)	0.9019×10^{-7}	8.351	19.991	0.191	10.809	10.810

Active constraints at optimum g_3, g_6 and g_8.
(P)—Poisson crossings, (V)—Vanmarcke failure rate.

Table 4.8   Cantilever beam with tip mass

$p_1 = 10^{-4}$ $D_1 = 5 \times 10^{-9}$	r	d_1	d_2	d_3	$W(\bar{d})$	$W(\bar{d}) + r \sum_i \frac{1}{g_i(\bar{d})}$
Initial design	0.8451	10.0	18.0	0.3	16.8	37.826
Optimum design (P)	0.8451×10^{-8}	8.995	19.677	0.205	11.781	11.784
Optimum design (V)	0.8451×10^{-9}	9.093	19.174	0.207	11.737	11.738
Initial design	1.1881	14.0	14.0	0.4	22.4	44.8
Optimum design (P)	1.1881×10^{-5}	9.026	19.580	0.206	11.796	11.797
Optimum design (V)	1.1881×10^{-6}	9.182	19.893	0.209	11.782	11.785

Active constraints at optimum g_3, g_6.

Table 4.9   Cantilever beam with tip mass

$p_1 = 10^{-4}$ $D_1 = 5 \times 10^{-8}$	r	d_1	d_2	d_3	$W(\bar{d})$	$W(\bar{d}) + r \sum_i \frac{1}{g_i(\bar{d})}$
Initial design	0.89502	14.0	14.0	0.35	19.6	40.11
Optimum design (P)	0.8952×10^{-9}	9.143	19.816	0.209	12.093	12.101
Optimum design (V)	0.8952×10^{-8}	9.126	19.876	0.208	12.089	12.090
Initial design	0.8293	10.0	18.0	0.300	16.800	37.746
Optimum design (P)	0.8293×10^{-6}	9.099	19.897	0.209	12.100	12.193
Optimum design (V)	0.8293×10^{-8}	9.094	19.992	0.207	12.081	12.082

Active constraints at optimum g_2, g_6 and g_8.

(2) A decrease in the failure probability level $p_1 = 10^{-3}$ to 10^{-4} increases the optimum weight as expected. In the particular problem considered the increase is about 9%.

(3) Fatigue damage constraint becomes active for the lower value of the damage level $D_1 = 5 \times 10^{-9}$ pushing up the optimal weight by 2%.

(4) In all the cases the depth of the beam d_2 almost takes its limiting value of $L/10$ at the optimum.

The same problem is again solved by Rao (1981) with the same method assuming the design variables and the material properties to be Gaussian random variables.

4.6.2 Uniform Cantilever Beam (Narayanan and Nigam, 1978)

Consider the minimum weight design of a uniform cantilever beam shown in Fig. 4.7b subjected to support white noise excitation. The response statistics of the beam used in estimating the relevant probabilities can be obtained by the normal mode method. The mean square responses at any location x along the length of the beam are given by (Narayanan, 1975)

$$E[w_r(x, t)^2] = 8\pi \sum_j \sum_k \frac{I_{jk} f_j(x) f_k(x) \zeta_j \omega_j}{m_j m_k [(\omega_j^2 - \omega_k^2)^2 + 8\zeta_j^2 \omega_j^2 (\omega_j^2 + \omega_k^2)]} \qquad (4.117)$$

$$E[S^2(x, t)] = 8\pi \left(\frac{Ed}{2}\right)^2 \sum_j \sum_k \frac{I_{jk} f_j''(x) f_k''(x) \zeta_j \omega_j}{m_j m_k [(\omega_j^2 - \omega_k^2)^2 + 8\zeta_j^2 \omega_j^2 (\omega_j^2 + \omega_k^2)]} \qquad (4.118)$$

where

$$f_j(x) = \frac{(\sin p_j L - \sinh p_j L)(\sinh p_j x - \sin p_j x)}{(\cosh p_j L - \sinh p_j L)} + (\cosh p_j x - \cos p_j x)$$

$$(4.119)$$

is the jth normal mode of the cantilever beam, $p_j = \{\omega_j/\sqrt{EI/m}\}^{1/2}$, ω_j is the jth natural frequency given by the characteristic equation $\cos p_j L = -1/\cosh p_j L$, m the mass per unit length of the beam, $w_r(x, t)$ the relative displacement and $S(x, t)$ the stress at location x and time t. $m_j = mL$ is the jth generalized mass and $\zeta_j = c/2m\omega_j$ is damping ratio in the jth mode and c is a uniform viscous damping coefficient per unit length of the beam,

$$I_{jk} = m^2 \Phi_0 \int_L \int_L f_j(x_1) f_k(x_2) dx_1 dx_2 \qquad (4.120)$$

and

$$f_j''(x) = \partial^2 f_j(x)/\partial x^2.$$

Having obtained the response statistics in closed form the structural optimization problem can be stated as

Find $\bar{d}*$, to minimize,

$$W(\bar{d}) = 2d_3(d_1 + d_2)$$

subject to

$$g_1(\bar{d}) = 1 - \omega_L/\omega_1(\bar{d}) \geq 0 \quad \text{(frequency constraint on first natural frequency)}$$

$$g_2(\bar{d}) = 1 - \frac{\omega}{2\pi\beta D_1} (\sqrt{2\sigma_s})^\alpha \, \Gamma(1 + \alpha/2) \geq 0 \quad \text{(fatigue damage constraint)}$$

$$g_3(\bar{d}) = 1.0 - \text{erfc} \, (S_1/\sqrt{2(\sigma_s^2 + \sigma_{s_y}^2)})/p_1 \geq 0, \quad \text{(stress constraint)}$$

$$g_4(\bar{d}) = 1 - \text{erfc} \, (C_1 L/\sqrt{2}\,\sigma_{w_r})/p_2 \geq 0, \quad \text{(tip deflection constraint)}$$

$$g_5(\bar{d}) = 1 - S_y/[k_1 E(d_3/d_2)^2] \geq 0, \quad \text{(local buckling constraint of web)}$$

$$g_6(\bar{d}) = 1 - S_y/[k_2 E(d_3/d_1)^2] \geq 0, \quad \text{(local buckling constraint of flange)}$$

$$
\left.
\begin{aligned}
g_7(\bar{d}) &= 1 - 10d_1/L \geq 0 \\
g_8(\bar{d}) &= 1 - 10d_2/L \geq 0
\end{aligned}
\right| \quad \text{(side constraints)}
$$

$$
\left.
\begin{aligned}
g_9(\bar{d}) &= d_1 \geq 0 \\
g_{10}(\bar{d}) &= d_2 \geq 0 \\
g_{11}(\bar{d}) &= d_3 \geq 0
\end{aligned}
\right| \quad \text{(non-negativity constraints)}
$$

The constraints have been formulated taking into consideration the following points:

1. In the fatigue damage constraint $g_2(\bar{d})$ for purposes of calculating the expected rate of fatigue damage by Palmgren-Miner law consideration has to be given for the fact that the stress history is composed of many frequency components corresponding to the normal modes. But in view of the observation that the magnitudes of the stress in the higher modes are very much less compared to the stress magnitude in the first mode, the rate of fatigue damage is calculated on the basis that the stress history is essentially narrow band with central frequency ω_1. But the mean square stress at the fixed end is considered to be contributed by the different modes and given by (4.118). Calculations reveal that the fatigue damage rate based on a simple superposition of the rates of damages in different modes is less than the rate of expected fatigue damage calculated by the above method.

$$\omega_1(\sqrt{2}\,\sigma_{s_1})^\alpha + \omega_2(\sqrt{2}\,\sigma_{s_2})^\alpha + \ldots < \omega_1(\sqrt{2}\,\sigma_s)^\alpha \tag{4.121}$$

Hence the fatigue damage constraint is expressed conservatively by $g_2(\bar{d})$.

2. The yield stress of the material is assumed to be a normal random variable with standard deviation σ_{s_y} and the stress constraint $g_3(\bar{d})$ is expressed in terms of expected fractional occupation time.

3. The tip deflection constraint $g_4(\bar{d})$ is introduced in terms of the expected fraction of the time the displacement stays above a specified level being less than a given value. The critical deflection level is taken to be a constant ratio of the cantilever length.

The optimization problem is solved for the following set of design data:

$\Phi_0 = 0.05$ g^2/Hz, $\zeta_1 = 0.02$, $L = 200$ in., $\omega_L = 10$ Hz, $p_1 = 10^{-3}$, $p_2 = 10^{-3}$, $S_y = 20000$ lbs/in^2., $\sigma_{s_y} = 1000$ lb/in^2., $E = 10.6 \times 10^6$ psi, $\alpha = 6.0$, $\beta = 6.4 \times 10^{31}$, $k_1 = 21.72$, $k_2 = 3.62$, $C_1 = 3/200$, $D_1 = 1.67 \times 10^{-8}$ and $\rho = 0.1$ lbs/in^3.

The converged optimal solutions starting from two design variables are shown in Table 4.10.

Table 4.10 Uniform cantilever beam

$p_1 = 10^{-4}$ $D_1 = 5 \times 10^{-9}$	r	d_1	d_2	d_3	$W(\bar{d})$	$W(\bar{d}) + r \sum_i \dfrac{1}{g_i(\bar{d})}$
Initial design	0.9804	10.0	16.0	0.4	20.8	41.6
Optimum design	0.9804×10^{-4}	4.68	12.496	0.116	4.0004	4.0005
Initial design	0.6815	10.0	16.0	0.3	15.6	31.290
Optimum design	0.6815×10^{-9}	4.679	12.495	0.116	4.0003	4.0005

Active constraints at optimum g_2 and g_5.

4.6.3 Sheet-Stringer Panel Subjected to Jet Noise Excitation
(Narayanan and Nigam, 1976)

Rows of sheet-stringer panels supported on frames are typical of wings, fuselage and control surfaces of flight vehicle structures. These panels are subjected to random pressure fields generated by jet noise and boundary layer turbulence. We shall consider in this section the minimum weight design of sheet-stringer panels subjected to jet noise excitation. The pressure field due to jet noise is treated as a stationary Gaussian random process and the optimum design problem is formulated within the reliability framework.

Consider a sheet-stringer panel shown in Fig. 4.8 between two adjacent frames distance a apart, supported by a periodic row of z-section stringers separated by the uniform distance b between them. In the design problem to be considered we assume the interframe distance a to be a constant. The design variables in the problem are the stringer spacing b, the web height of the stringer h, the flange width of the stringer d, the thickness of the plate t_s, and the thickness of the stringer t. Thus each possible design is defined by the components of the vector

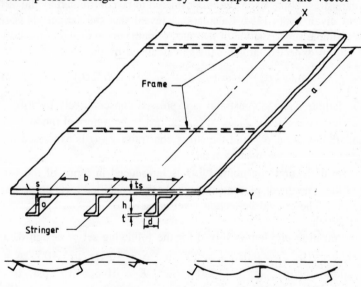

Fig. 4.8. Sheet-stringer combination.

$$\bar{d}^{\mathrm{T}} = \{d_1, d_2, d_3, d_4, d_5\}^{\mathrm{T}} \equiv \{b, h, d, t_2, t\}^{\mathrm{T}} \tag{4.122}$$

where the superscript T denotes the transpose of the vector.

OBJECTIVE FUNCTION: The design objective is to minimize the weight of the sheet-stringer panel. Since the distance a is assumed to be fixed in the design, and the stringers are uniformly spaced we consider only the weight of the panel between two adjacent stringers along with a stringer. Moreover, we assume the material of the panel and the stringer to be the same (aluminum). Hence minimizing the weight of the sheet-stringer combination is equivalent to minimizing the area of cross-section of the sheet stringer panel. Hence the objective function is

$$W(\bar{d}) = bt_s + t(h + 2d) \tag{4.123}$$

FREQUENCY CONSTRAINT: The natural frequencies of the periodic structure such as the sheet stringer combinsation group in distinct bands and there are as many number of natural frequencies as the number of bays within a frequency band. It has been shown by Lin (1962) and Mead (1970) that the limiting frequencies in each band of the multi-panel system correspond to the natural frequencies of a single panel between two stringers. In general for typical dimensions of the sheet-stringer combination the lower bounding frequency corresponds to the stringer torsion mode when adjacent panels vibrate out of phase and the upper bounding frequency corresponds to the stringer bending mode with adjacent panels vibrating in phase (Fig. 4.8). Clarkson (1962, 1968a, 1968b) has outlined certain general procedures for the design of structures to resist jet noise fatigue. These procedures specify conditions on the natural frequency which have to be satisfied by adjusting the parameters of the panel, such as stringer spacing, their torsional and bending stiffness so as to reduce response levels. In the frequency range 400–1000 Hz, the jet noise pressure correlation is high and positive upto about two feet. If the stringer pitch is of the order of six inches or so, at least four adjacent panels will have pressures in phase. Thus the stringer torsion mode in which adjacent panels vibrate out of phase will not be appreciably excited by the jet noise pressure field as compared to the stringer bending mode. To reduce response levels, use can therefore be made of the pressure spectrum to ensure that the frequency of the sheet corresponding to the stringer bending mode is well above the frequency of maximum sound energy. This .condition can be satisfied if the lower bounding frequency itself is above the frequency of maximum sound energy. This requirement takes the form of the frequency constraint

$$\omega_t(\bar{d}) \geq \omega_L \tag{4.124}$$

where $\omega_t(\bar{d})$ is the panel natural frequency corresponding to the stringer torsion mode and ω_L is a lower bound on that frequency corresponding to the frequency of maximum noise in the jet noise spectrum.

To ensure that the panel natural frequency corresponding to the stringer torsion mode ω_t does not in any case exceed that corresponding to stringer bending mode ω_b we require another constraint on the frequency

$$\omega_b(\bar{d}) \geq \omega_t(\bar{d}) \tag{4.125}$$

STRESS CONSTRAINT: The probability of failure constraint based on the stress response can be expressed considering individually the stresses at critical locations on the panel exceeding specified levels or as a overall failure probability constraint. To estimate the failure probabilities the statistics corresponding to the stress response have to be estimated. We describe briefly the random vibration analysis of the sheet-stringer panel to jet noise excitation adopted in Appendix 4A.

The critical stresses identified in the panel are the bending stresses along the x and y directions at the stringer location midway between the frames and the stress along y direction at the center of the panel. These stresses are denoted by the random processes $S_x(a/2, b/2, t)$, $S_y(a/2, b/2\ t)$, $S_y(a/2, 0, t)$, which are also mean zero, stationary Gaussian random processes because of the linearity of the structure. The stress constraints can be expressed in terms of the standard deviation of the stress response in one of the following three ways. In view of the difficulty of obtaining the statistics of the time derivative of the stress response in this example the stress constraints are expressed in terms of the fractional occupation time, instead of the first excursion failure probability.

(i) Failure based on individual failure modes

$$\text{erfc}\,\{S/(\sqrt{2}\sigma_{S_x}\,(a/2, b/2))\} \le p_1 \qquad (4.126)$$

$$\text{erfc}\,\{S/(\sqrt{2}\sigma_{S_y}\,(a/2, b/2))\} \le p_1 \qquad (4.127)$$

$$\text{erfc}\,\{S/(\sqrt{2}\sigma_{S_y}\,(a/2, 0))\} \le p_1 \qquad (4.128)$$

where S is a safe stress level which is taken to be the yield stress of the material and σ_{S_x}, σ_{S_y} are the standard deviations of the subscripted quantities.

(ii) Failure based on inequality (4.8)

$$\text{erfc}\,\{S/\sqrt{2}\sigma_{S_x}\,(a/2, b/2)\} + \text{erfc}\,\{S/\sqrt{2}\sigma_{S_y}\,(a/2, b/2)\}$$

$$+\ \text{erfc}\,\{S/\sqrt{2}\sigma_{S_x}\,(a/2, 0)\} \le p_2 \qquad (4.129)$$

(iii) Failure based on inequality (4.16)

$$a_1\,\text{erfc}\,\{S/\sqrt{2}\sigma_{S_y}\,(a/2, 0)\} + a_2\,\text{erfc}\,\{S/\sqrt{2}\sigma_{S_y}\,(a/2, b/2)\}$$

$$+\ a_3\,\text{erfc}\,\{S/\sqrt{2}\sigma_{S_x}\,(a/2, b/2)\} \le p_2 \qquad (4.130)$$

where $a_1 = 1$

$$a_2 = 1 - \text{erfc}\,(C_1)/\text{erfc}\,\{S/\sqrt{2}\sigma_{S_y}\,(a/2, b/2)\} \qquad (4.131)$$

$$a_3 = \min\left[\frac{1 - \text{erfc}\,(C_2)}{\text{erfc}\,\{S/\sqrt{2}\sigma_{S_y}\,(a/2, b/2)\}},\ \frac{1 - \text{erfc}\,(C_3)}{\text{erfc}\,\{S/\sqrt{2}\sigma_{S_x}\,(a/2, b/2)\}}\right] \qquad (4.132)$$

with

$$C_1 = \max\left\{\frac{S}{\sqrt{2}\sigma_{S_y}(a/2, b/2)},\ \frac{S\sigma_{S_y}(a/2, b/2)}{\sqrt{2}\sigma_{S_y}^2(a/2, 0)\,|\,\rho_{S_y S_y}(a/2, b/2; a/2, 0)\,|}\right\} \qquad (4.133)$$

$$C_2 = \max \left\{ \frac{S}{\sqrt{2}\,\sigma_{S_x}(a/2,\ b/2)}, \frac{S\sigma_{S_x}(a/2,\ b/2)}{\sqrt{2}\,\sigma_{S_y}(a/2,\ 0)\ |\ \rho_{S_x S_y}(a/2,\ b/2;\ a/2,\ 0)\ |} \right\}$$

(4.134)

and

$$C_3 = \max \left\{ \frac{S}{\sqrt{2}\,\sigma_{S_x}(a/2,\ b/2)}, \frac{S\sigma_{S_x}(a/2,\ b/2)}{\sqrt{2}\,\sigma_{S_y}^2(a/2,\ b/2)\ |\ \rho_{S_x S_y}(a/2,\ b/2;\ a/2,\ b/2)\ |} \right\}$$

(4.135)

and $\rho_{S_x S_y}$ etc. represent the correlation coefficient between the stresses corresponding to the subscripts and the location indicated in the parenthesis.

Fatigue Damage Constraint
The expected rate of fatigue damage is calculated on the basis of the critical stresses discussed earlier using the Palmgren-Miner cumulative damage rule. For the crossing rate of the stress cycles an average frequency between the stringer torsion mode and stringer bending mode is used. The fatigue damage constraint is expressed as

$$\frac{1}{2\pi\beta} \left(\frac{\omega_t + \omega_b}{2} \right) \sqrt{2} \ \{\sigma_{S_x}(a/2,\ b/2)\}^\alpha \ \Gamma(1 + \alpha/2) \le D_1$$

(4.136)

$$\frac{1}{2\pi\beta} \left(\frac{\omega_t + \omega_b}{2} \right) \sqrt{2} \ \{\sigma_{S_y}(a/2,\ b/2)\}^\alpha \ \Gamma(1 + \alpha/2) \le D_1$$

(4.137)

$$\frac{1}{2\pi\beta} \left(\frac{\omega_t + \omega_b}{2} \right) \sqrt{2} \ \{\sigma_{S_y}(a/2,\ 0)\}^\alpha \ \Gamma(1 + \alpha/2) \le D_1$$

(4.138)

The rest of the design specifications are the same in both the cases. These are: $a = 20$ in., $\zeta = 0.02$, $S = 20000$ lbs/in^2., $\alpha = 6.0$, $\beta = 6.4 \times 10^{31}$, $v = 1100$ ft/sec, $\theta = \pi/4$ radians, $\rho = 0.1$ lbs/in^3. The quantities Φ_o, v and θ which correspond to the jet noise power spectrum and correlation and ρ the density of the material occur in the response analysis discussed in Appendix 4A. The results of the optimization are given in Tables 4.11 and 4.12.

Table 4.11(a) gives the optimum values when the failure probability constraint based on the stress response is specified individually for the three locations as in inequalities (4.126) to (4.128). Table 4.11 (b) to (e) give the optimum value when the failure probability constraint based on the stress response is as in inequality (4.129). In Table 4.12(b) the failure probability constraint based on stress response is as in inequality (4.130). For the results presented in Table 4.11(a), the stress constraint based on $S_y(a/2, 0)$ (Inequality (4.128)) and the side constraint giving a lower bound on the stringer spacing are active at the optimum. For the results of Table 4.11(b), the fatigue damage constraint and the stringer spacing constraint are active. For the results of Table 4.11(c), the fatigue damage constraint, the stress constraint and the stringer spacing constraint are active. For the results of Table 4.11(d), the stress constraint and the stringer spacing constraint are active and for the results of Table 4.11(e), the fatigue damage constraint, the stringer

Table 4.11 Optimum design parameters for sheet-stringer combination

Optimum design for $\Phi_o = 0.4 \times 10^{-4}$ (lbs/in^2.)/(rad/sec), $\omega_L = 1884$ rad/sec. Initial design $d_1 = 5.0$, $d_2 = 2.4$, $d_3 = 0.6$, $d_4 = 0.12$, $d_5 = 0.08$. $W(\bar{d}) = 0.888$. $W(\bar{d})/b = 0.176$

	p_1 or p_2	D_1	d_1	d_2	d_3	d_4	d_5	$W(\bar{d})$	$W(\bar{d})/b$
(a)	$10^{-4}(p_1)$	5×10^{-6}	3.546	1.770	0.518	0.0606	0.0634	0.393	0.110
(b)	$10^{-3}(p_2)$	5×10^{-6}	3.621	1.801	0.516	0.0610	0.0597	0.390	0.108
(c)	$10^{-4}(p_2)$	5×10^{-6}	3.536	1.765	0.516	0.0604	0.064	0.393	0.111
(d)	$10^{-5}(p_2)$	5×10^{-6}	3.706	1.853	0.523	0.0676	0.069	0.451	0.121
(e)	$10^{-5}(p_2)$	1×10^{-6}	3.051	2.018	0.522	0.0779	0.0694	0.528	0.13

Table 4.12 Optimum design parameters for sheet-stringer combination

Optimum design for $\Phi_o = 0.7 \times 10^{-5}$ (lbs/in^2.)^2/(rad/sec), $\omega_L = 942$ rad/sec. Initial design $d_1 = 8.0$, $d_2 = 2.0$, $d_3 = 1.0$, $d_4 = 0.08$, $d_5 = 0.08$. $W(\bar{d}) = 0.96$. $W(\bar{d})/b = 0.12$

	p_1 or p_2	D_1	d_1	d_2	d_3	d_4	d_5	$W(\bar{d})$	$W(\bar{d})/b$
(a)	$10^{-3}(p_1)$	1×10^{-6}	2.897	1.447	0.501	0.035	0.361	0.190	0.0655
(b)	$10^{-3}(p_2)$	1×10^{-6}	3.044	1.521	0.501	0.0361	0.0321	0.111	0.0655

spacing constraint are active. For the results of Table 4.11(a) and (b) the fatigue damage constraint, the stringer spacing constraint and the side constraint on the flange width of the stiffener are active.

Thus we observe that the optimum design is sensitive to the constraints on the specified probability of failure in the stress constraints or the fatigue damage level. One or the other or both constraints become active at the optimum depending on the relative stringency of the constraint. We also observe that the side constraint on the stringer spacing is active in all cases. This is so because for a single panel as the stringer spacing reduces the area of the sheet also reduces without adversely affecting the other constraints. For a general structural system, such as a fuselage, minimum weight of a panel does not always imply minimum weight of the total system. It is meaningful, therefore to consider the area of cross-section of a single panel per unit length of the stringer spacing as the objective function

$$W(\bar{d}) = t_s + t(h + 2d)/b \tag{4.144}$$

Results of minimizing such an objective function is shown in Table 4.13. In both the cases of Table 4.13(a) and (b) the fatigue damage constraint and the side constraint on the stringer thickness become active. The optimal stringer spacing is no longer twice the web height of the stringer. We also observe that results based on this objective function are even better as compared to the results in Table 4.12(b).

Table 4.13. Optimum design parameters for sheet-stringer combination with modified objective funciton

Optimum design for $\Phi_o = 0.7 \times 10^{-5}$ (lbs/in^2.)^2/(rad/sec), $\omega_L = 942$ rad/sec. $W(\bar{d}) = t_s + (2d + h)/b$. Initial design $d_1 = 8.0$, $d_2 = 2.0$, $d_3 = 1.0$, $d_4 = 0.08$, $d_5 = 0.08$ $W(\bar{d}) = 0.12$

	p_2	D_1	d_1	d_2	d_3	d_4	d_5	$W(\bar{d})$
(a)	10^{-3}	1×10^{-6}	5.954	1.702	1.06	0.0403	0.023	0.05505
(b)	10^{-4}	1×10^{-6}	6.7	1.822	1.276	0.042	0.02	0.05518

Another factor to note is that the optimal design is not very sensitive to the way in which the probability of failure constraint based on stress response is specified either on the basis of individual stresses at the critical locations or on a overall probability of failure. This will be so if the stress constraint is inactive at the optimum. Even in cases when the stress constraint becomes active at the optimum, such insensitivity is noticed because, the mean square value of the stress at one location is significantly larger as compared to other locations, and hence in any case the stress levels become critical essentially at one location.

4.7   Design for Critical Excitation

In this chapter several examples have been presented on the design of structures operating in random vibration environment including optimization. In all the examples it has been assumed that the probabilistic descriptions of the random excitations are available. However, in many real situations such complete description of the random excitation processes may not be available. In such cases it is meaningful to implement the design for excitations for which the structural response is maximum in some sense. Such excitations are called critical excitations and the design for critical excitation is assumed conservative and safe. A brief introduction to design for critical excitation is presented in Appendix 4B.

Sheet Stringer Panels—Random response analysis
(Narayanan, 1975)

Consider the single panel between a pair of stringers and the two frames $x = 0$ and $x = a$ as shown in Fig. 4.8. The equation of motion for the panel is

$$D\left[\frac{\partial^4 w}{\partial x^4} + \frac{2\partial^4 w}{\partial x^2 \partial y^2} + \frac{\partial^4 w}{\partial y^4}\right] + c\dot{w} + m\ddot{w} = p(x, y, t) \qquad (4A.1)$$

where w is the transverse displacement of the panel, $D = Et_s^3/12(1 - v^2)$, the flexural rigidity, m mass per unit area, c damping coefficient per unit area assumed uniform over the plate and $p(x, y, t)$ is the random pressure field due to jet noise.

We assume the panel to be simply supported at the frames. We also assume only the first mode in the x direction makes significant contribution to the response. This assumption is reasonable as the jet noise pressure field is assumed to be fully correlated along the x direction, in which case the joint acceptance function representing the pressure field-structural modal coupling vanishes for even number of modes and progressively decreases in magnitude for odd number of modes (Bozich, 1964).

Express $w(x, y, t)$ in normal modes as

$$w(x, y, t) = \sum_j \sin\frac{\pi x}{a}\, \eta_j(y) q_j(t) \qquad (4A.2)$$

In generalized coordinates the equations of motion can be represented as

$$M_j\ddot{q}_j + C_j\dot{q}_j + K_j q_j = p_j(t) \qquad (4A.3)$$

where

$$M_j = m\int_0^b \int_0^a \left(\sin^2\frac{\pi x}{a}\, dx\right)\eta_j^2(y)\, dy \qquad \text{(generalized mass)}$$

$$C_j = c\int_0^b \int_0^a \left(\sin^2\frac{\pi x}{a}\, dx\right)\eta_j^2(y)\, dy \qquad \text{(generalized damping coefficient)}$$

$$K_j = D\int_0^b \int_0^a \left(\frac{\pi^4}{a^4}\sin^2\frac{\pi x}{a}\, dx\right)\eta_j^2(y)\, dy$$

$$+ \left\{\left(-\frac{\pi^2}{a^2}\right)\sin^2\frac{\pi x}{a}\, dx\right\}\eta_j''(y)\,\eta_j(y)\, dy$$

$$+ \left\{\sin^2\frac{\pi x}{a}\, dx\right\}\eta_j^{iv}(y)\,\eta_j(y)\, dy \qquad \text{(generalized stiffness)}$$

$$p_j(t) = \int_0^b \int_0^a p(x, y, t)\sin\frac{\pi x}{a}\,\eta_j(y)\, dx\, dy \qquad \text{(generalized force)}$$

We also assume that only the modes corresponding to the two bounding frequencies in the first frequency band contribute significantly to the response. Those are the stringer torsion mode and the stringer bending mode.

Following Lin (1960), the frequency equation and the modes for the stringer torsion mode and stringer bending mode are given by

Stringer torsion mode:

$$\cosh \frac{k_1 b}{2} \left(k_2 \beta \ \sin \frac{k_2 b}{2} + 2 D k_2^2 \ \cos \frac{k_2 b}{2} \right)$$

$$+ \cos \frac{k_2 b}{2} \left(k_1 \beta \ \sinh \frac{k_1 b}{2} + 2 D k_1^2 \ \cosh \frac{k_1 b}{2} \right) = 0 \qquad (4A.4)$$

$$\eta_t(y) = \cosh k_1 y - \alpha \ \cos k_2 y \qquad (4A.5)$$

Stringer bending mode:

$$k_1 \sinh \frac{k_1' b}{2} \left(\gamma \cos \frac{k_2' b}{2} - 2 D k_2'^3 \ \sin \frac{k_2' b}{2} \right)$$

$$+ k_2' \sin \frac{k_2' b}{2} \left(\gamma \cosh \frac{k_1' b}{2} - 2 D k_1'^3 \ \sinh \frac{k_1' b}{2} \right) = 0 \qquad (4A.6)$$

$$\eta_b(y) = \cosh k_1' y - \alpha' \ \cos k_2' y \qquad (4A.7)$$

The quantities used in (4A.4) to (4A.7) are

$$k_1 = [(\omega_t^2 m/D)^{1/2} + \pi^2/a^2]^{1/2}; \ k_2 = [(\omega_t \ m/D)^{1/2} - \pi^2/a^2]^{1/2}$$

$$k_1' = [(\omega_b^2 m/D)^{1/2} + \pi^2/a^2]^{1/2}; \ k_2' = [(\omega_b^2 \ m/D)^{1/2} - \pi^2/a^2]^{1/2}$$

$$\beta = E C_{ws} \ \pi^4/a^4 + G C \pi^2/a^2 - \rho I_s \omega^2$$

$$\gamma = E I \pi^4/a^4 - \rho A \omega^2$$

$$\alpha = \frac{\cosh (k_1 b/2)}{\cosh (k_2 b/2)}; \ \alpha' = \frac{k_1' \sinh (k_1' b/2)}{k_2' \cos (k_2' b/2)}$$

where ω_t—stringer torsion frequency; ω_b—stringer bending frequency; A—area of cross-section of the stringer; C—St. Venant torsion constant of stringer cross-section; C_{ws}—warping constant of stringer cross section about point of attachment; I_s—polar moment of inertia of stringer cross section about s; I_0—area moment of inertia of stringer cross section about horizontal axis; ρ—mass density of stringer material; v—Poisson's ratio; and E, G—Young's modulus and shear modulus of stringer material.

Jet Noise Pressure Field
The power spectral density of jet noise pressure field is of the form (Wallace, 1965)

$$\Phi_{pp}(x_1, y_1; x_2, y_2; \omega) = \Phi_o \cos \left\{ \frac{\omega \cos \theta}{v} (y_1 - y_2) \right\} \qquad (4A.8)$$

where v is the speed of sound in air and θ the angle of predominant direction of the jet noise with respect to y axis.

The power spectral density functions of the displacement and the stresses are given by

$$\Phi_{ww}(x_1, y_1; x_2, y_2; \omega) = \sum_j \sum_k f_j(x_1, y_1) f_k(x_2, y_2) H_j(\omega) H_k^*(\omega) I_{jk}(\omega)$$

$$\Phi_{S_x S_x}(x_1, y_1; x_2, y_2; \omega) = \sum_j \sum_k g_{jx}(x_1, y_1) \, g_{kx}(x_2, y_2) \, H_j(\omega) \, H_k^*(\omega) I_{jk}(\omega)$$

$$\Phi_{S_y S_y}(x_1, y_1; x_2, y_2; \omega) = \sum_j \sum_k g_{jy}(x_1, y_1) \, g_{ky}(x_2, y_2) \, H_j(\omega) \, H_k^*(\omega) I_{jk}(\omega)$$

$$\Phi_{S_x S_y}(x_1, y_1; x_2, y_2; \omega) = \sum_j \sum_k f_{jx}(x_1, y_1) \, g_{ky}(x_2, y_2) \, H_j(\omega) \, H_k^*(\omega) I_{jk}(\omega) \quad \text{(4A.9)}$$

where j, k are taken through the stringer torsion and stringer bending modes, that is

$$f_1(x, y) = \sin \frac{\pi x}{a} \, \eta_t(y) \quad \text{and} \quad f_2(x, y) = \sin \frac{\pi x}{a} \, \eta_b(y) \quad \text{(4A.10)}$$

$$g_{jx} = \frac{-Et_s}{1 - v^2} \left\{ \frac{\partial^2 f_j}{\partial x^2} + v \frac{\partial^2 f_j}{\partial y^2} \right\} \quad \text{and} \quad g_{jy} = \frac{-Et_s}{1 - v^2} \left\{ \frac{\partial^2 f_j}{\partial y^2} + v \frac{\partial^2 f_j}{\partial x^2} \right\} \quad \text{(4A.11)}$$

and

$$H_j(\omega) = \frac{1}{-M_j \omega^2 + iC_j \omega + K_j} \quad \text{(4A.12)}$$

The various response statistics are obtained by integrating the corresponding spectral density function by contour integration.
For example,

$$\sigma_{S_x}^2(x, y) = \int_{-\infty}^{\infty} \Phi_{S_x S_x}(x, y; x, y; \omega) \, d\omega \quad \text{(4A.13)}$$

and so on.

APPENDIX 4B: Design Through Critical Exicitation(s)

Design of systems to specified exicitation(s) involves deterministic or probabilistic constraints on one, or more, response quantities. For the class of problems, where the response can be determined analytically, given a 'complete' or 'precise' deterministic or probabilistic description of exication(s) the design process, including optimization, can be implemented as discussed in this chapter. In many real life problems, however, a complete description of exication(s) is not available. In such cases, the design process, must be implemented on the basis of available 'incomplete' or 'partial' description of the exicitation(s) in terms of one, or more, parameters that would characterize all excitations which can be reasonably expected. Amongst the class of excitations, consistent with the specified partial description, the excitation which maximizes a response quantity is called the 'critical excitation' or 'worst-input'. A system designed to be safe against such an excitation is expected to be safe against all excitations which can be reasonably expected. While this approach provides a formal basis for design against uncertainities, the degree of conservatism introduced in establishing the critical excitation(s) and an upper-bound on the response, determines whether the approach is far too pessimistic to be practical (Drenick, 1970, 1973, 1977; Iyengar, 1970, 1972; Iyengar and Manohar 1984, 1985, 1987 and Shinozuka 1970). Application of this approach to a sesmic design (Drenick, 1973) show that the approach is practical as it leads to design margins of the order of 2 against observed earthquakes, and 1.3 against linear combination of such earthquakes which are comparable to the factors of safety used in the design of structures. In the sequel we discuss a procedure to establish critical excitations and design based on such excitations.

4B.1 Deterministic Critical Excitation

Consider a system represented by a linear time-invariant operator L which relates the excitation $F(t)$ to the response $X(t)$ through the relation

$$L\,[X(t)] = F(t) \tag{4B.1}$$

Let $h(t)$ and $H(\omega)$ represent the impulse and the frequency response functions of the system respectively. Further, let $\hat{F}(\omega)$ and $\hat{X}(\omega)$ represent the Fourier transforms of $F(t)$ and $X(t)$ respectively.

We consider a class-C of excitations $F(t)$ satisfying the bound on total energy

$$C : E = \int_{-\infty}^{\infty} F^2(t)\,dt = \frac{1}{2\pi} \int_{-\infty}^{\infty} |\hat{F}(\omega)|^2\,d\omega \le M^2 \tag{4B.2}$$

Further, let

$$\int_{-\infty}^{\infty} h^2(t)\,dt = \frac{1}{2\pi} \int_{-\infty}^{\infty} |H(\omega)|^2\,d\omega \le N^2 \tag{4B.3}$$

The response of the system can be expressed in the frequency domain by the relation

$$X(t) = \frac{1}{2\pi} \int_{-\infty}^{\infty} H(\omega)\,\hat{F}(\omega)\,e^{i\omega t}\,d\omega \tag{4B.4}$$

We note that

$$| X(t) | = \left| \frac{1}{2\pi} \int_{-\infty}^{\infty} H(\omega) \hat{F}(\omega) e^{i\omega t} d\omega \right| \le \frac{1}{2\pi} \int_{-\infty}^{\infty} | H(\omega) | | \hat{F}(\omega) | d\omega = I$$

(4B.5)

and from Schwarz inequality, (4B.2) and (4B.3)

$$\frac{1}{2\pi} \int_{-\infty}^{\infty} | H(\omega) | | \hat{F}(\omega) | d\omega$$

$$\le \left[\frac{1}{2\pi} \int_{-\infty}^{\infty} | H(\omega) |^2 d\omega \right]^{1/2} \left[\frac{1}{2\pi} \int_{-\infty}^{\infty} | \hat{F}(\omega) |^2 d\omega \right]^{1/2} \le MN$$ (4B.6)

Combining (4B.5) and (4B.6)

$$\| X(t) \| = \max_t | X(t) | \le I \le MN$$

(4B.7)

in which $\| X(t) \|$ is a norm of $X(t)$ representing the maximum value of $| X(t) |$ in the time interval $-\infty < t < \infty$.

It can be shown that the critical excitation $F_c(t)$ belonging to the class-C (4B.2), which maximizes $| X(t) |$ is

$$F_c(t) = \pm \frac{M}{N} h(- t)$$

(4B.8)

and the upper-bound on $\| X(t) \|$ corresponding this excitation is given by (4B.7) (Drenick, 1970).

Drenick (1970) has shown that besides the constraint on total energy (4B.2), if the excitation is assumed to have a finite duration T,

$$\| X(t) \| = \max_{0 \le t \le T} | X(t) | \le MN_T$$

(4B.9)

where

$$N_T^2 = \int_0^T h^2(t_0 - \tau) d\tau$$

(4B.10)

and the critical excitation

$$F_c(t) = \pm \frac{M}{N_T} h(t_0 - t)$$

(4B.11)

In (4B.10) and (4B.11), $0 \le t_0 \lesssim T$ represents the time at which the integral on the r.h.s. of (4B.10) assumes the maximum value.

Drenick (1970) has also shown that besides the constraint on total energy (4B.2), if the excitation is assumed to be band-limited, that is $| \omega | \le \omega^L$

$$\| X(t) \| = \max_t | X(t) | \le MN_{\omega^L}$$

(4B.12)

where

$$N_{\omega^L}^2 = \frac{1}{2\pi} \int_{-\omega^L}^{\omega^L} | H(\omega) |^2 d\omega$$

(4B.13)

Shinozuka (1970) has shown that for excitations, such as, earthquakes the envelope of the Fourier transform may be readily established from the available information. If the excitation is specified in terms of such an envelope, instead of constraint on total energy (4B.2), an improved estimate can be made of the upper-bound on response.

Let $$\hat{F}_l(\omega) \geq |\hat{F}(\omega)| \qquad (4B.14)$$

From (4B.5)

$$\|X(t)\| = \max_t |X(t)| \leq \frac{1}{2\pi} \int_{-\infty}^{\infty} |\hat{F}(\omega)| |H(\omega)| \, d\omega = I_l \qquad (4B.15)$$

which represents the lower amongst the two bounds in (4B.7).

Multiple Excitations
The above results are limited to a single excitation. We can extend these to multiple excitation and response. Let $\bar{F}(t)$ represent a m-dimensional vector and $\bar{X}(t)$ a n-dimensional response vecotr. Let $H_{ij}(\omega)$, $i = 1, 2 \ldots m$; $j = 1, 2, \ldots n$, represent ijth term of the $m \times n$ frequency response matrix. It can be shown that

$$\|X_j(t)\| = \max_t |X_j(t)| \leq \frac{1}{2\pi} \int_{-\infty}^{\infty} \left| \sum_{i=1}^{m} H_{ij}(\omega) \hat{F}_j(\omega) \right| d\omega \qquad (4B.16)$$

which is an extension of (4B.5).

4B.2   Random Critical Excitation
A probabilistic formulation of design problems involves estimation of failure probabilities. Since failures are rare events, the estimates of failure probabilities are determined by the shape of tails, which are least known portions of the underlying distributions. The estimates of failure probabilities are therefore susceptible to large errors. Under the circumstances, a combination of probabilistic and worst-case analysis offers a viable option for design (Drenick, 1977).

Following Drenick (1977), let the response $X(t)$ be treated as a random process. If the available information on the random process is complete, it would imply that the probability measure on the sigma algebra, Σ, of all (measurable) sets of sample functions of $X(t)$ is specified. Generally, the information will be incomplete in the sense that the probability measure will be specified only on the sets of the family $\Sigma' \in \Sigma$. Let C be a set in Σ' which consists of, or contains, all sample functions that are of interest in a particular problem. We now consider the probability that some amongst those sample functions cause failure, that is the probability of the event

$$E : \{X : \|X(t)\| > R\} \cap \{X(t) \in C\} \qquad (4B.17)$$

Clearly,

$$0 \leq P\{\|X(t)\| > R \mid X(t) \in C\}P(C) \leq P(C) \qquad (4B.18)$$

The upper and the lower bounds in (4B.18) correspond respectively to

$$P\{\|X(t)\| > R \mid X(t) \in C\} = 1 \text{ or } 0 \qquad (4B.19)$$

the former implying that all sample functions of $X(t) \in C$, exceed the failure limit

R—which may be extremely conservative, and the latter implying that none do—which may happen only rarely.

The combined probabilistic and worst-case analysis procedure proposed by Drenick (1977) may be implemented by first choosing the family Σ' as fine as possible by utilizing all information considered reliable enough. The intersection in the event (4B.17) is then examined and assumed to be zero, if empty, and 1, if otherwise. Thus the probability theory is invoked to estimate the measured $P(C)$ and the worst-case analysis to choose the probabilities 0 or 1 in (4B.19).

We illustrate the procedure with an example. For the class of excitations defined by (4B.2)

$$P(C) = P(E \le M^2) \tag{4B.20}$$

which can be determined from the distribution the energy (Iyengar, 1972). The probability in (4B.19) is 0, if $MN < R$, and 1, if otherwise.

Ensemble of Critical Random Excitation

A uniformly modulated random process represents a nonstationary model of several excitations such as earthquakes, rocket exhaust. Iyengar and Manohar (1987) have developed a method to establish the psd of critical excitation which can be used to simulate an ensemble of such excitations. In their formulation, the class of admissible random excitations is characterized by specified variants, duration and the deterministic envelope. The psd of the associated stationary random process is determined from amongst this class to maximize the system response variance. Given the psd of the critical excitation, an ensemble of sample functions can be generated by any of the methods discussed in Chapter 2.

Let the excitation be modelled as a uniformly modulated nonstationary random process

$$F(t) = A(t)\, Y(t) \tag{4B.21}$$

where $A(t)$ is a deterministic envelope function, and $Y(t)$ is a Gaussian stationary random process with zero mean and known variance. Let T be the duration of the excitation.

Let $G(\omega)$ represent the one-sided psd of $Y(t)$, and $H(\omega, t)$ the frequency response like function of the system such that the response variance is expressed as

$$\sigma_X^2(t) = \int_0^\infty G(\omega)\, H(\omega, t)\, d\omega \tag{4B.22}$$

The critical excitation, $F_C(t)$ is defined as the input which maximizes $\sigma_X^2(t)$ within a class of random functions satisfying the constraints

$$C : \sigma^2 = \int_0^\infty G(\omega)\, d\omega, \quad 0 \le t \le T \tag{4B.23}$$

To determine $F_c(t)$, we have to find $G(\omega)$ such that $\sigma_X^2(t)$ is maximized. A particular solution is obtained by expanding $\sqrt{G(\omega)}$ in the form of a series

$$\sqrt{G(\omega)} = \sum_{i=1}^\infty a_i\, \phi_i(\omega) \tag{4B.24}$$

where $\phi_i(\omega)$, $i = 1, 2, \ldots$, are a set of known orthonormal functions.

Combining (4B.22) and (4B.23) we define the Lagrangian to be maximized as

$$\mathscr{L}[a_1, a_2, \ldots, \lambda(t)] = \int_0^\infty (\sum_i a_i \phi_i)^2 H(\omega, t)\, d\omega - \lambda(t)\,(\Sigma a_i^2 - \sigma^2) \quad (4B.25)$$

where $\lambda(t)$ is the Lagrangian multiplier.

Setting $\dfrac{\partial \mathscr{L}}{\partial a_i} = 0$, $i = 1, 2, \ldots$ and $\dfrac{\partial \mathscr{L}}{\partial \lambda} = 0$, we get

$$\sum_{i=1}^\infty a_i I_{ij}(t) - \lambda(t) a_j = 0, \quad j = 1, 2, \ldots, \qquad (4B.26)$$

$$\sum_{i=1}^\infty a_i^2 = \sigma^2 \qquad (4B.27)$$

where

$$I_{ij}(t) = \int_0^\infty \phi_i(\omega)\,\phi_j(\omega)\,H(\omega, t)\, d\omega \qquad (4B.28)$$

Equations (4B.22) and (4B.23) represent an algebraic eigenvalue problem which can be solved using standard techniques, yielding the eigen values $\lambda_i(t)$ and the corresponding eigen vectors $\bar{a}_i(t)$, $i = 1, 2, \ldots$. Substitution of (4B.26) in (4B.22) yields

$$\sigma_x^2(t) = \lambda(t)\, \sigma^2 \qquad (4B.29)$$

Hence, the largest eigenvalue for every $0 \le t \le T$, and the corresponding eigen vector yields the highest variance through (4B.29), and the psd of the critical excitation through (4B.24).

Example 4B.1: Critical Excitation for sdf System Subjected to Seismic Load
(Iyengar and Manohar, 1987)

Consider a linear sdf system

$$\ddot{X} + 2\zeta\omega_0 \dot{X} + \omega_0^2 X = -F(t) = -A(t)\ddot{Y}(t) \qquad (4B.30)$$

where $F(t)$ is the ground acceleration, $T = 10$ secs, the envelope function

$$A(t) = \exp(-0.5t) - \exp(-t) \qquad (4B.31)$$

and

$$\sigma^2 = \frac{0.01 g^2 T}{\displaystyle\int_0^T A^2(t)\, dt} \qquad (4B.32)$$

The system parameters are: $\zeta = 0.05$, and $f_0 = \omega_0/2\pi = 4$ Hz. The orthonormal functions in (4B.24) are assumed to be a linear combination of exponential functions

$$\phi_1(\omega) = \sqrt{2p}\, \exp(-p\omega) \qquad (4B.33)$$

$$\phi_2(\omega) = 2\sqrt{p}\, [2\exp(-p\omega) - 3\exp(-2p\omega)]$$

$$\vdots$$

with $p = 0.008$ and ϕ_i, $i = 1, 2, \ldots, 7$.

The psd of the critical excitation for the above data is shown in Fig. 4B.1(a); and a sample function of the ensemble of critical excitation is shown in Fig. 4B.1(b). It may be remarked that the sample function resembles real earthquake records.

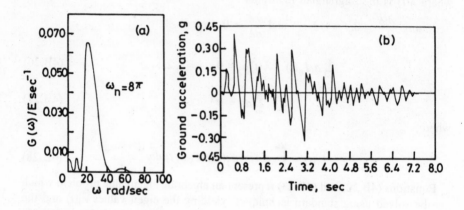

Fig. 4.B1   (a) Power spectral density; and (b) Sample function of the critical excitation.

REFERENCES

Allen, G.R., 1975. Evaluation of exposure to whole-body vibration below 1 Hz. U.K. group meeting on human response to vibration. University of Southampton.

Ang, A.H.S., 1972. Structural safety, a literature review, by Task Committee on structural safety of the Administrative Committee of analysis and design of structural division ASCE. J. of Str. Div. 98: 845–854.

Ang, A.H.S., 1973. Structural risk analysis and reliability based design. ASCE. J. Str. Div. 99: 1891–1910.

Ang, A.H.S. and A.M. Amin, 1968. Reliability of structures and structural systems. ASCE. J. of Engg. Mech. Div. 94: 559–583.

Ang, A.H.S. and A.M. Amin, 1969. Safety factors and probability in structural design. ASCE. J. Str. Div. 95: 1389–1405.

Ayorinde, E.O., and G.E. Warburton, 1980. Minimizing structural vibrations with absorbers. Int. J. Earthquake Engg. Str. Dyn. 8: 219–236.

Bender, E.K., 1967. Optimum linear random vibration isolation. JACC preprints. 135–143.

Bozich, D.J., 1964. Spatial correlation in acoustic structural coupling. J. Acous. Soc. Am. 36(1): 52–58.

Cartwright, D.E., 1958. On estimating the mean energy of sea waves from the highest waves in a record. Proc. Roy. Soc. Ser. A 247 22–48.

Cederkvist, J., 1982. Design of beams subjected to random loads. J. Str. Mech. 10: 49–65.

Clarkson, B.L., 1962. The design of structures to resist jet noise fatigue. J. Roy. Aero. Soc. 66: 603–616.

Clarkson, B.L., 1968a. Stresses in skin panels subjected to random acoustic loading. J. Roy. Aero. Soc. 72: 1000–1010.

Clarkson, B.L., 1968b. Design of fatigue resistant structures. In Noise and acoustic fatigue in aeronautics. Richards, E.J., and Mead, D.J. (Eds) John Wiley & Sons Ltd. New York. 354–371.

Crandall, S.H., and Mark, W.D., 1963. Random vibration in mechanical systems. Academic Press, New York.

Cornell, C.A., 1967. Bounds on the reliability of structural systems. ASCE. J. Str. Div., 93: 171–200.

Dahlberg, T., 1977. Parametric optimization of a 1-DOF vehicle travelling on a randomly profiled road. J. Sound Vib. 55: 245–253.

Dahlberg, T., 1978. Ride comfort and road holding of a 2-DOF vehicle travelling on a randomly profiled road. J. Sound Vib. 58: 179–187.

Dahlberg, T., 1979a. An optimized speed-controlled suspension of a 2-DOF vehicle travelling on a randomly profiled road. J. Sound Vib. 62: 541–546.

Dahlberg, T., 1979b. Optimization criteria for vehicles travelling on a randomly profiled road—a survey. Veh. Sys. Dyn. 8: 239–252.

Dahlberg, T., 1980. Comparison of ride comfort criteria for computer optimization of vehicle travelling on randomly profiled roads. Veh. Sys. Dyn. 9: 291–307.

Davenport, A.G., 1961a. The application of statistical concepts to the wind loading of structures. Proc. Inst. Civil Engrs. London. 19: 449–473.

Davenport, A.G., 1961b. The spectrum of horizontal gustiness near the ground in high winds. Quart. J. Roy. Met. Soc. 87: 194–211.

Davenport, A.G., 1964. Note on the distribution of the largest value of a random function with application to gust loading. Proc. Inst. Civil Engrs. London. 28: 187–196.

Davenport, A.G., 1967. Gust loading factors. ASCE J. Str. Div. 93: 1–34.

Davidson, J.W., L.P. Felton, and G.C. Hart, 1977. Reliability based optimization for dynamic loads. ASCE. J. Str. Div. 103: 2021–2035

Davidson, J.W., L.P. Felton, and G.C. Hart, 1980. On reliability based structural optimization for earthquakes. Computers and Structures. 12: 99–105.

Den Hartog, J.P. 1956. Mechanical Vibrations. McGraw-Hill, New York.

Drenick, R.F., 1970. Model free design of seismic structures. J. of Engg. Mech. Div., Proc. ASCE, 96 (EMA): 483–493.

Drenick, R.F., 1973. Aseismic design by way of critical excitation. J. of Engg. Mech. Div., Proc. ASCE, 99 (EMA): 649–667.

Drenick, R.F., 1977. On a class of non-robust problems in stochastic dynamics. Stochastic Problems in Dynamics (ed) B.L. Clarkson. London, Pitman: 237–255.

Fiacco, A.V. and G.P. McCormick, 1968. Nonlinear programming: Sequential Unconstrained minimization techniques. John Wiley & Sons Inc., New York.

Fletcher, R. and M.J.D. Powell, 1963. A rapidly convergent descent method for minimization. Computer Journal 6: 163–168.

Freudenthal, A.M., 1956. Safety and the probability of structural failure. Trans. ASCE 121: 1337–1375.

Freudenthal, A.M., 1962. Safety, Reliability and structural design. Trans. ASCE 127: Part II, 304–323.

Freudenthal, A.M. and J.M. Garrets, and M. Shinozuka, 1966. The analysis of structural safety. ASCE J. Str. Div. 92: 267–325.

Fujiwara, N., and Y. Murotsu, 1974. Optimum design of vibration isolators for random excitations. Bull. JSME. 17: 68–72.

Haugen, E.B., 1968. Probabilistic approaches to design. John Wiley & Sons. New York.

Haugen, E.B., 1980. Probabilistic mechanical design. John Wiley & Sons. New York.

Iyengar, R.N., 1970. Matched inputs. Center of Applied Stochastics. Purdue University, Rep 47, Series J.

Iyengar, R.N. 1972. Worst inputs and a bound on the highest peak statistics of a class of nonlinear systems. J. of Sound and Vibration. 25(1): 29–37.

Iyengar, R.N. and C.S. Manohar. 1984. Extreme seismic excitations. Symposium on Earthquake Effects on Plant and Equipment. Hyderabad, India.

Iyengar, R.N. and C.S. Manohar. 1985. System dependent critical stochastic excitations. Eighth international conference on SMIRT, Brussels, Belgium.

Iyengar, R.N. and C.S. Manohar. 1987. Nonstationary random critical seismic excitations. J. of Engg. Mech. Div., Proc. ASCE, 113 (EMA): 529–540.

Jacquot, R.G. and D.L. Hoppe, 1973. Optimal random vibration absorbers. ASCE. J. Engg. Mech. Div. 99: 612–616.

Kadam, V.R., 1983. Computer aided optimal design of the suspension system of a vehicle traversing a rough road. M. Tech. thesis, I.I.T. Madras.

Karnopp, D.C., 1968. On the optimization of saturating servo mechanisms subjected to random disturbances. J. Frank. Inst. 285: 475–482.

Karnopp, D.C. and A.K. Trikha, 1969. Comparative study of the optimization techniques for shock and vibration isolation. J. Engg. Ind. 91: 1128–1132.

Lin, Y.K., 1960. Free vibrations of continuous skin-strainger panels. J. App. Mech. 27: 669–676.

Lin, Y.K., 1962. Stresses in continuous skin-stiffner panels under random loading. J. Aero. Sci. 29: 67–76.

Maestrello, L., 1968. Design criterion of panel structure excited by turbulent boundary layer. J. Aircraft 5: 321–328.

Mains, R.M., 1958. Mechanical design for random loading. Random Vibration. Crandall, S.H. ed. M.I.T. Press, Cambridge, Mass.

Malhotra, A.K. and J. Penzien, 1970. Nondeterministic analysis of off-shore structures. J. Engg. Mech. Div. 96: 985–1003.

Mead, D.J., 1970. Free wave propagation.in periodically supported infinite beams. J. Sound Vib. 11: 181–197.

Moses, F. and J.D. Stevenson, 1970. Reliability based structural design. ASCE J. Str Div. 96: 221–244.

Narayanan, S., 1975. Structural optimization in random vibration environment. Doctoral Thesis, I.I.T. Kanpur.

Narayanan, S. and N.C. Nigam, 1976. Optimum structural design of sheet-stringer panels subject to jet noise excitation. Invited paper presented at IUTAM Symposium on stochastic problems in dynamics. University of Southampton. Also in stochastic problems in dynamics. Clarkson, B.L. Ed. 1977: 487–514.

Narayanan, S. and N.C. Nigam, 1978. Optimum structural design in random vibration environments. Engg. Optimization. 3: 97–108.

Newmark, N.M. and E. Rosenblueth, 1971. Fundamentals of Earthquake engineering. Prentice Hall Inc. Englewood Cliffs.

Nigam, N.C. 1983. Introduction to Random vibrations. The M.I.T. Press, Cambridge, Mass.

Racicot, R.L., 1969. Random vibration analysis, applications to wind loaded structures. Doctoral thesis, Case Western University, Ohio.

Rao, S.S., 1981. Reliability based optimization in random vibration environment. Computers and Structures. 14: 345–355.

Rosenblueth, E., 1964. Probabilistic design to resist earthquakes. ASCE J. Engg. Mech. Div. 90: 229–252.

Sarpkaya, T. and M. Isaaison, 1981. Mechanics of wave forces in off-shore structures. Van Nostrand Rheinhold, New York.

Sekiguchi, H. and K. Iida, 1970. The isolation of random vibrations. Bull. JSME. 13: 248–257.

Sekiguchi, H. and K. Iida, 1973. The isolation of random vibrations. Bull. JSME. 16: 1278–1288.

Shinozuka, M., 1964. Probability of structural failure under random loading. ASCE J. Engg Mech. Div. 90: 147–170.

Shinozuka, M., 1970. Maximum structural response to seismic excitation. J. of Engg. Mech. Div., Proc. ASCE, 96 (EMS): 729–738.

Shinozuka, M., 1974. Safety against dynamic forces. ASCE J. Str. Div. 100: 1821–1826.

Stevenson, J.D. and Moses, F., 1970. Reliability analysis of framed structures. ASCE. J. Str. Div. 96: 2409–2427.

Trikha, A.K. and Karnopp, D.C., 1969. A new criterion for optimizing linear vibration isolator systems subject to random input. ASME J. Engg Ind. 91: 1005–1010.

Vanmarcke, E.H., 1969. First passage and other failure criteria in narrowband random vibration. A discrete state approach. Doctoral thesis. MIT, Cambridge, Massachussets.

Vanmarcke, E.H., 1972. Matrix formulation of reliability analysis and reliability based design. Paper presented at the National Symposium on Computerized Structural Analysis and Design. George Washington University,

Wallace, C.E., 1965. Stress response and fatigue life of acoustically excited panels. Acoustical Fatigue in Aerospace Structures. Trapp, W.J. and D.M. Forrey. ed. Syracuse University Press, New York, 225–244.

Warburton, G.B., 1981. Optimum absorber parameters for minimizing vibration response. Earthquake Engg. Str. Dyn. 9: 251–262.

Warburton, G.B., 1982. Optimum absorber parameters for various combinations of response and excitation parameters. Earthquake Engg. Str. Dyn. 10: 381–401.

Wirsching, P.H. and Campbell, G.W., 1974. Minimal structural response under random excitation using the vibration absorber. 2: 203–312.

Yang, J.N. and Heer, E., 1971. Reliability of randomly excited structures. AIAA Journal, 9: 1262–1268.

5. Response of Aerospace Vehicles to Gust, Boundary Layer Turbulence and Jet Noise

5.1 Introduction

Aircrafts and aerospace vehicles are subjected to diverse source of random dynamic excitation during their service life. Such excitations include: (i) the ground loads induced during taxi, takeoff and landing of an aircraft, or during the transportation of a launch vehicle to the launch pad, (ii) the gust excitation caused by atmospheric turbulence, (iii) the aerodynamic excitation due to boundary layer turbulence and fluctuating wake forces, and (iv) the excitation due to jet and rocket noise. Dynamic response analysis for the design of aerospace structural systems operating in such environments involves the application of random vibration theory. Realistic estimation of response statistics requires the construction of suitable stochastic models of the environment based on measured data of the random phenomena. The response to ground induced excitation will be covered in Chapter 6. In this chapter we shall consider the random response analysis of aerospace vehicles to gust, boundary layer turbulence and jet noise excitations.

5.2 Characteristics of Atmospheric Turbulence

Atmospheric turbulence is caused by the differential heating of the earth's atmosphere by the sun which affects the air flow over it. These flows are aggravated by the earth's rotation through a Coriolis force effect. Atmospheric turbulence in the upper atmosphere is of concern to the aircraft designer because of its influence on the static design strength and fatigue life of the aircraft and in view of the control upset problem.

Atmospheric turbulence can be classified into two types: (i) that caused by wind shear and (ii) that due to convective phenomenon. Turbulence due to wind shear may be generated by friction of ground obstructions and irregularities, by the vertical gradient of the wind velocity profile, due to the presence of mountains along the wind path and the motion of a cold front of air encountering a warm current of air. Convective atmospheric turbulence is mainly due to the lifting and overturning of large air masses by nonuniform heating of the atmosphere, like in thunderstorms.

Atmospheric turbulence is generally assumed to be isotropic with nearly uniform directional characteristics. It implies almost no average rate of momentum transfer across shearing surfaces. The velocity fluctuations in the three orthogonal directions can essentially be considered as random processes with little correlation between them. Press (1957), and Houbolt, Steiner and Pratt (1964) have presented power spectral density (psd) curves from flight measurements for up and cross winds which show the similarity of the statistical properties of the turbulence in different directions. From several traverses of a test airplane over gusts corresponding to cumulus clouds and thunderstorms, it has been shown that the psds of the lateral and vertical velocity components have similar shapes with almost equal mean square velocities. Figure 5.1(a) shows the psds of the vertical and transverse gust

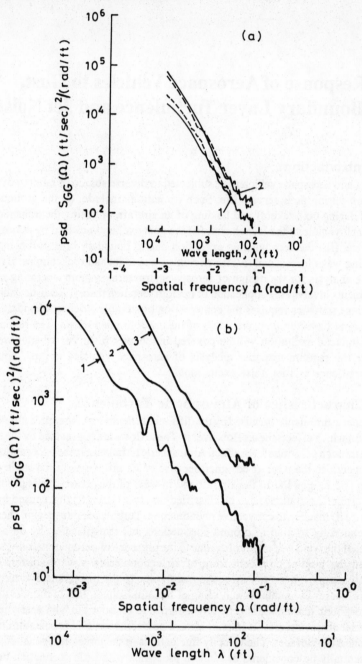

Fig. 5.1. (a) Psds of vertical and transverse gust velocity components. _____ vertical component; (1) standard deviation $\sigma = 7.46$ (ft/sec); (2) $\sigma = 13.38$ (ft/sec). - - - - transverse component; (1) $\sigma = 9$ (ft/sec); (2) $\sigma = 13.17$ (ft/sec) (Houbolt, Steiner and Pratt, 1964). (b) Psds of gust velocity for different turbulent conditions. (1) clear air, $\sigma = 4.48$ (ft/sec). (2) cumulus cloud, $\sigma = 6.14$ (ft/sec), (3) severe storm, $\sigma = 13.38$ (ft/sec) (Houbolt, Steiner and Pratt, 1964).

velocity components for two traverses of a T33 airplane through a thunderstorm at an altitude of 40000 ft, which clearly shows the isotropic nature of the gust. Typical spectra for clear air, cumulus clouds and thunderstorms are shown in Fig. 5.1(b) showing the similarity of the slopes and the variation of the intensity of turbulence with weather conditions.

To determine the aircraft response to atmospheric turbulence we assume that the turbulence could be considered as momentarily frozen in space relative to the airplane. This assumption is justified on the basis of the rapid traversing of the airplane through the gust. Such an assumption enables the transformation from time to space coordinates through the relation $s = Ut$, where U is the forward speed of the aircraft along the s direction and t denotes time. This assumption is known as Taylor's hypothesis. Another assumption again with respect to aircraft response to atmospheric turbulence is that the gusts are random in the flight direction only but uniform in the spanwise direction; this assumption is based on the notion that the scale of turbulence L is large compared to the air plane dimensions.

Earlier investigations on aircraft response to gusts considered atmospheric turbulence to be a homogeneous and stationary Gaussian random process. Routine response calculations are facilitated by use of the psd approach. But in as much as the intensity of the turbulence is dependent on weather conditions, which are known to change with time, the stationarity assumption can apply only in a limited sense. For large scale patterns of air movement, the stationarity and homogeneity conditions might be expected to apply within regions around 160 km and time durations of the order of an hour. In the case of severe turbulence such as that encountered in thunderstorms, these times and distances would be much shorter. In Fig. 5.2 is shown the psd of the vertical gust velocity obtained from two

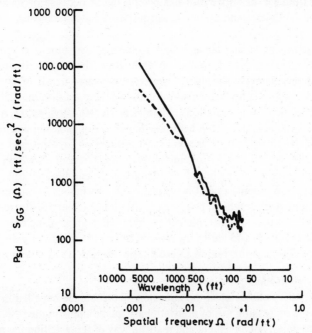

Fig. 5.2 Psd of vertical gust velocity. _____ $\sigma_1 = 15.74$(ft/sec);
- - - - $\sigma_1 = 13.38$ (ft/sec) (Houbolt, Steiner and Pratt, 1964).

successive half-sections of a particular traverse of the T33 airplane covering approximately a distance of 20 km or 100 seconds of time. The two spectra agree in general form considering that differences in both time and space are involved. Thus some degree of stationarity and homogeneity appear to be present in severe storm turbulence. It has also been tested that for the above traverse the vertical gust velocity is essentially Gaussian in character (Houbolt et al, 1964).

Dryden and von Karman Spectrum

Two forms of the psd of the gust velocity are commonly used for the stationary random process. They are respectively the Dryden and von Karman spectra. The expressions for the psd functions in these two cases are given by
Dryden Spectrum:

$$\Phi_{GG}(\omega) = \frac{\sigma^2 L}{2\pi U} \left\{ \frac{1 + (3\omega L/U)^2}{1 + (\omega L/U)^2} \right\} \tag{5.1}$$

von Karman Spectrum:

$$\Phi_{GG}(\omega) = \frac{\sigma^2 L}{2\pi U} \left\{ \frac{1 + 8/3(1.34\,\omega L/U)^2}{(1 + 1.34\,\omega L/U)^{11/6}} \right\} \tag{5.2}$$

where $G(t)$ represents the gust velocity, U the speed of aircraft, σ^2, variance of the gust velocity and $L = 2 \int_0^\infty R_{GG}(x)\,dx/\sigma^2$ is the scale of turbulence which is the measure of the average size of the eddies, and $R_{GG}(x)$ represents the autocorrelation of the gust velocity. It has been shown that the von Karman spectrum fits the data better than the Dryden spectrum. From the point of view of aircraft response calculations the Dryden spectrum has been popular because of its simpler form. The form of the Dryden and the von Karman spectra for different scales of turbulence are shown in Fig. 5.3.

The stationary and Gaussian assumptions for the gust velocity component enable the use of Rice's (1944) formulae in estimating the level crossings and peak statistics of the response variables. However, measured levels of gust loading on the aircraft indicate a consistent underestimation of these loads in the higher frequency ranges based on these assumptions. The discrepancies between the measured and predicted gust loads have often been attributed to nongaussian and nonstationary character of the gust velocity (Gaonkar, 1980). Based on experimental and analytical investigations, Piersol (1969) concluded that the discrepancies were primarily due to the nonstationarity of the turbulent data. Medium and high altitude atmospheric turbulence measurements reveal significant and continuous variations in the gust velocity intensity throughout patches of turbulence. Specifically measured time histories indicate that the gust intensity is light when an airplane enters a turbulent region, then increases and varies throughout the region and finally decreases as the aircraft passes out of the region (Verdon and Steiner, 1973; Ryan et al, 1971).

Nonstationary Models for Gust

The nonstationary character of the atmospheric turbulence have led in the past two decades to the increasing use of nonstationary gust models in the response analysis of aircraft to gusts. A convenient and realistic way of representing the nonstationary character of the gust velocity is to model it in separable form given by

$$W(t) = c(t)\ G(t) \tag{5.3}$$

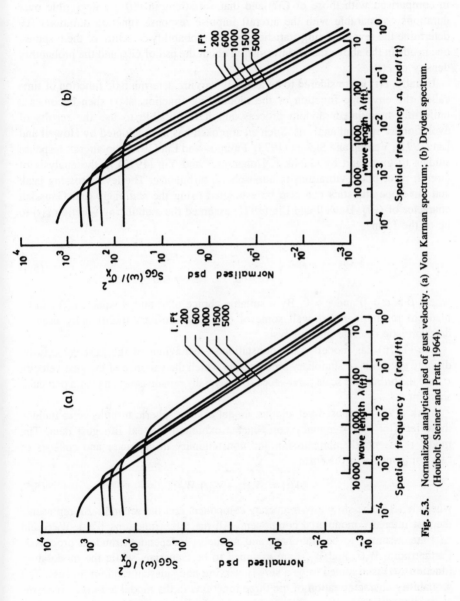

Fig. 5.3. Normalized analytical psd of gust velocity. (a) Von Karman spectrum; (b) Dryden spectrum. (Houbolt, Steiner and Pratt, 1964).

where $G(t)$ is a mean zero stationary Gaussian random process and $c(t)$ is a nonnegative function of time that may be regarded as being either deterministic or stochastic. The separable form of the gust velocity in (5.3) has been used by Press et al (1955), Houbolt (1973) in the aircraft dynamic response analysis. The above model is generally known as the Press model of turbulence. The studies of aircraft response using the Press model implicitly assume that fluctuations in $c(t)$ occur slowly in comparison with those of $G(t)$ and that variations in $c(t)$ are negligible over durations comparable with the aircraft impulse response function durations. To determine the level crossing statistics and the probability structure of the response one needs in the above model information about the psd of $G(t)$ and the probability density of $c(t)$.

Usually, $c(t)$ is considered to be a slowly varying deterministic function of time called the envelope function or the modulating function. $W(t)$ then becomes a uniformly modulated random process and it is possible to use the results of evolutionary spectral analysis. Such an approach has been adopted by Howell and Lin (1971), Verdon and Steiner (1973), Fujimori and Lin (1973a) in aircraft response studies to gusts and by Gaonkar, Hohenmser and Yin (1972) in the analysis of aircraft rotor blade vibrations to atmospheric turbulence. The level crossing peak and envelope statistics can also be estimated using the stationary and Gaussian character of $G(t)$. Howell and Lin (1971) assumed the modulating function $c(t)$ to be of the form

$$
c(t) = \begin{cases} c(e^{-\alpha t} - e^{-\beta t}), & t > 0 \\ 0, & t \leq 0 \end{cases} \tag{5.4}
$$

where $\beta > \alpha > 0$ and $c > 0$. By a suitable choice of α and β values, $c(t)$ can be made to resemble rather well some of the 'gust profiles' traditionally used in deterministic discrete gust analysis.

Lee (1976) has modelled the nonstationary behaviour of the gust velocity by discrete segments of stationary processes, in which the variance of the gust velocity σ^2 and the turbulence scale parameter L/U essentially remain constant over a particular segment.

Mark (1981) proposed yet another model of atmospheric turbulence to include the effects of low frequency components exhibited by real life gust data. The model is essentially nongaussian and nonstationary in character and consists of two components in the form

$$
W(t) = W_s(t) + c_f(t) \, W_f(t) \tag{5.5}
$$

where $W_s(t)$ is the slow low frequency component and the second term represents the fast intensity modulated component. All the three functions $W_s(t)$, $W_f(t)$ and $c_f(t)$ are assumed to be stationary and mutually independent random processes. Furthermore, $W_s(t)$ and $W_f(t)$ are assumed to be Gaussian, while the modulating function $c_f(t)$ is assumed to be a slowly varying nongaussian random process. The probability characterization of the three functions in the model are given in terms of the integral scale of appropriate transverse or longitudinal von Karman power spectrum for the component $W_f(t)$ and a series expansion for the autocorrelation functions of $W_s(t)$ and $c_f(t)$. The aircraft response statistics has been carried out by obtaining a series expansion for the probability density function of the response in terms of a Gram-Charlier series in conjunction with the use of Hermite polynomials. Other nongaussian models for atmospheric turbulence with emphasis on low wavelengths or low frequency contributions can be found in the works of Reeves (1974), Reeves et al (1974, 1975) and Murrow and Rhyne (1974) and Rhyne (1976).

5.3   Aircraft Response to Atmospheric Turbulence

In early aeronautical work during the 1930s and 40s gusts were modelled as deterministic functions of time and the gust loads were determined on the basis of a single or discrete gust encounter. The gust shape was considered as sharp edged but later modified to incorporate a specific profile. The peak vertical accelerations that were measured on airplanes flying through gusts were analyzed to estimate the maximum design speed and derive gust gradient distances, as though each acceleration value arose from an isolated discrete gust of fixed shape. These 'gusts' were then used as a means of establishing the design loads in terms of an increment to the load factor.

The inadequacy of the discrete gust approach as a basis for airplane structural design and the arbitrariness involved in the assumption of a specific profile for the gust velocity and the continuous evolution of the airplane in terms of configuration, shape, speeds and altitudes of operation etc. have necessitated a more realistic and rational model for the gust phenomenon. Atmospheric turbulence is a continuous phenomenon in which the airplane is subjected to repeated gusts. A realistic way of treating the continuous and variable character of the phenomenon is to consider it as a random process. As already observed the early attempts in this direction treated the gust velocity as a stationary Gaussian random process and the power spectral techniques in the frequency domain were used extensively in computing the dynamic response of rigid and flexible airplanes to atmospheric turbulence. These can be found in the works of Clementson (1950), Liepmaan (1952), Press and Mazelsky (1953), Fung (1953), Bisplinghoff et al (1951).

5.3.1   Stationary Response of a Rigid Wing to Stationary Gust

Quasi-steady State Theory
Consider the vertical response of a rigid wing of uniform chord width moving through turbulent air with a forward velocity U. Let $Z(t)$ be the vertical displacement of the wing measured positive downwards. The airfoil thickness and amplitude of vertical translation are assumed small in comparison with the chord. The fluctuating velocity components, u, v, w respectively in the s_1, s_2 and s_3 directions are assumed to be small compared with U. We also assume that the turbulence field does not change appreciably during the time $(2b/U)$ required for an air particle to pass the lifting surface of the vehicle, where b is the semichord length. Then with respect to a coordinate system fixed to the vehicle as shown Fig. 5.4, the gust velocity as experienced by the wing is a function of $s_1 - Ut$, s_2 and s_3, that is, $W = W(s_1 - Ut, s_2, s_3)$.

We assume that W is a mean zero Gaussian random process and uniform in the spanwise direction. The dependence of W on $(s_1 - Ut)$ implies that the turbulence field is 'frozen', that is, we assume that the Taylor's hypothesis holds. With the pitching motion of the airplane neglected, the aircraft can be considered as a mass point.

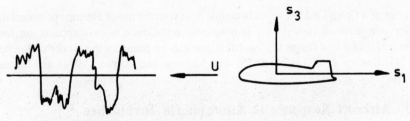

Fig. 5.4. Airplane entering a frozen pattern of gust field.

The vertical motion of the wing under these assumptions is given by

$$M\ddot{Z}(t) = -F(t) \tag{5.6}$$

where M is the wing mass and $F(t)$ is the lift force positive upwards. The number of dots represent the order of derivative with respect to time.

According to the quasi-steady state aerodynamic theory,

$$F(t) = \frac{1}{2}\rho US\left(\frac{dC_L}{d\alpha}\right)[W(t) + \dot{Z}(t)] \tag{5.7}$$

where ρ is the density of air, S the wing area, and $dC_L/d\alpha$ is the lift curve slope.

Equation (5.6) can be expressed in terms of the vertical velocity of the wing $V(t) = \dot{Z}(t)$ as

$$\dot{V}(t) + aV(t) = -aW(t) \tag{5.8}$$

where $a = \rho US(dC_L/d\alpha)/2M$.

Under the stationary and Gaussian assumption for $W(t)$ and the form of Dryden spectrum of (5.1) for the gust velocity the psd of the stationary response $V(t)$ is

$$\Phi_{vv}(\omega) = |H_v(\omega)|^2 \frac{\sigma^2 L}{2\pi U}\left\{\frac{1 + (3\omega L/U)^2}{[1 + (\omega L/U)^2]^2}\right\} \tag{5.9}$$

where

$$H_v(\omega) = \frac{a}{a + i\omega} \tag{5.10}$$

is the complex frequency response function representing the amplitude of the vertical velocity of the wing to a unit harmonic gust. The psd of the acceleration $A = \dot{V}(t)$ is,

$$\Phi_{AA}(\omega) = \frac{\omega^2 a^2}{a^2 + \omega^2} \frac{\sigma^2 L}{2\pi U}\left\{\frac{1 + (3\omega L/U)^2}{[1 + (\omega L/U)^2]^2}\right\} \tag{5.11}$$

The response variances can be obtained by contour integration in the complex plane and the results are,

$$\sigma_V^2 = \int_{-\infty}^{+\infty} \Phi_{VV}(\omega)\,d\omega = \frac{\sigma^2 aL(1 + 2aL/U)}{2U(1 + aL/U)^2} \tag{5.12}$$

and

$$\sigma_A^2 = \int_{-\infty}^{+\infty} \Phi_{AA}(\omega)\,d\omega = \frac{\sigma^2 a^2(2 + 3aL/U)}{2(1 + aL/U)^2} \tag{5.13}$$

Example 5.1 Lee (1976)

For a typical set of test data (HICAT) obtained by an U2 plane, the relevant quantities are, ρ (at 53230 ft) = 0.985×10^{-2} lb/ft^3, $S = 600$ ft^2, $dC_L/d\alpha = 7.4$ rad^{-1} for $M = 0.75$, $U = 700$ ft/sec, $M = 14300$ lb (2/3 fully fuelled U2 weight). The value of a corresponding to the above data $a \cong 1.07$ and $K = 0.9$. From (5.12) and (5.13) we get for the above data $\sigma_V^2 = 0.366\,\sigma^2$ and $\sigma_A^2 = 0.726\,\sigma^2$, where $K = L/U$, L being the scale of turbulence.

Unsteady Aerodynamic Theory

According to the unsteady aerodynamic theory the lift force $F(t)$ in (5.6) is given by

$$F(t) = \int_{\text{Span}} F_W\, dy + \int_{\text{Span}} F_M\, dy \tag{5.14}$$

where F_W is the lift per unit span due to the gust and F_M is the lift per unit span due to the wing motion.

For a sinusoidal gust of the form $W = \overline{W} \exp\{i\omega(t - s/U)\}$ and according to the strip theory (Bisplinghoff, Ashley, Halfmann, 1955), the lift per unit span is given by

$$F_W = 2\pi\,\rho U b\,\overline{W}\{C(k)[J_0(k) - iJ_1(k)] + iJ_1(k)\}e^{ik\tau} \tag{5.15}$$

where $C(k)$ is the Theodersen's function, J_0 and J_1 are the Bessel functions of first kind of order 0 and 1 respectively, $k = \omega b/U$ and $\tau = Ut/b$ is a nondimensional time parameter.

The lift per unit span due to the wing motion is given by

$$F_M = -\pi\rho U^2 Z''(\tau) - 2\pi\rho U^2 C(k)Z'(\tau) \tag{5.16}$$

where the prime indicates order of differentiation with respect to τ.

Substitute (5.15) and (5.16) in (5.6) to obtain

$$(2\lambda + 1)Z'' + 2C(k)Z' = \frac{2b}{U}\,\overline{W}K(k)e^{ik\tau} \tag{5.17}$$

where $\lambda = m/(\pi\rho Sb)$ is a nondimensional mass parameter and

$$K(k) = C(k)[J_0(k) - iJ_1(k)] + i\,J_1(k) \tag{5.18}$$

Equation (5.17) enables the evaluation of complex frequency response functions corresponding to the displacement Z, velocity $V = \dot{Z}(t)$ and the acceleration $A = \ddot{Z}(t)$ of the wing. They are,

$$H_Z(\omega) = \frac{b}{U}\left\{\frac{2K(k)}{k[2iC(k) - (2\lambda + 1)k]}\right\}$$

$$H_V(\omega) = \frac{b}{U}\left\{\frac{2K(k)}{ik(2\lambda + 1) + 2C(k)}\right\}$$

$$H_A(\omega) = -\frac{Uk}{b}\left\{\frac{2K(k)}{2iC(k) - (2\lambda + 1)k}\right\} \tag{5.19}$$

Assume the form of Dryden spectrum of (5.1) to get

$$\Phi_{ZZ}(\omega) = |H_Z(\omega)|^2 \, \Phi_{WW}(\omega)$$

$$\Phi_{VV}(\omega) = |H_V(\omega)|^2 \, \Phi_{WW}(\omega) \qquad (5.20)$$

and

$$\Phi_{AA}(\omega) = |H_A(\omega)|^2 \, \Phi_{WW}(\omega)$$

(Note that $\Phi_{WW}(\omega) = \Phi_{GG}(\omega)$ of (5.1).)

The mean square responses can be obtained by integration of the corresponding psds over the frequency domain.

Closed form integrations are difficult. However, with Liepmann's (1952) approximation of the Sear's function

$$|K(k)|^2 \cong \frac{1}{1 + 2\pi k} \qquad (5.21)$$

and with the quasi steady-state approximation

$$|2iC(k) - (2\lambda + 1)k|^2 \cong 4 + k^2(2\lambda + 1)^2 \qquad (5.22)$$

Fung (1953) has obtained the mean square acceleration by closed form integration.

$$\sigma_A^2 = \sigma^2 \left[\frac{4U^2}{\pi b^2 (2\lambda + 1)^2} \right] \left\{ \frac{a^2(3a^2 - 1)}{(a^2 - 1)^2} I_1(a_1, r) \right.$$

$$\left. - \frac{5a^2 - 3}{(a^2 - 1)^2} I_1(1, r) + \frac{2}{a^2 - 1} I_2(1, r) \right\} \qquad (5.23)$$

where

$$I_1(a, r) = \frac{2\pi r}{1 + 4\pi^2 a^2 r^2} \left(\ln 2\pi r + \ln a + \frac{1}{4ar} \right)$$

$$I_2(a, r) = \frac{8\pi^3 r^3}{(1 + 4\pi^2 a^2 r^2)^2} \left(\ln 2\pi r + \ln a + \frac{1}{4ar} \right) + \frac{\pi r}{a^2(1 + 4\pi^2 r^2 a^2)} \left(\frac{1}{4ra} - 1 \right)$$

$r = b/L$ and $a_1 = 2/r(2\lambda + 1)$.

The result of (5.23) is plotted in Fig. 5.5 as a function of r for two values of a. It can be observed from (5.23) that the mean square acceleration tends to zero for both $r \to 0$ and $r \to \infty$. This trend seems reasonable because when the wing chord is very large in comparison with the turbulence scale, the gusts are smoothed out and the effects of many minute uncorrelated disturbances on a large wing tend to cancel each other. On the other hand, when the turbulence scale is large compared with the wing chord the wing behaves in a quasi-stationary manner and hence experiences very little acceleration.

5.3.2 Transient Response of a Rigid Airplane to Stationary Gust

Quasi Steady State Theory

The response statistics given in (5.12) and (5.13) of the rigid airplane represent the stationary response after an elapse of time. If the gust encounter lasts for only a

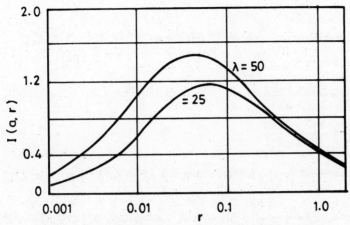

Fig. 5.5.   The scale factor, $I(a, r)$ in gust response (Fung, 1953).

short duration it may be of importance to estimate the instantaneous transient response statistics.

The autocorrelation function of the velocity $V(t)$ in the plunging mode of the airplane can be obtained from (5.8) as,

$$R_{VV}(t_1, t_2) = \int_0^{t_1} \int_0^{t_2} h_v(t_1 - \tau_1) \, h_v(t_2 - \tau_2) \, R_{WW}(\tau_1 - \tau_2) d\tau_1 d\tau_2 \qquad (5.24)$$

where $h_V(t)$ is the impulse response function corresponding to the velocity response

$$h_V(t) = e^{-at}, \qquad t > 0$$
$$= 0, \qquad \text{otherwise} \qquad (5.25)$$

and $R_{WW}(\tau)$ is the autocorrelation function of the gust. Equation (5.24) can be alternatively expressed in frequency domain as,

$$R_{VV}(t_1, t_2) = \int_{-\infty}^{+\infty} B(\omega, t_1) B^*(\omega, t_2) \, \Phi_{WW}(\omega) \, e^{i\omega(t_1 - t_2)} d\omega \qquad (5.26)$$

where

$$B(\omega, t) = -a \int_0^t e^{-(a+i\omega)u} \, du = -\frac{a[1 - e^{-(a+i\omega)t}]}{a + i\omega} \qquad (5.27)$$

the asterisk representing complex conjugate. (Henceforth the asterisk will always represent complex conjugate for a scalar and both complex conjugate and transpose for a vector or a matrix).

Substitute the form of Dryden spectrum form (5.1) in (5.26) and integrate over a contour using Cauchy's residue theorem. We obtain the relevant response statistics as,

$$\sigma_V^2(t) = \frac{\sigma^2 a^2 K}{2\pi} \left[(1 + e^{-2at}) \, I_1(K) - 2e^{-at} \, I_3(K, t) \right]$$

$$\sigma_A^2(t) = \frac{\sigma^2 a^2 K}{2\pi} \left[a^2 e^{-2at} \, I_1(K) + I_2(K) - 2ae^{-at} \, I_4(K, t) \right]$$

and

$$R_{VA}(t) = \frac{\sigma^2 a^2 K}{2\pi} \left[- ae^{-2at} I_2(K) + ae^{-at} I_3(K, t) + e^{-at} I_4(K, t)\right] \quad (5.28)$$

where

$$I_1(K) = \frac{\pi(1 + 2aK)}{a(1 + aK)^2}$$

$$I_2(K) = \frac{\pi(2 + 3aK)}{K(1 + aK)^2}$$

$$I_3(K, t) = \frac{\pi}{a(1 - a^2K^2)^2} \left[e^{-at}(1 - 3a^2K^2) + e^{-t/K}(2a^3K^3 + at(1 - a^2K^2))\right]$$

$$I_4(K, t) = \frac{\pi}{a(1 - a^2K^2)^2} \left[e^{-at}(1 - 3a^2K^2) + \frac{e^{-t/K}}{K}(3a^2K^2 - 1 + t(1 - a^2K^2))\right]$$

$$(5.29)$$

where $K = L/U$, L being the scale of turbulence.

It can be observed from (5.28) and (5.29) that as $t \to \infty$ the results approach the stationary values given by (5.12) and (5.13). The transient response covariances are shown in Fig. 5.6 for the data of example 5.1 (Lee, 1976).

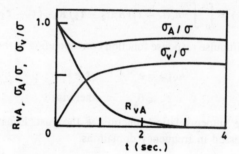

Fig. 5.6.   Transient gust response statistics in plunging mode (Lee, 1976).

5.3.3   Response of Rigid Airplane to Nonstationary Gust

Evolutionary Spectral Analysis

It has been verified experimentally that even though the gust velocity is approximately Gaussian distributed, nonstationary characteristics are exhibited by low altitude atmospheric turbulence over rough terrain. (Jones, Jones and Manson, 1969a, 1969b). Therefore for a realistic analysis of the airplane response to gusts one should consider a nonstationary model for the gust velocity. One way of treating the nonstationary character of the gust velocity is to consider the mean square gust velocity to be a random variable (Houbolt, Steiner and Pratt, 1964).

Another convenient way of modelling the nonstationary character of the gust is through the evolutionary spectral analysis. Howell and Lin (1971) were the first to use the evolutionary spectral density approach to obtain the plunging mode statistics of a rigid wing to nonstationary gust. The gust velocity is assumed to be a uniformly modulated random process given by

$$W(t) = c(t)\, G(t) \tag{5.30}$$

where $c(t)$ is a deterministic function of time and $G(t)$ a weakly stationary Gaussian random process The modulating function $c(t)$ is given by (5.4). The constant c is determined from the normalizing condition

$$\underset{t}{\text{Sup}}\, |\, c(t)\, | = 1 \tag{5.31}$$

Equation (5.30) can be expressed as a Fourier Stieltjes integral

$$W(t) = \int_{-\infty}^{+\infty} c(t)\, e^{i\omega t}\, d\tilde{G}(\omega) \tag{5.32}$$

where $\tilde{G}(\omega)$ is an orthogonal random process with

$$E[d\tilde{G}(\omega_1) d\tilde{G}^*(\omega_2)] = 0 \quad \omega_1 \neq \omega_2$$

$$E[|\, dG(\omega)\, |^2] = \Phi_{GG}(\omega)\, d\omega \tag{5.33}$$

In (5.33) $\Phi_{GG}(\omega)$ is the power spectral density of the stationary random process $G(t)$.

The autocorrelation function and the mean square value of $W(t)$ can be obtained using the evolutionary spectral analysis as

$$R_{WW}(t_1, t_2) = \int_{-\infty}^{+\infty} c(t_1)\, c(t_2)\, e^{i\omega(t_1-t_2)}\, \Phi_{GG}(\omega)\, d\omega$$

$$E[W^2(t)] = \int_{-\infty}^{+\infty} c^2(t)\, \Phi_{GG}(\omega)\, d\omega \tag{5.34}$$

Plunging Mode Response of Airplane: Quasi-steady state approach
Now consider the equation of motion of the airplane in its plunging mode as given by (5.8). With the gust velocity given by (5.32), the vertical displacement response $Z(t)$ can be expressed as a Fourier-Stieltjes integral

$$Z(t) = \int_{-\infty}^{+\infty} B_Z(t, \omega)\, e^{i\omega t}\, dG(\omega) \tag{5.35}$$

where

$$B_Z(t, \omega) = -a \int_0^t h_z(u)\, c(t - u)\, e^{-i\omega u}\, du \tag{5.36}$$

and $h_z(u)$ is the impulse response function associated with the displacement response and given by

$$\left.\begin{array}{ll} h_z(t) = \dfrac{1}{a}\, [1 - e^{-at}], & t > 0, \\[2mm] \quad\ = 0 & \text{otherwise} \end{array}\right\} \tag{5.37}$$

and $a = 2U/[b(2\lambda + 1)]$ wih $\lambda = M/(\pi\rho S b)$ is the nondimensional mass parameter, S, the wing area and ρ the density of air.

The autocorrelation function and the mean square value of the response are

$$R_{ZZ}(t_1, t_2) = E[Z(t_1)Z^*(t_2)] = \int_{-\infty}^{+\infty} B_Z(t_1, \omega) B_Z^*(t_2, \omega) \Phi_{GG}(\omega) e^{i\omega(t_1 - t_2)} \, d\omega$$

and

$$\sigma_Z^2(t) = E[Z^2(t)] = \int_{-\infty}^{+\infty} |B_Z(t, \omega)|^2 \, \Phi_{GG}(\omega) \, d\omega \qquad (5.38)$$

Likewise the autocorrelation function of the velocity and acceleration responses are

$$R_{VV}(t_1, t_2) = \int_{-\infty}^{+\infty} \{\dot{B}_Z(t_1, \omega) \dot{B}_Z^*(t_2, \omega) + \omega^2 B_Z(t_1, \omega) B_Z^*(t_2, \omega)$$

$$+ i\omega [B_Z(t_1, \omega) \dot{B}_Z^*(t_2, \omega)$$

$$- \dot{B}_Z(t_1, \omega) B_Z^*(t_2, \omega)]\} \, \Phi_{GG}(\omega) e^{i\omega(t_1 - t_2)} \, d\omega \qquad (5.39)$$

$$R_{AA}(t_1, t_2) = \int_{-\infty}^{+\infty} \{\ddot{B}_Z(t_1, \omega) + 2i\omega \dot{B}_Z(t_1, \omega) - \omega^2 B_Z(t_1, \omega)\}$$

$$\times \{\ddot{B}_Z^*(t_2, \omega) - 2i\omega \dot{B}_Z^*(t_2, \omega)$$

$$- \omega^2 B_Z^*(t_2, \omega)\} \, \Phi_{GG}(\omega) e^{i\omega(t_1 - t_2)} \, d\omega \qquad (5.40)$$

The mean square velocity and acceleration statistics can be obtained by considering directly the impulse reponse function $h_V(t)$ and $h_A(t)$ corresponding to the velocity and acceleration response respectively, namely,

$$h_V(t) = e^{-at}, \qquad t > 0$$

$$= 0, \qquad \text{otherwise} \qquad (5.41)$$

and

$$h_A(t) = \delta(t) - a \, e^{-at} u(t) \qquad (5.42)$$

where $\delta(t)$ and $u(t)$ represent the Dirac delta function and the Heaviside unit step function respectively. The relevant autocorrelation and cross correlation functions can now be represented as

$$R_{ZV}(t_1, t_2) = \int_{-\infty}^{+\infty} B_Z(t_1, \omega) B_V^*(t_2, \omega) \Phi_{GG}(\omega) \, e^{i\omega(t_1 - t_2)} \, d\omega$$

$$R_{VV}(t_1, t_2) = \int_{-\infty}^{+\infty} B_V(t_1, \omega) B_V^*(t_2, \omega) \Phi_{GG}(\omega) \, e^{i\omega(t_1 - t_2)} \, d\omega$$

$$R_{VA}(t_1, t_2) = \int_{-\infty}^{+\infty} B_V(t_1, \omega) B_A^*(t_2, \omega) \Phi_{GG}(\omega) \, e^{i\omega(t_1 - t_2)} \, d\omega$$

and

$$R_{AA}(t_1, t_2) = \int_{-\infty}^{+\infty} B_A(t_1, \omega) B_A^*(t_2, \omega) \Phi_{GG}(\omega) \, e^{i\omega(t_1 - t_2)} \, d\omega \qquad (5.43)$$

where

$$B_V(t, \omega) = - a \int_0^t h_V(u) \, c(t - u) e^{-i\omega u} \, du$$

and

$$B_A(t, \omega) = - a \int_0^t h_A(u) \, c(t - u) e^{-i\omega u} \, du$$

$$(5.44)$$

The mean square velocity and the mean square accelerations are

$$\sigma_V^2(t) = \int_{-\infty}^{+\infty} | B_V(t, \omega) |^2 \, \Phi_{GG}(\omega) \, d\omega$$

$$\sigma_A^2(t) = \int_{-\infty}^{+\infty} | B_A(t, \omega) |^2 \, \Phi_{GG}(\omega) \, d\omega \qquad (5.45)$$

For a modulating function of the form (5.31) with $c = 1$ the various complex frequency response functions are

$$B_Z(t, \omega) = e^{-i\omega t} \left[\frac{1}{\beta - i\omega} - \frac{1}{\alpha - i\omega} + \frac{e^{at}}{\alpha - a - i\omega} - \frac{e^{-at}}{\beta - a - i\omega} \right]$$

$$+ e^{-at} \left[\frac{1}{\alpha - i\omega} - \frac{1}{\alpha - a - i\omega} \right] - e^{-\beta t} \left[\frac{1}{\beta - i\omega} - \frac{1}{\beta - a - i\omega} \right]$$

$$B_V(t, \omega) = a e^{-(a+i\omega)t} \left[\frac{1}{\beta - a - i\omega} - \frac{1}{\alpha - a - i\omega} \right]$$

$$+ a \left[\frac{e^{-\alpha t}}{\alpha - a - i\omega} - \frac{e^{-\beta t}}{\beta - a - i\omega} \right]$$

and

$$B_A(t, \omega) = a(e^{-\beta t} - e^{-\alpha t}) + a^2 e^{-(a+i\omega)t} \left[\frac{1}{\alpha - a - i\omega} - \frac{1}{\beta - a - i\omega} \right]$$

$$- a^2 \left[\frac{e^{-\alpha t}}{\alpha - a - i\omega} - \frac{e^{-\beta t}}{\beta - a - i\omega} \right] \qquad (5.46)$$

It can be observed that if we substitute $c(t) = 1$, in the first equation of (5.44) it reduces to (5.27) and the transient response to a stationary gust can be recovered with $\alpha \to 0$ and $\beta \to \infty$ from the relevant response statistics.

The integrations in (5.38), (5.39), (5.40), (5.43) and (5.45) can be performed by contour integration in the complex plane. An approximate closed form expression for the mean square value of the acceleration can be obtained under the condition that the constant a is much larger than both α and β, by replacing both $a - \alpha$ and $a - \beta$ by $a - (\alpha + \beta)/2$. With this approximation integrating the second equation of (5.45) with $\Phi_{GG}(\omega)$ given by the Dryden spectrum of (5.1) gives, (Howell and Lin, 1971)

$$\sigma_A^2(t) \cong \frac{\sigma^2 a^2}{2} c^2(t) \left\{ \frac{2 - \Delta L/U - 4(\Delta L/U)^2 + 3(\Delta L/U)^3}{[1 - (\Delta L/U)^2]^2} \right\} \qquad (5.47)$$

where $\Delta = a - (\alpha + \beta)/2$. In deriving (5.47) the approximation $a\Delta \cong a^2$ is also used.

With the assumption of the gust velocity to be a Gaussian random process, the nonstationary exceedance rate of the plunging velocity response of level ξ can be expressed as

$$N^+(\xi, t) = \frac{\sigma_A}{2\pi\sigma_V} (1 - \rho_{VA}^2)^{1/2} e^{-\xi^2/2\sigma_V^2} [\exp(-\eta^2) + \sqrt{\pi}\ \eta(1 + \operatorname{erf}\eta)] \quad (5.48)$$

where $\eta = \xi\rho_{VA}/(\sigma_V\sqrt{[2(1 - \rho_{VA}^2)]})$ is the correlation coefficient between the velocity and acceleration considered at the same time, that is,

$$\rho_{VA} = R_{VA}(t, t)/[\sigma_V(t)\ \sigma_A(t)] \quad (5.49)$$

The peak distribution of the displacement response is given by

$$p_p(\xi, t) = \frac{1}{\sqrt{2\pi}} [\alpha_{11}^{1/2} - \alpha_{13}^2/(\alpha_{11}^{1/2}\alpha_{33})]\{\exp(-\alpha_{11}\xi^2/2)$$

$$+ \xi(\pi/2)^{1/2}(\alpha_{13}/\alpha_{33}^{1/2})$$

$$\exp[-(\alpha_{11} - \alpha_{13}^2/\alpha_{33})\xi^2/2][1 + \operatorname{erf}(\alpha_{13}\xi/(2\alpha_{33})^{1/2})]\} \quad (5.50)$$

where α_{jk} is the jkth element of the inverse of the covariance matrix between Z, V and A.

Example 5.2: Response of an aircraft to gust (Howell and Lin, 1971)
The following data for the aircraft are used in the response evaluation. $L = 500$ ft, $U = 200$ ft/sec $b = 3.0$ ft, $\lambda = 59.5$, $\sigma = 1$ ft/sec, $c(t) = e^{-\alpha t} - e^{-\beta t}$, $\alpha = 0.009$, $\beta = 0.04$ and $a = 1.11$ sec^{-1}. The response statistics computed using (5.38) and (5.43) with $t_1 = t_2 = t$ are shown in Fig. 5.7. The nonstationary character of the

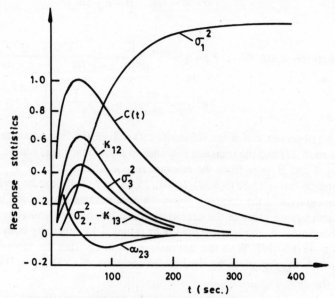

Fig. 5.7. Nonstationary gust response statistics.
$\sigma_1^2 = E[Z^2(t)]$ ft^2; $K_{12} = E[Z(t)\dot{Z}(t)]$ (ft^2/sec);
$\sigma_2^2 = E[\dot{Z}^2(t)]$ (ft/sec)^2; $K_{13} = E[Z(t)\ddot{Z}(t)]$ (ft^2/sec^2);
$K_{23} = E[\dot{Z}(t)\ddot{Z}(t)]$ (ft^2/sec^3) (Howell and Lin, 1971).

response is clearly observed with the variances being functions of time and the covariances being generally nonzero. As the modulating function tends to zero all the statistics of response vanish except σ_Z^2, which approaches a non-zero constant. This implies that a certain amount of uncertainty prevails in the aircraft altitude after it has passed through a gust field.

For the above data with $\alpha = 0$, that is, for a gust modulating function $c(t) = 1 - \exp(-\beta t)$, the mean square acceleration is shown in Fig. 5.8 for different times and for different values of β. It can be observed that for all values of β as t increases, that is, $c(t) \to 1$ the mean square acceleration approaches the stationary solution. However, in a stationary analysis the transient overload is not detected as we see in Fig. 5.8.

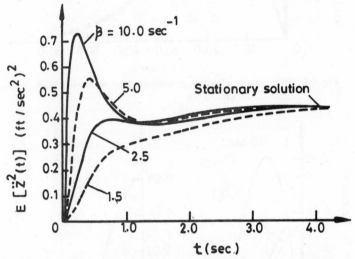

Fig. 5.8 Mean square plunging acceleration due to nonstationary gust (Howell and Lin, 1971).

The mean square acceleration calculated exactly by integration and the approximation of (5.47) are shown in Fig. 5.9 for the data $U = 125$ fps, $b = 4$ ft, $\lambda = 50.0$, $L = 500$ ft, $a = 0.611$ sec^{-1} and $\sigma = 1.0$ fps. The agreement between them is very good.

The peak density functions of the aircraft displacement response for the aircraft data of example 5.2 calculated from (5.50) corresponding to two times, $t = 30$ sec and $t = 150$ sec, are shown in Fig. 5.10. Also shown in the figure are Gaussian and Rayleigh probability densities of the displacement corresponding to a reference variance σ_Z^2 calculated at these times. The peak distribution is much closer to the Gaussian distribution indicating the broad band nature of the displacement response of the aircraft.

Mixed Unsteady and Quasi-steady State Approach
It is generally recognized that the quasi-steady aerodynamic theory is adequate when the scale of turbulence L is much larger than the semichord length b. An approximate mixed unsteady and quasi-steady approach has also been used in conjunction with the evolutionary spectral analysis for the response of a rigid

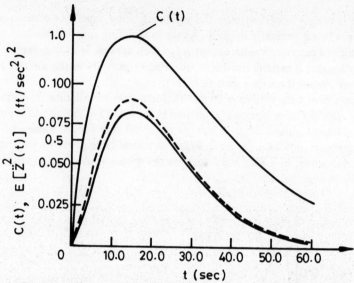

Fig. 5.9   Mean square plunging acceleration due to nonstationary gust (Howell and Lin, 1971).

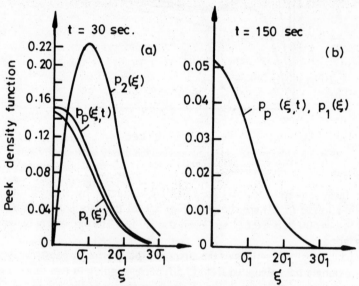

Fig. 5.10.   Peak distribution of displacement response to nonstationary gust. $p_p(\xi)$—actual; $p_1(\xi)$—Gaussian; and $p_2(\xi)$—Rayleigh (Howell and Lin, 1971).

airplane to nonstationary gust (Howell and Lin, 1971). In the sequel we present the important results of their analysis.

For a passive system, it is generally possible to establish an exponential bound for the impulse response function $h_Z(t)$, that is, it is possible to find a γ such that $|h_Z(\tau)| \le \frac{1}{a} e^{-\gamma \tau}$. Considering the response at instance of $t \gg 1/\gamma$ and assuming the modulating function $c(t)$ to be a slowly varying function such that $|c(t-\tau) - c(t)| \ll |c(t)|$ for $|\tau|$ less than about $3/\gamma$, (5.36) can be approximated by

$$B_Z(t, \omega) \cong - a \int_0^\infty h(\tau) c(t) e^{-i\omega \tau} \, d\omega \cong ac(t) H_Z(\omega) \tag{5.51}$$

where $H_Z(\omega) = 1/(ia\omega - \omega^2)$.

The mean square displacement with this approximation is

$$\sigma_Z^2(t) \cong c^2(t) \int_{-\infty}^{+\infty} \Phi_{GG}(\omega) \, | H_Z(\omega) |^2 \, d\omega$$

$$\sigma_V^2(t) \cong c^2(t) \int_{-\infty}^{+\infty} \omega^2 \Phi_{GG}(\omega) \, | H_Z(\omega) |^2 \, d\omega \tag{5.52}$$

and

$$R_{ZV}(t, t) \cong c(t) \dot{c}(t) \int_{-\infty}^{+\infty} \Phi_{GG}(\omega) \, | H_Z(\omega) |^2 \, d\omega$$

The preceding expressions imply that the nonstationary variances are approximated by the stationary variances multiplied by the square of the modulating function. Priestley (1967) has also suggested essentially the same approximation for slowly varying modulating functions.

Howell and Lin (1971) have used the frequency response function for the acceleration

$$H_A(\omega) = H_1(\omega) \, H_2(\omega) \tag{5.53}$$

where $H_1(\omega)$ is the admittance of the lift to a sinusoidal gust and $H_2(\omega)$ is the frequency response of the acceleration to a sinusoidal lift.

For an incompressible flow

$$H_1(\omega) = \sqrt{2} \, \pi \rho b U \, K(k) \tag{5.54}$$

and the quasi-steady state approximation for $H_2(\omega)$ is

$$H_2(\omega) = \sqrt{2} \, k/\{\pi \rho b [k(1 + 2\lambda) - 2i]\} \tag{5.55}$$

where $K(k)$ is the Sear's function.

For the Sear's function $K(k)$ the following approximation is used.

$$K(k) \cong [0.065/(0.13 + ik) + 0.5/(1 + ik)] \, e^{1.1k} \tag{5.56}$$

This enables the estimation of the variance of the derivative of the aircraft acceleration and the cross correlation between the aircraft acceleration and its derivative which are useful in the calculation of the nonstationary threshold crossing rate of the acceleration response. It may be mentioned here that the use of Liepmann's approximation for the Sear's function $| K(k) |^2 \cong 1/(1 + 2k)$, which was used in the stationary gust response would have yielded unbounded variance for the derivative of the acceleration response.

Example 5.3: Response of an aircraft to gust (Howell and Lin, 1971)

The following data for the aircraft are used in the response evaluation using the mixed unsteady and quasi-steady approach. $L = 200$ ft, $U = 200$ ft/s, $b = 3.5$ ft, $\sigma^2 = 1.0$ ft^2/sec^2, $c(t) = 1 - e^{-0.05t}$, $\lambda = 59.5$, $a = 1.11$ sec-1. Fig. 5.11 shows the

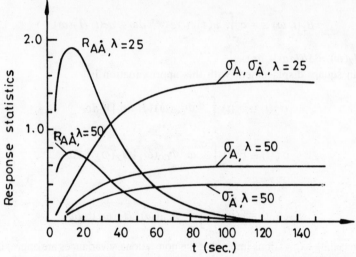

Fig. 5.11. Variances and covariances of acceleration response and rate of change of acceleration: $A = \ddot{Z}(t)$ and $\dot{A} = \dot{Z}(t)$ (Howell and Lin, 1971).

acceleration statistics of σ_A^2, $\sigma_{\dot{A}}^2$ and $R_{A\dot{A}}$. The threshold crossing statistics of the acceleration response for threshold levels of 1.0, 2.0 and 3.0 ft/sec² are shown in Fig. 5.12.

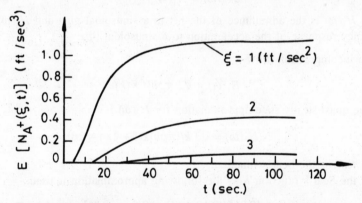

Fig. 5.12. Expected rate of threshold crossings of the acceleration response (Howell and Lin, 1971).

Other Approaches to Nonstationary Gust Response Analysis
Verdon and Steiner (1973) use a similar model as in (5.30) with the modulating envelope function $c(t)$ represented by a Fourier series which models the arbitrary intensity variations of the gust. A rectangular step funcion with the modulating function models a finite patch length, that is

$$c(t) = \left(\sum_{k=-\infty}^{\infty} a_p e^{ik\omega t} \right)[u(t) - u(t - s)] > 0 \qquad (5.57)$$

where $u(t)$ is the unit step function. A quasi-steady state aerodynamic theory is used in their analysis and the response statistics, threshold crossing rates are obtained by means of a time domain analysis.

Lee (1976) has used a discrete nonstationary gust model for the plunging mode response of an aircraft where the turbulent gusts are modelled by a series of discrete stationary processes with different mean square gust velocities and scales of turbulence. He retains however, the Gaussian assumption so that the exceedance statistics can be obtained by the use of Rice's formula.

Mark (1981) has proposed a nonstationary and nongaussian model to account for low frequency components exhibited by real gust data. In his model the gust velocity is assumed to be a sum of a stationary and a nonstationary random process, where the nonstationary random process is modelled similar to the evolutionary random process, with the exception that the modulating function $c(t)$ is considered as a stationary nongaussian random process. An analysis considering two degrees of freedom of rigid body motion pitching and plunging has been presented by Peele and Steiner (1970) with the psd of the gust response given by the von Karman spectrum. Trevino (1981) has considered the effects of inhomogeneities in atmospheric turbulence on the dynamic response of an aircraft.

5.3.4 Response of Aircraft Including Wing Bending Flexibility to Nonstationary Gust

In Sections 5.2 and 5.3 we have discussed the response analysis of a rigid wing to atmospheric turbulence for stationary and evolutionary nonstationary gusts. The only degree of freedom considered was the vertical plunging mode of the airplane. Both quasi-steady state and unsteady aerodynamic theories were used to obtain the impressed random forces on the wing due to the gust. Practical considerations, however, require that other degrees of the freedom of the airplane such as pitching, wing bending modes, fuselage sweep angle effects are considered in the response calculations. This is especially true of large airplanes.

In the deterministic discrete gust response analysis the flexibility effects of the airplane are usually considered by means of the normal mode approach. A similar approach can be extended to stationary gust response analysis (Bisplinghoff et al, 1955). Lin (1965) has used the transfer matrix approach in the gust response analysis of a flexible aircraft when the structure was considered as an assemblage of periodic elements. The gust velocity in his analysis was assumed to be a stationary Gaussian random process. Fujimori and Lin (1973a, 1973b) have analyzed aircraft response including the effects of wing bending flexibility to nonstationary gusts. The normal mode approach was used in the analysis. This approach is general and uses unsteady aerodynamic theory in determining the lift forces. We present in the sequel the essential features of the approach and some pertinent results.

Normal Mode Approach with Unsteady Aerodynamic Forces
Consider the equations of motion of a two dimensional wing of an airplane

$$M_j \ddot{q}_j + \beta_j \dot{q}_j + M_j \omega_j^2 q_j = Q_j \qquad (5.58)$$

where q_j represents the jth normal coordinate, M_j the generalized mass, ω_j the natural frequency, β_j the damping coefficient and Q_j the generalized force corresponding to the jth natural mode. The generalized mass M_j and the generalized force Q_j can be expressed as

$$M_j = \iint_s m(s_1, s_2)\, u(s_1, s_2)\, ds_1 ds_2 \qquad (5.59)$$

$$Q_j = \iint_s [F_W(s_1, s_2, t) + F_M(s_1, s_2, t)] u_j(s_1, s_2) \, ds_1 ds_2] \qquad (5.60)$$

where $u_j(s_1, s_2)$ is the jth normal mode and $F_W(s_1, s_2, t)$ and $F_M(s_1, s_2, t)$ respectively represent the lift per unit area due to the gust and the wing motion and m is the mass per unit area of the wing. s_1 represents the body coordinate along the longitudinal axis of the aircraft assumed positive in the opposite direction of the airplane motion and s_2 an orthogonal body coordinate in the plane of the wing with origin at the center of gravity of the airplane. F_W and F_M can be obtained using unsteady aerodynamic theory and use of Taylor's hypothesis of 'frozen' turbulence field as

$$F_W(s_1, s_2, t) = \pi \rho U \int_0^t W(s_1 - U\tau, s_2) \, \dot{\psi}(t - \tau) \, d\tau$$

$$F_M(s_1, s_2, t) = - \pi \rho b \ddot{Z}(s_1, s_2, t)/2 - \pi \rho U \int_0^t \ddot{Z}(s_1, s_2, \tau) \varphi(t - \tau) \, d\tau \quad (5.61)$$

where $Z(s_1, s_2, t) = \Sigma u_j(s_1, s_2) q_j$ is the vertical motion of the wing and $\varphi(t)$ and $\psi(t)$ are respectively the Wagner and Küssner functions which can be approximated by

$$\left. \begin{aligned} \varphi(\vartheta) &= 1 - 0.615 e^{-0.045\vartheta} - 0.355 e^{-0.3\vartheta} \\ \psi(\vartheta) &= 1 - 0.5 e^{-0.13\vartheta} - 0.5 e^{-\vartheta} \end{aligned} \right\} \qquad (5.62)$$

where $\vartheta = Ut/b$ is the nondimensional time parameter in terms of the wing chord.

Substitute (5.62) into the generalized force expression (5.60) and use (5.61) to obtain

$$Q_j(t) = Q_{W_j}(t) + Q_{M_j}(t)$$

where

$$Q_{W_j}(t) = \pi \rho U \int_0^t \dot{\psi}(t - \tau) \left\{ \iint_s W(s_1 - U\tau, s_2) u_j(s_1, s_2) \, ds_1 ds_2 \right\} d\tau$$

$$Q_{M_j}(t) = \pi \rho U \sum_r \left[(b/2U) \ddot{q}_r(t) + \Delta_{jr} \int \ddot{q}_r(\tau) \varphi(t - \tau) \, d\tau \right] \qquad (5.63)$$

with

$$\Delta_{jr} = \iint_s u_j(s_1, s_2) u_r(s_1, s_2) \, ds_1 ds_2$$

Substituting (5.63) into (5.58) it is observed that the equations of motion are coupled in the generalized coordinates due to the interaction of the structural motion and the fluid flow.

Take Laplace transform of (5.58), the resulting equation can be expressed in matrix form

$$C(p)\bar{q}(p) = \bar{Q}_W(p) \qquad (5.64)$$

where $\bar{\bar{q}}(p)$ and $\bar{\bar{Q}}(p)$ are vectors representing respectively the Laplace transforms of the generalized coordinate vector $\bar{q}$ and the $\bar{Q}_W$ vector whose jth elements are q_j and Q_{W_j}. A typical element $C_{jr}(p)$ of the matrix $C(p)$ is given by

$$C_{jr}(p) = \{M_{jr}\, p^2 + \beta_{jr}\, p + M_{jr}\, \omega_{jr}^2 + \pi\rho U\Delta_{jr}\, p^2 [\hat{\varphi}(p) + b/2U]\} \tag{5.65}$$

where

$$\hat{\varphi}(p) = \frac{1}{p} - \frac{0.165}{p + 0.0455U/b} - \frac{0.335}{p + 0.3U/b} \tag{5.66}$$

is the Laplace transform of the Wagner's function and $M_{jr} = M_j\,\delta_{jr}$, $\beta_{jr} = \beta_j\,\delta_{jr}$ and $\omega_{jr} = \omega_j\,\delta_{jr}$ with δ_{jr} being the Kronecker delta.

Take the inverse Laplace transform of (5.64) to obtain

$$q_j(t) = \sum_r \left[\int_0^{t'} h_{jr}(t - \tau)Q_{W_r}(\tau)\, d\tau \right] \tag{5.67}$$

where

$$h_{jr}(t) = L^{-1}[C_{jr}^{-1}]$$

with C_{jr}^{-1} being the jrth element of the matrix C^{-1}. Equation (5.67) represents the input output relationship of a linear system where the $h_{jr}s$ play the role of impulse response functions. Hence the statistics of $q_j s$ can be obtained from the statistics of $Q_{W_r}s$. In general the cross-correlation function is

$$R_{q_j q_k}(t_1, t_2) = E[q_j(t_1)q_k(t_2)]$$

$$= \sum_r \sum_r \int_0^{t_1} \int_0^{t_2} E[Q_{W_r}(\tau_1)Q_{W_s}(\tau_2)]h_{jr}(t_1 - \tau_1)h_{ks}(t_2 - \tau_2)\, d\tau_1\, d\tau_2 \tag{5.68}$$

Use the first of (5.63) to obtain the explicit input output relation between the gust and the aircraft response

$$R_{q_j q_k}(t_1, t_2) = (\pi\rho U)^2 \sum_r \sum_s \int_0^{t_1} h_{jr}(t_1 - \tau_1) \int_0^{t_2} h_{ks}(t_2 - \tau_2)$$

$$\times \int_0^{\tau_1} \dot{\psi}(\tau_1 - \sigma_1) \int_0^{\tau_2} \dot{\psi}(\tau_2 - \sigma_2) E[W(s_1^1 - U\sigma_1, s_2^1)W(s_1^2 - U\sigma_2, s_2^2)]$$

$$\times u_r(s_1^1, s_2^1)u_s(s_1^2, s_2^2)ds_1^2\, ds_2^2\, ds_1^1\, ds_2^1\, d\sigma_2\, d\sigma_1\, d\tau_2\, d\tau_1 \tag{5.69}$$

where σ_i and τ_i, $i = 1, 2$ are dummy time variables and s_1^1, s_2^1, s_1^2, and s_2^2 are dummy space variables.

The gust velocity W is assumed to be a uniformly modulated evolutionary random process having the form

$$W(s_1 - Ut, s_2) = c(s_1 - Ut, s_2)\, G(s_1 - Ut, s_2) \tag{5.70}$$

with $c(s_1 - Ut, s_2)$ a deterministic envelope function and $G(s_1 - Ut, s_2)$, a stationary random process having the following cross spectral density function

$$\Phi_{GG}(\omega, \xi, \eta) = \frac{\sigma^2 L e^{-i\omega\xi/U}}{2a^2\pi U} \{(\eta\omega/U)^2 K_2(a\eta/L)$$

$$+ a\eta/L[3K_1(a\eta/L) - (a\eta/L)K_2(a\eta/L)]\} \qquad (5.71)$$

where $\xi = s_1^1 - s_1^2$ and $\eta = s_2^1 - s_2^2$, σ^2 is the mean square value of the gust velocity, L the scale of turbulence and $a = [1 + (\omega L/U)^2]^{1/2}$. K_1 and K_2 are modified Bessel functions of second kind of orders 1 and 2 respectively. The form of the exponential correlation of the gust field along the s_1 direction and Bessel function correlation along the s_2 direction was adopted by Lin (1967) assuming an isotropic and spatially homogeneous turbulence field. The spectral density of the gust velocity at a single location is obtained by setting $\xi = 0$ and $\eta = 0$. It can be verified noting that as $z \to 0$, $zK_1(z) \to 1$ and $z^2 K_2(z) \to 2$ that for $\xi = 0$ and $\eta = 0$, (5.71) reduces to the Dryden spectrum of (5.1).

Evolutionary Spectral Analysis

$W(s_1 - Ut, s_2)$ given by (5.70) admits a Fourier-Stieltjes integral representation

$$W(s_1 - Ut, s_2) = \int_{-\infty}^{+\infty} c(s_1 - Ut, s_2) e^{i\omega t} d\tilde{G}(\omega, s_1, s_2) \qquad (5.72)$$

with

$$E[d\tilde{G}(\omega_1, s_1^1, s_2^1) d\tilde{G}^*(\omega_2, s_1^2, s_2^2)] = 0, \qquad \omega_1 \neq \omega_2$$

and

$$E[d\tilde{G}(\omega, s_1^1, s_2^1) d\tilde{G}^*(\omega, s_1^2, s_2^2)] = \Phi_{GG}(\omega, \xi, \eta) d\omega$$

As in section 5.3.3 we assume the form of the modulating function to be

$$c(s_1 - Ut, s_2) = c_0[e^{\alpha(s_1 - Ut)} - e^{\beta(s_1 - Ut)}]u(Ut - s_1) \qquad (5.73)$$

where c_0 is the normalizing constant, α and β arbitrary constants with $0 < \alpha < \beta$ and $u(\cdot)$ is the Heaviside unit step function. In (5.73) the variation of the modulating function is assumed uniform in the s_2 direction which implies that the statistical variation of the gust field over a distance of the order of the normal airplane along the s_2 direction can be neglected.

Airplane Response Statistics

Introduce the following notations

$$\mathscr{J}_{rs,\alpha\beta} = \iint_s \iint_s \Phi_{GG}(\omega, \zeta, \eta) e^{\alpha s_1^1 + \beta s_1^2} u_r(s_1^1, s_2^1) u_s(s_1^2, s_2^2) \, ds_1^2 \, ds_2^2 \, ds_1^1 \, ds_2^1 \qquad (5.74)$$

and

$$B_{jr,\alpha}(t, \omega) = e^{-\alpha Ut} \int_0^t e^{-(i\omega - \alpha U)\tau} h_{jr}(\tau) \left[\int_0^\tau \dot{\psi}(\tau_1) e^{-(i\omega - \alpha U)\tau_1} \, d\tau_1 \right] d\tau \qquad (5.75)$$

and substitute in (5.69), we obtain

$$R_{q_j q_k}(t_1, t_2) = \int_{-\infty}^{+\infty} \Phi_{jk}(t_1, t_2, \omega) e^{i\omega(t_1 - t_2)} d\omega \qquad (5.76)$$

where

$$\Phi_{jk}(t_1, t_2, \omega) = (\pi \rho U)^2 c_0^2 \sum_r \sum_s [J_{rs,\alpha\alpha}(\omega) B_{jr,\alpha}(t_1, \omega) B^*_{ks,\alpha}(t_2, \omega)$$

$$+ J_{rs,\beta\beta}(\omega) B_{jr,\beta}(t_1, \omega) B^*_{ks,\beta}(t_2, \omega)$$

$$- J_{rs,\alpha\beta}(\omega) B_{jr,\alpha}(t_1, \omega) B^*_{ks,\beta}(t_2, \omega)$$

$$- J_{rs,\beta\alpha}(\omega) B_{jr,\beta}(t_1, \omega) B^*_{ks,\alpha}(t_2, \omega)] \qquad (5.77)$$

The quantity defined in (5.77) is the evolutionary cross spectral density of the jth and kth generalized coordinates at times t_1 and t_2. The mean square value of each modal response can be evaluated by setting $t_1 = t_2 = t$ and $j = k$ in (5.77). The cross-correlation function of the total response at (s_1^1, s_2^1) and time t_1 and at s_1^2, s_2^2 and time t_2 is,

$$E[Z(s_1^1, s_2^1, t_1) Z(s_1^2, s_2^2, t_2)] = \sum_j \sum_k u_j(s_1, s_2) u_k(s_1^2, s_2^2) R_{q_j q_k}(t_1, t_2) \qquad (5.78)$$

The cross-correlation of the derivatives of the modal response can be obtained as

$$R_{\dot{q}_j \dot{q}_k}(t_1, t_2) = \frac{\partial^2}{\partial t_1 \partial t_2} R_{q_j q_k}(t_1, t_2) = \int_{-\infty}^{+\infty} \left\{ \omega^2 \Phi_{jk}(t_1, t_2, \omega) + i\omega \frac{\partial \Phi_{jk}}{\partial t_2}(t_1, t_2, \omega) \right.$$

$$\left. - i\omega \frac{\partial \Phi_{jk}}{\partial t_1}(t_1, t_2, \omega) + \frac{\partial^2 \Phi_{jk}}{\partial t_1 \partial t_2}(t_1, t_2, \omega) \right\} e^{i\omega(t_1 - t_2)} d\omega \qquad (5.79)$$

Example 5.4: Response of an aircraft to gust (Fujimori and Lin, 1973a, b)
Consider an idealized airplane with a swept wing as shown in Fig. 5.13. It is characterized by the following parameters:

Λ: sweep angle

EI: uniform flexural rigidity of the wing

m_F: distributed mass per unit length along the longitudinal axis assumed attached to the wing but does not contribute to wing rigidity.

$m - 2b\rho_W \cos \Lambda$: uniform mass distribution of the wing per unit length of its span

l: span length

$M_F - bm_F$: one half of fuselage mass

$M_T = (2M_F + 2ml)$: total mass of airplane

$R = M_F/ml$: ratio of fuselage mass to wing mass

b: semichord of the wing

ρ_W: distributed wing mass per unit area.

The coordinate axes are also shown in Fig. 5.13. x is the longitudinal coordinate positive opposite to the direction of motion of the wing and y an orthogonal coordinate with origin at the centre of gravity of the airplane. y' is a spanwise coordinate and

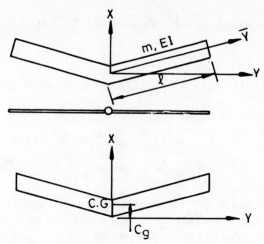

Fig. 5.13. Idealized model of an airplane and coordinate systems.

x' is orthogonal to it. The x, y and x', y' coordinates are related to each other by the following transformations.

$$\begin{Bmatrix} x' \\ y' \end{Bmatrix} = \begin{bmatrix} \cos \Lambda & -\sin \Lambda \\ \sin \Lambda & \cos \Lambda \end{bmatrix} \begin{Bmatrix} x \\ y \end{Bmatrix} \quad \text{for } y > 0$$

$$= \begin{bmatrix} \cos \Lambda & \sin \Lambda \\ -\sin \Lambda & \cos \Lambda \end{bmatrix} \begin{Bmatrix} x \\ y \end{Bmatrix} \quad \text{for } y < 0 \qquad (5.80)$$

The following three degrees of freedom are considered:

plunging: $u_1'(x, y) = 1$

pitching: $u_2'(x, y) = -x$

fundamental wing bending:

$$u_3'(x, y) = A_1 e^{-\Omega_3 y'/l} + A_2 \cos \Omega_3 y'/l + A_3 \sinh \Omega_3 y'/l + A_4 \sin \Omega_3 y'/l \quad (5.81)$$

where Ω_3 the nondimensional frequency parameter which is related to the wing natural frequency in bending ω_3 by

$$\omega_3 = (\Omega_3/l)(EI/M)^{1/2} \qquad (5.82)$$

The plunging and pitching modes being rigid body modes, the natural frequencies corresponding to these modes are $\omega_1 = \omega_2 = 0$. They are normalized according to

$$u_i(x, y) = N_i u_i'(x, y) \qquad (5.83)$$

such that

$$\iint_s N_1^2 u_1'^2(x, y) \rho_s(x, y) dx \, dy = M_T \qquad (5.84)$$

where ρ_s is the mass per unit area of the airplane including wing and fuselage.

For convenience of calculation, shift the origin of the coordinates to the front tip of the wing. The distance c_g of the centre of gravity from the tip is

$$c_g = b + [l \sin \Lambda/2(R + 1)] \tag{5.85}$$

The modal coefficients Δ_{jr} of (5.63) are

$$\Delta_{jr} = N_j N_r \iint_s u_j'(x - c_g, y) u_r'(x - c_g, y) \, dx \, dy \tag{5.86}$$

Substitute (5.86) in (5.65), the Laplace transformations of the elements of the inverse matrix can be shown to be the ratio of two polynomial functions of the form

$$C_{jr}^{-1}(p) = \sum_{k=1}^{K_{jr}} D_{jr,k} \, p^{k-1} / p \sum_{n=1}^{11} D_n p^{n-1} \tag{5.87}$$

where $K_{jr} = \begin{cases} 10 & j = r \\ 9 & j \neq r \end{cases}$ and $D_{jr,1} = 0$ with either $j = 3$ or $r = 3$ or both. The

denominator in (5.87) is the determinant of matrix C and the numerator represents the cofactor of element C_{jr}. The impulse response functions of (5.68) are of the general form

$$h_{jr}(\tau) = \sum_{l=1}^{L_{jr}} E_{jr,l} \, \exp(\gamma_l \tau) \tag{5.88}$$

where $\gamma_1, \gamma_2, \ldots \gamma_{11}$, are the characteristic roots of the matrix $\mathbf{C}$.

Letting $\gamma_{11} = 0$,

$$E_{jr,l} = \sum_{k=1}^{K_{jr}} D_{jr,k} \, \gamma_l^{k-1} \bigg/ \left[\frac{d}{dp} \left(p \sum_{n=1}^{11} D_n p^{n-1} \right) \right]_{p=\gamma_l} \tag{5.89}$$

$L_{jr} = 10$ when either $j = 3$ or $r = 3$ or both, and for any other combination $L_{jr} = 11$. The roots $\gamma_1, \gamma_2 \ldots \gamma_{11}$ are either real or in complex conjugate pairs. If γ_l and γ_{l+1} are a pair of conjugate roots, the corresponding coefficients $E_{jr,l}$ and $E_{jr,l+1}$ are a complex conjugate pair so that the sum in (5.88) is real.

Substitute the derivative of the Küssner's function

$$\dot{\psi}(\tau) = (U/2b)\{0.13 \exp[-0.13(U/b)\tau] + \exp[-U/b\tau]\} \tag{5.90}$$

in (5.75) and integrate. The resulting expression is

$$B_{jr,\alpha}(t, \omega) = \frac{U}{2b} \sum_{l=1}^{L_{jr}} E_{jr,l} \, \{I_1 e^{-(0.13U/b+i\omega)t} + I_2 e^{-(U/b+i\omega)t}$$

$$+ (I_3 + I_4 - I_1 - I_2) e^{(-i\omega+\gamma_l)t} - (I_3 + I_4) e^{-\alpha Ut}\} \tag{5.91}$$

where the I_1s are functions of U/b, ω, αU, γ_l given by

$$I_1(U/b, \omega, \alpha U, \gamma_l) = 0.13 \, \{0.13U/b + i\omega - \alpha U] \, [0.13U/b + \gamma_l]\}^{-1}$$

$$I_2(U/b, \omega, \alpha U, \gamma_l) = \{(U/b + i\omega - \alpha U) \, (U/b + \gamma_l)\}^{-1}$$

$$I_3(U/b, \omega, \alpha U, \gamma_i) = 0.13 \{[0.13U/b + i\omega - \alpha U] [- i\omega + \alpha U + \gamma_i]\}^{-1}$$

$$I_4(U/b, \omega, \alpha U, \gamma_i) = \{(U/b + i\omega - \alpha U)(- i\omega + \alpha U + \gamma_i)\}^{-1} \qquad (5.92)$$

The computation of the functions $J_{rs,\alpha\beta}(\omega)$ involves the modified Bessel functions K_1 and K_2 which occur in (5.71). It can be seen from the psd expression that as the scale of turbulence L increases indefinitely or decrease to zero, that is, $L \to \infty$ or $L \to 0$ $\Phi_{GG}(\omega, \xi, \eta) \to 0$. This implies as in the analysis of response to stationary gust, that a forcing field whose correlation distances are very large compared with the length of the airplane causes little disturbance to the airplane. On the other hand when the correlation distances are negligibly small the minute uncorrelated disturbance on a large plane tend to cancel each other. In most practical applications one is interested in cases where L is finite but $| \eta/L | \leq 1$. From the properties of the modified Bessel functions as $z \to 0$, $zK_1(z) \to 1$ and $z^2 K_2(z) \to 2$ and for negligibly small $| \eta/L |$ one can approximate

$$\Phi_{GG}(\omega, \xi, \eta) \cong \Phi_{GG}(\omega)e^{-i\omega\xi/U} \qquad (5.93)$$

where $\Phi_{GG}(\omega)$ is the Dryden spectrum given in (5.1) and

$$J_{rs,\alpha\beta}(\omega) = \Phi_{GG}(\omega) J_{r,\alpha} J_{s,\beta}^* \qquad (5.94)$$

Substituting (5.93) and (5.94) into (5.78) and (5.79) and using (5.77) the airplane response statistics can be evaluated.

The response relations are applied to the case of an airplane having following data: $l = 42$ ft, $b = 7$ ft, $\Lambda = 30°$, $U = 586.6$ ft/sec, $EI = 1.2 \times 10^9$ lb/ft^2, $\rho_W = 2.5$ slugs/ft^2, $R = 3$, $\Omega_3 = 1.95$, $L = 750$ ft, $\alpha = 0.1 \times 10^{-3}$ ft^{-1}, $\beta = 0.4 \times 10^{-3}$ ft^{-1}, $\omega = 13.63$ rad/sec.

The results of the numerical computations are shown in Figs. 5.14 and 5.15. The displacement evolutionary spectra corresponding to the plunging and pitching modes resemble each other. The evolutionary spectral density of the displacement corresponding to the wing fundamental bending mode exhibits two dominant peaks one around 1 rad/sec and another at the fundamental natural frequency of wing bending. The peak at the lower frequency corresponds to the high energy content at lower frequencies of the gust energy spectrum.

We have presented in the preceding sections, the response analysis of an aircraft to atmospheric turbulence. For the sdf system plunging mode response of the airplane simple relations between the response and gust statistics are presented both using quasi-steady state and unsteady aerodynamic theories and both for stationary and evolutionary nonstationary gusts. The last example considered is a rather complicated one including the wing bending flexibility. Even for a three degree of freedom model, the response evaluation is involved. We have considered in all the examples the Dryden spectrum for the stationary part of gust velocity mainly because of its simplicity. The inclusion of other form of spectral densities for the gust velocity like the von Karman spectrum does not in any way alter he analysis procedures, except that some of the integrations have to be performed numerically.

5.4 Structure of Boundary Layer Pressure Fluctuations

The flow of a turbulent boundary layer over the external surfaces of aerospace vehicles is important both in terms of the structural response and the sound radiated

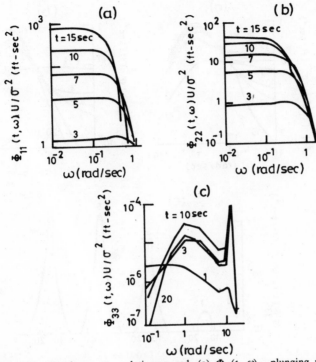

Fig. 5.14. Displacement evolutionary psd. (a) $\Phi_{11}(t, \omega)$—plunging mode; (b) $\Phi_{22}(t, \omega)$—pitching mode; (c) $\Phi_{33}(t, \omega)$—wing fundamental bending mode (Fujimori and Lin, 1973a).

on either side of the structure. If the structure is rigid, the turbulent pressure fluctuations radiate sound directly into the adjacent air. When the structure is flexible and vibrates as a result of the boundary layer excitation, it acts as an efficient sound radiator and gives rise to an additional sound field in the fluid which may have a much greater intensity than that associated with the flow over rigid surface. Further as the boundary layer flow over the surface lasts throughout the period of flight, the vibration induced by the turbulent pressure fluctuations cause fatigue stresses. Thus acoustic fatigue is one of the important consideration in the design of high speed aircrafts excited by turbulent flow. The calculation of the structural response and radiated sound fields requires a knowledge of the character and statistical properties of the wall pressure fluctuations induced by the turbulent boundary layer.

A turbulent boundary layer can be considered as being a collection of randomly sized eddies or vortices which are formed taking energy from the shear in the boundary layer between the vehicle and the surrounding air and the turbulent flow past the vehicle's skin. Kraichnan (1956) presented a theoretical analysis of the nature of the pressure fluctuations in a turbulent flow over a flat plate. He attributed the pressure fluctuations as being caused by the interaction of turbulence with the mean shear at the wall. Because the skin is almost in direct contact with the convected eddies which characterize the turbulent boundary layer, the pressure fluctuations at the skin are actually the dynamic pressures due to the velocity fluctuations normal to the skin. Hence pressure fluctuations in Kraichanan's analysis

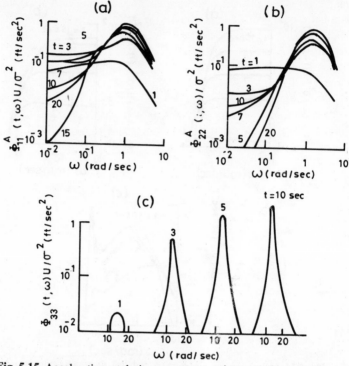

Fig. 5.15. Acceleration evolutionary psd. (a) $\Phi_{11}^{A}(t, \omega)$—plunging mode; (b) $\Phi_{22}^{A}(t, \omega)$—pitching mode; (c) $\Phi_{33}^{A}(t, \omega)$—wing fundamental bending mode (Fujimori and Lin, 1973a).

were expressed in terms of the velocity fluctuations which were functions of the mean shear stress at the wall. Kraichanan estimated the root mean square pressure p_{rms} to be $6\tau_{w}$, where τ_{w} is the mean shear stress at the wall. Measurements under different experimental conditions varying from laboratory to flight tests made by many investigators, Willmarth and Woolridge (1962), Lilley (1960), Bull (1963) indicate the consistent set of results for the ratio p_{rms}/τ_{w} in the range of 2.5 to 3. Alternatively, p_{rms} is expressed in terms of the free stream dynamic pressure q. The ratio p_{rms}/q shows a range of variation from 1.5×10^{-3} to 12×10^{-3}, but a more generally accepted result for incompressible flow is $p_{rms} \cong 0.006\, q$.

While the ratio p_{rms}/τ_{w} is more or less constant for different speeds within the subsonic range, experimental results for supersonic flow indicate a gradual increase in its value with Mach number with values of 3 to 4.5 in the range $1.2 < M < 1.6$ and even values of 5 to 6 for $M = 5$ (Kistler and Chen, 1962). Variations of p_{rms}/τ_{w} with changes in Reynolds's number had also been reported in the subsonic range while in the supersonic range it remained more or less constant with respect to changes in Reynold's number.

Spectral Density and Correlations of the Pressure Field

For estimating the structural response to the boundary layer excitation, the nature of the wall pressure fluctuations on the structure should be ascertained. Measurements on wind tunnel models have shown that the pressure fluctuations constitute almost a stationary and homogeneous random process. The psd of the fluctuating pressures in boundary layer is usually obtained through wind tunnel scale model tests over the Mach number range of interest. Measurement of cross-spectra of a turbulent pressure field usually show some decay in the statistical correlation in addition to convection at a characteristic velocity. The computation of spectral response to such a random excitation is more tedious than would be the case when the pressure field is convected without decay. In this case, it can be considered to be convected as a frozen pattern and the Taylor's hypothesis can be used. The relevant acceptance functions representing the coupling between the pressure field and the structural motion corresponding to the case of convection with decay have to be in most cases evaluated by numerical integration.

Dyer (1959) approximated the cross-correlation function of the boundary layer pressure field over a flat plate by

$$R_{pp}(s_1^1, s_2^1, t_1; s_1^2, s_2^2, t_2) = E[p(s_1^1, s_2^1, t_1) \, p(s_1^2, s_2^2, t_2)]$$

$$= p_{rms}^2 \, e^{-|\tau|/\vartheta} \, e^{-\alpha[(\xi - U_c \tau)^2 + \eta^2]^{1/2}} \tag{5.95}$$

where $\xi = s_1^1 - s_1^2$, $\eta = s_2^1 - s_2^2$ are the spatial separations, $\tau = t_1 - t_2$ is the time lag, ϑ is the life time of the turbulent eddies and $2/\alpha$ is the correlation distance in the radial direction and U_c is the mean convection speed of the turbulent flow. Equation (5.95) implies that the boundary layer pressure field is a weakly stationary and homogeneous random process convecting along the direction of flow. The pressure correlation decays exponentially with time lag. With further assumption that the correlation distances are small compared with the structural dimensions, the cross correlation function can be replaced by delta functions of the form

$$R_{pp}(\xi, \eta, \tau) = A p_{rms}^2 \, \delta(\xi - U_c \tau) \, \delta(\eta) e^{-|\tau|/\vartheta} \tag{5.96}$$

where $A = 2\pi/\alpha^2$ is the correlation area and $\delta(\cdot)$ is the Dirac delta function. The convection velocity U_c is assumed to be about 0.8 times the free stream velocity $U_c \cong 0.8 U_\infty$.

The delta function correlation of (5.96) has been used by many early investigators (Ribner, 1956; Dyer, 1959; El Baroudi, 1964; and Lin, 1967) in estimating the random response of panels and the associated sound radiation characteristics. The delta function correlation is analytically simple but not very realistic in representing the actual pressure field.

Experimentally measured cross spectral densities of pressure fluctuations in a turbulent boundary layer with respect to a fixed frame of reference have the general form (Corcos, 1963)

$$\Phi_{pp}(\xi, \eta, \omega) = \Phi_{pp}(\omega) e^{-(i\omega\xi/U_c)} R_{pp}(\xi, 0, \omega) R_{pp}(0, \eta, \omega) \tag{5.97}$$

where Φ_{pp} is the wall pressure psd, $R_{pp}(\xi, 0, \omega)$, and $R_{pp}(0, \eta, \omega)$ are longitudinal and lateral correlations respectively which are usually approximated by negative exponential functions of the longitudinal and lateral Strouhal numbers $\omega\xi/U_c$ and $\omega\eta/U_c$. The form of the separable correlation function of (5.97) is generally attributed to Corcos (1963) and provides accurate empirical fits for experimental data. The usual form considered for the normalized correlation function are

$$R_{pp}(\xi, 0, \omega) = e^{-\alpha_{s1}|\omega\xi/U_c|}, R_{pp}(0, \eta, \omega) = e^{-\alpha_{s2}|\omega\eta/U_c|} \tag{5.98}$$

where α_{s_1} and α_{s_2} are empirical constants that fit the experimental data. Implicit in the form of (5.97) is that the turbulence field is not a frozen one as the spatial correlation functions decay exponentially. The pressure field reduces to a plane wave field if $R_{pp}(0, \eta, \omega) = 1$ and to a frozen turbulence field if $R_{pp}(\xi, 0, \omega) = 1$.

Bull, Wilby and Blackman (1963), Wilby (1967) and Bull (1967) proposed the following approximation:

$$\Phi_{pp}(\xi, \eta, \omega) = \Phi_{pp}(\omega) \exp(-0.1\omega \mid \xi \mid /U_c)$$
$$\exp(-0.75\ \omega \mid \eta \mid /U_c) \cos(\omega\xi/U_c) \tag{5.99}$$

to fit their experimental data. The form of the separable spatial correlations in the longitudinal and lateral directions with slight variations in the empirical constants has been used by Crocker (1969), Strawdermann and Brand (1969), Strawdermann (1969), Davies (1969), Bolotin and Elishakoff (1971), Elishakoff and Khromatoff (1971), Chyu and Au Yang (1972) and Elishakoff (1975), in the response analysis of structures to boundary layer turbulence and the attended sound radiation. Another assumption implied in the above formulation is that the convection velocity U_c is constant. But in actual conditions U_c is a slowly varying function of frequency. Another drawback regarding the correlation function represented by (5.97) and (5.98) is that they represent a nondifferentiable random process. Willmarth and Roos (1965) proposed for the longitudinal and lateral correlation an analytic approximation corresponding to a differentiable random process, which does not differ materially from Corcos' approximation. The cross spectral density function in this case is of the form.

$$\Phi_{pp}(\xi, \eta, \omega) = \Phi_{pp}(\omega) \exp\left(-\frac{i\omega\xi}{U_c}\right)\left\{\exp\left(-0.1455 \left| \frac{\omega\xi}{U_c} \right|\right)\right.$$

$$\left. + 0.1455 \left| \frac{\omega\xi}{U_c} \right| \exp\left(-2.5 \left| \frac{\omega\xi}{U_c} \right|\right)\right\}$$

$$\times \left\{\exp\left(-0.092 \left| \frac{\omega\eta}{U_c} \right|\right) + 0.7 \exp\left(-0.789 \left| \frac{\omega\eta}{U_c} \right|\right)\right.$$

$$+ 0.145 \exp\left(-2.916 \left| \frac{\omega\eta}{U_c} \right|\right)$$

$$\left. + 0.99 \left| \frac{\omega\eta}{U_c} \right| \exp\left(-4 \left| \frac{\omega\eta}{U_c} \right|\right)\right) \tag{5.100}$$

Based on extensive measurements of the statistical parameters of the wall pressure fluctuations and velocity fluctuations in a boundary layer flow Maestrello (1965a, 1965b, 1967) in a series of papers has proposed two models for the nature of the wall pressure fluctuations that reflect the essential characteristics of the boundary layer turbulence. In both the models the pressure field is assumed to be stationary and convecting along the direction of free stream. The temporal and spatial dependence of the correlation are of a form that retains the characteristics of the superposition of a wave system with phase and amplitude associated with wave number and frequency spectrum. In the first model, the convection velocity is assumed to be constant implying a semi frozen turbulence pattern. In the other model consideration is given to the unsteady nature of the convection velocity with temporal fluctuations based on the experimental observations of Fisher and Davies (1964). The normalized cross-correlation function of the wall pressure fluctuations in the two models are given by

$$R_{pp}(\xi, \eta, \omega) = \frac{e^{-|\xi|/U_c\vartheta}}{\sum\limits_{n=1}^{3} A_n/K_n} \left\{ \sum_{n=1}^{3} \frac{A_n K_n}{K_n^2 + (1/FU_c)^2[(\xi - U_c\tau)^2 + \eta^2]} \right\} \quad (5.101)$$
(semi frozen pattern)

$$R_{pp}(\xi, \eta, \tau) = \{A_1(e^{-|\tau|/\vartheta_1} + a_1(e^{-|\tau|/\vartheta_1'} \sin \omega_1 |\tau|)\} \{e^{-[(\xi - U_c\tau)^2 + \eta^2]/2\sigma_1^2}\}$$
(unsteady convection)

$$+ \{A_2(e^{-|\tau|/\vartheta_2} + a_2 (e^{-|\tau|/\vartheta_2'} \cos \omega_2 |\tau|)\} \{e^{-[(\xi - U_c\tau)^2 + \eta^2]/2\sigma_2^2}\}$$
$$(5.102)$$

where K_ns and A_ns are constants fitting the measured data and $F = \delta^*/U$, where δ^* is the boundary layer displacement thickness and U_∞ is the free stream velocity. The other notations correspond to the same as in (5.95). The temporal fluctuation in the convection velocity for the unsteady convection model are expressed in terms of oscillatory decaying functions as in (5.102) with $\vartheta_1', \vartheta_2'$; and a_1, a_2; and ω_1, ω_2 representing respectively the decays, amplitudes and frequencies that may be caused by distortions of the components of the eddies as they interact with the mean shear.

An equivalent representation of the turbulent pressure field in terms of wave number and frequency spectra can be obtained by taking the triple Fourier transform of the cross-correlation function

$$\Phi_{pp}(k_1, k_2, \omega) = \frac{1}{(2\pi)^3} \int\!\!\!\int\!\!\!\int\limits_{-\infty}^{+\infty} R_{pp}(\xi, \eta, \tau) e^{-i(k_1\xi + k_2\eta + \omega\tau)} \, d\xi \, d\eta \, d\tau \quad (5.103)$$

where k_1 and k_2 are the wave numbers. For example, the normalized cross spectral density function corresponding to (5.101) for zero spatial separation along s_2 direction $\eta = 0$ is

$$\Phi_{pp}(k_1, \omega) = \frac{\vartheta \left[\dfrac{FU_c}{2\pi} \right] \sum\limits_{n=1}^{3} A_n e^{-|\omega| K_n F}}{\left(\sum\limits_{n=1}^{3} A_n/K_n \right) [1 + \vartheta^2(\omega + k_1 U_c)^2]} \quad (5.104)$$

The excellent fit afforded by the model represented by (5.101) and (5.104) can be seen from Fig. 5.16, which shows the normalized power spectrum of the wall pressure fluctuations for various ranges of free stream velocities, Reynold's number and wall shear stress. Measured normalized power spectra for supersonic turbulent boundary layer flow after applying corrections for attenuation due to the finite size of transducer used is shown in Fig. 5.17. The measured spectra by Bull (1967), Kistler and Chen (1963) and Willmarth and Woolridge (1963) are also shown in the figure. It is seen from the figure that the cross spectral density has a peak corresponding to $\omega\delta^*/U_\infty \cong 2$. Maestrello (1969) proposed another model for the boundary layer pressure fluctuation in supersonic flow which is similar in form

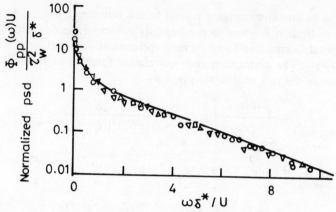

Fig. 5.16. Comparison of dimensionless psd of boundary layer wall pressure fluctuations with experimental data (Maestrello, 1967).

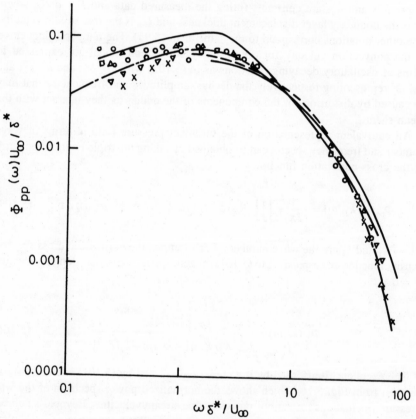

Fig. 5.17. Comparisons of dimensionless psds of boundary layer wall pressure fluctuations in supersonic flow of different models with experimental data. 1—Bull (1967), 2—Kistler and Chen (1963), 3—Willmarth and Woolridge (1963), 4—Maestrello (1969).

represented by (5.101) and (5.104), except that it takes into consideration the appearance of the peak in the power spectrum of a boundary layer developed over a flat smooth plate at zero pressure gradient. The normalized cross spectral density function in this model is of the form

$$\Phi_{pp}(\xi, \eta, \omega) = \frac{1}{2} \frac{\delta^*}{U_\infty} \{\sum_{n=1}^{4} A_n \, e^{-K_n(|\omega|\delta^*/U_\infty)}\} \, e^{-|\xi|/\alpha_1\delta^*} \, e^{-|\eta|/\alpha_2\delta^*} \qquad (5.105)$$

The values of the constants that fit the experimental data are $A_1 = 4.4 \times 10^{-2}$, $A_2 = 7.5 \times 10^{-2}$, $A_3 = 9.3 \times 10^{-1}$, $A_4 = 2.5 \times 10^{-2}$, $K_1 = 5.78 \times 10^{-2}$, $K_2 = 2.43 \times 10^{-1}$, $K_3 = 1.12$, $K_4 = 11.57$, $\alpha_2 = 0.26$ and $\alpha_1 = 3$. The value of the convection velocity ranges from $U_c = 0.78 \, U_\infty$ at $M = 1.98$ to $U_c = 0.72 \, U_\infty$ at $M = 3.03$. The fit obtained by this model is also shown in Fig. 5.17, which shows very good agreement with experimental results especially in the supersonic flow regime. A similar form of the cross spectral density function with variations in the constants has been adopted by Maekawa and Lin (1977) in subsonic flow while estimating the response statistics of a periodic structure to boundary layer turbulence.

Note that (5.105) is very similar to the separable form of (5.97). As we have observed, the models proposed by Maestrello show good agreement between theoretical and experimental results especially in light fluid media (air). These models also facilitate the incorporation of realistic fluid and structural data, greater generality in analysis and resultant vibratory response. It has been demonstrated by Leibowitz (1971, 1972, 1975) that the Maestrello's models can be modified and extended further, emphasizing the practicality of using them for the prediction of the turbulence induced vibratory response of a panel in both light and heavy fluids.

5.5   Structural Response to Boundary Layer Pressure Fluctuations

The response analysis of aerospace structural systems to boundary layer turbulence and the associated problem of sound radiation are quite complicated, if the structure were to be treated in its entirety and the complete sound-structure interaction effects were to be included in the analysis. Hence most of the theoretical investigations in this respect have been concentrated mainly on simple structural configurations like rectangular panels and membranes or sheet-stiffener arrays. Even the experimental studies of Maestrello (1967, 1969), Dowell (1969), El Baroudi (1964) and Bies (1966) have been confined mainly to simple model structures in the laboratory which have been used to compare with the theoretical results. However, Wilby and Gloyna (1972) have presented extensive results based on experimental investigations of response measurements of a representative aircraft fuselage to boundary layer turbulence.

Real structural configurations tend to be more complex and it is somewhat difficult to extrapolate the results of simplified models to practical structural systems. However, such results provide valuable information regarding the nature of the problem and the effects of the interaction of structural motion with the external random pressure field and also furnish pertinent design data. Moreover, with the rapid advances in the finite element method in which complex structural systems are considered as an assemblage of simple structural elements, the methods of the random response analysis of panel like structures can be used to advantage in conjunction with the overall finite element method.

5.5.1   Response of Rectangular Panels to Turbulent Boundary Layer Excitation

Subsonic Flow
Consider the rectangular panel of uniform thickness h with sides of length a and

b held in a rigid baffle and coordinate axes as shown in Fig. 5.18. The upper side of the panel is exposed to a convected fluid moving in the positive s_1 direction. The displacement X of the panel, the turbulent pressure p in the boundary layer are taken positive in the downward direction.

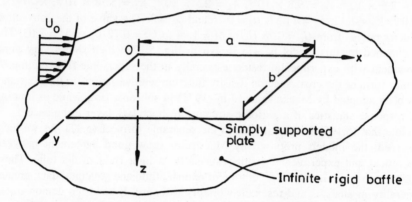

Fig. 5.18. Panel held in rigid baffle subjected to boundary layer pressure fluctuation.

The excitation force on the panel arises both from the pressure fluctuations in the boundary layer and also from the unsteady pressure distribution arising from the vibration of the panel in air. The contribution to the exciting force due to panel motion can be expressed as a complicated integral expression of X, which when substituted into the equation of motion of the panel, leads to an integro-differential equation for the response and sound radiation problem (Maestrello, Yen, Padula, 1975). To make the problem analytically tractable, the sound-structure interaction effects can be accounted in terms of a virtual mass and an acoustic damping coefficient. For panels vibrating in air, the virtual mass effect can be neglected. For a rigidly oscillating infinite panel the acoustic damping coefficients has the value $2\rho c$, where ρ is the equilibrium air density and c is the speed of sound in air. This value needs some modifications for finite panels (Liebowitz, 1975). For a flow being convected at subsonic velocities, the radiation loading is not of much significance. However, in the case of supersonic convection velocity, the acoustic damping has to be taken into consideration as then the supersonic convection velocity excites strong coupling modes to the acoustic field. We also assume that the panel motion does not modify the turbulent boundary layer. With these assumptions the equation of motion for the panel can be expressed as

$$D \nabla^4 X(s_1, s_2, t) + \beta \frac{\partial X}{\partial t} (s_1, s_2, t) + \mu \frac{\partial^2 X}{\partial t^2} (s_1, s_2, t) = p(s_1, s_2, t) \quad (5.106)$$

where $D = Eh^3/12(1 - v^2)$ is the effective plate flexural rigidity, β is the panel damping per unit area including acoustic damping, μ is the plate mass per unit area, v the poisson's ratio and $p(s_1, s_2, t)$ is the random excitation corresponding to the turbulent wall pressure fluctuations only.

Assume the cross spectral density function of the wall pressure fluctuation to be given by (Strawdermann, 1969)

$$\Phi_{pp}(\xi, \eta, \omega) = \Phi_{pp}(\omega) \exp(-0.115 \,|\omega\xi/U_c|)$$
$$\exp(-0.7 \,|\omega\eta/U_c|) \exp(-i\omega\xi/U_c) \tag{5.107}$$

where

$$\Phi_{pp}(\omega) = \begin{cases} A\alpha^2, & \omega \le 1.932U_c/\delta^* \\ 2A(\omega\delta^*/U_\infty)^{-3}\alpha^2, & \omega > 1.932U_c/\delta^* \end{cases} \tag{5.108}$$

and $A = 0.75 \times 10^{-5}\rho^2 U_\infty^3\, \delta^*$ and α is a constant which for air can be taken as 3.

Normal Mode Method

The normal mode method can be used to estimate the response statistics. The cross spectral density of the displacement response corresponding to two locations using stationary random vibration analysis is,

$$\Phi_{XX}(s_1^1, s_2^1; s_1^2, s_2^2; \omega) = \sum_m \sum_n \sum_r \sum_s u_{mn}(s_1^1, s_2^1)\, u_{rs}(s_1^2, s_2^2)\, H_{mn}(\omega)\, H_{rs}^*(\omega)\, j_{mnrs}^2(\omega) \tag{5.109}$$

where

$$j_{mnrs}^2(\omega) = \int_0^a \int_0^b \int_0^a \int_0^b u_{mn}(s_1^1, s_2^1)\, u_{rs}(s_1^2, s_2^2)\, \Phi_{pp}(s_1^1, s_2^1, s_1^2, s_2^2, \omega)\, ds_2^2\, ds_1^2\, ds_2^1\, ds_1^1 \tag{5.110}$$

is similar to the cross-acceptance functions defined by Powell (1959), $u_{mn}(s_1, s_2)$ is the mnth normal mode and

$$H_{mn}(\omega) = (M_{mn}(\omega_{mn}^2 - \omega^2 + 2i\zeta_{mn}\omega_{mn}\omega))^{-1} \tag{5.111}$$

is the complex frequency response function corresponding to the mnth mode and M_{mn}, ω_{mn} and ζ_{mn} are respectively the generalized mass, natural frequency and damping ratio corresponding to the mnth natural mode given by

$$M_{mn} = \int_0^a \int_0^b \mu\, u_{mn}^2(s_1, s_2)\, ds_1\, ds_2$$

$$\omega_{mn}^2 = \frac{1}{M_{mn}} \int_0^a \int_0^b D[\nabla^4 u_{mn}(s_1, s_2)] u_{mn}(s_1, s_2)\, ds_1\, ds_2$$

and

$$\zeta_{mn} = \frac{1}{2M_{mn}\omega_{mn}} \int_0^a \int_0^b \beta\, u_{mn}^2(s_1, s_2)\, ds_1\, ds_2 = \frac{\beta}{2\mu\omega_{mn}} \tag{5.112}$$

Example 5.5: Response of a rectangular panel (Strawdermann, 1969)
Consider a steel plate with $a = 3.5$ in., $b = 3.5$ in. and $h = 0.01$ in. The edges of the panel are assumed to be simply supported. For the simply supported panel,

$$u_{mn}(s_1, s_2) = \sin\left(\frac{m\pi s_1}{a}\right) \sin\left(\frac{n\pi s_2}{b}\right)$$

$$M_{mn} = \mu ab/4, \quad \omega_{mn}^2 = \frac{D}{\mu}\left[\left(\frac{m\pi}{a}\right)^2 + \left(\frac{n\pi}{b}\right)^2\right]^2 \tag{5.113}$$

$$2\xi_{mn}\,\omega_{mn} = \beta/\mu$$

Substitute (5.113) in (5.110) with $\Phi_{pp}(s_1^1, s_2^1; s_1^2, s_2^2; \omega)$ from (5.107) and integrate to obtain

$$j_{mnrs}^2(\omega) = \Phi_{pp}(\omega)\, I_{mr}(\gamma_{s_1}, \delta_{s_1})\, I_{ns}(\delta_{s_2}) \tag{5.114}$$

where

$$I_{mr}(\gamma_{s_1}, \delta_{s_1}) = mr^2\pi\, \frac{\{1 - (-1)^m\, e^{-(\delta_{s_1}+i\gamma_{s_1})a} - (-1)^r\, e^{-(\delta_{s_1}-i\gamma_{s_1})a} + (-1)^{m+r}\}}{a^2\left[(i\gamma_{s_1} - \delta_{s_1})^2 + \left(\dfrac{r\pi}{a}\right)^2\right]\left[(\delta_{s_1} + i\gamma_{s_1})^2 + \left(\dfrac{m\pi}{a}\right)^2\right]}$$

and

$$I_{ns}(\delta_{s_2}) = -\,\frac{ns\pi^2\{1 - e^{-\delta}\, s_2^b\,[(-1)^n + (-1)^s] + (-1)^{s+n}\}}{b^2\left[\delta_{s_2}^2 + \left(\dfrac{n\pi}{a}\right)^2\right]\left[\delta_{s_2}^2 + \left(\dfrac{s\pi}{b}\right)^2\right]} \tag{5.115}$$

with $\delta_{s_1} = 0.115\,|\omega/U_c|$ and $\gamma_{s_1} = |\omega/U_c|$ and $\delta_{s_2} = 0.7\,|\omega/U_c|$.
The other parameters used in the response analysis are $U = 539$ ft/sec and $\delta^* = 0.172$ in.

Using the above data the cross spectral density of the displacement response can be computed from (5.109). Figure 5.19 shows the psd of the displacement by setting $s_1^1 = s_1^2 = s_1$ and $s_2^1 = s_2^2 = s_2$. In computing the response the modal damping

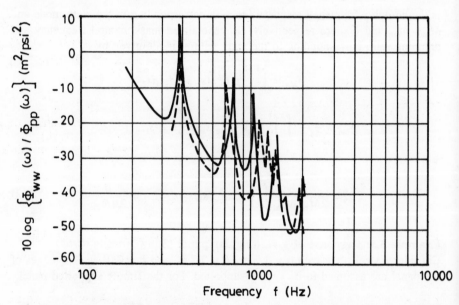

Fig. 5.19. Ratio of displacement and pressure psds. $U_0 = 539$ (ft/sec), $\delta^* = 0.172$ in.; ——— computed (Strawdermen, 1969); - - - - experimental (Bull et al, 1963).

is assumed constant for all the modes, which is 0.5% of critical damping. This value of damping corresponds to the measured value of damping (Bull 1963) for the second natural mode of the panel. In Fig. 5.19 is also shown the experimental results due to Bull et al, (1963). While comparing the two results it should be noted that the experimental reults correspond to clamped edge conditions. Hence the computed displacement spectral density differs somewhat from the experimental results especially at low frequencies. However, at the higher panel modes, the spectral levels are of the same order of magnitude, even though the natural frequencies differ because of the different types of boundary conditions in the experimental and theoretical investigations.

Supersonic Flow
The previous example considered a simple idealized panel simply supported in all its edges to boundary layer turbulence corresponding to subsonic flow. Crocker (1969) used the normal mode method in predicting the response of practical structures representative of fuselage structures of a large supersonic transport aircraft. Even though the final models considered reduce the structure to equivalent flat panels, the dimensions of such panels relate to the actual structure. The effect of the mass and stiffness distributions due to the presence of stringers and frames are taken into consideration in the modelling of the equivalent flat panel.

Three views of a typical supersonic transport aircraft considered in the analysis is shown in Fig. 5.20. The modal pattern of vibration of the entire fuselage will be extremely complex and hence only a part of the fuselage is considered. In the

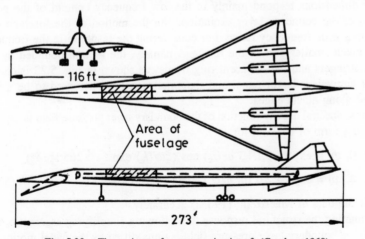

Fig. 5.20.   Three views of a supersonic aircraft (Crocker, 1969).

fuselage there are several vibrational discontinuities at the ring frames, the cabin floor-fuselage joint etc., which act as nodal points. The part of the fuselage between two heavy ring frames as in Fig. 5.21 is chosen for the analysis as it is of fairly uniform construction. This part of fuselage consists of titanium skin of 0.045 in thick and is unbroken by access doors, although there are discontinuities due to the presence of windows. The skin is reinforced by light titanium stringers as shown in Fig. 5.22. The covered fuselage panel is idealized to a dynamically similar flat plate by allowing for an increase in mass and stiffness, over that of the skin alone,

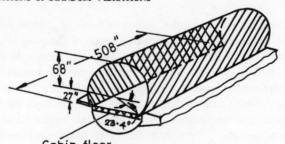

Fig. 5.21. Large fuselage section (Crocker, 1969).

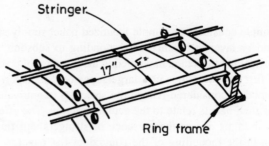

Fig. 5.22. Typical small panel bay of fuselage section (Crocker, 1969).

caused by the presence of stringers and ring frames. The idealized structure with its large dimensions respond mainly to the low frequency content of the power spectrum of the boundary layer excitation. But the motion of the fuselage is of interest in a wide frequency range. For considering the response in the frequency range a much smaller part of the fuselage, namely, the section bounded by two adjacent stringers and two adjacent ring frames as shown in Fig. 5.22 is chosen. Both the equivalent larger panel and the small panel are assumed to be simply supported along all its edges.

The cross spectral density function of the boundary layer pressure field is assumed to be of the form of (5.97) with

$$R_{pp}(\xi, 0, \omega) = \exp(-0.1 |\xi| \omega/U_c) \cos(\xi\omega/U_c) \exp(-0.265 |\xi|/\delta^*)$$

$$R_{pp}(0, \eta, \omega) = \exp(-2.0 |\eta| \omega/U_c) \exp(-2.0 |\eta|/\delta^*) \qquad (5.116)$$

which fit the data of Maestrello (1965a) accurately.

In estimating the structural response the following simplifying assumptions are made: (i) the boundary layer pressure field is unmodified by the panel motion and (ii) the modal damping factor ζ_{mn} in all the modes is a constant. Thus the radiation loading on the panel is not explicitly taken into account. The equation of motion for the panel is the same as (5.107). Apart from these assumptions the cross-modal coupling terms in (5.109) have been neglected. The spectral density of the displacement response corresponding to a location (s_1, s_2) on the panel is of the form

$$\Phi_{xx} = \frac{\Phi_{xx}(s_1, s_2, \omega)}{\Phi_{pp}(\omega)} = \sum_m \sum_n f_{mn}^2(s_1, s_2) |H_{mn}(\omega)|^2 j_{mn}^2(\omega) \qquad (5.117)$$

where

$$j_{mn}^2(\omega) = \int_0^a \int_0^b \int_0^a \int_0^b R_{pp}(s_1^1, s_2^1; s_1^2, s_2^2; \omega) u_{mn}(s_1^1, s_2^1) u_{mn}(s_1^2, s_2^2) ds_1^1 ds_2^1 ds_1^2 ds_2^2$$

$$(5.118)$$

A slight modification of the expression for j_{mn}^2 from that given in (5.114) and (5.115) is required in view of the change in the normalized correlation function in (5.116) and the factoring out of $\Phi_{pp}(\omega)$. Inserting the normal modes (5.113), (5.118) can be integrated to yield,

$$j_{mn}^2(\omega) = \frac{4 I_{s_1}(\gamma_{s_1}, \delta_{s_1}) I_{s_2}(\delta_{s_2}) a^2 b^2}{m^2 n^2 \pi^4} \qquad (5.119)$$

where

$$I_{s_1}(\gamma_{s_1}, \delta_{s_1}) = \frac{1}{\Delta_{s_1}^2} \left\{ q_{s_1}[1 - (-1)^m \exp(-\delta_{s_1}) \cos \gamma_{s_1}] \right.$$

$$\left. + 4(-1)^m r_{s_1} \exp(-\delta_{s_1}) \sin \gamma_{s_1} + \frac{m\pi}{2} s_{s_1} \Delta_{s_1} \right\}$$

with

$$\Delta_{s_1} = [1 + (\delta_{s_1}/m\pi)^2 + (\gamma_{s_1}/m\pi)^2]^2 - 4(\gamma_{s_1}/m\pi)^2$$

$$q_{s_1} = [1 + \delta_{s_1}/m\pi]^2 - (\gamma_{s_1}/m\pi)^2 - 4(\gamma_{s_1}/m\pi)^2(\delta_{s_1}/m\pi)^2$$

$$r_{s_1} = (\delta_{s_1}/m\pi)(\gamma_{s_1}/m\pi)[1 + (\delta_{s_1}/m\pi)^2 - (\gamma_{s_1}/m\pi)^2]$$

$$s_{s_1} = (\delta_{s_1}/m\pi)[1 + (\delta_{s_1}/m\pi)^2 + (\gamma_{s_1}/m\pi)^2] \qquad (5.120)$$

$$I_{s_2}(\delta_{s_2}) = \frac{1}{\Delta_{s_2}^2} \left\{ q_{s_2}[1 - (-1)^n \exp(-\delta_{s_2})] + \frac{n\pi}{2} s_{s_2} \Delta_{s_2} \right\}$$

where Δ_{s_2}, q_{s_2}, r_{s_2} and s_{s_2} are obtained from Δ_{s_1}, q_{s_1}, r_{s_1} and s_{s_1} by replacing s_1 and s_2 and m by n and taking $\gamma_{s_2} = 0$. In (5.120) $\delta_{s_1} = 0.1\omega a/U_c + 0.265a/\delta^*$, $\delta_{s_2} = 2(\omega b/U_c + b/\delta^*)$ and $\gamma_{s_1} = \omega a/U_c$.

Example 5.6: Response of panels (Crocker, 1969)

The panel dimensions are (i) large panel: $a = 508$ in., $b = 242$ in., $h = 0.045$ in. (ii) small panel: $a = 17$ in., $b = 4$ in., $h = 0.045$ in. and the material is titanium. Crocker (1969) computed for the above panels the acceleration spectral density by multiplying (5.117) by ω^4. Figure 5.23 shows the acceleration response of the large fuselage panel for two convection speeds $U_c = 20600$ in./sec and $U_c = 6210$ in./sec and two values of the modal damping $\zeta_{mn} = 0.02$ and 0.1. About 400 modes are included in the response computation. It is seen from the figure that corresponding to the low frequency response of the large panel for a particular value of damping and for both the convection velocities, the response statistics are not very much

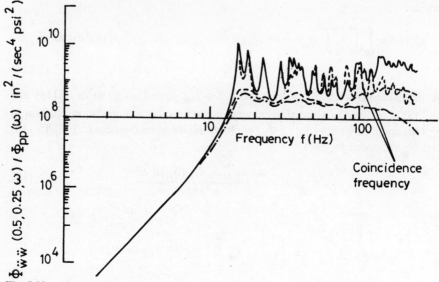

Fig. 5.23. Normalized acceleration psd of large fuselage section to boundary layer noise. at $s_1/a = 0.5$; $s_2/b = 0.25$. _____ $U_c = 20600$ in./sec ($M = 2.7$), $Q = 25$; - - - - $U_c = 6210$ in./sec ($M = 0.6$), $Q = 25$; – – – – $U_c = 20600$ in./sec ($M = 2.7$), $Q = 5$; —·—·— $U_c = 6210$ in./sec ($M = 0.6$), $Q = 5$ (Crocker, 1969).

different upto the frequency corresponding to the coincidence frequency of the panel. Beyond the coincidence frequency the response levels reduce significantly for the case of the flow with lower convection velocity. For the high frequency response of the smaller panel a similar trend is observed as shown in Fig. 5.24. In this case about 200 modes are included in the response computation. The coincidence speed is defined as the speed at which the component of the mean flow convection velocity U_c is equal to the speed of propagation of bending waves of the panel in the longitudinal direction. The frequency at which this velocity matching occurs is called the coincidence frequency (Ribner, 1956) and is a significant parameter as at this frequency large vibratory response of the panel can be expected due to resonance or matching in the wavenumber-frequency domain.

For the dimensions and the material of the large panel with the effect of the inertia and stiffness of the stringers taken into account the coincidence frequency corresponding to a convection speed of $U_c = 6210$ in./sec turns out to be 121 Hz and that corresponding to $U_c = 20600$ in./sec is 1350 Hz. The coincidence frequencies for the small panel bay turns out to be 2280 Hz and 27200 Hz corresponding to $U_c = 6210$ in./sec and $U_c = 20600$ in./sec respectively.

5.5.2 Response of Rectangular Panel Including Sound Structure Interaction Effects

In the previous section the response of a panel held in a rigid baffle to boundary layer excitation was presented in some detail under rather simplifying assumptions. The main assumptions which reduced the analysis procedure to a considerable extent are: (i) simply supported panel edge conditions, (ii) the noninclusion of the

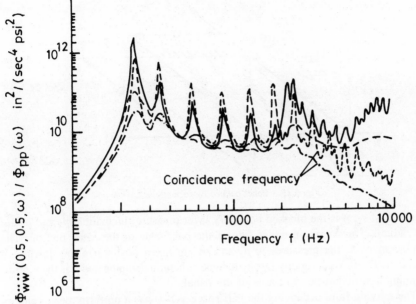

Fig. 5.24. Normalized acceleration psd of small panel bay to boundary layer excitation.
_____ U_c = 20600 in./sec (M = 2.7), Q = 25; - - - - U_c = 6210 in./sec (M = 0.6),
Q = 5; ———— U_c = 20600 in./sec (M = 2.7), Q = 25; —·—·— U_c = 6210 in./sec
(M = 0.6) Q = 5 (Crocker, 1969).

radiation loading in the panel associated with the vibratory interaction of the
structure and the surrounding fluid and the subsequent transmission of acoustical
energy through the fluid.

Davies (1971) considered the problem of structural response and sound radiation
of a panel excited by turbulent boundary layer by including the radiation pressure
loading due to the motion of the panel. The radiation loading on the panel, is
determined through a solution of the acoustic wave equation. Liebowitz (1975) has
summarized the different approaches by which the radiation loading can be
determined.

In this section we shall consider the response of a finite rectangular panel to
supersonically convected turbulence. The upper face of the panel is in contact with
a supersonically convected flow moving in the positive s_1 direction. The lower face
of the panel is in contact with the fluid in a rectangular cavity (Fig. 5.25). The
radiation loading on either side of the panel due to the panel motion is taken into
consideration in the analysis. The panel edges are assumed to be clamped adding
to the analytical complexity of the problem. This problem is of interest in determining
the interior cabin noise levels and the structural response. We follow closely the
approach of Yen, Maestrello and Padula (1975, 1980) who used a Ritz-Galerkin
procedure for the solution of the coupled vibroacoustic problem.

The equation of motion for the panel is

$$D\nabla^4 X + \beta \dot{X} + \mu \ddot{X} = p(s_1, s_2, t) + p_1(s_1, s_2, t) - p_2(s_1, s_2, t) \qquad (5.121)$$

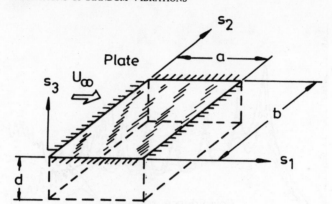

Fig. 5.25. Panel backed by rectangular cavity.

where $p(s_1, s_2, t)$ is the blocked boundary layer pressure fluctuation, p_1 and p_2 are the radiation pressures on either side of the panel due to the panel motion. The displacement X, the pressures p, p_1 and p_2 are taken positive in the downward direction. The flexural rigidity D may include structural damping and β is the viscous damping coefficient per unit area of the panel.

The moving fluid occupying the half space $s_3 > 0$ has a uniform mean velocity of flow U with velocity fluctuations governed by the linearized acoustic wave equation

$$\nabla^2 \varphi_1 - \frac{1}{c^2} \left(\frac{\partial}{\partial t} + U \frac{\partial}{\partial s_1} \right)^2 \varphi_1 = 0, \quad s_3 > 0 \tag{5.122}$$

where $\varphi_1 = \varphi_1(s_1, s_2, s_3, t)$ is the acoustic velocity potential and c is the speed of sound. The boundary conditions at the fluid-solid interface are

$$\frac{\partial X}{\partial t}(s_1, s_2, t) = \frac{\partial \varphi_1}{\partial s_3}(s_1, s_2, s_3, t)\bigg|_{s_3=0} \tag{5.123}$$

and

$$p_1(s_1, s_2, t) = \rho \left(\frac{\partial}{\partial t} + U \frac{\partial}{\partial s_1} \right) \varphi_1(s_1, s_2, s_3, t)\bigg|_{s_3=0} \tag{5.124}$$

for $0 \le s_1 \le a$ and $0 \le s_2 \le b$ and ρ is the fluid density.

For the clamped edges of the panel, the boundary conditions are

$$X(0, s_2, t) = X(a, s_2, t) = X(s_1, 0, t) = X(s_1, b, t) = 0,$$

$$\frac{\partial X}{\partial s_1}(s_1, s_2, t)\bigg|_{s_1=0} = \frac{\partial X}{\partial s_1}(s_1, s_2, t)\bigg|_{s_1=a}$$

$$= \frac{\partial X}{\partial s_2}(s_1, s_2, t)\bigg|_{s_2=0} = \frac{\partial X}{\partial s_2}(s_1, s_2, t)\bigg|_{s_2=b} = 0 \tag{5.125}$$

The pressures p_1 and p_2 induced by the panel motion are governed by linear integral operators

$$p_1 = \mathcal{L}_1(X) \qquad \text{and} \qquad p_2 = \mathcal{L}_2(X) \tag{5.126}$$

The specific forms of the operators $\mathcal{L}_1(.)$ and $\mathcal{L}_2(\cdot)$ are presented in the sequel.

Denote the Fourier transform of all the time dependent variables by a overhat. For example, the Fourier transform $X(s_1, s_2, t)$ is $\hat{X}(s_1, s_2, \omega)$ and defined by the relation

$$X(s_1, s_2, t) = \int_{-\infty}^{+\infty} \hat{X}(s_1, s_2, \omega) e^{-i\omega t} \, d\omega \tag{5.127}$$

In terms of the Fourier transforms (5.121) to (5.126) become

$$D\nabla^4 \hat{X} - \beta i\omega \hat{X} - \mu\omega^2 \hat{X} = \hat{p} + \hat{p}_1 - \hat{p}_2 \tag{5.128}$$

$$\nabla^2 \hat{\varphi}_1 = -\frac{1}{c^2}\left(-i\omega + U\frac{\partial}{\partial X}\right)\varphi_1 = 0, \quad s_3 > 0 \tag{5.129}$$

$$-i\omega\hat{X} = \frac{\partial\hat{\varphi}_1}{\partial s_3}(s_1, s_2, s_3, \omega)\bigg|_{s_3=0} \tag{5.130}$$

$$\hat{p}_1(s_1, s_2, \omega) = \rho\left(-i\omega + U\frac{\partial}{\partial s_1}\right)\hat{\varphi}_1(s_1, s_2 \, s_3, \omega)\bigg|_{s_3=0} \tag{5.131}$$

$$\hat{X}(0, s_2, \omega) = \hat{X}(a, s_2, \omega) = \hat{X}(s_1, 0, \omega) = \hat{X}(s_1, b, \omega) = 0$$

$$\frac{\partial\hat{X}}{\partial s_1}(s_1, s_2, \omega)_{s_1=0} = \frac{\partial\hat{X}}{\partial s_1}(s_1, s_2, \omega)_{s_1=a} = \frac{\partial\hat{X}}{\partial s_2}(s_1, s_2, \omega)_{s_2=0}$$

$$= \frac{\partial\hat{X}}{\partial s_1}(s_1, s_2, \omega)_{s_2=b} = 0 \tag{5.132}$$

$$\hat{p}_1 = \mathcal{L}_1(\hat{X}) \text{ and } \hat{p}_2 = \mathcal{L}_2(\hat{X}) \tag{5.133}$$

The Fourier transform of the cavity pressure can be expressed as

$$\hat{p}_2 = \mathcal{L}_2(\hat{X}) = -\rho\omega^2 \int_0^b\int_0^a \hat{h}_2(s_1^1, s_2^1; s_1^2, s_2^2)\,\hat{X}(s_1^2, s_2^2)ds_1^2 ds_2^2 \tag{5.134}$$

where $\hat{h}_2$ is the Fourier transform of the Green's function for the rectangular cavity given by (Morse and Feshbach, 1953)

$$\hat{h}_2(s_1^1, s_2^1; s_1^2, s_2^2; \omega) = \frac{4}{ab}\sum_{s=0}^{\infty}\sum_{s=0}^{\infty}\frac{\cos\left(\dfrac{r\pi s_1^1}{a}\right)\cos\left(\dfrac{s\pi s_2^1}{b}\right)\cos\left(\dfrac{r\pi s_1^2}{a}\right)\cos\left(\dfrac{s\pi s_2^2}{b}\right)}{k_{rs}\sin(k_{rs}d)}$$

$$\tag{5.135}$$

with $k_{rs}^2 = k^2 - (r\pi/a)^2 - (s\pi/b)^2$, where $k = \omega/c$ is the wave number and d is the depth of the cavity.

The Fourier transform of p_1 is

$$\hat{p}_1 = \hat{\mathscr{L}}_1(\hat{X}) = \left(-i\omega + U \frac{\partial}{\partial s_1^1}\right) \int_0^a \int_0^b \hat{h}_1(s_1^1 - s_1^2, s_2^1 - s_2^2)$$

$$\times \left(-i\omega + U \frac{\partial}{\partial s_1^2}\right) \hat{X}(s_1^2, s_2^2) \, ds_1^2 \, ds_2^2 \qquad (5.136)$$

where $\hat{h}_1$ is the Fourier transform of the acoustic Green's function which can be determined in the following manner:

$$\hat{h}_1(s_1^1 - s_1^2; s_2^1 - s_2^2; \omega) = \hat{\psi}_1(s_1^1 - s_1^2; s_2^1 - s_2^2; \omega)\bigg|_{s_3=0} \quad \begin{matrix} 0 < s_1^2 < a \\ 0 < s_2^2 < b \end{matrix} \qquad (5.137)$$

and $\hat{\psi}_1$ is the transform of the acoustic potential that satisfies the transformed convective wave equation

$$\nabla^2 \hat{\psi}_1 - \frac{1}{c^2}\left(-i\omega + U \frac{\partial}{\partial s_1}\right)\hat{\psi}_1 = 0\bigg|_{s_3=0}$$

with

$$\frac{\partial \hat{\psi}_1}{\partial s_3} = \begin{cases} \delta(s_1^1 - s_1^2)\,\delta(s_2^1 - s_2^2) & \text{for } (s_1^1, s_2^1) \text{ in } A \\ 0, & \text{otherwise} \end{cases}$$

where A denotes the region $0 < s_1 < a$, $0 < s_2 < b$ and (s_1, s_2) is in A.

Yen, Maestrello and Padulla (1975, 1980) have shown that in the case of supersonic flow $U/c = M > 1$, $\hat{\psi}_1$ must vanish outside the Mach cone with vertex at $(s_1^2, s_2^2, 0)$. $\hat{\psi}_1$ is determined by taking a Fourier transform with respect to $s_2^1 - s_2^2$ and Laplace transform with respect to $s_1^1 - s_1^2$ using the fact that $\hat{\psi}_1 = 0$ for $s_1^1 - s_1^2 < 0$. The resulting equation in s_3 is solved and after applying the radiation condition as $s_3 \to \infty$ and taking the inverse Fourier transform by contour integration, $\hat{\psi}_1$ is obtained

$$\hat{\psi}_1(s_1^1 - s_1^2; s_2^1 - s_2^2; s_3) = -\frac{1}{\pi} \exp\left(\frac{iMk(s_1^1 - s_1^2)}{M^2 - 1}\right)$$

$$\times \frac{\cos\{[k/(M^2 - 1)]\sqrt{(s_1^1 - s_1^2)^2 - (M^2 - 1)[(s_2^1 - s_2^2)^2 + s_3^2]}\}}{\sqrt{(s_1^1 - s_1^2)^2 - (M^2 - 1)[(s_2^1 - s_2^2)^2 + s_3^2]}}$$

$$\text{if } (s_1^1 - s_1^2) > \{(M^2 - 1)[(s_2^1 - s_2^2)^2 + s_3^2]\}^{1/2}$$

$$= 0 \text{ otherwise} \qquad (5.138)$$

$\hat{h}_1(s_1^1 - s_1^2; s_2^1 - s_2^2; \omega)$ is obtained from (5.138) by setting $s_3 = 0$.

Ritz-Galerkin Procedure

A Ritz-Galerkin procedure is used to solve the integro-differential equation for X.

For the clamped edge conditions the eigen functions are assumed to be uncoupled in the s_1 and s_2 coordinates and can be approximated by the appropriate beam functions with clamped end conditions. In the analysis to follow, only a single mode in the s_2 direction is considered. The Fourier transform of the displacement of the panel can be expressed as,

$$\hat{X}(s_1, s_2) = \sum_{j=1}^{n} \hat{q}_j u_j(s_1) v_1(s_2) = \bar{U}^T \bar{Q} \tag{5.139}$$

where $\bar{U}$ and $\bar{Q}$ are vectors with elements

$$U_j(s_1, s_2) = u_j(s_1)\, v_1(s_2) \text{ and } Q_j = \hat{q}_j \tag{5.140}$$

The normalized beam functions are given by

$$u_j(s_1) = \frac{1}{\sqrt{a}} \left\{ \left(\frac{\sin k_j + \sinh k_j}{\cos k_j - \cosh k_j} \right) [\sin (k_j s_1/a) - \sin (k_j s_1/a)] \right.$$

$$\left. + \cos (k_j s_1/a) - \cosh (k_j s_1/a) \right\} \tag{5.141}$$

$$v_1(s_2) = \frac{1}{\sqrt{b}} \left\{ \left(\frac{\sin k_1 + \sinh k_1}{\cos k_1 - \cosh k_1} \right) [\sin (k_1 s_2/b) - \sinh (k_1 s_1/b) \right.$$

$$\left. + \cos (k_1 s_2/b) - \cosh (k_1 s_2/b) \right\} \tag{5.142}$$

where k_j, $j = 1, 2, \ldots$ are the roots of the equation $\cos (k) \cosh (k) = 1$, with $k_1 = 4.73$, $k_2 = 7.853$ and $k_j \cong (2j + 1)\pi/2$ for $j > 2$.

Substitute (5.139) in (5.128), multiply the resulting equation by $u_r(s_1)v_1(s_2)$ and integrate over the area of the panel. Confining the number of modes (say n) we get the following matrix equation

$$\mathbf{B}\bar{Q} = \bar{P}(\omega) \tag{5.143}$$

where $\mathbf{B} = \mathbf{B}(\omega)$ is a square matrix given by

$$\mathbf{B} = D/a^4 \mathbf{K} + 2Dn_b^2 \mathbf{N} + [(D/b)^4 k_1^4 - i\omega\beta - \mu\omega^2]\mathbf{I} - \mathbf{H}_1 + \mathbf{H}_2 \tag{5.144}$$

where $\mathbf{K}$ is a diagonal matrix with elements

$$K_{jj} = k_j^4 \tag{5.145}$$

$$n_b^2 = \int_0^b v_1'(s_2)\, v_1'(s_2)\, ds_2 \tag{5.146}$$

$\mathbf{N}$ is a square matrix with elements,

$$N_{ij} = \int_0^a u_i'(s_1)\, u_j'(s_1)\, ds_1 \tag{5.147}$$

and $\mathbf{I}$ is the identity matrix.

H_1 and H_2 are square matrices defined by

$$H_i \bar{Q} = \int_0^a \int_0^b \bar{U} \, \hat{\mathscr{L}}_i [\bar{U}^T \bar{Q}] \, ds_1 \, ds_2 \tag{5.148}$$

and $\bar{P}(\omega)$ is a vector with elements

$$P_j(\omega) = \int_0^a \int_0^b \hat{p}(\omega) \, u_j(s_1, s_2) \, ds_1 \, ds_2 \tag{5.149}$$

In (5.145) to (5.149) use has been made of the orthogonal properties of the beam functions. The primes indicate the order of differentiation with respect to s_1 or s_2 as the case may be. All the matrices are of order $n \times n$ and all the vectors are of dimension n.

We mention here, that in (5.143) if we set the generalized force vector $\bar{P}(\omega) = 0$, the solution for nontrivial $\bar{Q}$ represents the free vibration problem or the panel flutter problem and the corresponding natural frequencies can be obtained by setting

$$\det \mathbf{B}(i\omega) = 0 \tag{5.150}$$

The interaction between the flutter and random vibration problem is considered in the next section.

Panel Response

The response statistics of the panel to the external boundary layer excitation which is a multidimensional random process of space and time can be estimated in the following manner. The random excitation is taken to be a weakly stationary and homogeneous random process. Consider the truncated Fourier transform of the different variables of interest defined by, for example

$$\hat{X}(s_1, s_2; \omega, T) = \int_{-T}^{T} X(s_1, s_2, t) e^{i\omega t} \, dt \tag{5.151}$$

We know that the cross spectral density function of X corresponding to two locations (s_1^1, s_2^1) and (s_1^2, s_2^2) on the panel is given by

$$\Phi_{XX}(s_1^1, s_2^1; s_1^2, s_2^2; \omega) = \underset{T \to \infty}{\text{Lt}} \frac{\pi}{T} E[\hat{X}(s_1^1, s_2^1, \omega, T) \hat{X}^*(s_1^2, s_2^2, \omega, T)] \tag{5.152}$$

Similarly the cross spectral density function of the boundary layer pressure field is

$$\Phi_{PP}(s_1^1, s_2^1; s_1^2, s_2^2; \omega) = \underset{T \to \infty}{\text{Lt}} \frac{\pi}{T} E[\hat{P}(s_1^1, s_2^1, \omega, T) \hat{P}^*(s_1^2, s_2^2, \omega, T)] \tag{5.153}$$

From (5.147), the cross spectral density matrix of the generalized forces is given by

$$\Phi_{PP}(\omega) = \int_0^a \int_0^b \int_0^a \int_0^b \bar{U}(s_1^1, s_2^1) \, \Phi_{pp}(s_1^1, s_2^1; s_1^2, s_2^2; \omega) \, \bar{U}^{*T}(s_1^2, s_2^2) \, ds_2^2 \, ds_1^2 \, ds_2^1 \, ds_1^1 \tag{5.154}$$

From (5.143) the cross spectral density matrix of the generalized displacements is

$$\Phi_{\bar{Q}\bar{Q}}(\omega) = S(\omega)\,\Phi_{\bar{P}\bar{P}}(\omega)\,S^{*T}(\omega) \tag{5.155}$$

where the matrix

$$S(\omega) = B^{-1}(\omega) \tag{5.156}$$

The cross spectral density of the displacement is then

$$\Phi_{XX}(s_1^1, s_2^1; s_1^2, s_2^2; \omega)$$

$$= \bar{U}^T(s_1^1, s_2^1)\,S(\omega)\,\Phi_{\bar{P}\bar{P}}(\omega)\,S^{*T}(\omega)\,U(s_1^2, s_2^2)$$

$$= \bar{U}^T(s_1, s_2)\,S(\omega)\left[\int_0^a\int_0^b\int_0^a\int_0^b \bar{U}(s_1^1, s_2^1)\,\Phi_{PP}(s_1^1, s_2^1; s_1^2, s_2^2; \omega)\right.$$

$$\left.\times \bar{U}^{*T}(s_1^2, s_2^2)\,ds_2^2\,ds_1^2\,ds_2^1\,ds_1^1\right]S^{*T}(\omega)\,\bar{U}(s_1^2, s_2^2) \tag{5.157}$$

Equation (5.157) represents the explicit relation between the cross spectral densities of the output displacement to the cross spectral density of the external excitation, namely the boundary layer pressure field including the interaction effects of the structural motion and the radiated pressure loadings on either side of the panel. It represents the most general form of the input-output relation, which could include any type of support conditions at the panel edges and also more than a single mode along the s_2 direction as has been the case considered in the present analysis. The only difference will be in the shape of the normal mode and the elements of the interaction matrix $S(\omega)$, which also depend on the normal modes.

Computational Procedure
We observe from (5.157), the major computational efforts required in the response analysis are in the determination of the cross spectral density matrix $\Phi_{\bar{P}\bar{P}}(\omega)$ and the fluid-structure interaction matrix $S(\omega)$.

$S(\omega)$ is the inverse of the matrix $B(\omega)$ defined by (5.144). Most of he efforts involved in estimating the elements of $B(\omega)$ are in estimating matrices $H_1(\omega)$ and $H_2(\omega)$ which involve complicated integral expressions of the acoustic Green's functions corresponding to the cavity on one side of the panel. A typical element of $H_1(\omega)$ is

$$H_1(\omega)_{ij} = -\int_0^a\int_0^a \left(i\omega + U\frac{d}{ds_1^1}\right)u_i(s_1^1)\left(-i\omega + U\frac{d}{ds_1^1}\right)u_j(s_1^2)$$

$$\times \left[\int_0^b\int_0^b \hat{h}_1(s_1^1 - s_1^2; s_2^1 - s_2^2; \omega)v_1(s_2^1)\,v_1(s_2^2)ds_2^1ds_2^2\right]ds_1^1ds_1^2 \tag{5.158}$$

Similarly one can obtain an expression for the elements of $H_2(\omega)$. The indicated integrations are extremely difficult to evaluate in closed form and numerical integration procedures have to be resorted to.

For the form of the cross spectral density function of the boundary layer pressure field given in (5.105) the elements of the matrix $\Phi_{\bar{p}\bar{p}}(\omega)$ can be obtained in closed form by integration. A typical element of the matrix $\Phi_{\bar{p}\bar{p}}(\omega)$ is

$$(\Phi_{\bar{p}\bar{p}}(\omega))_{ij} = \int_0^a \int_0^b \int_0^a \int_0^b u_i(s_1^1)\, v_1(s_1^1)\, \Phi_{PP}(s_1^1 - s_1^2; s_2^1 - s_2^2; \omega)$$

$$\times u_j(s_2^1)\, v_1(s_2^2)\, ds_2^2\, ds_1^2\, ds_2^1\, ds_1^1 \tag{5.159}$$

which after substitution for u_i and v_1 from (5.141) and (5.142) and for $\Phi_{\bar{p}\bar{p}}(s_1^1 - s_1^2; s_2^1 - s_2^2; \omega)$ from (5.105) and integration gives

$$(\Phi_{\bar{p}\bar{p}}(\omega))_{ij} = \Phi_{pp}(\omega)\, \chi_{ij}\, \psi \tag{5.160}$$

$$\chi_{ij} = [(A^4 - k_i^4)(A^4 - k_j^4)]^{-1} [\exp(-Aa)\{Au_i''(0) - u_j'''(0)\}$$

$$- \{Au_i''(a) - u_i'''\,)\}\{Au_j''(a) - u_j'''(a)]$$

$$+ [(A^{*4} - k_i^4)(A^{*4} - k_j^4)]^{-1}[\exp(-A^*a)\{A^*u_i''(0) - u_i'''(0)\}$$

$$- \{A^*u_j''(a) - u_j'''(a)\}\{A^*u_i''(a) - u_i'''(a)\}]$$

$$+ [A^4 - k_i^4]^{-1}[A^3\delta_{ij} + A^2 U_{ij}' + A U_{ij}'' + U_{ij}''']$$

$$+ [A^{*4} - k_j^4]^{-1}[A^{*3}\delta_{ij} + A^{*2} U_{ij}' + A^* U_{ij}'' + U_{ij}'''] \tag{5.161}$$

with

$$A = \frac{1}{\alpha_1\delta} + \frac{i\omega}{U_\infty},\ A^* = \frac{1}{\alpha_1\delta^*} - \frac{i\omega}{U_\infty}$$

$$U_{ij}' = \int_0^a u_i'(s_1)\, u_j(s_1)\, ds_1$$

$$U_{ij}'' = \int_0^a u_i''(s_1)\, u_j(s_1)\, ds_1$$

$$U_{ij}''' = \int_0^a u_i'''(s_1)\, u_j(s_1)\, ds_1$$

$$\psi = 2[A_1^4 - k_i^4]^{-2} [\exp(-A_1 b)\{A_1 v_1''(0) - v_1'''(0)\}$$

$$- \{A_1 v_1''(b) - v_1'''(b)\}]^2 + 2(A_1^4 - k_1^4)^{-2}(A_1^3 - n_b^2 A_1)] \tag{5.162}$$

and $A_1 = 1/(\alpha_2\delta^*)$.

Example 5.7: (Yen, Maestrello and Padula, 1980)
Numerical results of the response calculations have been presented for the following data by Yen, Maestrello and Padula, (1980).

Panel Material: Titanium, $E = 1.1. \times 10\ \text{N/m}^2$, $v = 0.3$, $\rho_s = 4450\ \text{kg/m}^3$
Panel dimensions: $a = 30.5$ cm, $b = 15.2$ cm, $h = 0.00157\ m$
Flow characteristics: Fluid-air, $\rho = 1.23\ \text{kg/m}^3$, $c = 211$ m/s, $U = 475$ m/s, $M = 3.03$.

Characteristics of boundary layer pressure fluctuations: Cross spectral density function of wall pressure fluctuations is given by (5.105) with the same parameters. $\delta = 3.45 \times 10^{-2}$ m and the mean square value of the wall pressure fluctuation $p_{rms}^2 = 1.19 \times 10^5$ N^2/m^4. The other derived properties for the panel are: $D = Eh^3/12(1 - v^2) = 39.5$ Nm, $m = \rho_s h$—mass/unit area of panel = 7.01 kg/m^2.

In calculating the response statistics the viscous damping has been neglected ($\beta = 0$) in comparison with acoustic damping. Also the effect of cavity on the panel response is neglected. The computed psd of the panel displacement corresponding to the location $s_1 = a/2$, $s_2 = b/2$ is shown in Fig. 5.26. A five term Ritz-Galerkin approximation was taken for $X(s_1, s_2)$ with $j = 1, 3, ..., 9$. Only the odd number modes are taken into consideration as by symmetry, the eigen functions $u_j(s_1)$ with j even vanish at $s_1 = a/2$. In Fig. 5.26 is also shown the experimental results. It is seen from the figure that general agreement exists between theory and experiment. Except at the peaks, the numerical results are generally lower than the experimental results. This can be attributed to the five term Galerkin approximation and to the fact that exact fixed end conditions might not have been achieved in the experimental setup. The acoustic damping in the model arises from the pressure terms p_1 and p_2 in (5.128) and are included in the terms of the matrices H_1 and H_2 in (5.142). The matrices H_1 and H_2, apart from being responsible for giving rise to damping may also have effects on the shifts of the resonant frequencies of the panel and on the aerodynamic virtual mass. The exact roles played by terms

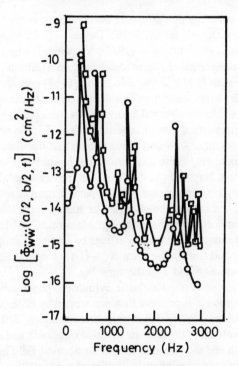

Fig. 5.26. Comparison of displacement psd at centre of panel to supersonic turbulent boundary layer excitation. □ —experimental; ○ — numerical (Yen et al, 1980).

of various order in ω in the elements of $\mathbf{H}_1$ and $\mathbf{H}_2$ are difficult to analyze. For a qualitative description of such effects readers are referred to the works of Liebowitz (1971, 1972).

The characteristics of the radiation sound field can be obtained using (5.134) and (5.136) in conjunction with the acoustic and Green's functions defined by (5.135) and (5.138) once the statistics of the displacement of the panel is established. For further details regarding this problem readers can refer to the works of Davies (1969), and Maestrello (1969).

5.5.3 Flutter and Random Vibration in Turbulence

The random vibration of elastic bodies to turbulent boundary layer excitation has often been considered as a forced vibration problem neglecting the fluid-structure interaction effects especially if the flowing fluid is air. These effects become important at supersonic speeds, in which case a portion of the vibration energy is dissipated in the medium as radiated noise. The response of rectangular panels presented in Section 5.5.1 did not consider such radiation loading, while that presented in Section 5.5.2 for supersonically convected turbulence included the radiation loading on either side of the panel. The acoustic field inside the rectangular cavity was also determined as a solution of the coupled sound-structure interaction problem.

The fluid-structure interaction effects in the context of random vibration analysis have been treated by Dowell (1969, 1970), Davies (1971), Maestrello and Linden (1971), Strawdermann and Christman (1972), Vaicaitis, Jan and Shinozuka (1972). The problem of nonlinear panel flutter under random excitation has been analysed by Easter and McIntosh (1971). Dzygaldo (1966, 1967) considered the problem of plate vibration under supersonic flow in a deterministic frame work assuming that the plate to be excited by an external harmonic pressure loading in addition to that generated by its own motion due to supersonic flow. Linear potential theory was assumed for the supersonic flow and it was shown that the plate undergoes a sharp resonance with infinite amplitude at a critical Mach number inspite of material and aerodynamic damping. Nagornov (1967) used the above formulation in the response of panels to boundary layer turbulence with a separable form of space-time correlation function neglecting the convective effects. The spatial correlation was assumed to be of delta function form in the spatial lag. Maestrello and Linden (1971) have taken into consideration the convective effects by having a coupled delta function correlation in space and time in the form $\delta(\xi - U_c\tau)$ along the flow direction where ξ is the spatial separation and τ is the time lag.

The random response of a viscoelastic cylindrical panel subjected to turbulent boundary layer excitation in supersonic flow was treated by Elishakoff and Khromatov (1971) using the Corcos (1963) form of correlation function. Bolotin and Elishakoff (1971) treated the problem of the random vibration of smooth and stiffened cylindrical shells of finite length and the determination of the acoustic field inside the cylindrical cavity. Elishakoff and his co-workers in a series of papers (Elishakoff and Efimtsoff, 1972; Elishakoff, 1973; Elishakoff, 1977(a); 1977(b); Elishakoff, Th. van Zanten and Crandall, 1979; Elishakoff and Livshits, 1984; 1989) have considered the random vibration to turbulent flow of beams, plates and shells, including such effects as cross-correlations between normal modes, orthotrophy. The problem of the random vibration of a two-span beam and closed form solutions for the response of Euler-Bernoulli and Timoshenko beams to turbulent flow have also been considered. Vaicaitis and Lin (1972), Lin et al (1977) have also treated the problem of the response of a periodic beam to supersonic boundary layer pressure fluctuations.

As usually, the damping characteristics of structural systems are determined in air, rather than in vacuum, it may be assumed that the radiation damping is automatically taken into account. At low flight speeds, the amount of radiation damping is relatively very small compared to the structural damping. At supersonic flow velocities the effect of radiation damping is significant. Moreover, with flexible structures aeroelastic instabilities may set in. The frequency characteristics of the structure alter considerably at these velocities and flutter type of dynamic instability due to fluid-structure interaction occur. The random response analysis is valid only for flow velocities which are lower than a critical velocity $U < U_{cr}$ at which these instabilities set in. It is therefore important to know apriori the condition under which flutter occurs. Elishakoff (1974) has presented a general formulation of the problem of interaction of flutter and random vibration of elastic structures in supersonic flow arising out of the combination of the external random pressure fluctuations of the turbulent boundary layer and the pressure loading induced on the structure due to its motion. He has applied the above formulation to the solution of the random response of plates and beams (Elishakoff, 1977a; 1983) below the critical velocity using the Galerkin method in conjunction with the in vacuo natural modes of the structure. The 'piston theory' (Ashley and Zartarian, 1956) used extensively in aeroelastic design in the supersonic regime was adopted in deriving the equation of motion. The conditions for onset of flutter depending on the number of modes taken in the analysis have been obtained. In the following we present briefly the interaction of flutter and random vibration of a beam in supersonic flow. Essentially the same analysis can be extended to the case of a plate (Elishakoff, 1977a; 1983).

Piston Theory
According to the piston theory (Ashley and Zartarian, 1956) the pressure perturbation on a vibrating surface with the supersonic flow oriented towards the s_1 direction is approximately given by

$$p_a \cong \frac{Kp_\infty}{c_\infty}\left(\dot{X} + U\frac{\partial X}{\partial s_1}\right) \tag{5.163}$$

where $X(s_1, s_2, t)$ is the displacement of the surface, p_∞ and c_∞ respectively are the unperturbed pressure and velocity of sound in the medium. U is the flow velocity and K is the polytropic exponent.

Equation (5.163) can be rewritten as,

$$p_a \cong c_\infty \rho_\infty \left(\dot{X} + U\frac{\partial X}{\partial s_1}\right) \tag{5.164}$$

where $\rho_\infty c_\infty$ represents the active component of the acoustic impedance at wavelengths sufficiently large compared with its characteristic dimension (Morse, 1948). Dowell (1970) has shown that at sufficiently high flow velocities the form of the pressure perturbation (5.163) in the analysis yields an accurate estimate of the critical flutter velocity.

Structural Motion

Considering that the correlation scale of turbulence is much larger than the characteristic dimension of the structure and that the standard deviation of the displacement is much smaller than the thickness, the response problem can be formulated in a linear setting. For a beam of length l and kept in supersonic flow along the s_1 direction, the equation of motion and boundary conditions are

$$EI \frac{\partial^4 X}{\partial s_1^4} + c\dot{X} + m\ddot{X} = p(s_1, t) - p_a(s_1, t) \tag{5.165}$$

where EI is the flexural rigidity, c is the structural damping coefficient assumed to be of viscous type, m is the mass, $p(s_1, t)$ is the external pressure loading due to boundary layer turbulence and $p_a(s_1, t)$ is the aeroelastic interaction loading given by (5.163) all per unit length of the beam. Substituting (5.163) into (5.165), we obtain

$$EI \frac{\partial^4 X}{\partial s_1^4} + 2\zeta_0 m\dot{X} + gmU \frac{\partial X}{\partial s_1} + m\ddot{X} = p(s_1, t) \tag{5.166}$$

where $\quad \zeta_0 = \frac{1}{2}\left(\frac{c}{m} + g\right)$ and $g = Kp_\infty/(mc_\infty)$

For the beam clamped at both ends the boundary conditions are

$$X(0, t) = X(l, t) = \frac{\partial X}{\partial s_1}(0, t), \frac{\partial X}{\partial s_1}(l, t) = 0$$

with the mode shapes given by the normalized beam functions (5.141) and natural frequencies $\omega_j^2 = EIk_j^2/(ml^4)$, $j = 1, 2 \ldots$ where k_js are as defined in Section (5.5.2). Express $p(s_1, t)$ and $X(s_1, t)$ in the form

$$p(s_1, t) = \sum_{j=1}^{N} \int_{-\infty}^{+\infty} \hat{P}_j(\omega) u_j(s_1) e^{i\omega t} d\omega$$

$$X(s_1, t) = \sum_{j=1}^{N} \int_{-\infty}^{+\infty} \hat{X}_j(\omega) u_j(s_1) e^{i\omega t} d\omega \tag{5.167}$$

where $\hat{P}_j(\omega)$ and $\hat{X}_j(\omega)$ respectively are the Fourier transform of the generalized force and displacement corresponding to the jth normal mode and N is the number of modes considered.

Substitute (5.167) in (5.166), multiply both sides of the resulting equation by $u(s_1)$, $k = 1, 2, \ldots N$ and integrate over the length of the beam to obtain the matrix equation

$$A \bar{\hat{X}} = \frac{1}{m} \bar{\hat{P}} \tag{5.168}$$

where $\bar{P}$ is the vector containing $\hat{P}_j(\omega)s$ and A is a matrix with elements

$$a_{jk} = (\omega_j^2 - \omega^2 + 2i\zeta_0\omega)\delta_{jk} + gUb_{jk} \tag{5.169}$$

with

$$b_{jk} = \left\{\int_0^l \frac{du_k(s_1)}{ds_1} u_j(s_1)\, ds_1\right\} \Big/ v_j^2; \; v_j^2 = \int_0^l u_j^2(s_1)\, ds_1 \tag{5.170}$$

For flow velocities for which A is nonsingular the cross spectral density of the beam displacement corresponding to positions s_1^1, s_1^2 is

$$\Phi_{XX}(s_1^1, s_1^2; \omega) = \frac{1}{m^2}\sum_{i=1}^N \sum_{j=1}^N \sum_{k=1}^N \sum_{l=1}^N (A_{ik}^{-1})^* A_{jl}^{-1} \Phi_{P_k P_l}(\omega) u_i(s_1^1) u_j(s_1^2) \tag{5.171}$$

where $(A_{ik}^{-1})^*$ is the complex conjugate of the ikth element and A_{jl}^{-1} is the jlth element of the inverse of matrix A, that is A^{-1} and

$$\Phi_{P_k P_l}(\omega) = \int_0^l \int_0^l \Phi_{pp}(s_1^1, s_1^2; \omega) u_k(s_1^1)\, u_l(s_1^2)\, ds_1^1 ds_1^2/(v_k v_l)^2 \tag{5.172}$$

being the cross spectral density between the generalized force P_k and P_l and $\Phi_{pp}(s_1^1, s_1^2; \omega)$ is the cross spectral density of the boundary layer pressure fluctuations.

Aeroelastic instability occurs for the flow velocity $U = U_{cr}$ at which A becomes singular. Consequently, the response calculations indicated are valid only in the range $U < U_{cr}$. U_{cr} can be obtained by setting det $A(\omega) = 0$, which gives the necessary and sufficient conditions for the onset of aeroelastic instability which are

$$\text{Re}\,\{\det A(\omega)\} = 0; \; \text{Im}\,\{\det A(\omega)\} = 0 \tag{5.173}$$

as the elements of A are generally complex valued.

It has been shown by Elishakoff (1974), that the linear system given by (5.166) is mean square stable, if the corresponding deterministic system is asymptotically stable with the right hand side of (5.166) is identically set to zero, which again reduces to the condition that $U < U_{cr}$ provided that

$$E[X^2(s_1)] = \int_{-\infty}^{+\infty} \Phi_{XX}(s_1, s_2; \omega)\, d\omega < \infty \tag{5.174}$$

It has also been shown by Elishakoff (1977) that the occurrence or non-occurrence of flutter and the flow velocity at which it occurs depend on the number of modes taken into consideration in the analysis.

One Mode Approximation

In this case, A becomes a scalar with det $A(\omega) = a_{11} = \omega_1^2 - \omega^2 + 2i\zeta_0\omega + gUb_{11} = 0$ and the critical velocity is obtained by setting $\text{Re}(a_{11}) = \omega_1^2 - \omega^2 + gUb_{11}$ and $\text{Im}(a_{11}) = 2i\zeta_0\omega = 0$. As the total damping factor ζ_0 is not zero, the second condition holds only for $\omega = 0$; which means that flutter type instability is apparently impossible with the one mode approximation. The condition that $\text{Re}(a_{11}) = 0$ is also not possible because $b_{11} = 0$ is impossible. Thus the single mode approximation often used in random vibration analysis does not cover the complete physical picture of the problem and leads to the incorrect conclusion of nonexistence of flutter type instability.

Two Mode Approximation
In this case, the determinant of the matrix $\mathbf{A}$ becomes with

$$\det \mathbf{A}(\omega) = (\omega_1^2 - \omega^2 + gUb_{11} + 2i\zeta_0\omega)(\omega_2^2 - \omega^2$$
$$+ gUb_{22} + 2i\zeta_0\omega) - g^2U^2b_{12}b_{21} \qquad (5.175)$$

$$\text{Re}\{\det \mathbf{A}(\omega)\} = C_0\omega^4 - C_2\omega^2 + C_4; \text{Im}\{\det \mathbf{A}(\omega)\}$$
$$= -C_1\omega^3 + C_3\omega \qquad (5.176)$$

where

$$C_0 = 1, C_1 = 4\zeta_0, C_2 = \omega_1^2 + \omega_2^2 + gU(b_{11} + b_{22}) + 4\zeta_0^2, C_3 = 2\zeta_0(C_2 - 4\zeta_0^2)$$

and

$$C_4 = (\omega_1^2 + gUb_{11})(\omega_2^2 + gUb_{22}) - g^2U^2b_{12}b_{21}$$

Setting $\text{Re}\{\det \mathbf{A}(\omega)\} = \text{Im}\{\det \mathbf{A}(\omega)\} = 0$ for the flutter condition, from the second of (5.176) we obtain

$$\omega_{1,\text{cr}} = 0; \omega_{2,\text{cr}} = C_3/C_1 = \frac{1}{2}[\omega_1^2 + \omega_2^2 + gU(b_{11} + b_{22})]$$

for which

$$\text{Re}\{\det \mathbf{A}(\omega)\} = \omega_1^2\omega_2^2 - g^2U_{\text{cr}}^2b_{12}b_{21} \qquad (5.177)$$

$$\text{Im}\{\det \mathbf{A}(\omega)\} = \frac{1}{C_1^2}[-C_3(C_1C_2 - C_0C_3) + C_4C_1^2] \qquad (5.178)$$

The critical flow velocity associated with $\omega_{1,\text{cr}}$ which is identically zero indicates static instability or *divergence* while that associated with $\omega_{2,\text{cr}}$ indicates dynamic instability or *flutter*. Setting each of (5.177) to zero, it is shown that it does not have any real solution because of $b_{12}b_{21} < 0$ indicating that divergence instability is not possible even with two modes taken into account in the analysis. The critical flutter velocity is obtained by setting (5.178) to zero as

$$U_{\text{flutter}} = \frac{1}{2g}\left\{\frac{-[(\omega_2^2 - \omega_1^2)^2 + 8\zeta_0(\omega_1^2 + \omega_2^2)]}{b_{12}b_{21}}\right\} \qquad (5.179)$$

when $U < U_{\text{flutter}}$, $\mathbf{A}$ is nonsingular and (5.168) has the following solutions

$$\hat{X}_1(\omega) = (a_{22}\hat{P}_1 - a_{12}\hat{P}_2)/(m\Delta); \hat{X}_2(\omega) = (a_{11}\hat{P}_1 - a_{21}\hat{P}_2)/(m\Delta) \qquad (5.180)$$

where $\Delta = a_{11}a_{22} - a_{12}a_{21}$. The mean square value of the displacement at s_1 is

$$E[X^2(s_1, t)] = \int_{-\infty}^{+\infty}[\Phi_{P_1P_1}\varphi_{11} + \Phi_{P_2P_2}\varphi_{22}$$
$$- 2\text{Re}(\Phi_{P_1P_2}\varphi_{12})]m^{-2}\Delta^{-2}d\omega \qquad (5.181)$$

where

$$\varphi_{11} = |a_{22}|^2 u_1^2(s_1) + |a_{21}|^2 u_2^2(s_1) + 2\mathrm{Re}\,(a_{21}^* a_{22})u_1(s_1)u_2(s_1)$$

$$\varphi_{22} = |a_{12}|^2 u_1^2(s_1) + |a_{11}|^2 u_2^2(s_1) + 2\mathrm{Re}\,(a_{11}^* a_{12})u_1(s_1)u_2(s_1)$$

$$\varphi_{12} = a_{22}^* a_{12} u_1^2(s_1) + a_{21}^* a_{11} u_2^2(s_1) + (a_{21}^* a_{12} + a_{22}^* a_{11})u_1(s_1)u_2(s_1) \qquad (5.182)$$

Elishakoff (1977a) has obtained similar approximations with three and four modes. Using the principal Hurwitz determinant he has shown the divergence is not possible in these cases also. He has derived the conditions for flutter instability and the equations to determine the same. The response analysis as mentioned earlier is valid only for $U < U_{\text{flutter}}$. He has assumed the cross spectral density of the turbulent boundary layer pressure fluctuations in the form

$$\Phi_{\text{pp}}(s_1^1, s_1^2; \omega) = \sigma_p^2\,\Phi_{\text{pp}}(\omega)\exp\left\{\left(-\frac{0.1\omega l}{U_c} + 0.265\,\frac{l}{\delta^*}\,\alpha\right)|\,\xi_1 - \xi_2\,|\right.$$

$$\left. -\frac{i\omega l}{U_c}|\,\xi_1 - \xi_2\,|\right\} \qquad (5.183)$$

where σ_p^2 is the mean square value of the pressure fluctuation, $\Phi_{\text{pp}}(\omega)$ is the normalized spectral density, δ^* is the displacement thickness of the boundary layer, its nondimensional value being $\delta^*/l = 0.32$. The convection velocity was taken as a constant $U_c = 0.8U$. For $\alpha = 0$, (5.183) reduces to Wilby's (1967) approximation and for $\alpha = 1$, it reduces to Crocker's approximation for the cross spectral density. The nondimensional cross spectral density of the maximum normal stress s_{max} is expressed in terms of the nondimensional frequency $\bar{\omega} = \omega l^2(m/EI)^{1/2}$ as

$$\bar{\Phi}_{s_{\text{max}}s_{\text{max}}} = \frac{m^2 h^4 \Phi_{s_{\text{max}}s_{\text{max}}}(\bar{\omega})}{36\,(EI)^2\,\sigma_p^2\,\Phi_{\text{pp}}(\bar{\omega})} \qquad (5.184)$$

where h is the thickness of the beam.

Example 5.8: Flutter of a plate (Elishakoff, 1977a; 1983)
Numerical data have been obtained for the following data and flow parameters:
$l = 0.55$ m, $\rho = 2700$ kg/m^3, $E = 7.05 \times 10^{10}$ N/m^2, $\zeta = 0.1$, $c_\infty = 303.3$ m/s, $\rho = 0.459$ kg/m^3, altitude = 10000 m, $h = 1.5 \times 10^{-3}$m, Poisson's ratio$v = 0.3$. Even though the analysis has been presented for a plate, only the variations with respect to the flow direction s_1 have been considered essentially reducing the problem to one dimension. The flutter Mach number for these data have been calculated as $M_{\text{flutter}} = 1.545$, 1.976 and 2.08 respectively for two mode, three mode and four mode approximations.

The nondimensional psd of the maximum stress $\Phi_{s_{\text{max}}s_{\text{max}}}$ is plotted against $\bar{\omega}$ for the location $\bar{s}_1 = s_1/l = 0$ in Fig. 5.27 for three Mach numbers. At $M = 1.05$ and $M = 1.3$ the plots contain two distinct maxima, the first occurring at a frequency higher than the first natural frequency and the second at a frequency which is lower than the second natural frequency. As the Mach number increases and

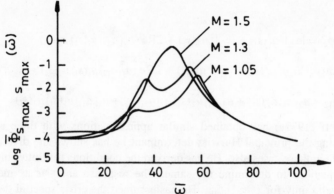

Fig. 5.27. Nondimensional cross spectral density of stress response at $x = 0$, $N = 2$.

approaches the flutter level at $M = 1.5$ the two maxima merge into a single maximum close to the critical flutter frequency $\bar{\omega}_{cr}$. Obviously for $M = M_{flutter}$, $\bar{\Phi}_{s_{max}s_{max}}$ which is valid only in the preflutter range has a pole at $\bar{\omega} = \bar{\omega}_{cr}$. Elishakoff (1983) has also plotted $\bar{\Phi}_{s_{max}s_{max}}$ as a function of $\bar{\omega}$ taking four modes in the analysis which is shown in Fig. 5.28, for $M = 1.4$, 1.6 and 1.8 which are less than $M_{flutter}$ for the four mode approximation. For the two mode approximation, the last two

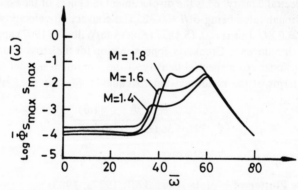

Fig. 5.28. Nondimensional cross spectral density of stress response for different Mach numbers at $x = 0$, $N = 4$.

Mach numbers of 1.6 and 1.8 are greater than the corresponding $M_{flutter} = 1.545$, illustrating how approximations below a minimum number of modes N^* are of only qualitative significance as with the two mode approximation the random response calculated for $M > M_{flutter}$ is not valid. In Fig. 5.29 is shown $\bar{\Phi}_{s_{max}s_{max}}(\bar{\omega})$ corresponding to $\bar{s}_1 = 0$ and 1 for $M = 1.4$ on an extended frequency scale. The plots show four maxima, the first of them occurring considerably above the first natural frequency and the remaining three close to the corresponding natural frequencies. It has also been observed that as the Mach number increased, the first two maxima draw closer together and eventually merge. This degenerate behaviour may be considered as an indication of the onset of possible flutter type instability. It has been found

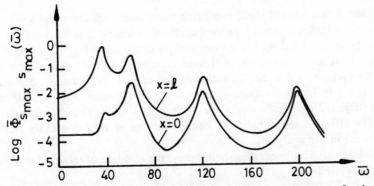

Fig. 5.29. Nondimensional cross spectral density of stress response at $x = 0$ and $x = l$; $N = 4$, $M = 1.4$.

that a minimum number of 12 modes are required for satisfactory convergence of results. Results for $N = 12$ have been plotted in Fig. 5.30 for $M = 1.4$. It can also be observed from Figs. 5.29 and 5.30 that the mean square maximal stress at $x = l$ exceeds that at $x = 0$.

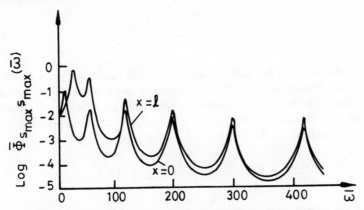

Fig. 5.30. Nondimensional cross spectral density of stress response at $x = 0$ and $x = l$; $N = 12$, $M = 1.4$.

5.5.4 Random Vibration of Elastic Shells in Turbulent Flow and Containing an Acoustic Medium

In Section 5.5.2, we presented the random vibration analysis of a panel backed by a rectangular acoustic cavity. In this section we present briefly the formulation and solution procedure of the problem of determining the random response of an elastic shell and the acoustic field inside the shell cavity when it is excited by turbulent boundary layer pressure fluctuations. We follow closely the work of Bolotin and Elishakoff (1971) which is also covered in Bolotin (1984). Similar works on the sound transmission through cylindrical shells due to random incident sound using a normal mode approach include those of Fedoroff (1963), Nemat-Nasser (1968), Hwang and Pi (1969), Kana (1971) and Narayanan and Shanbhag (1984). A different approach to an analogous problem using statistical energy analysis can be found in the works of White (1966), Fahy (1970) and Pope (1971).

Consider a two layered shell, consisting of an outer stiff layer and a less rigid inner insulating layer enclosing the shell cavity of volume V. A is the middle surface of the exterior shell and A_H is the inner surface of the insulating shell which is of thickness H. A set of curvilinear coordinates denoted by $\bar{s} = \{s_1, s_2\}$ on the surface A and another set of coordinates including the normal coordinate S_3 denoted by $\bar{r} = \{s_1, s_2, s_3\}$ to locate positions inside the cavity are used. The exterior of the shell is subjected to a random pressure loading $p_e(\bar{s}, t)$ due to turbulent boundary layer. The effect of sound radiation into the interior of shell due to its vibration is neglected in the analysis.

Assuming classical linear thin shell theory, the equation of motion for the shell can be expressed as

$$\mathscr{L}[X(\bar{s}, t)] = p_e(\bar{s}, t) - \mathscr{L}_{11}[X(\bar{s}, t)] - \mathscr{L}_{12}[X_H(\bar{s}, t)] \tag{5.185}$$

where X and X_H respectively are the normal displacements of the outer and inner shells, $\mathscr{L}(\cdot)$ is the linear differential operator representing inertial and damping forces of the outer shell and $\mathscr{L}_{11}[\cdot]$ and $\mathscr{L}_{12}[\cdot]$ are linear differential operators representing the elastic forces of the outer and inner shells respectively.

The acoustic field inside the shell cavity is governed by the wave equation

$$\nabla^2 \varphi(\bar{r}, t) - \frac{1}{c^2} \frac{\partial^2 \varphi}{\partial t^2}(\bar{r}, t) = 0 \tag{5.186}$$

where ∇^2 is the Laplace operator (used mainly in the cylindrical polar coordinates in the present context), φ is the acoustic potential and c is the velocity of sound in the medium. The boundary conditions representing equality of normal velocity of the fluid and structure at the interface and relating acoustic pressure to shell displacements similar to (5.123) and (5.124) are given by

$$\left. \frac{\partial \varphi}{\partial s_3} \right|_{A_H} = \frac{\partial X_H}{\partial t}; \quad -\rho \left. \frac{\partial \varphi}{\partial t} \right|_{A_H} = L_{21}[X] + L_{22}[X_H] \tag{5.187}$$

where ρ is the density of the acoustic medium and $\mathscr{L}_{21}[\cdot]$ and $\mathscr{L}_{22}[\cdot]$ are appropriate linear differential operators. Other boundary conditions on the outer shell depending on its fixity conditions and on the acoustic potential depending on whether a portion of the inner surface A_H is acoustically stiff or soft, that is $\left. \dfrac{\partial \varphi}{\partial s_3} \right|_{A_0} = 0$ if the portion of A_0 of A_H is acoustically stiff or $\left. \dfrac{\partial \varphi}{\partial t} \right|_{A_0} = 0$ if it is acoustically soft are also supplemented.

The random pressure loading $p_e(\bar{s}, t)$ is assumed to be stationary with repect to time and in general nonhomogeneous with respect to spatial coordinates. Without loss of generality p_e is assumed to have zero mean value as a consequence of which the mean values of X, X_H and φ are also zero. If $p_e(\bar{r}, t)$ is the acoustic pressure inside the cavity its mean square value within a frequency band (ω_1, ω_2) can be obtained by integration as

$$E[p_a^2(\bar{r}; \omega_1, \omega_2)] = \int_{\omega_1}^{\omega_2} \Phi_{p_a p_a}(\bar{r}^1, \bar{r}^2, \omega)\, d\omega \qquad (5.188)$$

where $\Phi_{p_a p_a}(\bar{r}^1, \bar{r}^2; \omega)$ is the cross spectral density of p_a corresponding to locations $\bar{r}^1$ and $\bar{r}^2$ and given by

$$\Phi_{p_a p_a}(\bar{r}^1, \bar{r}^2; \omega) = \frac{1}{2\pi} \int R_{p_a p_a}(\bar{r}^1, \bar{r}^2; \tau)\, e^{-i\omega\tau}\, d\tau \qquad (5.189)$$

with $R_{p_a p_a}(\bar{r}_1, \bar{r}_2; \tau)$ being the cross-correlation function. The transmitted noise inside the shell in decibels is given by

$$L_{p_a}(\bar{r}; \omega_1, \omega_2) = 10 \log \left\{ \frac{E[p_a^2(\bar{r}; \omega_1, \omega_2)]}{p_{\text{ref}}^2} \right\} \qquad (5.190)$$

where $p_{\text{ref}} = 2 \times 10^{-5}$ N/m^2. The determination of L_{p_a} thus involves the solution for $\Phi_{p_a p_a}(\bar{r}^1, \bar{r}^2; \omega)$ which is obtained by solving the coupled structure-acoustic random vibration problem (stochastic boundary value problem) given by (5.185) to (5.187) with the other boundary conditions. We shall consider now the solution procedure.

We expand $X(\bar{s}, t)$, $X_H(\bar{s}, t)$ and $\varphi(\bar{r}, t)$ in terms of the invacuo structural normal modes and the cavity acoustic modes and in view of the stationarity with respect to time of $p_e(\bar{s}, t)$ and consequent stationarity of the response process, express them in Fourier-Stieltjes integral representation in frequency domain as

$$X(\bar{s}, t) = \sum_j u_j(\bar{s}) \int_{-\infty}^{+\infty} e^{i\omega t}\, dS_{q_j}(\omega) \qquad (5.191)$$

$$X_H(\bar{s}, t) = \sum_j u_j(\bar{s}) \int_{-\infty}^{+\infty} e^{i\omega t}\, dS_{q_{Hj}}(\omega) \qquad (5.192)$$

$$\varphi(\bar{r}, t) = \sum_l v_l(\bar{s}) \int_{-\infty}^{+\infty} z_l(s_3, \omega)\, e^{-i\omega t}\, dS_{\eta_l}(\omega) \qquad (5.193)$$

where $u_j(\bar{s})$ is the jth structural mode which is complete and orthonormal on A and satisfies all the necessary boundary conditions and $v_l(\bar{s}) Z_l(s_3, \omega) = \chi_l(\bar{r}, \omega)$ is the lth acoustic normal mode satisfying the Helmholtz equation $\nabla^2 \chi - (\omega^2/c^2)\chi = 0$ satisfying either the rigid wall boundary condition (acoustically hard) or flexible wall boundary condition (acoustically soft) as the case may be and of boundedness within the volume V. Note that the acoustic mode is expressed in separable form in the coordinate $\bar{s}$ and s_3.

$$q_j(t) = \int_{-\infty}^{\infty} e^{i\omega\tau}\, dS_{q_j}(\omega),\ q_{Hj}(t) = \int_{-\infty}^{\infty} e^{i\omega t}\, dS_{q_{Hj}}(\omega),\ \eta_l(t) = \int_{-\infty}^{\infty} e^{i\omega\tau}\, dS_{\eta_l}(\omega)$$

are generalized coordinates corresponding to X, X_H and φ, with the corresponding orthogonal random process in the frequency domain $S_{q_j}(\omega)$ etc., satisfying relations of the form (Nigam 1983)

$$E[dS_{q_j}(\omega_1) dS_{q_k}^*(\omega_2)] = \Phi_{q_j q_k}(\omega_1)\, \delta(\omega_2 - \omega_1)\, d\omega_1 d\omega_2 \qquad (5.194)$$

$$E[dS_{q_{Hj}}(\omega_1) dS_{q_{Hk}}^*(\omega_2)] = \Phi_{q_{Hj} q_{Hk}}(\omega_1)\, \delta(\omega_2 - \omega_1)\, d\omega_1 d\omega_2 \qquad (5.195)$$

$$E[dS_{\eta_l}(\omega_1) dS_{\eta_m}^*(\omega_2)] = \Phi_{\eta_l \eta_m}(\omega_1)\, \delta(\omega_2 - \omega_1)\, d\omega_1 d\omega_2 \qquad (5.196)$$

where $\Phi_{q_j q_k}(\omega)$ etc., are the corresponding cross spectral densities, $\delta(\cdot)$ is the Dirac delta function and the asterisk represents the complex conjugate.

Also expanding the external pressure loading $p_e(s, t)$ in terms of the orthonormal modes

$$p_e(\bar{s}, t) = \sum_j u_j(\bar{s}) \int_{-\infty}^{\infty} e^{i\omega t}\, dS_{Q_j}(\omega) = \sum u_j(\bar{s})\, Q_j(t) \qquad (5.197)$$

where $Q_j(t)$ can be considered as the generalized force in the jth normal mode and substituting (5.194) to (5.197) in (5.185) to (5.187) it can be shown that

$$[\hat{\mathscr{L}}(\omega) + \hat{\mathscr{L}}_{11}(\omega)] dS_{q_j}(\omega) + \hat{\mathscr{L}}_{12}(\omega) dS_{q_{Hj}}(\omega) = dS_{Q_j}(\omega) \qquad (5.198)$$

$$\alpha_l'(\omega)\, dS_{\eta_l}(\omega) = i\omega \sum \gamma_{lj}\, dS_{q_{Hj}}(\omega) \qquad (5.199)$$

$$-i\omega\, \rho\alpha_l(\omega)\, dS_{\eta_l}(\omega) = \sum_j \gamma_{lj} [\hat{\mathscr{L}}_{21}(\omega)\, dS_{q_j}(\omega) + \hat{\mathscr{L}}_{22}(\omega)\, dS_{q_{Hj}}(\omega)] \qquad (5.200)$$

where

$$\alpha_l(\omega) = Z_l(s_3, \omega)\,|_{s_3 = -H} \text{ and } \alpha_l'(\omega) = \left.\frac{\partial Z_l(s_3, \omega)}{\partial s_3}\right|_{s_3 = -H} \qquad (5.201)$$

and

$$\gamma_{lj} = \frac{\displaystyle\int_{A_H} v_l(\bar{s})\, u_j(\bar{s})\, d\bar{s}}{\displaystyle\int_{A_H} v_l^2(\bar{s})\, d\bar{s}} \qquad (5.202)$$

which can be considered as acousto-structural modal coupling coefficients. $\hat{\mathscr{L}}(\omega)$, $\hat{\mathscr{L}}_{11}(\omega)$ etc., represent the differential operators in the frequency domain.

To calculate the cross spectral density of the acoustic pressure inside the cavity corresponding to two positions $\bar{r}^1$ and $\bar{r}^2$, namely $\Phi_{p_a p_a}(\bar{r}^1, \bar{r}^2; \omega)$, it is essential to establish a relation between the acoustic potential spectral function $dS_{\eta_l}(\omega)$ and the generalized force spectral function $dS_{Q_j}(\omega)$. Such a relation can be expressed in the form

$$dS_{\eta_l}(\omega) = \sum_j H_{lj}(i\omega)\, dS_{Q_j}(\omega) \qquad (5.203)$$

where $H_{lj}(\omega)$ can be considered as an acousto-structural modal coupling transfer function which depends on γ_{lj}.

We shall now establish relations between the cross spectral densities

$\Phi_{XX}(\bar{s}^1, \bar{s}^2; \omega)$, $\Phi_{X_H X_H}(\bar{s}^1, \bar{s}^2; \omega)$, $\Phi_{p_a p_a}(\bar{s}^1, \bar{s}^2; \omega)$, $\Phi_{p_e p_e}(\bar{s}^1, \bar{s}^2; \omega)$. From (5.198) to (5.202) it can be readily shown that the cross spectral relations are given by

$$\Phi_{q_{Hj} q_{Hk}}(\omega) = \beta_{2l}(\omega) \beta_{2l}^{*}(\omega) \, \Phi_{Q_j Q_k}(\omega) \tag{5.204}$$

$$\Phi_{q_j q_k}(\omega) = \beta_{1l}(\omega) \beta_{1l}^{*}(\omega) \beta_{2l}(\omega) \beta_{2l}^{*}(\omega) \, \Phi_{Q_j Q_k}(\omega) \tag{5.205}$$

$$\Phi_{\eta_l \eta_m}(\omega) = \sum_j \sum_k H_{lj}(\omega) H_{mk}^{*}(\omega) \, \Phi_{Q_j Q_k}(\omega) \tag{5.206}$$

where

$$\beta_{1l}(\omega) = \frac{1}{\hat{\mathscr{L}}_{21}(\omega)} \left\{ \frac{\rho \omega^2 \alpha_l(\omega)}{\alpha_l'(\omega)} - \hat{\mathscr{L}}_{22}(\omega) \right\} \tag{5.207}$$

$$\beta_{2l}(\omega) = \{ [\hat{\mathscr{L}}(\omega) + \hat{\mathscr{L}}_{11}(\omega)] \beta_{1l}(\omega) + \hat{\mathscr{L}}_{12}(\omega) \}^{-1} \tag{5.208}$$

$$H_{lj}(i\omega) = \frac{i\omega \, \gamma_{lj} \beta_{2l}(\omega)}{\alpha_l'(\omega)} \tag{5.209}$$

Multiplying (5.197) by $u_k(\bar{s})$ and integrating over the domain of the structure A, we get

$$\int_A p_e(\bar{s}, t) u_k(\bar{s}) d\bar{s} = \int_A \{ \sum_j u_j(\bar{s}) Q_j(t) \} u_k(\bar{s}) d\bar{s} = Q_k(t) \tag{5.210}$$

where use has been made of the orthogonality of the normal modes and the normalizing condition $\int_A u_j^2(\bar{s}) d\bar{s} = 1$. From (5.110) the cross spectral density of the generalized forces in the jth and kth modes is given by

$$\Phi_{Q_j Q_k}(\omega) = \int_A \int_A u_j(\bar{s}^1) u_k(\bar{s}^2) \Phi_{p_e p_e}(\bar{s}^1, \bar{s}^2; \omega) d\bar{s}^1 d\bar{s}^2 \tag{5.211}$$

Equations (5.204) to (5.209) together with (5.211) and (5.191) to (5.193) constitute the solution of the random vibration problem of the shell containing an acoustic medium and subjected to the external pressure loading which can be assumed to be due to turbulent boundary layer excitation with $\Phi_{p_e p_e}(\bar{s}^1, \bar{s}^2; \omega)$ given by an appropriate function discussed in this chapter. For the actual solution of the problem it is necessary to evaluate the γ_{lj}s in (5.202).

A class of problem for which $v_l(\bar{s}) = u_l(\bar{s})$, that is the forces of shell natural modes coincide with those of the surface acoustic modes $H_{lj}(\omega)$ can be readily evaluated as in this case $\gamma_{lj} = \delta_{lj}$, where δ_{lj}, is the Kronecker delta. Bolotin and Elishakoff (1971) suggest an approximation when $v_l(\bar{s}) \neq u_l(\bar{s})$ by disregarding the X_H terms from (5.185) in which case

$$H_{lj}(i\omega) = \frac{i\omega \, \hat{\mathscr{L}}_{21}(\omega) \, \gamma_{lj}}{[\rho \omega^2 \alpha_l(\omega) - \hat{\mathscr{L}}_{22}(\omega) \alpha_l'(\omega)][\hat{\mathscr{L}}(\omega) + \hat{\mathscr{L}}_{11}(\omega)]} \tag{5.212}$$

To evaluate the mean square acoustic pressure inside the cavity the corresponding psd has to be integrated over the frequency range of interest. Since there will be a large number of cavity and structural resonances the function $H_{ij}(i\omega)$ will have a large number of poles leading to lot of difficulties in the numerical evaluation of the integral. An approximate method of calculating the integral using a steepest descent method has been proposed by Bolotin and Elishakoff (1971) and Bolotin (1984).

5.6 Jet and Rocket Noise

Jet noise and rocket noise are other main sources of random dynamic excitation of aircraft and launch vehicle structural systems. The exhaust from the jet engine and rocket motors generate severe acoustic environments in the wake of the aircraft and launch vehicle. The severity of the acoustic environment generally increases with increasing vehicle size and performance. It is estimated that on an average 0.4 to 0.8% of the total power developed by a conventional rocket motor is converted into acoustic power.

As exhaust gases from the nozzle exit plane of a rocket motor flow into the atmospheric air at high speeds, a zone of turbulent flow develops on the boundaries of the exhaust flow because of the large shear present between the exhaust stream and the surrounding air. This turbulent zone is characterized by a collection of eddies or vortices formed because of the shear and generates a sound field in the surrounding air. Thus the basic mechanism of jet noise generation is the turbulent mixing of the hot exhaust gas with the surrounding air. The thickness of the turbulent mixing zone increases continuously with the distance from the nozzle. The spectra of the acoustic pressure fluctuations radiated by points along the length of jet stream show different frequency concentration. In the fully developed turbulent region of the exhaust flow away from the nozzle, low frequency noise but relatively of higher intensity is generated. But in general the acoustic pressure due to the jet exhaust is characterized by a broad spectral distribution over the audible and sub-audible frequency range. Even though the structural resonance frequencies are usually in the low frequency range, the modes are not strongly excited as the wavelengths corresponding to the low frequency noise are very much larger compared to the dimensions of the structure. However, at high frequencies of the order of 100–10000 Hz where the skin and component resonances are concentrated, the structural dimensions are no longer small compared with the acoustic wavelengths which are of the order of 10–0.1 ft. Consequently at the resonance frequencies, skin and substructures absorb considerable amount of acoustic energy resulting in significant vibratory response.

Both theoretical and experimental investigations have been undertaken on the noise generating mechanism and the associated pressure field of jet noise. The fundamental works of Lighthill (1952, 1954) on the subject show the mechanism of noise generation in the jet stream to be due to the fluctuating Reynold's shear stress. The essence of Lighthill's theory of aerodynamic noise generation is the formulation of an acoustic analogy in which the complicated process of sound generation by turbulence is modelled in terms of an acoustically equivalent set of sound sources embedded in an otherwise uniform medium at rest. He concluded from his analysis that the far field radiation field of the jet noise is caused by a string of quadrapole sound sources along the jet stream.

Using simple order of magnitude studies and dimensional analysis, Lighthill (1952) showed that the total radiated acoustic power in the far field can be expressed as

$$\pi_{rad} \propto \rho_c^2 l_c^2 U_j^8 / (\rho_0 c_0^5) \qquad (5.213)$$

where ρ_c is the characteristic jet density, U_j is the characteristic jet velocity, ρ_0 is the ambient density, l_c a characteristic dimension, c_0 is the ambient speed of sound in the surrounding atmosphere. This is the famous eighth power law of Lighthill for sound radiation in subsonic flows. Callaghen and Coles (1957) have observed excellent correlation between their experiments in the subsonic flow regime, and the modified form of the relationship, namely,

$$\pi_{rad} \propto \rho_0 A U_j^8 / c_0^5 \qquad (5.124)$$

where A is the nozzle exit area. The eighth power law does not predict the acoustic power in a supersonic jet as it assumes that almost the entire mechanical power is radiated as sound. Consequently, several empirical relationships have been developed based on the experiments to predict the acoustic power of a supersonic jet. For supersonic jets the acoustic power radiation can be approximately given by (Chandramani et al, 1967).

$$\pi_{rad} \cong 0.0025 \, (\rho_0 A/2) U_j^3 \qquad (5.215)$$

5.6.1   Statistical Characteristics of Jet Noise Excitation

To estimate the structural response statistics subjected to jet noise and rocket noise excitation, it is essential to know the statistical characteristics of the turbulent pressure field. The far field radiation characteristics of the jet noise are of importance as a source of community noise problem and the parameters on which it depends have been now fairly well understood. But from the point of view of structural response, the loads that the engine excites on the vehicle close to the jet boundary cannot be so simply related to the different parameters of the engine. They depend on the details of the vehicle and engine configuration in a complicated manner.

In predicting the structural loads one has to depend almost entirely on experimental results and empirical fits. Howes et al, (1957), Mayer, Landford and Hubbard (1959), Lawrence (1956) have experimentally investigated the characteristics and correlation pattern in the near field as well as in the far field of the jet noise. The determination of acoustic loads for structural response applications is best illustrated by the work of Clarkson (1965). In two particular studies on the geometric zone pressure field of the Comet engine and in the mockup of the Caravelle tail section, he has obtained data on the scaling laws of pressure correlations between jets of varying sizes. From the point of view of predicting structural loading the region around the jet efflux is divided into two regions, the hydrodynamic turbulent field and the geometric field.

In the hydrodynamic field which extends to about two diameters outward from the jet boundary and downstream to a plane about twelve diameters from the nozzle exit plane, Clarkson (1965) measured the longitudinal and lateral pressure correlations patterns. In the longitudinal direction the pressure correlations are similar to those of boundary layer pressure fluctuation, convecting along the centre line, but in this case the convection velocity depends on the frequency in a strong manner, the dependence being approximately of the form $U_c \propto f^{1/2}$, where U_c is the convection velocity and f is the frequency in Hz. In the near geometric field which extends to about ten diameters out from the jet boundary, twenty diameters downstream and ten diameters upstream of the nozzle exit plane, the longitudinal correlation pattern is directional and depends on the angle which the correlation traverse line makes with the line joining the reference point to the source. It can be said as a result of experimental measurements the correlation function along the axial direction on the structural surfaces of the noise field is approximately sinusoidal. The lateral correlations are also similar except that the extent of the correlation in the circumferential direction is somewhat larger than that in the axial direction further emphasizing the fact that the jet noise pressure field is a highly directional source.

The spatial correlation functions along the axial direction can be represented by fitting curves to the experimental data. For example Callaghan, Howes and Coles (1957) have presented two empirical forms of correlation functions along the axial given by

$$R_{pp}(\xi, \omega) = p_{rms}^2 \cos\left(\frac{\omega \cos(\theta)}{c} \xi\right) \tag{5.216}$$

$$R_{pp}(\xi, \omega) = p_{rms}^2 \left(1 - 1/2(\omega \cos\theta/c)^2 \xi^2 + 1/24(\omega \cos\theta/c)^4 \xi^4\right) \tag{5.217}$$

where θ is the angle between the s_1 coordinate and the predominant direction of sound propagation, ξ is the spatial separation along the s_1 direction and c is the speed of sound.

It has been observed that by using suitable scaling laws the psd functions of the jet noise pressure field for all types of jet flows can be collapsed into a normalized power spectrum which can be expressed as a function of the nondimensional frequency parameter, the Strouhal number fD/U. Eldred (1960) has shown that by such a technique good fits can be obtained for the power spectra of turbojets, rockets and subsonic cold or hot jets over a wide frequency range. The power spectral density obtained by Clarkson (1957) from an octave band analysis of sound measurement in a ground running test on a Comet 1 aircraft is shown in Fig. 5.31. The power spectrum exhibits a typical peak characteristic of the jet noise excitation. Kanematsu (1971) has fitted an empirical curve for the one sided power spectrum in the form of a filtered white noise spectrum

$$G_{pp}(\omega) = \frac{0.0307(\omega/1650)^2}{[1 - (\omega/1600)^2]^2 + (1.6\omega/1650)^2} \tag{5.218}$$

which is also shown in the same figure. This gives a close approximation to the measured spectra.

It could be said in conclusion that the structural excitation loads due to acoustic noise from jets, subsonic or supersonic must be based on direct measurements of

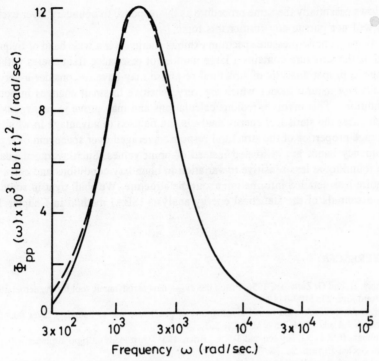

Fig. 5.31 Jet noise power spectrum _____ Measured (Clarkson, 1970);
- - - - eq. (5.218) (Kanematsu, 1971).

the loads on similar or scaled vehicles. Theoretical analysis of sound generation
and propagation are at best useful guides in the prediction of frequency scales and
overall levels of pressure fluctuations.

5.7   Structural Response to Jet Noise Excitation

Once the excitation field of the jet noise is specified in terms of the psd function
and the correlation functions, the response statistics can be estimated by the use
of the normal mode approach. Clarkson and Ford (1962) and Clarkson (1962) have
presented a response analyis of typical aircraft structures to jet noise excitation
based on many simplifying asumptions. In their analysis the jet noise is approximated
to be a stationary Gaussian random process. The local white noise assumption is
made with regards to the exciting function. Only a few normal modes which are
in the vicinity of the frequency of maximum power spectral values of the pressure
are taken into consideration in the modal summation. The contributions from cross
modal coupling terms to the total response are neglected under the assumption of
low damping and widely separated frequencies. Other examples of structural response
of a simple panel, sandwich panel and a sheet stiffener combination to jet noise
excitation using the correlation functions of (5.166) have been presented by Nigam
(1983), Wallace (1965) and Narayanan (1975). An example of the response of a
sheet stringer combination to jet noise excitation is presented in Chapter 4 in a
general optimization problem. As the response analysis by normal mode method

follows essentially the same procedure as the response to boundary layer excitation we will not pursue any further this topic.

As the jet noise pressure spectrum exhibits energy over a wide band of frequencies and as the structure contains a large number of resonance frequencies within this range, a proper analysis of structural response may have to consider a very large number of natural modes which are very sensitive to small changes in boundary conditions. This involves tedious calculations and may prove to be expensive. In such cases the statistical energy analysis can be used to advantage in which gross average properties of the structural response averaged over space and over certain frequency bands are estimated instead of point values. Such averaged quantities are found to be less sensitive to variation in boundary conditions and calculations require less detailed information about the structure. We shall treat in some detail the essentials of the statistical energy analysis (SEA) method in Chapter 10.

REFERENCES

Ashley, H. and G. Zartarian, 1956. Piston theory—a new aerodynamic tool for the aeroelastician. J. Aero. Sci. 23: 1109–1118.

Bies, D.A., 1966. A wind tunnel investigation of panel response to boundary layer fluctuation at Mach 1.4 and Mach 3.5 NACA CR 501.

Bisplinghoff, R.L., G. Isakson and T.F. O' Brien, 1951. Gust loads on rigid airplanes with pitching neglected J. Aero. Sc. 18: 33–42.

Bisplinghoff, R.L., H. Ashley and R.L. Halfman, 1955. Aeroelasticity, Addison-Wesley, Reading, Massachussets.

Bolotin, V.V. and I.Elishakoff, 1971. Random oscillations in elastic shells containing an acoustic medium. Mech. Solids. 5: 99–106.

Bolotin, V.V., 1984. Random Vibrations of elastic systems. Martinus Nijhoff publishers. Hague.

Bull, M.K., 1963. Properties of the fluctuating wall-pressure field of a turbulent boundary layer. University of Southampton. AASU Report No. 234.

Bull, M.K., J.F. Wilby and D.R. Blackman, 1963. Wall pressure fluctuations in boundary layer flow and response of simple structures to random pressure fields. University of Southampton. AASU Report No. 243.

Bull, M.K., 1967. Wall pressure fluctuations associated with subsonic turbulent boundary layer flow. J. Fluid. Mech. 28: 719–754.

Callaghen, E.E., W.L. Howes and W.D. Coles, 1956. Near noise field of a jet engine exhaust, II. Cross correlation and sound pressure. NACA. TN. 3764.

Callaghen, E.E. and W.D. Coles, 1957. Far noise field of air jets and jet engines. NACA Report 1329.

Chandiramani, K.L., S.E. Widnall, R.H. Lyon and P.A. Franken 1966. Structural response to inflight acoustic and aerodynamic environments. BBN Report 1417.

Chyu, W.J. and M.K. Au Yang, 1972. Random response of rectangular panels to the pressure fields beneath a turbulent boundary layer in subsonic flows. NACA TND-6970.

Clarkson, B.L., 1957. Stresses produced in aircraft structures by jet efflux. J. Roy. Aero. Sci. 61: 110–112.

Clarkson, B.L., 1962. The design of structures to resist jet noise fatigue. J. Roy. Aero. Soc. 66: 603–616.

Clarkson, B.L. and R.D. Ford, 1962. The response of a typical aircraft structure to jet noise. J. Roy. Aero. Soc. 66: 603–616.

Clarkson, B.L., 1965. Scaling of near field pressure correlation patterns around a jet exhaust. Acoustical Fatigue in Aerospace Structures, W.J. Trapp and D.M. Forney, ed. Syracuse University Press N.Y., pp. 19–38.

Clementson, G.C., 1950. An investigation of the power spectral density of atmospheric turbulence. SCD Thesis M.I.T. Cambridge, Massachussets.

Corcos, M.J., 1963. Resolution of pressure in turbulence. J. Acous. Soc. Am. 35: 192–199.

Crocker, M.J., 1969. The response of a supersonic transport fuselage to boundary layer and to reverberant noise, J. Sound. Vib. 9: 6–20.

Davies, H.G., 1969. Acoustic radiation from fluid loaded rectangular plates. M.I.T. Acoustic and Vibration Laboratory Report 71476-1.

Davies, H.G., 1971. Sound from turbulent boundary layer excited panel. J. Acous. Soc Am. 49: 878–889.

Dowell, E.H., 1969. Transmission of noise from a turbulent boundary layer through a flexible plate into a closed cavity. J. Acous. Soc. Am. 46: 238–252.

Dowell, E.H., 1970. Panel flutter: A review of aeroelastic stability of plates and shells. AIAA. J. 8: 385–399.

Dyer, I., 1959. Response of plates to a decaying and convecting random pressure field. J. Acous. Soc. Am. 31: 922–928.

Dzygaldo, Z., 1966. Forced vibration of a plate on hinged supports in supersonic flow. Proc. Vib. Problems. 7: 121–134.

Dzygaldo, Z., 1967. Forced vibration of a plate of a finite length in plane supersonic flow. Proc. Vib. Problems. 8: 61–77.

Easter, F.E. and S.C. McIntosh. Jr, 1971. Analysis of non linear panel flutter and response under random excitation or non linear aerodynamic loading. AIAA. J. 9: 411–418.

ElBaroudi., 1964. Turbulence induced panel vibrations. UTIA Report 98, University of Toronto.

Eldred, K.M., 1960. Review of the noise generation of rockets and jets, J. Acous. Soc. Am. 32: 1502.

Elishakoff, I. and V. Khromatoff, 1971. Statistical analysis of the vibrations of a panel in supersonic flow. Mech. Solids. 5: 41–45.

Elishakoff, I. and B. Efimtsoff, 1972, Vibrations of unbounded plates in a random force field. Soviet App. Mech. 8: 103–107.

Elishakoff, I., 1973. Random vibration of a two span beam. J. Israel Inst. Tech. 11: 317–320.

Elishakoff, I., 1974. Mean square stability of elastic bodies in supersonic flow. J. Sound. Vib. 33: 67–78.

Elishakoff, I., 1975. Turbulent flow excited vibration of a shallow cylindrical shall. AIAAJ, 13: 1109–1118.

Elishakoff, I., 1977a. Flutter and radom vibration in plates. Stochastic Problems in Dynamics. B.L. Clarkson. ed. Pitmen. London. 390–411.

Elishakoff., I., 1977b. On the role of cross-correlations in random vibrations of shells. J. Sound. vib. 50: 239–252.

Elishakoff, I., 1977c. Random vibrations of orthotropic plates with all edges clamped or simply supported. Acta Mechanica. 28: 165–176.

Elishakoff, I., 1983. Probabilistic methods in the theory of structures. John Wiley & Sons. New York.

Elishakoff, I., Th. Van Zanten, and S.H. Crandall, 1979. Wide band random axisymmetric vibration of cylindrical shells. Trans. ASME. J. App. Mech. 96: 417–423.

Elishakoff, I. and D. Livshits, 1984. Some closed form solutions in random vibrations of Bernoulli-Euler beams. Int. J. Eng. Sci. 22: 1291–1302.

Elishakoff, I. D. Livshits, 1989. Some closed form solutions in random vibrations of Timoshenko beams. J. Prob. Mech. 4: 47–54.

Fahy, F.J., 1970. Response of cylinder to random sound in the contained fluid. J. sound. vib. 13: 171–194.

Federoff, Yu., 1963. Vibration of closed circular cylindrical shell in a field of random acoustic pressures. Inzherernyi zhurnal 3: 498–503.

Fisher, M.J. and P.O.A.L. Davies, 1964. Correlation measurement in a non-frozen pattern of turbulence. J. Fluid. Mech. 18: 97–116.

Fujimori, Y. and Y.K. Lin, 1973a. Analysis of airplane response to nonstationary turbulence including wing bending flexibility. AIAA. J. 11: 334–339.

Fujimori, Y. and Y.K. Lin, 1973b. Analysis of airplane response to nonstationary turbulence including wing bending flexibility-II, AIAA. J. 11: 1343–1345.

Fung, Y.C., 1953. Statistical aspects of Dynamic loads. J. Aero. Sci. 20: 317–330.

Gaonkar, G.J., K.H. Hohenmser and S.K. Yin, 1972. Random gust response statistics for coupled torsion-flapping rotor blade vibrations. J. Aircraft, 9: 726–729.

Gaonkar, G.J., 1980. Review of nonstationary gust-response of flight vehicles. 21st Struct. Dyn. Mat. Conf. AIAA/ASME/ASCE/AHS/Seattle, Washington.

Houbolt, J.C., R. Steiner, and K.G. Pratt, 1964. Dynamic responses of airplanes to atmospheric turbulence including flight data on input and response, NACA. TR. R. 199.

Houbolt, J.C., 1973. Atmospheric turbulence, AIAA. J. 11: 421–437.

Howell, L.J. and Y.K. Lin, 1971. Response of flight vehicles to nonstationary atmospheric-turbulence. AIAA. J. 9. 2201–2207.

Howes, W.L., E.E. Callaghan, W.D. Coles, and H.R. Mull, 1957. Near noise field of a jet engine exhaust, NACA Report 1338.

Hwang, C. and W.S. Pi, 1969. Random acoustic response of a cylindrical shell. AIAA. J. 7: 2204–2210.

Jones, G.W., J.W. Jones and K.R.A Monson, 1969a. Interim analysis of low altitude atmospheric turbulence (Lolocat) data, ASD-TR-69-7, Wright Patterson Air Force Base, Ohio.

Jones, G.W., J.W. Jones and K.R.A. Monson, 1969b. Low altitude atmospheric turbulence lo-locat Phase III Interim Report, AFFOL-TR 69-63 vol. I and II Wright Patterson Air Force Base, Ohio.

Kana, D.D., 1971. Response of a cylindrical shell to random acoustic excitation. AIAA. J. 9: 425–431.

Kanematsu, H., 1971. Random vibration of thin elastic plates and shallow shells. Mean square response of a non-linear system to non-stationary random excitation. Ph.D. Thesis, University of Massachussets.

Kistler, A.L. and W.S. Chen, 1962. The fluctuating pressure field in a supersonic turbulent boundary layer. J. Fluid Mech. 16: 41–64.

Kraichnan, R.H., 1956. Pressure fluctuations in a turbulent flow over a flat plate, J. Acous. Soc. Am. 28: 378–390.

Lawrence, J.C., 1956. Intensity, scale and spectra of turbulence in mixing region of free subsonic jet. NACA TR. 1292.

Lee, J., 1976. The plunging mode response of an idealized airplane to atmospheric turbulence. J. Sound. Vib. 44: 47–62.

Leibowitz, R.C., 1971. Methods for computing fluid loading and the vibratory response of fluid loaded finite rectangular plates subject to turbulence excitation option. 3. NSRDC Report 2976 C.

Leibowitz, R.C., 1972. Methods for computing radiation damping and the vibratory response of fluid loaded and acoustically radiating finite rectangular plates subject to turbulence excitation. Option 4. NSRDC Report 2976 D.

Leibowitz, R.C., 1975. Vibroacoustic response of turbulence excited thin rectangular finite plates in heavy and light fluid media, J. Sound. Vib. 40: 441–495.

Liepmann, H.W., 1952. On the application of statistical concepts to the buffeting problem. J. Aero. Sci. 19: 793–812.

Lighthill, M.J., 1952. On sound generated aerodynamically, Part I. General Theory. Proc. Roy. Soc. London, vol. A 221: 564–587.

Lighthill, M.J., 1954. On Sound generated aerodynamically, Part II. Turbulence as a source of Sound, Proc. Roy. Soc. London. vol. A 222: 1–32.

Lilley, G.M., 1960. Pressure fluctuations in an incompressible turbulent boundary layer. College of Aeronautics, Cranfield Report. No. 133.

Lin, Y.K., 1965. Transfer matrix representation of flexible airplanes in gust response study. J. Aircraft. 2: 116–121.

Lin, Y.K., 1967. Probabilistic Theory of Structural Dynamics. McGraw-Hill Book Company, New York.

Lin. Y.K., S. Maekawa, H. Nijim and L. Maestrello, 1977. Response of a periodic beam to supersonic boundary layer pressure fluctuation in stochastic problems in dynamics. ed. B.L. Clarkson Pitman. London. 468–485.

Maekawa, S and Y.K. Lin, 1977. Vibroacoustic response of structures and perturbation Reynold's stress near structure turbulence interface, NACA CR. 2876.

Maestrello, L., 1965a. Measurement of noise radiated by boundary layer excited panels. J. Sound. Vib. 2: 100–115.

Maestrello, L., 1965b. Measurement and analysis of the response field of turbulent boundary layer excited panels. J. Sound. Vib. 2: 270–292.

Mastrello, L., 1967. Use of turbulent model to calculate the vibration and radiation responses of a panel, with practical suggestions for reducing sound level. J. Sound. Vib. 15: 407–448.

Maestrello, L., 1969. Radiation from panel response to a supersonic turbulent boundary layer. J. Sound. Vib. 10: 261–295.

Maestrello, L. and T.L.J. Linden, 1971. Response of an acoustically loaded panel excited by supersonically convected turbulence. J. Sound. Vib. 16: 365–384.

Mark, W.D., 1981. Characterization, parameter estimation and aircraft response statistics of atmospheric turbulence. NASA CR 3463.

Mayer, W.H., W.E. Landford and H.H. Hubbard, 1959. Near field and far field noise surveys of solid fuel rocket engines for a range of nozzle exit pressure, NACA TN D-21.

Morse, P.M., 1948. Vibration and Sound 2nd ed. McGraw-Hill, New York.

Morse, P. and H. Feshback, 1953. Methods of Theoretical Physics. Vol. 1, McGraw-Hill Book Company, New York.

Murrow, H.N. and R.H. Rhyne, 1974. Atmospheric turbulence measurements with emphasis on long wave lengths. Proc. 6th Conf. Aero. Met. Soc. 313–316.

Nogornov, L.N., 1967. On the random vibrations of plates in supersonic turbulent flow of gas. Sci. Conf. Moscow. Ener. Ins. Collection 251–263.

Narayanan, S., 1975. Structural optimization in random vibration environment, Ph.D. thesis. Indian Institute of Technology, Kanpur.

Narayanan, S. and R.L. Shanbhag, 1984. Sound transmission through layered cylindrical shells with applied damping treatment. J. Sound. Vib. 92: 541–558.

Nemat-Nasser, S., 1968. On the response of shallow thin shells to random excitations. AIAA. J. 6: 1327–1331.

Nigam, N.C. 1983. Introduction to Random Vibrations, M.I.T. Press, Cambridge, Massachussets.

Peele, E.L. and R. Steiner, 1970. A simplified method of estimating the response of light aircraft to continuous atmospheric turbulence, J. Aircraft. 7: 402: 407.

Piersol, A.G., 1969. Investigation of the statistical properties of atmospheric turbulence data. Measurement analysis Corporation, Technical Report 28032–07, California.

Pope, L.D., 1971. On the transmission of sound through finite closed shells, statistical energy analysis, modal coupling and non-resonant transmission. J. Acous. Soc. Am. 50: 1004–1018.

Press, H., 1957. Atmospheric turbulence environment with special reference to continuous turbulence, AGARD Report 115.

Press, H. and B. Mazelsky, 1953. A study of power spectral methods of generalized harmonic analysis to gust loads on airplanes, NACA. TN 3312.

Press, H., M.T. Meadows and I. Hadlock, 1955. Estimates of probability distribution of root mean square gust velocity of atmospheric turbulence from operational gust-load data by random process theory. NACATN 3312.

Priestley, M.B., 1967. Power spectral analysis of non-stationary random processes, J. Sound. Vib. 6: 86–97.

Reeves, P.M., 1974. A non-Gaussian model of continuous atmospheric turbulence for use in aircraft design. Ph.D. Dissertation, University of Washington.

Reeves, P.M., G.S. Campbell, V.M. Ganzer and R.G. Joppa, 1974. Development and application of a non-gaussian atmospheric turbulence model for use in flight simulators, NASA CR-2451.

Reeves, P.M., R.G. Joppa and V.M. Ganzer, 1975. A non-gaussian model of continuous atmospheric turbulence proposed for use in aircraft design. 13th AIAA. Aero. Sp. Sci. Paper No. 75–31, Pasadena, California.

Rhyne, R.H., 1976. Flight assessment of an atmospheric turbulence measurements system with emphasis on long wave lengths, NASA TN-D 8315.

Ribner, H.S., 1956. Boundary layer induced noise in the interior of aircraft, UTIA Report 37, University of Toronto.

Rice, S.O., 1944. Mathematical analysis of random noise. Bell System Tech J. 23: 282–332.

Ryan, J.P., A.P. Berns, A.C. Robertson, R.J. Dominie and K.C. Rolle, 1971 Medium altitude critical atmospheric turbulence (MEDCAT) Data Processing and Analysis. AFFDL-TR-71-82. Wright Patterson Air Force Base, Ohio.

Strawderman, W.A., 1969. Turbulence induced plate vibrations: An evaluation of finite and infinite plate modes. J. Acous. Soc. Am. 46: 1294–1307.

Strawderman, W.A. and R.S. Brand, 1969. Turbulent flow excited vibration of a simply-supported rectangular flat plate. J. Acous. Soc. Am. 45: 177–192.

Strawderman, W.A. and R.A. Christman, 1972. Turbulent induced plate vibrations: Some effects of fluid loading on finite and infinite plates. J. Acou. Soc. Am. 52: 1537–1552.

Trevino, G., 1981. Effects of inhomogeneities in atmospheric turbulence on the dynamic response of a rigid aircraft to nonstationary atmospheric turbulence, AIAA J. 11: 1087–1092.

Vaicaitis, R., C.M. Jan, and M. Shinozuka, 1972. Non-linear panel response from a turbulent boundary layer. AIAA. J. 10: 895–899.

Vaicaitis, R. and Y.K. Lin, 1972. Response of finite periodic beam to turbulent boundary layer excitation. AIAA. J. 16: 1020–1024.

Verdon, J.M. and R. Steiner, 1973. Response of a rigid aircraft to nonstationary atmospheric turbulence. AIAA J. 1087–1092.

Wallace, C.E., 1965. Stress response and fatigue life of acoustically excited panels. Acoustical Fatigue in Aerospace Structures. Trapp, W.J. and Forney, D.M. ed. Syracuse university Press, New York, 225–244.

White, P.H., 1966. Sound transmission through a finite closed cylindrical shell. J. Acous. Soc. Am. 40: 1121–1130.

Wilby, J.F., 1967, The response of simple panels to turbulent boundary layer excitation. AFFDL-TR-67–70.

Wilby, J.F. and F.L. Gloyna, 1972. Vibration measurements of an airplane fuselage structure. I. Turbulent boundary layer excitation. J. Sound. Vib. 23: 443–466.

Willmarth, W.W. and C.E. Woolridge, 1962. Measurements of the fluctuating pressure at the wall beneath a thick turbulent boundary layer. J. Fluid Mech. 14: 187–210.

Willmarth, W.W. and F.W. Roos, 1965. Resolution and structure of the wall pressure field beneath a turbulent boundary layer. J. Fluid Mech. 22: 81–94.

Yen, D.H.Y., L. Maestrello and S. Padula, 1975. On an Integro-differential equation model for the study of the response of an acoustically coupled panel. 2nd AIAA Aero. Acous. Conf. Hampton, Virginia.

Yen, D.H.Y., L. Maestrello and S. Padula, 1980. Response of a panel to a supersonic turbulent boundary studies on a theoretical model. J. Sound Vib. 71: 271–282.

6. Response of Vehicles to Guideway Unevenness

6.1 Introduction

The vehicles constituting the modern transport system can be divided into three groups:

(i) Road vehicles, which include two wheel vehicles—bicycles, motor cycles; four wheel vehicles—cars, buses and trucks; and articulated vehicles—truck-trailer combinations;

(ii) Rail vehicles, which include locomotives, passenger coaches and wagons; and

(iii) Aircrafts, which include small and large passenger and military aircrafts.

Each of these vehicles operate on prepared tracks, namely, roads, rails and runway/taxiway. We shall use the generic word 'guideway' to refer to these prepared tracks. Even the best of guideways exhibit random spatial unevenness about a mean level and are the source of random vibration in a moving vehicle. With an increasing trend towards higher operating speeds and complex multi-body, multi-wheel vehicles, the dynamic aspects of vehicle behaviour have become important. The major consideration in the design and operation of vehicles include:

(i) passenger comfort and cargo safety;
(ii) ride safety and speed potential of a guideway;
(iii) design of suspension system and other components for fatigue and strength; and
(iv) stability and controlability of vehicles.

The dynamic response of vehicle to guideway induced vibration depends on the nature of guideway unevenness, vehicle motion, guideway-vehicle contact and the dynamic characteristics of the vehicle. To carry out a random vibration analysis of a vehicle, it is necessary to construct a stochastic model of the guideway profile. Depending on the vehicle motion and the nature of wheel-guideway contact, the guideway unevenness is transformed to time-dependent random excitation(s) at each contact point. If a mathematical model of the vehicle is constructed, the response of vehicle-guideway system is reduced to a set of differential equations with random forcing function. The theory of random vibration can be used to determine the distribution of the response and the vehicle design can be implemented in a reliability framework.

The significant aspects of random vibration analysis of vehicles include; nonlinear behaviour of vehicles, particularly the suspension system, and rail-wheel contact forces in the railway systems; and nonstationary response during acceleration and deceleration phases of vehicle motions specially in aircrafts. The method of equivalent linearisation has been extensively used to treat the nonlinear behaviour analytically. The state space formulation of vehicle equations of motion in the time and space domains has proved very effective in the analytical treatment of nonstationary response. Finally, the design requirements of ride safety and ride comfort often impose conflicting demands on vehicle design. These can be resolved by formal optimization procedures subject to operational and functional constraints.

In this chapter, we shall first discuss the physical features of the guideways—roads and railway tracks—and their stochastic models. We then consider the linear and nonlinear vehicle models, including the models of guideway-wheel contact forces. The stationary and nonstationary random vibration of vehicles moving on a guideway is then treated under a unified approach. We conclude with a discussion on the design of vehicle systems, including optimization, in a reliability framework.

Several examples are included to illustrate the application of the methods discussed. The discussion in the chapter is confined to vehicles with passive suspension elements. During the past few years, considerable interest has been generated in the use of active suspension elements to control the vehicle behaviour specially with regard to ride comfort and ride safety. This extends the random vibration analysis to include control theory. We give a brief review of the state-of-the-art of vehicles with active suspension systems in Appendix 6B.

6.2 Stochastic Models of Guideways

6.2.1 Roads/Runways

A road/runway consists of a prepared two-dimensional surface of finite width, a nominal camber and grade. Since road/runway pavements are considerably more rigid than vehicle tyres, we assume that they are perfectly rigid, and, therefore, the profiles measured under no-load conditions represent the dynamic *roughness*. The measured surface elevations of guideways exhibit random fluctuations about the nominal geometry, and can be treated as a two-dimensional random process, $\tilde{Z}(s_1, s_2)$, with space coordinates s_1 and s_2 as the indexing parameters as shown in Fig. 6.1. If isolated large fluctuations, such as, potholes are disregarded, the fluctuations of the guideway surface can be treated as a homogeneous, Gaussian random process with zero mean (Parkhilovskii, 1968; Dodds and Robson, 1973). Under these

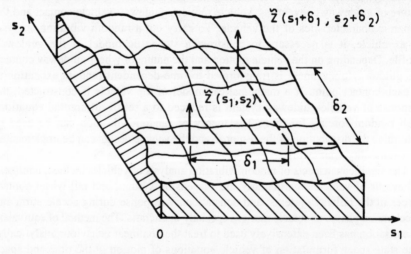

Fig. 6.1. Two-dimensional guideway roughness.

assumptions, the probability structure of guideway profile is completely defined by the two-dimensional auto-correlation function:

$$R(\delta_1, \delta_2) = E[\tilde{Z}(s_1, s_2)\, \tilde{Z}(s_1 + \delta_1, s_2 + \delta_2)] \qquad (6.1)$$

or two dimensional power spectral density (psd) $\Phi(k_1, k_2)$, where k represents the wave number (spatial frequency). To determine the two-dimensional auto-correlation function, or psd, it is necessary to measure the surface profile with respect to a reference plane at a large number of guideway cross-sections. This is a formidable task, both in data acquisition and numerical computation. Efforts have, therefore, been made to simplify guideway models (Parkhilovskii, 1968; Dodds, 1974; Dodds and Robson, 1973).

Single Profile Model
The simplest model describes the guideway surface as a cylindrical surface defined by a single longitudinal profile $\tilde{Z}(s)$. Assuming $\tilde{Z}(s)$ to be a zero mean, homogeneous, Gaussian random process, its probability structure can be defined by the auto-correlation function, or by the psd. The psd can be obtained by a spectral analysis of the guideway profile measured along any longitudinal section (usually centre-line). Figure 6.2 shows the psd of two parallel tracks along three different types of roads as a function of wave number. It may be noted that the psd of two tracks are quite close, justifying the assumption of a cylindrical surface.

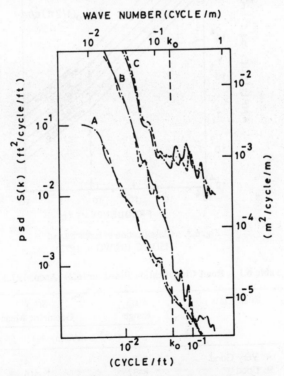

Fig. 6.2.   Spectral densities of two parallel tracks along three different types of motor roads. A—motorway; B—minor road; C—paved road (Dodds and Robson, 1973).

For a random vibration analysis of vehicles moving on a guideway, it is necessary to fit an analytical expression to the measured psd (La Barre and Forbes, 1969; Robson and Dodds, 1970, and Anon, 1972). Following analytical description has been proposed to fit the measured psd

$$S(k) = S(k_0)\left(\frac{k}{k_0}\right)^{-\gamma_1} \qquad k \leq k_0$$

$$= S(k_0)\left(\frac{k}{k_0}\right)^{-\gamma_2} \qquad k \geq k_0 \qquad (6.2)$$

where $S(k)$ is the one-sided spatial power spectral density (m²/c/m); $S(k_0)$ is the roughness coefficient; $k_0 = 1/2\pi$ (c/m); and $\gamma_1 = 2.0$ and $\gamma_2 = 1.5$ are waviness parameters. Figure 6.3 shows the classification of roads based on (6.2) and the values of roughness coefficient given in Table 6.1. A simpler expression over the

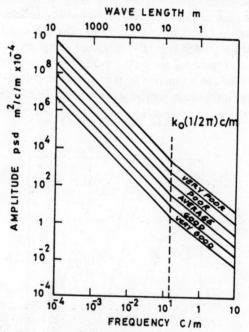

Fig. 6.3. Classification of roads by psd
(ISO/TC 108/WG-9, 1972).

Table 6.1 **Road Classification Based on (6.2)** (Anon, 1972a)

Road Class	$S(k_0)$ Range	$S(k_0)$ Geometric Mean
	$(10^{-6} - $ m²/c/m$)$	
A: Very Good	2–8	4
B: Good	8–32	16
C: Average	32–128	64
D: Poor	128–512	256
E: Very Poor	512–2028	1024

entire frequency range

$$S(k) = S(k_0)\left(\frac{k}{k_0}\right)^{2.5} \tag{6.3}$$

may also be used.

The mathematical expression (6.2) for the psd is simple and convenient for computational and design purposes. However, it creates mathematical difficulties because $S(k)$ becomes infinite at $k = 0$. This difficulty can be circumvented by filtering the psd through a high-pass filter with transfer function

$$H(ik) = \frac{i\lambda_c(\Omega)}{1 + i\lambda_c(\Omega)} \tag{6.4}$$

where $\Omega = 2\pi k$; and λ_c is the largest wave length of interest. This leads to a finite value of psd at $k = 0$. Yadav and Nigam (1978) proposed the following analytically well behaved expression for psd:

$$S(\Omega) = \frac{C}{\Omega^2 + a^2} \qquad \Omega \leq \Omega_c$$

$$= 0, \qquad\qquad \Omega < \Omega_c \tag{6.5}$$

where Ω_c is the cut-off frequency; and C the roughness constant. For $\Omega_c \to \infty$, the auto-correlation function corresponding to (6.5) is

$$R(\delta) = \frac{\pi C e^{-a|\delta|}}{a} \tag{6.6}$$

Road profile corresponding to (6.6) may be generated by filtering white-noise through the following first-order linear filter

$$d\tilde{Z}/ds + a\tilde{Z} = W(s) \tag{6.7}$$

where $W(s)$ is white noise with one-sided psd $S_0 = 2C$.

Double Profile Model

The single-profile stochastic model of the guideway is adequate for two-wheel vehicles, such as, bicycles or motor cycles. For vehicles, such as, car, tractor-trailer, aircraft having two or more wheel per axle, the model does not account for different excitations at different points across the width of the guideway. We now consider a stochastic model of the guideway for vehicles having a pair of wheels moving on two parallel tracks separated by a constant distance $2b$, as shown in Fig. 6.4. The longitudinal profile along the two tracks is represented by the random processes $\tilde{Z}_i(s)$, $i = 1, 2$, obtained from the guideway surface $\tilde{Z}(s_1, s_2)$. From (6.1),

$$R_{11}(\delta) = E(\tilde{Z}_1(s)\,\tilde{Z}_1(s + \delta)] = R(\delta, 0)$$

$$R_{22}(\delta) = E[\tilde{Z}_2(s)\,\tilde{Z}_2(s + \delta)] = R(0, \delta) \tag{6.8}$$

Since $\tilde{Z}(s_1, s_2)$ is homogeneous,

$$R_{11}(\delta) = R_{22}(\delta) \tag{6.9}$$

and the cross-correlation functions

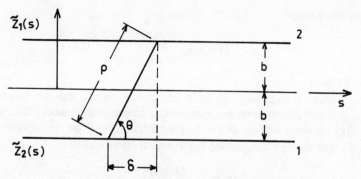

Fig. 6.4. Two parallel tracks—isotropic guideway model.

$$R_{12}(\delta) = E[\tilde{Z}_1(s)\,\tilde{Z}_2(s+\delta)]$$

$$= E[\tilde{Z}(s,-b)\,\tilde{Z}(s+\delta,b)]$$

$$= R(\delta, 2b) \tag{6.10a}$$

$$R_{21}(\delta) = R(\delta, -2b) \tag{6.10b}$$

Since $R(\delta_1, \delta_2)$ is an even function of δ_1 and δ_2,

$$R_{12}(\delta) = R_{21}(\delta) = R_{12}(-\delta) \tag{6.11}$$

We now assume that $\tilde{Z}(s_1, s_2)$ is an isotropic random surface (Dodds and Robson, 1973, Kamash and Robson, 1977; 1978). Isotropy requires that for any profile making an angle θ with x-axis (Fig. 6.4):

$$R(\rho\cos\theta, \rho\sin\theta) = R(\rho) \tag{6.12}$$

From (6.8), (6.9) and (6.12), it follows that

$$R_{11}(\delta) = R_{22}(\delta) = R(\delta) \tag{6.13}$$

$$R_{12}(\delta) = R_{21}(\delta) = R(\rho) = R(\sqrt{\delta^2 + 4b^2}) \tag{6.14}$$

Since the autocorrelation function and psd form a Fourier transform pair,

$$R(\delta) = \frac{1}{2}\int_0^\infty S(k)\,e^{i\pi 2k\delta}\,dk \tag{6.15}$$

$$S(k) = 2\int_0^\infty R(\sqrt{\delta^2 + 4b^2})\,e^{-i2\pi k\delta}\,d\delta \tag{6.16}$$

Thus, for an isotropic random surface, once a single-profile spectrum, $S(k)$, is defined, the cross-spectral densities $S_{12}(k)$ of the profiles of any parallel pair of tracks can be deduced. Note that, since $R_{12}(\delta)$ are even, $S_{12}(k)$ are real.

Dodds and Robson (1973) have tested the hypothesis of isotropy by careful analysis of experimental data and have concluded that it provides a valid model for road surface description. Kamash and Robson (1977, 1978) have examined the mathematical implications of isotropy and application of isotropy to road surface modelling. An alternative model of two parallel tracks was proposed by Parkhilovskii

(1968). This model assumed that the two profiles, $\tilde{Z}_i(s)$, $i = 1, 2$, are derived from two uncorrelated random processes $\tilde{Z}(s)$ and $\theta(s)$ (Fig. 6.4), that is

$$\tilde{Z}_1(s) = \tilde{Z}(s) + b\theta(s)$$

$$\tilde{Z}_2(s) = \tilde{Z}(s) - b\theta(s) \qquad (6.17a)$$

$$\tilde{\bar{Z}}(s) = \mathbf{T}\bar{U}(s) \qquad (6.17b)$$

where $\tilde{\bar{Z}}(s) = [\tilde{Z}_1, \tilde{Z}_2]^T$, $\bar{U}(s) = [\tilde{Z}(s), \theta(s)]^T$,

and

$$\mathbf{T} = \begin{bmatrix} 1 & b \\ 1 & -b \end{bmatrix} \qquad (6.18)$$

It follows that

$$\mathbf{S}_{\tilde{Z}}(k) = \mathbf{T}\mathbf{S}_U(k)\mathbf{T}^T \qquad (6.19a)$$

or

$$S_{ii}(k) = S_{\tilde{Z}}(k) + b^2 S_\theta(k), \, i = 1, 2$$

$$S_{ij}(k) = S_{\tilde{Z}}(k) - b^2 S_\theta(k), \, i \neq j = 1, 2 \qquad (6.19b)$$

since

$$\mathbf{S}_U(k) = \begin{bmatrix} S_{\tilde{Z}}(k) & 0 \\ 0 & S_\theta(k) \end{bmatrix} \qquad (6.19c)$$

Thus, direct and cross-psd can be expressed in terms of direct psds of $Z(s)$ and $\theta(s)$. Robson (1978) has examined the physical and mathematical basis of Parkhilovskii model. He has shown that it can be made compatible with isotropic model for a profile-pair description of guideway, and may be used where isotropy assumption is not valid.

6.2.2 Guideway Induced Excitation

As a vehicle traverses a guideway, the wheels follow the profile and transmit a time-dependent vertical displacement to the vehicle at each contact point. The temporal nature of the transmitted excitations depends on the velocity of the vehicle. In this regard, two cases are distinguished: (i) vehicles moving with constant velocity V; and (ii) vehicle moving with a variable velocity. In the first case, $s = Vt$ and the guideway profile is linearly transformed from space to the time domain. Since the guideway profile is modelled as a homogeneous, Gaussian random process in the space domain, it is transformed to a stationary, Gaussian random process in the time domain. In the second case, the excitation is nonstationary and its characteristics are determined by the nature of the vehicle velocity.

Motion with Constant Velocity
Consider a vehicle moving with constant velocity $\dot{s} = V$. Let $Z(t)$ represent the vertical displacement transmitted to the vehicle at a contact point. Then, for a guideway whose vertical spatial profile is represented by $\tilde{Z}(s)$,

$$Z(t) = \tilde{Z}(Vt) \qquad (6.20)$$

The temporal auto-correlation function

$$R(\tau) = E[Z(t)\,Z(t + \tau)] = E\left[\tilde{Z}\left(\frac{s}{V}\right)\tilde{Z}\left(\frac{s+\delta}{V}\right)\right] = R\left(\frac{\delta}{V}\right)$$

and the one-sided psd of $Z(t)$

$$G(f) = \frac{1}{V}\,S\left(\frac{f}{V}\right) \tag{6.21a}$$

where $f = kV$. The one-sided psd in terms of $\omega = 2\pi f$

$$G(\omega) = \frac{1}{2\pi V}\,S\left(\frac{\omega}{2\pi V}\right) \tag{6.21b}$$

and the two-sided psd

$$\Phi(\omega) = \frac{1}{4\pi V}\,S\left(\frac{\omega}{2\pi V}\right) \tag{6.22}$$

Most vehicles have two, or more, wheels separated by a fixed distance called the *wheel-base*. Let us first consider a vehicle with two wheels following each other at a fixed distance l. Let $Z_1(t)$ and $Z_2(t)$ denote the excitation transmitted by the front and rear wheels, respectively. Then,

$$Z_1(t) = Z(t)$$

$$Z_2(t) = Z\left(t - \frac{l}{V}\right)$$

and

$$R_{12}(\tau) = E[Z_1(t)\,Z_2(t + \tau)] = R\left(\tau - \frac{l}{V}\right)$$

$$R_{21}(\tau) = E[Z_2(t)\,Z_1(t + \tau)] = R\left(\tau + \frac{l}{V}\right)$$

The one-sided cross-psd are, therefore,

$$G_{12}(\omega) = G(\omega)\,e^{-i\omega l/V}$$

$$G_{21}(\omega) = G(\omega)\,e^{i\omega l/V} \tag{6.23}$$

The psd matrix of the vector $Z(t) = [Z_1, Z_2]^T$ can, therefore, be expressed as

$$\mathbf{G}(\omega) = G(\omega)\begin{bmatrix} 1 & e^{-i\beta} \\ e^{i\beta} & 1 \end{bmatrix} \tag{6.24}$$

where $\beta = \omega l/V$. Note that $\mathbf{G}(\omega)$ is hermitian.

The above result can be generalized to m-wheels following each other. Let $Z_j(t)$, $j = 1, 2, \ldots m$, represent the excitation transmitted by the jth wheel, and let l_{jk} be the distance between the jth and kth wheel. Then,

$$G_{jj}(\omega) = G(\omega),\ j = 1, 2, \ldots m \text{ (no sum)}$$

$$G_{jk}(\omega) = G(\omega) \, e^{-i\beta_{jk}}, j \neq k = 1, 2, \dots m$$

$$G_{kj}(\omega) = G_{jk}^{*}(\omega) = G(\omega) \, e^{i\beta_{jk}}, j \neq k = 1, 2, \dots m \qquad (6.25)$$

where $\beta_{jk} = \omega l_{jk}/V$.

The above treatment applies to single profile guideway models. For double profile guideway models, psd matrix for multi-wheel trains separated by a constant distance can be similarly derived for isotropic, or Parkhilovskii models using (6.15, 6.16) or (6.19, 6.20), respectively.

Motion with Variable Velocity

Consider a vehicle moving with variable velocity $\dot{s} = V(t)$. Let $Z(t)$ represent the vertical displacement transmitted to the vehicle at a contact point. If the vehicle is moving on a profile defined by (6.7), the excitation transmitted to the vehicle may be obtained by changing the independent variable s to t, that is

$$\dot{Z} + a\dot{s}Z = \dot{s} \, W(s(t)) \qquad (6.26)$$

Clearly, $Z(t)$ is a nonstationary random process.

The forcing function $W(s(t))$ in (6.26) is a parameterised white process, and

$$E[W(s(t_1)) \, W(s(t_2))] = \pi C \, \delta(s(t_2) - s(t_1)); \; t_2 \geq t_1 \qquad (6.27)$$

where $\delta(\cdot)$ is a delta function, which has the following property (Hoskins, 1979; Hammond and Harrison, 1981)

$$\delta(s(t_2) - s(t_1)) = \frac{\delta(t_2 - t_1)}{|\dot{s}(t_1)|} = \frac{\delta(t_2 - t_1)}{\dot{s}(t_1)} \qquad (6.28)$$

assuming $\dot{s}(t) > 0$. (6.27) can now be expressed as

$$E[W(s(t_1)) \, W(s(t_2))] = \pi C \, \frac{\delta(t_2 - t_1)}{\dot{s}(t_1)} \qquad (6.29a)$$

We now define a nonstationary random process $W_1(t)/\sqrt{\dot{s}}$, in which $W_1(t)$ is a white noise process with $E[W_1(t_1)W_1(t_2)] = \pi C\delta(t_2 - t_1)$.
Hence,

$$E\left[\frac{W_1(t_2)}{\sqrt{\dot{s}(t_1)}} \frac{W_1(t_2)}{\sqrt{\dot{s}(t_2)}}\right] = \frac{\pi C\delta(t_2 - t_1)}{\dot{s}(t_1)} \qquad (6.29b)$$

Processes $W(s(t))$ and $W_1(t)/\sqrt{\dot{s}}$ are, therefore *equivalent in a covariance sense*. We can generate a random process $Z_1(t)$ by replacing $W(s(t))$ in (6.26) by $W_1(t)/\sqrt{\dot{s}}$, that is

$$\dot{Z}_1 + a\dot{s} \, Z_1 = \sqrt{\dot{s}} \, W_1(t) \qquad (6.30)$$

Clearly, $Z(t)$ and $Z_1(t)$ are different random processes, but they are covariance equivalent. Since they are assumed to be Gaussian, they are equivalent in the stochastic sense.

The random process $Z_1(t)$ defined by (6.30), has a significant mathematical advantage, as it is generated by filtering a modulated white noise process which admits Stieltjes integral representation and also leads to other analytical simplifications. We shall discuss this aspect in Section 6.4.2.

6.2.3 Railway Track

Figure 6.5 shows the cross-section of a railway track consisting of two rails R_1 and R_2. Let subscript $i = 1, 2$, refer to right and left rails respectively. Further, let $\tilde{Z}_i(s)$ and $\tilde{Y}_i(s)$ denote the vertical and lateral rail profiles respectively.

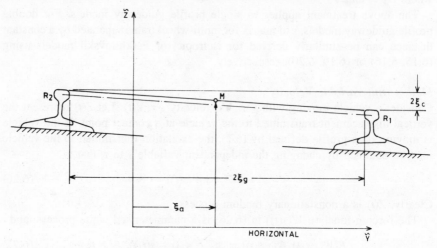

Fig. 6.5 Cross-section of a railway track

Following four irregularities are generally used to define the track geometry

$$\xi_V = \frac{\tilde{Z}_1 + \tilde{Z}_2}{2} \; ; \; \xi_c = \frac{\tilde{Z}_1 - \tilde{Z}_2}{2}$$

$$\xi_a = \frac{\tilde{Y}_1 + \tilde{Y}_2}{2} \; ; \; \xi_g = \frac{\tilde{Y}_1 - \tilde{Y}_2}{2} \tag{6.31}$$

where ξ_V is the vertical profile; $2\xi_c$ is the cross level; ξ_a is the alignment; and $2\xi_g$ is the gauge. Inverting (6.31), we have

$$\tilde{Z}_1 = \xi_V + \xi_c \; ; \; \tilde{Z}_2 = \xi_V - \xi_c$$

$$\tilde{Y}_1 = \xi_a + \xi_g \; ; \; \tilde{Y}_2 = \xi_a - \xi_g \tag{6.32}$$

Based on the experimental results, it is assumed that all irregularities (ξ_V, ξ_c, ξ_a and ξ_g) are homogeneous, Gaussian random processes with zero mean and are mutually uncorrelated (Garivaltis, et al, 1980).

For vehicle motion with constant velocity V, the vertical and lateral track induced excitations are represented by

$$Z_i(t) = \tilde{Z}_i(Vt), i = 1, 2$$

$$Y_i(t) = \tilde{Y}_i(Vt), i = 1, 2 \tag{6.33}$$

Since the track irregularities are assumed to be uncorrelated, the cross correlation functions $R_{Vc} = R_{ag} = R_{ac} = R_{Vg} = 0$, and

$$R_{Z_i}(\tau) = R_V(\tau) + R_c(\tau), i = 1, 2$$

$$R_{Y_i}(\tau) = R_a(\tau) + R_g(\tau), i = 1, 2$$

$$R_{Z_1 Z_2}(\tau) = R_{Z_2 Z_1}(\tau) = R_V(\tau) - R_c(\tau)$$

$$R_{Y_1 Y_2}(\tau) = R_{Y_2 Y_1}(\tau) = R_a(\tau) - R_g(\tau)$$

and

$$R_{Z_i Y_j}(\tau) = 0, i = 1, 2, j = 1, 2 \qquad (6.34)$$

Taking the Fourier transform of (6.34), we have the elements of the psd matrix

$$G_{Z_i}(\omega) = S_V(\omega) + S_c(\omega), i = 1, 2$$

$$G_{Y_i}(\omega) = S_a(\omega) + S_g(\omega), i = 1, 2$$

$$G_{Z_1 Z_2}(\omega) = G_{Z_2 Z_1}(\omega) = S_V(\omega) - S_c(\omega)$$

$$G_{Y_1 Y_2}(\omega) = G_{Y_2 Y_1}(\omega) = S_a(\omega) - S_g(\omega)$$

$$G_{Z_i Y_j}(\omega) = 0, i = 1, 2, j = 1, 2 \qquad (6.35)$$

Following expression is commonly used to model the one-sided psd of the track:

$$S(k) = \frac{C}{k^r} \qquad (6.36)$$

where C is a constant representing track roughness, and the exponent $r = 2$. Other values of r (Ahlbeck, et al, 1974), and different values over various frequency ranges have also been used (Corbin and Kaufman, 1975). Hedrick has proposed following expressions for the psd of track irregularities:

$$S_V(k) = S_a(k) = \frac{2\pi A_V k_c^2}{(k^2 + k_c^2)k^2} \qquad (6.37)$$

and

$$S_c(k) = S_g(k) = \frac{8\pi A_c k_c^2}{(k^2 + k_c^2)(k^2 + k_s^2)} \qquad (6.38)$$

It is suggested (Hooks and Pearce, 1974) that the magnitude constants A_v and A_c may be increased in (6.37) and (6.38) as the track deteriorates.

The random process model of the railway track irregularities described above assumes a frozen profile. In reality, the track profile is dynamic and is closely coupled with the vehicle dynamics. The dynamic behaviour of track-vehicle system is very complex, and to simplify the treatment the track is generally modelled as a beam supported on resilient foundation. Several models have been proposed to represent the dynamic behaviour of railway track. Jaiswal and Iyengar (1991) have reviewed these models, and have proposed models, which include the effect of foundation inertia, integration of observed track irregularities with the track dynamics and inelastic foundation behaviour.

Multi-wheel-axle Vehicles
Railway vehicles consist of a train of wheel-axle sets. Each set transmits four

input excitations, two vertical and two lateral. Let us consider a system with m-axles. The total number of output excitations will be $4m$. Let the inputs be numbered, such that

(i) vertical inputs are denoted by

$$Z_j, Z_p; \quad j, p = 1, 3, 5 \ldots$$

(ii) lateral inputs are denoted by

$$Y_k, Y_q; \quad k, q = 2, 4, 6 \ldots$$

From (6.35), the elements of the $4m \times 4m$ input psd matrix can be expressed as

$$G_{jk}(\omega) = 0, j = 1, 3, 5, \ldots, k = 2, 4, 6, \ldots$$

$$G_{jj}(\omega) = G_{Z_1}(\omega) = G_{Z_2}(\omega), \, j = 1, 3, 5, \ldots \text{(no sum)}$$

$$G_{kk}(\omega) = G_{Y_1}(\omega) = G_{Y_2}(\omega), k = 2, 4, 6, \ldots \text{(no sum)} \qquad (6.39a)$$

The cross-spectra of vertical inputs are

$$G_{Z_j Z_p}(\omega) = G_{Z_i}(\omega)\, e^{-\omega(s_j - s_p)/V}, \; i = 1 \text{ or } 2 \text{ and } (j \neq p, \text{ on the same rail}),$$

$$= G_{Z_1 Z_2}(\omega)\, e^{-\omega(s_j - s_p)/V} \; (j \neq p, \text{ on different rails}) \qquad (6.39b)$$

Similarly, for the lateral inputs

$$G_{Y_k Y_q}(\omega) = G_{Y_i}(\omega)\, e^{-\omega(s_k - s_q)/V}, \; i = 1 \text{ or } 2, (k \neq q, \text{ on the same rail}),$$

$$= G_{Y_1 Y_2}(\omega)\, e^{-\omega(s_k - s_q)/V}, \; (k \neq q, \text{ on different rails}), \qquad (6.39c)$$

where s denotes the location of the wheel on the track. Further, matrix $\mathbf{G}$ is hermitian.

In deriving the above expressions, we have assumed a point contact between the wheel and the guideway which follows the guideway profile. In reality, particularly, in vehicles with tyres, there is a finite contact area called 'footprint' (Newland, 1986), which modifies the profile transmitted by the wheel. We discuss this aspect in the next section.

6.3   Vehicle System Models

Aircraft, road and rail vehicles represent a wide range of complex dynamic systems. To study the guideway induced vibration of these systems, it is necessary to construct mathematical models of the vehicle. The first step in this direction is usually a physical model incorporating the idealised inertia, damping and stiffness properties. A mathematical statement of physical laws then leads to a set of equations relating the vehicle response to the excitation through a mathematical operator. As a rule the model should be as simple as possible, and consistent with the aims of the study. This requires considerable engineering judgement and an understanding of the significant features of the system, its dynamic behaviour, and operating constraints.

We shall divide the vehicles into three groups and briefly discuss their significant characteristics.

ROAD VEHICLES include motor cycles, passenger cars, trucks/buses, and articulated road vehicle systems, such as, truck/tractor-trailer combinations. Major components of these vehicles are: passenger-seat system; cargo; body; chasis; suspension system; and wheels with pneumatic tyres.

RAILWAY VEHICLES include coaches, wagons and locomotive. Major components of these vehicles are: passenger-seat system, cargo; car body; truck frames; suspension-system; wheel-axle sets and flexible track. The rail-wheel contact forces depend, amongst other parameters, on creep, gravit tional stiffness and vehicle velocity.

Both, road and rail vehicles move mostly with a constant velocity. The major design considerations for these vehicles include: ride comfort; cargo safety; fatigue and fracture of structural systems and mechanical components; limits on travel of moving parts; and ride safety, which is determined by the dynamic fluctuations of the contact force between the wheel and the guideway. The dynamic behaviour of contact force is also of interest for the design of pavement and railway track. In railway systems, stability problems, such as, hunting, wheel hopping are also encountered.

AIRCRAFTS include civilian and military aircraft. Major components of aircrafts are: fuselage, wings/tail, engines and the landing gear, including nose and main wheel systems. The landing gear usually consists of a complex oleo-pneumatic system which has a pneumatic spring action, orifice damping and coulomb friction.

During taxi, an aircraft moves with a constant velocity, but during take-off and landing runs its velocity is variable. The landing impact constitutes a critical loading condition for the landing gear. The major design considerations include: pilots' comfort; fatigue and fracture of structural systems and mechanical components, and limits on travel of moving parts.

The size and the details of a model are largely determined by the aims of the investigation. For example, if the ride comfort of the driver is to be investigated, the model must incorporate the driver-seat subsystem in greater detail, include the pitch degree of freedom to estimate the fore-and-aft vibration of the driver (back slapping, head snapping), but may include only a simple model of the body and the suspension system. On the other hand, if the investigation is for the design of the suspension system, the passenger-body sub-system may be modelled as a rigid body, but the suspension should be included in detail. For a preliminary investigation, a simple sdf model is often adequate.

The major source of complexity and analytical difficulty in the vibration of vehicle systems is the nonlinear behaviour of suspension elements. The damping in suspenion systems is provided by coulomb friction and by shock absorbers with nonlinear characteristics. The tyre and the leaf and coil springs also exhibit nonlinear behaviour for large deflections. In railway vehicles, the contact forces between wheel and rail also contribute to nonlinear behaviour. Due to analytical difficulties in obtaining the response of nonlinear systems, linear mathematical models are generally constructed either by using linearised damping and stiffness elements in the physical model, or by linearizing the governing equations. We shall first treat the linear models, and then discuss the nonlinear aspects, including various methods of linearization.

With the increase in the speed of the vehicles, there is a growing trend to reduce the vibration level through active elements in the suspension system using feed-back controls. We shall discuss the vibration of vehicles with active suspension systems, in Appendix 6B.

Linear Models

The elements representing inertia, damping the stiffness characteristics constitute the basic components of dynamic systems. In vehicles, the inertia properties are generally separated into sprung and unsprung mass. The sprung mass includes the passengers, seats, car body and other subsystems supported by the suspension system. It may be modelled by one, or more, masses, and may represent heave, pitch, roll, yaw and other degrees of freedom depending upon the nature of the model. The unsprung mass includes equivalent mass of the wheel-axle and suspension systems. In the dynamic analysis of aircraft, the flexibility of wing-fuselage system is some times incorporated in the model by one, or more, spring-mass-dashpot systems representing the first and higher modes (Kirk and Perry, 1971). The damping and stiffness are generally represented through discrete linear elements. Figures 6.6, 6.10, 6.14, 6.16 and 6.22 show several discrete models of vehicle systems incorporating varying levels of detail.

The equations of motion of vehicle systems for small oscillations can be derived using the Lagrange equations. Let $\bar{X}$ be a set of n independent generalized coordinates which completely specify the configuration of the system measured from the equilibrium position. Then the kinetic energy T, potential energy U, and dissipative energy D, functions can be expressed as:

$$T = \frac{1}{2}\dot{\bar{X}}^{\mathrm{T}}\mathbf{M}\dot{\bar{X}}$$

$$U = \frac{1}{2}\bar{X}^{\mathrm{T}}\mathbf{K}\bar{X}$$

$$D = \frac{1}{2}\dot{\bar{X}}^{\mathrm{T}}\mathbf{C}\dot{\bar{X}} \tag{6.40}$$

where $\mathbf{M}$, $\mathbf{K}$ and $\mathbf{C}$ are the mass, stiffness and viscous damping matrices, respectively.

The Lagrange equations of motion are:

$$\frac{d}{dt}\left(\frac{\partial T}{\partial \dot{X}_j}\right) + \frac{\partial D}{\partial \dot{X}_j} + \frac{\partial U}{\partial X_j} = 0, j = 1, 2, \ldots n \tag{6.41}$$

Equations (6.40) and (6.41) yield a set of simultaneous, nonhomogeneous, second order differential equations with constant coefficients

$$\mathbf{M}\ddot{\bar{X}} + \mathbf{C}\dot{\bar{X}} + \mathbf{K}\bar{X} = \mathbf{C}_f\dot{\bar{Z}} + \mathbf{K}_f\bar{Z} = \bar{F}(t) \tag{6.42}$$

where $\mathbf{C}_f$ and $\mathbf{K}_f$ are $n \times m$ forced damping and stiffness matrices, respectively, and $\bar{Z}$ is m dimensional excitation vector representing guideway induced displacements. Since $\bar{Z}(t)$ represents a vector of random processes, $\bar{F}(t)$ is a random vector process.

Defining $\bar{Y}(t) = [\dot{\bar{X}} \mid \bar{X}]^{\mathrm{T}}$, and $\bar{\xi}(t) = [\dot{\bar{Z}} \mid \bar{Z}]^{\mathrm{T}}$, (6.42) can also be expressed as

$$\dot{\bar{Y}}(t) + \mathbf{A}\bar{Y}(t) = \mathbf{B}\bar{\xi}(t) = \bar{P}(t) \tag{6.43}$$

where

$$\underset{(2n \times 2n)}{\mathbf{A}} = \left[\begin{array}{c|c} \mathbf{M}^{-1}\mathbf{C} & \mathbf{M}^{-1}\mathbf{K} \\ \hline -\mathbf{I} & 0 \end{array}\right]; \quad \underset{(2n \times 2m)}{\mathbf{B}} \left[\begin{array}{c|c} \mathbf{M}^{-1}\mathbf{C}_f & \mathbf{M}^{-1}\mathbf{K}_f \\ \hline 0 & 0 \end{array}\right] \tag{6.44}$$

and

$$\bar{P}(t) = \left\{ \frac{\mathbf{M}^{-1}\bar{F}(t)}{\bar{0}} \right\} = \left\{ \frac{\mathbf{M}^{-1} \mid \mathbf{C}_f\mathbf{K}_f \mid \bar{\xi}}{\bar{0}} \right\} \tag{6.45}$$

Equations (6.42) and (6.43) represent a linear mathematical model of vehicle systems.

6.3.1 Nonlinear and 'Equivalent' Linear Models

In vehicles, nonlinearities arise primarily due to nonlinear force-deformation characteristics of damping and stiffness elements. In railway vehicles, the rail wheel contact force and geometry also contribute to the nonlinear behaviour. If the nonlinear behaviour is included, the mathematical model of a vehicle is represented by the system of nonlinear differential equations

$$\dot{\bar{X}} + \bar{g}(\bar{X}, \dot{\bar{X}}) = \bar{F}(t) \tag{6.46}$$

or

$$\dot{\bar{Y}} + \bar{g}(\bar{Y}) = \bar{P}(t) \tag{6.47}$$

If the elements of the vehicle exhibit hysteretic behaviour, the terms $\bar{g}(\bar{X}, \dot{\bar{X}})$ and $\bar{g}(\bar{Y})$ are replaced by $\bar{g}(\bar{X}, \dot{\bar{X}}, t)$ and $\bar{g}(\bar{Y}, t)$, respectively.

An analytical solution of nonlinear equations is complicated by the fact that the principle of superposition does not apply and, therefore, the time and frequency domain approaches based on convolution integral are not valid. Further, the response of nonlinear systems to Gaussian excitation is non-Gaussian. Exact solutions of nonlinear systems are available for only a few simple problems. The response is, therefore, obtained either by approximate methods, or by Monte-Karlo procedure. A comprehensive treatment of the response of nonlinear systems to random excitation is contained in Nigam (1983).

An approximate response of a vehicle to guideway induced random excitation can be obtained, if it is represented by an 'equivalent' linear model. An equivalent linear model can be constructed, either by replacing the individual nonlinear elements by linear elements, or by linearizing the equations of motion. We shall first discuss the equivalent linear models of individual damping and stiffness elements.

Damping

The damping in vehicles may be represented by three types of nonlinear models: Coulomb, orifice and hysteretic. The equivalent linear viscous model is constructed by equating the energy dissipated by the nonlinear model over a cycle (for harmonic excitation), or over a given period of time, with the energy dissipated by the linear viscous model.

Let $X(t)$ denote the relative displacement of the ends of a damping element. Then, for *Coulomb* damping, the damping force is given by

$$F_D = \mu N \, \text{sgn}\,(\dot{X}) \tag{6.48}$$

where μ is the coefficient of friction and N is the normal contact force. For the equivalent linear viscous model (Kirk and Perry, 1971)

$$F_D = C_e \dot{X} \tag{6.49}$$

where

$$C_e = \frac{4\mu N}{\pi A \omega} \tag{6.50}$$

for harmonic excitation with amplitude, A, and frequency ω.

For Gaussian random excitation,

$$C_e = \mu N \sqrt{\frac{2}{\pi}} / \sigma_{\dot{x}} \tag{6.51}$$

For *Orifice* type damping,

$$F = C |\dot{X}| \dot{X} \tag{6.52}$$

and

$$C_e = \frac{8 C A \omega}{3\pi} \tag{6.53}$$

for harmonic excitation.

For Gaussian random excitation,

$$C_e = 2 \sqrt{\frac{2}{\pi}} C \sigma_{\dot{X}} \tag{6.54}$$

For *hysteretic* damping, we can write (Wen, 1980)

$$g(X, \dot{X}, t) = g_1(X, \dot{X}) + C v(t) \tag{6.55}$$

where $g_1(X, \dot{X})$ is the non-hysteric component; and $v(t)$ is hysteretic component, which satisfies the following nonlinear differential equation

$$\dot{v} = A \dot{X} - \gamma |\dot{X}| |v|^{n-1} v - \beta \dot{X} |v|^n \tag{6.56}$$

in which parameters γ and β control the shape of the hysteresis loop; A the strength and stiffness; n the smoothness of transition from elastic to plastic regions. A nearly elasto-plastic system may be modelled by

$$g(X, \dot{X}, t) = \alpha k X + (1 - \alpha) k v \tag{6.57}$$

in which k is the pre-yielding stiffness; and α the ratio of pre-yielding stiffness to post-yielding stiffness. Wen (1980) has shown that for $n = A = 1$, and zero mean Gaussian excitation (6.56) can be replaced by the following linear equation

$$\dot{v} + C_1 \dot{X} + C_2 v = 0 \tag{6.58}$$

in which

$$C_1 = \sqrt{\frac{2}{\pi}} \left[\gamma \frac{E[\dot{X} v]}{\sigma_{\dot{X}}} + \beta \sigma_v \right] - A$$

and

$$C_2 = \sqrt{\frac{2}{\pi}} \left[\gamma \sigma_{\dot{X}} + \beta \frac{E[\dot{X} v]}{\sigma_v} \right] \tag{6.59}$$

Stiffness

The nonlinear restoring force can be expressed as a polynomial function of relative displacement, given by

$$F_S = \sum_n k_n X^n, n = 1, 3, 5 \tag{6.60}$$

For random excitation, it can be linearized to

$$F_S = k_e X, \tag{6.61}$$

where

$$k_e = \sum_{n=1,3,\ldots} k_n \frac{E[X^{n+1}]}{E[X^2]} \tag{6.62}$$

If X is assumed to be Gaussian, (6.52) reduces to

$$k_e = \sum_{n=1,3,\ldots} k_n (1, 3 \ldots n) \sigma_X^{n-1} \tag{6.63}$$

If each nonlinear element of the vehicle is replaced by an 'equivalent' linear element using the relations given above, a linear mathematical model can be constructed as discussed in the preceding section. It must be noted that the expressions for equivalent linear coefficients involve moment functions of response of each element. An iterative procedure is, therefore, required while using the equivalent linear models.

A linear model can also be constructed by linearizing the equations of motion (6.46) or (6.47), using the method of equivalent linearization (Caughey, 1963; Iwan and Yang, 1972; Atalik and Utku, 1976). For Gaussian random excitation, and some fairly general conditions on the nature of nonlinear functions (Nigam, 1983), (6.46) can be reduced to (6.42), with the elements of matrices **C** and **K** given by (10.110*).

$$c_{ij} = E\left[\frac{\partial g_i(\bar{X}, \dot{X})}{\partial \dot{X}_j}\right] \tag{6.64}$$

and

$$k_{ij} = E\left[\frac{\partial g_i(\bar{X}, \dot{X})}{\partial X_j}\right], i, j = 1, 2, \ldots n \tag{6.65}$$

An equivalent linear form of (6.47) can be similarly constructed. The elements of the coefficient matrices of the linearized models are functions of response moments, and the comment regarding iterative procedure made earlier again applies.

Amongst the class of vehicles considered in this chapter, the landing gears of an aircraft constitute a complex and strongly nonlinear suspension system. We discuss the behaviour of landing gear in Appendix 6A.

6.3.2 Contact Forces between Wheels and Guideway

The contact forces between the wheel and the guideway are the source of vibration for the vehicle-guideway system. In road vehicles with pneumatic tyres, the contact forces are relatively simple and depend on the stiffness and damping characteristics of the tyre, the shape and size of the contact area (footprint), and the guideway profile. In railway vehicles, however, the contact forces are complex and nonlinear, and include the effects of gravity, creep due to difference in strain rates of wheel and rails, besides vertical and lateral rail profiles. We shall discuss the nature of the contact forces for road and rail vehicles separately.

Road Vehicles

The pneumatic tyre of a road vehicle can be modelled by a linear spring and a dashpot in parallel. The contact force is, therefore, given by

$$F(t) = k_T(X - Z') + C_T(\dot{X} - \dot{Z}') \tag{6.66}$$

where k_T and C_T are stiffness and damping constants of the tyre; and $(X - Z')$ is the relative displacement. Z' is the 'perceived' vertical guideway profile as transmitted by the wheel and depends on the shape and size of the contact area, called the 'footprint'. The input at each wheel is assumed to be represented by a weighted average of the road surface height within the footprint. The form of the weight function depends on the characteristics of the tyre.

Let the road profile be represented as a two dimensional surface $\tilde{Z}(s_1, s_2)$ (Fig. 6.1). Let the weight function $W(s_1, s_2)$ of the position vector $\bar{s} = [s_1, s_2]^T$, be normalized such that

$$\iint\limits_{-\infty}^{\infty} W(\bar{s}) \, d\bar{s} = 1 \tag{6.67}$$

The smoothed surface profile $Z'(\bar{s})$, as perceived by the wheel, is then given by

$$Z'(\bar{s}) = \iint\limits_{-\infty}^{\infty} \tilde{Z}(\bar{s} + \bar{u}) \, W(\bar{u}) \, d\bar{u} \tag{6.68}$$

Newland (1986) has shown that the two-dimensional psd of $Z'(\bar{s})$ and $\tilde{Z}(\bar{s})$ are related through the Fourier transform of the weight function

$$\Phi_{Z'Z'}(\overline{\Omega}) = \Phi_{ZZ}(\overline{\Omega})(2\pi)^4 \tilde{W}^*(\overline{\Omega}) \, \tilde{W}(\overline{\Omega}), \tag{6.69}$$

where $\overline{\Omega} = 2\pi\bar{k} = [\Omega_1, \Omega_2]^T$ is the spatial frequency vector,

and

$$\overline{W}(\overline{\Omega}) = \frac{1}{(2\pi)^2} \iint\limits_{-\infty}^{\infty} W(\bar{s}) \, e^{-\overline{\Omega} \cdot \bar{s}} \, d\bar{s} \tag{6.70}$$

is the two-dimensional Fourier transform of $W(\bar{s})$.

The displacement input at each wheel corresponds to a single-point contact with smoothed surface $Z'(s)$ whose psd is given by (6.69). Note that, if $W(\bar{s}) = \delta(\bar{s})$, the delta function, $Z'(s)$ reduces to $Z(s)$. Newland (1986a) has shown that smoothing filters out the high frequency components in the response.

Railway Vehicles

The rail-wheel contact forces in railway vehicles are determined by a complex dynamic interaction between the wheelset and the rails. A wheelset consists of two profiled wheels rigidly connected by an axle. It has four degrees of freedom:

vertical Z^a; lateral Y^a; yaw ψ^a; and roll ϕ^a. Due to the conocity of the wheel, the c.g. of the wheelset gets vertically displaced due to a lateral displacement developing a gravitational restoring force, which is the source of gravitational stiffness. Further, the contact between the wheel and the rails causes local deformation, and longitudinal, lateral and spin creep, giving rise to longitudinal and lateral creep forces and a creep moment about the normal axis. These forces are nonlinear and depend upon the contact ellipticity, normal load, elastic constants and surface contamination. Kalker (1967, 1979) has shown that for small creepage and spin, force-creepage relations are linear.

Let the generalized displacements of the wheelset be denoted by $\bar{X}^a = [Z^a, Y^a, \phi^a, \psi^a]^T$. Let $\bar{F}^a(t)$ denote the generalized forces between the wheelset and the rails. The contributions to $\bar{F}^a(t)$ come from two sources—the rail profile and the creep. Hence,

$$\bar{F}^a(t) = \bar{F}_r^a + \bar{F}_{cp}^a \tag{6.71}$$

where subscripts r and cp refer to rail and creep, respectively. The linearized relations between these forces and rail input displacements can be expressed as

$$\bar{F}_r^a(t) = (\mathbf{T}^r)^T \mathbf{K}_r \bar{Z}^r \tag{6.72}$$

in which $\mathbf{T}^r$ and $\mathbf{K}_r$ are the transfer matrix and the stiffness matrix of the track; and $\bar{Z}^r$ is the vector of rail input displacements Z_i and Y_i, $i = 1, 2$: and

$$\bar{F}_{cp}^a(t) = \mathbf{K}_g (T^r)^T \bar{X}'^r + \mathbf{C}_g \dot{\bar{X}}^a \tag{6.73}$$

where $\bar{X}'^r$ is a vector denoting relative displacement between the wheelset and rails. Matrices $\mathbf{K}_g$ and $\mathbf{C}_g$ represent the effects of gravity and creep, and are given by (Garivaltis et al, 1980)

$$\mathbf{K}_g = \begin{bmatrix} 0 & 0 & 0 & 0 \\ 0 & -k_g & 0 & 2f_L \\ 0 & -k_g r & 0 & 2f_L r - 2f_T \dfrac{\lambda b}{r} \\ 0 & \dfrac{2f_{s23}}{r}\varepsilon & 0 & k_{gw} \end{bmatrix} \tag{6.74}$$

$$\mathbf{C}_g = \begin{bmatrix} 0 & 0 & 0 & 0 \\ 0 & \dfrac{-2f_L}{V} & 0 & \dfrac{-2f_{s23}}{V} \\ 0 & \dfrac{-2f_L}{V} & 0 & 0 \\ 0 & 0 & 0 & \dfrac{-2f_T b^2}{V} - \dfrac{2f_{s23}}{V} \end{bmatrix} \tag{6.75}$$

in which k_g is the lateral gravitational stiffness for a wheelset; k_{gw} is gravitational stiffness for a wheelset; f_L, f_T are lateral and tangential creep coefficients; f_{s23} and

f_{s33} are lateral and longitudinal spin creep coefficients; r is the wheel tread radius; ε is the rate of change of contact plane slope with respect to the lateral displacement of wheelset; λ is the effective conocity; and b is the half distance between contact points of wheel treads and rails in lateral direction.

It may be noted that matrix $\mathbf{C_g}$ depends on the velocity of the vehicle. If the railwheel contact forces (6.71) for each axle are included in (6.42), equation of motion of a railway vehicle can be expressed in the following general form (Wickens and Gilchrist, 1977)

$$\mathbf{M}\ddot{\bar{X}} + [\mathbf{C} + \mathbf{C_g}(V)]\dot{\bar{X}} + [\mathbf{K} + \mathbf{K_g}]\bar{X} = \mathbf{K_f}\bar{Z}(t) = \bar{F}(t) \qquad (6.76)$$

Note that, damping matrix includes yaw gravitational stiffness which is a function of vehicle speed. This is similar to the aeroelastic damping and leads to a critical speed.

A recent literature survey by Sankar and Samaha (1986) provides a comprehensive review of the state of art in rail vehicle dynamics.

Influence Function Models

We have so far constructed the mathematical models of the vehicles in terms of equations of motion. For linear systems, mathematical models can also be constructed in terms of impulse, or frequency, influence function matrices $\mathbf{h}(\bar{r}, \bar{s}, t)$ or $\mathbf{H}(\bar{r}, \bar{s}, \omega)$, in which $\bar{r}$ and $\bar{s}$ are the position vectors of a point on the vehicle and track, respectively, in a coordinate frame moving with the vehicle (Newland, 1986b). The response of the vehicle at any point, and the correlation between the response at any two points, can be expressed in terms of influence functions (Nigam, 1983, Newland, 1986b).

Since vehicles are a complex dynamic system, it is not always easy to write down the equations of motion using the Lagrangian formulation. Symbolic formalism, such as NEWEUL, have been developed recently, which make it possible to write down the equations of motion with the aid of a digital computer treating the vehicle as a multi-body system (Schiehlen, 1982; Haug, 1984).

6.4 Random Vibration Response of Vehicles

In Section 6.2, we have seen that the guideway profile may be modelled by a two-dimensional, or by a set of one, or more, one-dimensional, homogeneous, Gaussian random processes. For a multi-wheel vehicle, the excitation is, therefore, represented by a random vector. The temporal nature of the excitation depends on the velocity of the vehicle. If the vehicle moves with a constant velocity, the excitation is stationary. However, if the velocity varies, the excitation becomes nonstationary. Road and rail vehicles operate with constant velocity for a major part of their service life. However, for aircrafts, variable velocity run during take-off and landing forms a significant part of their ground operation phase. The primary interest in vehicle vibration is, therefore, confined to stationary response. We shall discuss both stationary and nonstationary behaviour of vehicles.

The design of vehicle systems and an analysis of their operational performance involve quantitative assessment of ride comfort, ride safety, fatigue and strength analysis, and travel of moving parts. A random vibration analysis of the vehicle in terms of psd, or covariance, of the response and its derivatives, provides the basis for a practical reliability based design and performance evaluation. In Section

6.3, we have discussed three class of vehicle models: (i) linear, (ii) 'equivalent' linear; and (iii) nonlinear. Classical frequency and time domain methods of random vibration theory apply to linear and 'equivalent' linear models for determination of stationary response. Since the coefficients of 'equivalent' linear models are functions of response moments, an iterative procedure is generally needed to determine the response statistics for these models. For nonstationary response of linear models, recent developments in evolutionary random processes, and a use of state space formulation provide elegant analytical framework for determination of response statistics. For a class of nonlinear models, response statistics can be determined by Markov vector formulation, or by approximate methods, such as, perturbation method and Gaussian closure (Nigam, 1983).

6.4.1 Stationary Response

For linear, or 'equivalent' linear models of vehicles, the equations of motion can be expressed as (6.42), (6.76)

$$\mathbf{M}\ddot{\bar{X}} + \mathbf{C}\dot{\bar{X}} + \mathbf{K}\bar{X} = \mathbf{C}_f\dot{\bar{Z}} + \mathbf{K}_f\bar{Z} = \bar{F}(t) \tag{6.77}$$

or, in the state space as (6.43)

$$\dot{\bar{Y}} + \mathbf{A}\bar{Y} = \mathbf{B}\bar{\xi} = \bar{P}(t) \tag{6.78}$$

in which $\bar{F}(t)$ and $\bar{P}(t)$ are stationary random vectors. The coefficients of the matrices in the above equations are constant. In case of equivalent linear models, they are functions of stationary moment functions.

The stationary response of a linear system represented by (6.77) or (6.78) can be obtained in time, or frequency, domains using the standard results of linear random vibration theory (Nigam, 1983). In vehicle vibration studies, frequency domain method is preferred, as it yields the psd of the response, which is directly related to design criteria for ride comfort, and through the spectral moments to excursion probability, fatigue and strength analysis. We reproduce below the well known relation between the input and output psd. From (6.77),

$$\Phi_{\bar{X}\bar{X}}(\omega) = \mathbf{H}_{\bar{X}\bar{Z}}(\omega)\, \Phi_{\bar{Z}\bar{Z}}(\omega)\, \mathbf{H}_{\bar{X}\bar{Z}}^*(\omega)^{\mathrm{T}} \tag{6.79}$$

where

$$\mathbf{H}_{\bar{X}\bar{Z}}(\omega) = [-\omega^2\mathbf{M} + i\omega\,\mathbf{C} + \mathbf{K}]^{-1}[i\omega\,\mathbf{C}_f + \mathbf{K}_f] \tag{6.80}$$

Alternatively, from (6.78)

$$\Phi_{\bar{Y}\bar{Y}}(\omega) = \mathbf{H}_{\bar{Y}\bar{\xi}}(\omega)\, \Phi_{\bar{\xi}\bar{\xi}}(\omega)\, \mathbf{H}_{\bar{Y}\bar{\xi}}^*(\omega)^{\mathrm{T}} \tag{6.81}$$

where

$$\mathbf{H}_{\bar{Y}\bar{\xi}}(\omega) = [i\omega\,\mathbf{I} + \mathbf{A}]^{-1}\mathbf{B} \tag{6.82}$$

Let $Q(t)$ be a response quantity of interest, and let

$$Q(t) = [\bar{a}_1^{\mathrm{T}} \mid \bar{a}_2^{\mathrm{T}}]\left\{\frac{\dot{\bar{X}}}{\bar{X}}\right\} + [\bar{b}_1^{\mathrm{T}} \mid \bar{b}_2^{\mathrm{T}}]\left\{\frac{\dot{\bar{Z}}}{\bar{Z}}\right\} \tag{6.83}$$

where $\bar{a}_1, \bar{a}_2$ represent $n \times 1$ vectors, and $\bar{b}_1, \bar{b}_2$ represent $m \times 1$ vectors. Then,

where
$$\Phi_{QQ}(\omega) = \underset{(1 \times m)}{\mathbf{H}_{Q\bar{Z}}(\omega)} \; \underset{(m \times n)}{\Phi_{\bar{Z}\bar{Z}}(\omega)} \; \underset{(m \times 1)}{\mathbf{H}^{*}_{Q\bar{Z}}(\omega)^{\mathrm{T}}} \qquad (6.84)$$

$$\underset{(1 \times m)}{\mathbf{H}_{Q\bar{Z}}(\omega)} = \underset{(1 \times n)}{[i\omega\bar{a}_1^{\mathrm{T}} + \bar{a}_2^{\mathrm{T}}]} \; \underset{(n \times m)}{\mathbf{H}_{\bar{X}\bar{Z}}(\omega)} + \underset{(1 \times m)}{[i\omega\bar{b}_1^{\mathrm{T}} + \bar{b}_2^{\mathrm{T}}]} \; \underset{(m \times m)}{\mathbf{H}_{\bar{Z}\bar{Z}}(\omega)} \qquad (6.85)$$

The spectral moments of $Q(t)$ are given by

$$\lambda_{Q_j} = \int_0^\infty \omega^j G_{QQ}(\omega)\, d\omega, j = 0, 1, 2, \ldots \qquad (6.86)$$

where $G_{QQ}(\omega) = 2\Phi_{QQ}(\omega)$ is one-sided psd of $Q(t)$, and

$$\Phi_{\dot{Q}\dot{Q}}(\omega) = \omega^2 \Phi_{QQ}(\omega)$$

$$\Phi_{\ddot{Q}\ddot{Q}}(\omega) = \omega^4 \Phi_{QQ}(\omega) \qquad (6.87)$$

The design criteria for vehicles require response characteristics, such as, level-crossing, peaks, fractional occupation time, distribution and moments of the maximum value of the response. These can be obtained in terms of the first three spectral moments of $Q(t)$ and its derivative processes (see Chapter 6*).

Influence Function Method
The response of linear vehicle models can also be obtained in terms of influence function of the vehicle and the psd of the guideway (Newland, 1986). Consider an orthogonal system of coordinates with unit vectors $\bar{e}_i$, $i = 1, 2, 3$, moving with the vehicle. Let the datum plane of the guideway be parallel to the plane defined by $\bar{e}_1$ and $\bar{e}_2$, and let the vehicle be moving in the direction $\bar{e}_1$ with constant velocity $V\bar{e}_1$. Let $\bar{s}_j, j = 1, \ldots m$, define the locations of m points of contact between the vehicle and the guideway surface. Let $Q(\bar{r}, t)$ denote the scalar response at a point identified by a position vector $\bar{r}$ from the moving origin. Let $Z'_j(t), j = 1, 2, \ldots m$ denote the 'smoothed' (6.68) profile at position $\bar{s}_j$.

We shall find the response at two points represented by position vectors $\bar{r}_1$ and $\bar{r}_2$. Let the dynamic model of the vehicle be represented by the frequency response function matrix $H_{ij}(\omega)$, where

$$H_{ij}(\omega) = H(\bar{r}_i, \bar{s}_j, \omega), i = 1, 2; j = 1, 2, \ldots m. \qquad (6.88)$$

Let $\Phi_{QQ}(\bar{r}_1, \bar{r}_2, \omega)$ denote the cross-spectral density of the responses $Q(\bar{r}_1, t)$ and $Q(\bar{r}_2, t)$, and let $\Phi_{Z'Z'}(\bar{s}_j, \bar{s}_k, \omega), j, k = 1, 2, \ldots m$, denote the temporal cross-spectral density for a pair of m displacement inputs. From the theory of random vibration, we have

$$\Phi_{QQ}(\bar{r}_1, \bar{r}_2, \omega) = \sum_{j=1}^{m} \sum_{k=1}^{m} H^*(\bar{r}_1, \bar{s}_j, \omega) H(\bar{r}_2, \bar{s}_k, \omega) \Phi_{Z'Z'}(\bar{s}_j, \bar{s}_k, \omega) \qquad (6.89)$$

The spectral density of the smoothed guideway profile Z' is related to the original profile Z through (6.69). Let

$$\bar{\Omega} = \Omega_1 \bar{e}_1 + \Omega_2 \bar{e}_2$$

Since the vehicle is moving in the direction $\bar{e}_1$ with constant velocity V, $\Omega_1 = \omega/V$, and

$$\bar{\Omega} = \frac{\omega}{V}\,\bar{e}_1 + \bar{\Omega}_2\bar{e}_2 \qquad (6.90)$$

Newland (1986) has shown that (6.89) can be expressed in the following form:

$$\Phi_{QQ}(\bar{r}_1, \bar{r}_2, \omega) = \frac{1}{V}\int_{-\infty}^{\infty} \Phi_{ZZ}(\bar{\Omega})\, g^*(\bar{r}_1, \bar{\Omega}, \omega)\, g(\bar{r}_2, \bar{\Omega}, \omega)\, d\Omega \qquad (6.91)$$

where

$$g(\bar{r}, \bar{\Omega}, \omega) = \sum_{j=1}^{m} H(\bar{r}, \bar{s}_j, \omega)\,(2\pi)^2\, W_j^*(\bar{\Omega})\, e^{i\bar{\Omega}\cdot\bar{s}_j} \qquad (6.92)$$

is the *sensitivity function* of the vehicle. $\bar{\Omega}$ and $W_j(\bar{\Omega})$ are defined by (6.90) and (6.70) respectively. $\Phi_{ZZ}(\bar{\Omega})$ is the two dimensional psd of the guideway.

The cross-spectral density of the response defined by (6.91) can be used to determine the response characteristics for purposes of design and performance evaluation. The influence function formulation is particularly suited for numerical computation of the response. Newland (1986) has illustrated the method for the motion of a two-wheeled trailer on a road surface.

Example 6.1: Response of tractor-semi-trailer train
(Dokainish and Elmadany, 1980)

A tractor-semi-trailer train is a complex, coupled dynamic system. A symmetrical model of the train is shown in Fig. 6.6. The tractor consisting of the cab, engine and chasis is treated as one rigid body with translational and rotational degrees of freedom. The semi-trailer with its chassis is treated as the second rigid body. The truck and semitrailer are coupled and constrained at the fifth wheel, and the combined system has three degrees of freedom. The unsprung mass of each wheel and axle assembly is represented by a mass having vertical translational freedom only. The driver and seat assembly are modelled as a sdf system attached to the tractor as shown. The discrete dynamical model of the tractor-semi-trailer-driver system thus has seven degrees of freedom. The springs and viscous dampers representing the stiffness and damping of the suspension system and the tires, are assumed to be linear.

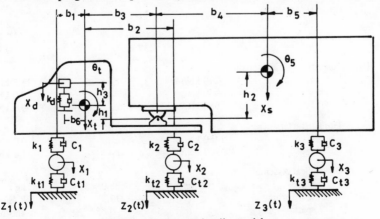

Fig. 6.6. Tractor-semi-trailer model.

The equation of motion of the above system is given by (6.77), in which $\bar{X}$ is 7-dimensional generalised coordinate vector, $\mathbf{M}$, $\mathbf{C}$ and $\mathbf{K}$ are 7×7 matrices, $\bar{Z}$ is 3-dimensional excitation vector, and $\mathbf{C_f}$ and $\mathbf{K_f}$ are 7×3 matrices. The frequency response matrix is given by (6.80), and the psd matrix of the response by (6.79).

The psd of the road surface is given by (6.2). Two classes of roads—A: motorway and B: minor roads are considered. The characteristics of the roads are given in Table 6.2.

Table 6.2   Road surface characteristics

Route	$S(k_0)$ (m³/c)	r_1	r_2
A : Motorway	13.0×10^{-6}	2.1	1.42
B : Minor Road	28.4×10^{-6}	3.3	1.48

The parameters of tractor-semitrailer system are given in Table 6.3. The response of the system is determined both for the loaded and unloaded conditions. The dynamic characteristics of the vehicle under these two conditions are given in Table 6.4.

Table 6.3   Tractor-semi-trailer parameters

I.	Tractor				
	(a)	General			
		sprung mass (M_1)		2400 kg	
		front unsprung mass (M_1)		440 kg	
		rear unsprung mass (M_2)		775 kg	
		pitch moment of inertia (I_1)		1660 kg m²	
	(b)	Dimensions			
		b_1	0.82 m	b_5	1.62 m
		b_2	1.84 m	h_1	− 0.15 m
		b_6	0.51 m	h_3	0.95 m
II.	Semitrailer				
	(a)	General		Laden	Unladen
		sprung mass (M)		14125 kg	1740 kg
		unsprung mass (M_3)		510 kg	510 kg
		(I)			
	(b)	Dimensions			
		b_3		2.55 m	2.35 m
		b_4		3.02 m	3.22 m
		h_2		1.37 m	0.15 m
III.	Suspension Characteristics				
		k_1	290 kN/m	c_1	11.5 kNs/m
		k_2	960 kN/m	c_2	29.0 kNs/m
		k_3	960 kN/m	c_3	29.0 kNs/m
IV.	Tire Characteristics				
		k_{11}	1450 kN/m	c_{11}	0.7 kNs/m
		k_{12}	2900 kN/m	c_{12}	1.2 kNs/m
V.	Driver-Seat Assembly				
		mass	(m_d)	88.0	kg
		seat spring	(k_d)	17.5	kNs/m
		seat damping	(c_d)	0.65	kNs/m

Table 6.4   Dynamic characteristics of tractor-semi-trailer

Case	Decay Rate (σ), 1/s	Damped Frequency (ω), 1/s	Undamped Frequency (ω), Hz	Mode Description
(a) Loaded vehicle	− 1.21	10.03	1.6	Tractor bouncing mode
	− 2.42	12.00	1.95	Driver-seat bouncing mode
	− 3.91	14.64	2.41	Semi-trailer pitching mode
	− 4.00	17.64	2.87	Tractor pitching mode
	− 15.52	66.24	9.87 ⎫	
	− 21.75	60.06	11.1 ⎬	Axle wheel hop modes
	− 31.05	80.32	13.7 ⎭	
(b) Unloaded vehicle	− 2.54	11.92	1.93	Driver-seat bouncing mode
	− 4.27	14.83	2.46	Tractor bouncing mode
	− 8.18	27.86	4.62	Tractor pitching mode
	− 10.6	45.27	7.43	Semi-trailer pitching mode
	− 15.58	60.18	9.89 ⎫	
	− 41.78	48.02	10.13 ⎬	Axle wheel hop modes
	− 33.45	66.4	11.83 ⎭	

The psd of the vertical and fore-aft acceleration at the drivers seat for Routes A (smooth) and B (rough) are shown in Figs. 6.7 and 6.8, respectively. The ISO criteria for one hour and eight hour reduced comfort boundaries are also plotted in the figures. The peaks corresponding to the dominant modes of vibration can be identified. Both, vertical and fore-aft, spectra exceed the 1 hour ISO criteria indicating ride discomfort due to excessive vibration even for the smooth road. For rough road, there is a general increase in the vibration level over a large part of the frequency range.

Figure 6.9 shows the vertical and fore-aft vibration level experienced by the driver as a function of the speed. Except in the range 25–55 km/hour, the vibration level increases with speed. A detailed analysis of the response suggests design modifications and operational guidelines to improve ride performance (Dokanish and Elmadany, 1980).

Example 6.2:   Response and fatigue life of a locomotive to random track input (Garivaltis, Garg and D' Souza, 1980b)

Locomotives are the most complex amongst railway vehicle systems. We shall consider the determination of the psd of the response and fatigue life of the suspension elements of a six-axle locomotive travelling on a tangent track. The three views of the locomotives showing the assumed degrees of freedom, dimensions and parameters of the dynamic model are shown in Fig. 6.10.

The model has 39 degrees of freedom:

Carbody: $\{X^b\} = [Y^b \; Z^b \; \phi^b \; \psi^b \; \Theta^b]^T$

Trucks: $\{X_j^t\} = [Y_j^t Z_j^t \phi_j^t \psi_j^t \Theta_j^t]$

Axles: $\{X_k^a\} = [Y_k^a Z_k^a \phi_k^a \psi_k^a]$

where Y, Z, ϕ, ψ, Θ represent vertical, lateral, roll, yaw and pitch degrees of freedom, respectively. The 39-dimensional generalized displacement vector:

$$\bar{X} = [\bar{X}^b; \bar{X}_j^t, j = 1, 2; \bar{X}_k^a, k = 1, 6]^T$$

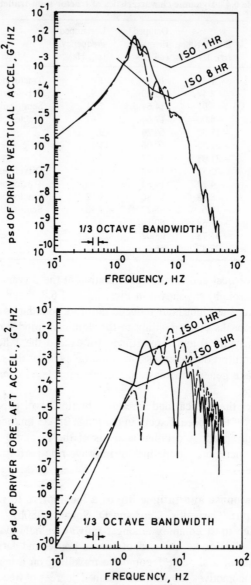

Fig. 6.7. Vertical and fore-aft acceleration spectra for route −A. (———) loaded vehicle; (—·—·—) unloaded vehicle. Speed 80 km/h (Dokainish and Elmadany, 1980).

The locomotive is assumed to travel with a constant velocity V. The equations of the locomotive are given by (6.77) with

$$\bar{F}(t) = \mathbf{R}\bar{Z}^{\mathrm{r}}$$

in which

$$\mathbf{R} = \left[\begin{array}{c|c} 0 & 0 \\ \hline 0 & R_{\mathrm{rs}} \end{array}\right], r, s = 1, 2, \ldots 24$$

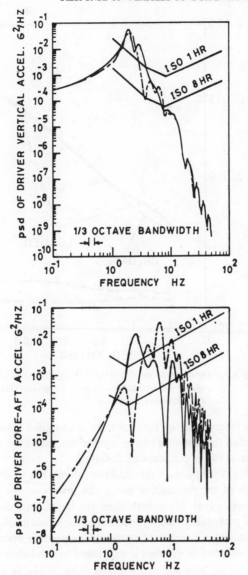

Fig. 6.8. Vertical and fore-aft acceleration spectra for route –B. (————) loaded vehicle; (— — — — —) unloaded vehicle. Speed 80 km/h (Dokanish and Elmadany, 1980).

and

$$\bar{Z}^r = [0 \ldots 0 \; Z^r_{16} \ldots Z^r_{39}]$$

represents the lateral and vertical track inputs at each wheel. The psd of the track irregularities are assumed to be given by (6.37) and (6.38). The input due to multi-wheel axles is determined by (6.39).

The stationary response of the locomotive is given by (6.79) to (6.87). The dominant frequencies of the locomotive are given in Table 6.5.

The psd of car body displacements is shown in Fig. 6.11. The computations were carried out over the frequency range 0.1–35 Hz for an operating speed of 80 mph.

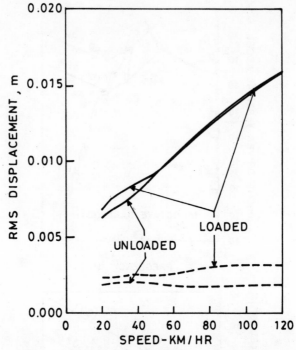

Fig. 6.9. Vertical and fore-aft displacement against speed for route –A. (——) vertical; (– – – –) fore-aft (Dokainish and Elmadany, 1980).

The theoretical frequencies in Table 6.5 are determined from the peaks in the psd curve. The psd of the response can be used to determine the variance and other response characteristics for purposes of design.

Garivaltis et al, (1980b) have estimated the fatigue life of the elements of the suspension system of the locomotive using the results of the random vibration analysis. The mean fatigue life calculations are based on Palmgren-Miner law, which requires the probability density function of the stress levels in the suspension elements and estimates of number of stress cycles per unit distance. The force in the springs of the suspension systems is expressed by a relation of the form (6.83) with $(\bar{a}_1 = \bar{b}_1 = \bar{b}_2 = 0)$, and psd is computed from (6.84). A parameter n called the stiffness ratio is defined as the ratio between the vertical secondary suspension stiffness and the vertical primary suspension stiffness per truck side. Figures 6.12(a), (b) show the psd of the force in the vertical springs of primary and secondary suspensions, respectively. Figure 6.13 shows the fatigue life in miles for the vertical spring of the primary and secondary suspensions as a function of stiffness ratio. The stiffness ratio $\simeq 3$ gives maximum expected fatigue life, which falls sharply as stiffness ratio increases. Such analysis can be imbedded in the optimum design of suspension elements.

6.4.2 Nonstationary Response
A vehicle is subjected to nonstationary excitation, if it moves on a guideway with a variable velocity, and its response is, therefore, nonstationary. The response of vehicle moving with constant velocity is also nonstationary if it suddenly encounters

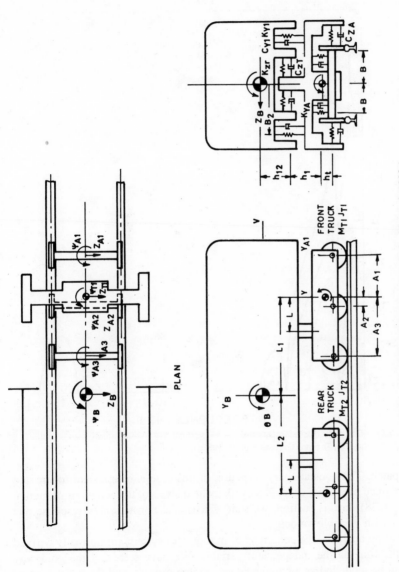

Fig. 6.10. Six-axle locomotive (Garivaltis et al, 1980a).

Table 6.5. Dominant frequencies, Hz

Degree of Freedom-Body	Test	Theory
Bounce	1.5–1.6	1.58
Lateral	0.5	0.5
Roll	0.9	0.6
Pitch	1.62–1.7	1.2

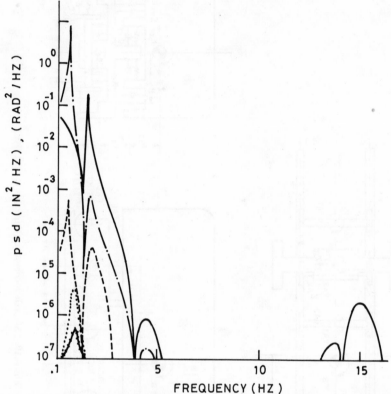

Fig. 6.11. Car body displacements psd. (——) vertical, (—·—·—) lateral; (– – –) roll; and (....) yaw (Garivaltis et al, 1980a).

a rough patch. The nonstationary response, in this case is of transient nature and can be obtained by a time, or frequency, domain analysis of the equations of motion (6.77) or (6.78). In this section, we shall discuss the nonstationary response due to motion with variable velocity.

In vibration analysis, the equations of motion of a system are normally derived treating the time as the independent variable. We have seen that the guideway profile can be modelled as a homogeneous random process, and the nonstationary excitation is introduced in the vehicle vibration problem due to motion with variable velocity. To retain the homogeneous property of the excitation this suggests a transformation of the equations of motion from time to space domain, with distance along the guideway as the independent variable. Such a transformation results in governing equations with variable (space-dependent) coefficients which make the response nonhomogeneous. However, in the space domain, it is possible to obtain

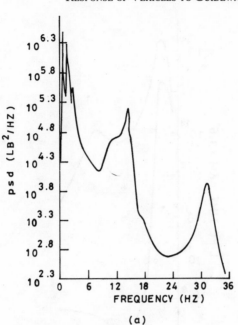

(a)

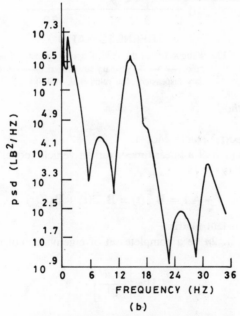

(b)

Fig. 6.12. Force psd for the vertical spring. (a) primary
suspension; (b) secondary suspension.
Stiffness ratio η = 12.62 (Garivaltis et al,
1980b).

the response statistics in a closed form, within the framework of evolutionary
spectral analysis (Yadav, 1976; Yadav and Nigam, 1974, 1978). We shall discuss
both the space and time domain methods.

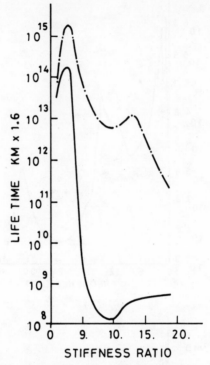

Fig. 6.13. Fatigue life of the vertical spring against stiff-ness ratio. (—.—.—) primary suspension; (———) secondary suspension. (Garivaltis et al, 1980b).

Space Domain Method

LINEAR TIME-INVARIANT VEHICLE MODEL
The equations of motion of a linear time-invariant vehicle model in the state space are given by (6.78) as

$$\dot{\bar{Y}} + \mathbf{A}\,\bar{Y} = \mathbf{B}\,\dot{\bar{\xi}}(t) = \mathbf{B}_1\dot{\bar{Z}}(t) + \mathbf{B}_2\bar{Z}(t) \tag{6.93}$$

in which **A** is a constant matrix.

Let α_j, $j = 1, 2, \ldots 2n$ be a complete set of eigenvalues of matrix **A**, and let

$$U = \left[\frac{\alpha_j\,\bar{u}^j}{\bar{u}^j}\right] \tag{6.94}$$

be the modal column matrix, in which $\bar{u}_j$ are n-dimensional modal column vectors. Then,

$$\mathbf{U}^{-1}\,\mathbf{A}\mathbf{U} = \mathbf{D} = \text{diag.}\,(\alpha_1, \alpha_2, \ldots \alpha_{2n}) \tag{6.95}$$

Using the transformation $\bar{Y}(t) = \mathbf{U}\bar{R}(t)$, (6.93) can be expressed in the canonical form

$$\dot{\bar{R}} - \mathbf{D}\bar{R} = \mathbf{U}^{-1}\mathbf{B}_1\dot{\bar{Z}} + \mathbf{U}^{-1}\mathbf{B}_2\bar{Z} = \tilde{\mathbf{B}}_1\dot{\bar{Z}} + \tilde{\mathbf{B}}_2\bar{Z} \tag{6.96a}$$

or

$$\dot{R}_j - \alpha_j R_j = \sum_{k=1}^{m} \tilde{b}_{1jk} \dot{Z}_k(t) + \sum_{k=1}^{m} \tilde{b}_{2jk} Z_k(t), \quad j = 1, 2, \ldots 2n \qquad (6.96b)$$

Let $V(s) = ds/dt$. Transforming (6.96) from time to space domain, we get

$$V(s) R_j' - \alpha_j R_j = \sum_{k=1}^{m} \tilde{b}_{1jk} V(s) Z_k'(s) + \sum_{k=1}^{m} \tilde{b}_{2jk} Z_k(s), \quad j = 1, 2, \ldots 2n \qquad (6.97)$$

where $(')$ denotes d/ds. Let

$$f(s) = \int_0^s \frac{ds}{V(s)} \qquad (6.98)$$

Solution of (6.96) can be expressed as

$$R_j(p) = C_j \exp\left[\alpha_j f(s)\right] + \int_0^s h_j(s, p) \left[\sum_{k=1}^{m} \{V(p) \tilde{b}_{1jk} Z_k'(p)\right.$$

$$\left. + \tilde{b}_{2jk} Z_k(p)\}\right] dp \qquad (6.99)$$

where

$$h_j(s, p) = \exp\left[\alpha_j\{f(s) - f(p)\}\right]/V(p) \qquad (6.100)$$

is the Greens' function of (6.97); and $Z_k(s)$, $k = 1, 2, \ldots m$, are a set of homogeneous random processes. Let

$$Z_k(s) = \int_{-\infty}^{\infty} \exp(i\Omega s) \, dF_k(\Omega) \qquad (6.101)$$

with

$$E[|dF_k(\Omega)|^2] = \Phi_{Z_k Z_k}(\Omega) \, d\Omega, \text{ and } E[dF_k(\Omega)] = 0 \qquad (6.102)$$

Substituting for $Z_k(s)$ from (6.101) in (6.99),

$$R_j(p) = C_j \exp\left[\alpha_j f(s)\right]$$

$$+ \int_{-\infty}^{\infty} \left[\sum_{k=1}^{m} \int_0^s \{h_j(s, p) \exp(i\Omega p) (V(p) \tilde{b}_{1jk} i\Omega + \tilde{b}_{2jk})\} \, dp \, dF_k(\Omega)\right]$$

$$= C_j \exp\left[\alpha_j f(s)\right] + \int_{-\infty}^{\infty} \sum_{k=1}^{2n} \{H_{jk}(\Omega, s) e^{i\Omega s} \, dF_k(\Omega)\} \qquad (6.103)$$

where

$$H_{jk}(\Omega, s) = \int_0^s \{h_j(s, p) \exp(i\Omega(p - s)) (V(p) \tilde{b}_{1jk} i\Omega + \tilde{b}_{2jk})\} \, dp \qquad (6.104)$$

The response of the vehicle can now be expressed as

$$X_i(s) = \sum_{j=1}^{2n} u_i^j C_j \cdot \exp(-\alpha_j f(s))$$

$$+ \sum_{j=1}^{2n} u_i^j \int_{-\infty}^{\infty} \sum_{k=1}^{m} \{H_{jk}(\Omega, s) e^{i\Omega s} \, dF_k(\Omega)\} \qquad (6.105)$$

and

$$\dot{X}_i(s) = \sum_{j=1}^{2n} \alpha_j u_i^j C_j \exp(-\alpha_j f(s))$$

$$+ \sum_{j=1}^{2n} \alpha_j u_i^j \int_{-\infty}^{\infty} \sum_{k=1}^{m} \{H_{jk}(\Omega, s) e^{i\Omega s} dF_k(\Omega)\} \tag{6.106}$$

It follows that the mean response

$$E[X_i(s)] = \sum_{j=1}^{2n} u_i^j C_j \exp(-\alpha_j f(s)) \tag{6.107}$$

$$E[\dot{X}_i(s)] = \sum_{j=1}^{2n} \alpha_j u_i^j C_j \exp(-\alpha_j f(s)) \tag{6.108}$$

and the cross-covariance response

$$K_{X_i X_e}(s_1, s_2) = \sum_{j=1}^{2n} \sum_{q=1}^{2n} u_i^j u_e^q I_{jq}(s_1, s_2) \tag{6.109}$$

$$K_{\dot{X}_i \dot{X}_e}(s_1, s_2) = \sum_{j=1}^{2n} \sum_{q=1}^{2n} \alpha_j \alpha_q^* u_i^j u_e^q I_{jq}(s_1, s_2) \tag{6.110}$$

where

$$I_{jq}(s_1, s_2) = \sum_{k=1}^{m} \sum_{r=1}^{m} \int_{-\infty}^{\infty} H_{jk}(\Omega, s_1) H_{qr}(\Omega, s_2) e^{i\Omega(s_1-s_2)} \Phi_{Z_k Z_r}(\Omega) d\Omega \tag{6.111}$$

We have obtained a closed form solution for the nonstationary response statistics of a vehicle moving with variable velocity. However, integral (6.111) requires numerical evaluation, except in a few simple cases. The response statistics can be converted into the time-domain using the relationship between s and t. Note that, response statistics for motion with constant velocity is a special case of the above results with $V(s) = V = $ constant, and $h(s, p) = h(s - p)$. We shall illustrate the application of the above method for the response of a simple model of the aircraft during taxi, take off and landing.

Example 6.3:   Response of aircraft to ground environment
(Nigam and Yadav, 1974; Yadav and Nigam, 1984)

An aircraft is modelled as a linear, two-degree-of-freedom system shown in Fig. 6.14. The equations of motion of the model are given by (6.77) with

$$\mathbf{M} = \begin{bmatrix} m_1 & 0 \\ 0 & m_2 \end{bmatrix}; \mathbf{C} = \begin{bmatrix} c_1 & -c_1 \\ -c_1 & c_1 \end{bmatrix}$$

$$\mathbf{K} = \begin{bmatrix} k_1 & k_1 \\ -k_1 & k_1 + k_2 \end{bmatrix}; \bar{F}(t) = \begin{Bmatrix} L(t) \\ k_2(Zt) \end{Bmatrix}$$

where $L(t) = C_L \dot{s}^2$ is the lift force. Table 6.6 gives the values of the model parameters.

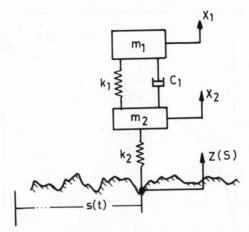

Fig. 6.14.   An aircraft model.

Table 6.6   **Vehicle parameter values for aircraft in Fig. 6.14**
(Nigam and Yadav, 1974)

$m_1 = 14{,}970$ kg sec^2/m	$m_2 = 230.2$ kg sec^2/m
$k_1 = 30.2 \times 10^6$ kg/m	$k_2 = 0.240 \times 10^6$ kg/m
$c_1 = 173.4 \times 10^3$ kg sec/m	
$a_1 = 0;\ a_2 = 1.2205125$	$b = 2$
$c = 1 \times 10^{-5};\ a = 0.001791$	

The motion of the aircraft during taxi, take off and landing runs is described by

$$s(t) = a_1 t + a_2 t^b$$

The psd of the runway/taxiway is assumed to be

$$\Phi_{ZZ}(\Omega) = \frac{c}{\Omega^2 + a^2}, \qquad |\Omega| \leq \Omega_c$$
$$= 0; \qquad \Omega > \Omega_c$$

The mean profile of the runway is assumed to be

$$E[Z(s)] = g_0 s + g \sin \frac{2\pi s}{L}$$

The response statistics of the aircraft during taxi, take off and landing runs is obtained from (6.107)–(6.110). The integral in (6.111) is evaluated numerically. The r.m.s. acceleration of the c.g. of the aircraft during take off run is shown in Fig. 6.15 (Nigam and Yadav, 1974).

The fatigue life of the wheel-axle of the nose landing gear of an aircraft is determined using the crack propagation approach. The fatigue damage is assumed to take place in three phases: crack initiation, crack propagation and fracture, which occurs when the stress level exceeds the residual strength value. The failure is modelled as a time-dependent one-sided barrier crossing problem. The statistics of the stress history during each ground-to-air cycle is used to determine the probability of failure due to cumulative fatigue damage. Figure 6.16 shows the results for crack growth and probability of failure for two values of sink velocities during landing impact (Yadav and Nigam, 1984).

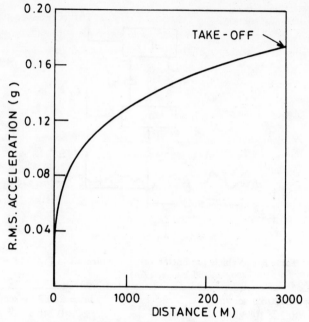

Fig. 6.15. Upper mass rms acceleration against distance (Nigam and Yadav, 1974).

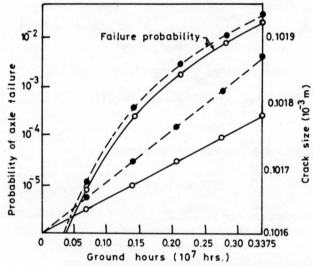

Fig. 6.16. Crack growth and probability of failure against ground hours. Sink velocity: 0.6 m/s (—o—); 1.8 m/s(--•--), (Yadav and Nigam, 1984).

Equivalent Linear Vehicle Model

The equations of motion of a nonlinear vehicle model can be expressed in the state space, as

$$\dot{\bar{Y}} + \mathbf{A}\,\bar{Y} + \bar{f}(\bar{Y}) = \mathbf{B}_1 \dot{\bar{Z}}(t) + \mathbf{B}_2 \bar{Z}(t) \qquad (6.112)$$

where $\bar{f}(\bar{Y})$ is a nonlinear vector function. Equation (6.112) can be reduced to an equivalent linear system using the method of equivalent linearization (Iwan and Mason, 1980), or by covariance analysis by describing function technique (CADET) (Gelb and Warren, 1972; Harris and Hammond, 1986b). The linearized equation of motion can be expressed as

$$\dot{\bar{Y}} + [A + \tilde{A}(K_{\bar{Y}\bar{Y}}(t))]\bar{Y} = B_1\dot{\bar{Z}}(t) + B_2\bar{Z}(t) \tag{6.113}$$

Since the response is nonstationary, the covariance matrix $K_{\bar{Y}\bar{Y}}$ in (6.113) is time dependent. The equivalent linear system has time dependent coefficients, and therefore, it cannot be reduced to a system of first order equations through a canonical transformation, as in the case of a linear time-invariant model. A determination of covariance response of (6.113) using the time, or frequency, domain state transition matrix, though feasible, is cumbersome. A well known result in linear system theory can, however, be used to determine zero-lag auto-covariance matrix $K_{\bar{Y}\bar{Y}}(t)$ through numerical integration of Lyapunov matrix equation.

In Section 6.2, we have seen that a homogeneous guideway profile can be generated by passing white noise through a linear filter. Let

$$\bar{Z}(s) = P\,\bar{g}(s) \tag{6.114}$$

and

$$\frac{d}{ds}\,\bar{g}(s) + J\,\bar{g}(s) = R\,\bar{W}(s) \tag{6.115}$$

where $\bar{W}(s)$ is a white noise vector with

$$E[\bar{W}(s)] = 0; \text{ and } E[\bar{W}(s_1)\,\bar{W}^T(s_2)] = G(s_1)\,\delta(s_2 - s_1) \tag{6.116}$$

and P, J, R are matrices of appropriate dimensions.

We now transform (6.116) from space to time domain. Since $d/ds = \frac{d}{dt}\frac{1}{\dot{s}}$, we have

$$\frac{d}{dt}\,\bar{g}(t) + \dot{s}\,J\,\bar{g}(t) = R\,\dot{s}\,\bar{W}(s(t)) \tag{6.117}$$

In Section 6.2, we have shown that a $W(s(t))$ can be replaced by a 'covariance equivalent' modulated white noise process $W_1(t)/\sqrt{\dot{s}}$. We replace (6.117) by

$$\frac{d}{dt}\,\bar{g}_1(t) + \dot{s}\,J\,\bar{g}_1(t) = R\sqrt{\dot{s}}\,\bar{W}_1(t) \tag{6.118}$$

Clearly $\bar{g}_1(t)$ and $\bar{g}(t)$ are covariance equivalent.

We now define $\bar{\eta}(t) = [\bar{Y}^T\,\bar{g}_1^T]$, and combine (6.113), (6.114) and (6.118) to write the augmented equation of motion

$$\dot{\bar{\eta}} + \hat{A}(t)\,\bar{\eta}(t) = \hat{B}(t)\,\bar{W}_1(t) \tag{6.119}$$

where

$$\hat{A}(t) = \left[\begin{array}{c|c} \hat{A} + A(K_{\bar{Y}\bar{Y}}(t)) & -\dot{s}\,B_1\,PJ + B_2 \\ \hline 0 & \dot{s}\,J \end{array}\right] \tag{6.120}$$

$$\hat{\mathbf{B}}(t) = \left[\frac{\sqrt{\dot{s}} \; \mathbf{B}_1 \mathbf{PR}}{\sqrt{\dot{s}} \; \mathbf{R}} \right] \tag{6.121}$$

The solution of (6.119) is given by

$$\bar{\eta}(t) = \Psi(t, t_0) \, \bar{\eta}(t_0) + \int_{t_0}^{t} \Psi(t, \tau) \, \hat{\mathbf{B}}(\tau) \, \overline{W}_1(\tau) \, d\tau \tag{6.122}$$

and the covariance matrix of $\bar{\eta}(t)$

$$\mathbf{K}(t, \tau) = \mathbf{K}(t) \, \Psi^{\mathrm{T}}(t, \tau) \tag{6.123}$$

The zero-lag covariance matrix $\mathbf{K}(t)$ satisfies the Lyapunov matrix equation (7.142*):

$$\dot{\mathbf{K}} + \mathbf{AK} + \mathbf{KA}^{\mathrm{T}} = \hat{\mathbf{B}} \mathbf{G}(t) \hat{\mathbf{B}}^{\mathrm{T}}; \; \mathbf{K}(t_0) = K_0 \tag{6.124}$$

$\Psi(t, \tau)$ is the state transition matrix of (6.119). For time-variant systems, numerical calculations are required to determine $\Psi(t, \tau)$.

In vehicle vibration studies, it is generally sufficient to know the zero-lag covariance matrix $\mathbf{K}(t)$. It can be obtained by numerical integration of (6.124) starting with the initial value K_0. (For a discussion on numerical integration see, Karnopp, 1978; Hammond and Harrison, 1981.) Since $\mathbf{K}$ is symmetric, (6.124) yields $N(N + 1)/2$ independent first-order equations, if $N \times N$ is size of covariance matrix.

The above formulation is applicable to a class of problems, in which the excitation is a white noise vector process. In vehicles, this includes problems, in which the model has a single excitation, or the wheels are laterally aligned (e.g. separated by an axle) so that the excitation vector consists of two or more random processes at the same instant of time. Most vehicles have a train of wheels longitudinally displaced at fixed distances. This gives rise to an excitation vector with time variable delays, which makes the problem much more complicated and requires a separate treatment.

Vehicles with Multiple Wheel Train
For simplicity, we consider a vehicle with two wheels separated by a constant distance L. The state-space equation of motion of the vehicle can be expressed as

$$\dot{\bar{\eta}} + \hat{\mathbf{A}}(t) \bar{\eta} = \mathbf{B}_1(t) \, \overline{W}(s(t)) + \mathbf{B}_2(t) \, \overline{W}(s(t) - L) \tag{6.125}$$

Harrison and Hammond (1986c) have shown that the Lyapunov equation associated with (6.125) can be expressed as

$$\dot{\mathbf{K}} + \hat{\mathbf{A}}(t) \mathbf{K} + \mathbf{K} \hat{\mathbf{A}}^{\mathrm{T}}(t) = \mathbf{B}_1(t) \, \mathbf{GB}_2^{\mathrm{T}}(t)/\dot{s}(t) + \mathbf{B}_2(t) \, \mathbf{GB}_2^{\mathrm{T}}(t)/\dot{s}(t)$$
$$+ \Psi(t, \tau) \, \mathbf{B}_1(t) \, \mathbf{GB}_2^{\mathrm{T}}(t)/\dot{s}(t) + \mathbf{B}_2(t) \, \mathbf{GB}_1^{\mathrm{T}}(t) \, \Psi^{\mathrm{T}}(t, \tau)/\dot{s}(\tau) \tag{6.126}$$

where $\tau = s^{-1}(s(t) - L)$.

To solve (6.126), an explicit evaluation of the time variable state transition matrix is required. This involves simultaneous numerical integration of (6.126), and computation of $\Psi(t, \tau)$. If the vehicle model is time-invariant, the space domain method described in the preceding section is applicable, and is more efficient. If the vehicle has more than two wheels following each other, additional delay terms should be added to the r.h.s. of (6.125). Clearly this will add more term to (6.126) and the numerical computations become very cubersome.

Narayanan (1987) has developed an approximate method to circumvent the complications due to time variable delay in multi-wheel vehicles. He has proposed a Taylor series expansion of the excitation at a follower wheel in terms of the excitation at the front wheel.

Let us consider a vehicle with two wheels, and let $Z_f(t)$ and $Z_r(t)$ denote the excitation at the front and rear wheels separated by a distance L. Noting that

$$\frac{d}{ds} = \frac{1}{\dot{s}}\frac{d}{dt}; \frac{d^2}{ds^2} = \left(\frac{1}{\dot{s}^2}\frac{d^2}{dt^2} - \frac{\ddot{s}}{\dot{s}^3}\frac{d}{dt}\right)$$

We have by Taylor series expansion

$$Z_r(t) = Z_f(s(t) - L) = Z_f(s(t)) - \frac{L}{\dot{s}}\frac{dZ_f}{dt} + \frac{1}{2!}\left(\frac{L}{\dot{s}}\right)^2\left(\frac{d^2Z_f}{dt^2} - \frac{\ddot{s}}{\dot{s}}\frac{dZ_f}{dt}\right) + 0\left[\left(\frac{L}{\dot{s}}\right)^2\right]$$

$$\simeq Z_f(t) - \left[\frac{L}{\dot{s}} + \frac{1}{2}\left(\frac{L}{\dot{s}}\right)^2\frac{\ddot{s}}{\dot{s}}\right]\dot{Z}_f(t) \tag{6.127a}$$

and

$$\dot{Z}_r(t) \simeq \left[1 + \left(\frac{L}{\dot{s}}\right)\frac{\ddot{s}}{\dot{s}}\right]\dot{Z}_f(t) \tag{6.127b}$$

The above approximation is obtained by neglecting terms of order higher than $(L/\dot{s})^2$, and the second and higher derivatives of $Z_f(t)$ in (6.127). The first approximation is justified if $L/\dot{s} \ll 1$. The wheel base L is usually small, and, for practical operating speeds the assumption is justified. It must be noted, however, that the r.h.s of (6.127) is singular at $\dot{s} = 0$, $L/\dot{s} > 1$ for small values of $\dot{s}$. This will cause difficulties for vehicle starting from rest. The difficulty may be circumvented by assuming that until $L/\dot{s} < 1$, the vehicle moves on a perfectly smooth surface. From practical considerations, it appears to be an acceptable approximation.

With the above approximation the excitation terms involving time delay in (6.126) may be replaced by terms without delay, and the equations of motion can be expressed by (6.113). Thus instead of (6.126), the zero-lag covariance matrix Eq. (6.124) is now applicable.

Example 6.4: Nonstationary response of a nonlinear vehicle (Narayanan, 1987) Consider a four-degree-of-freedom vehicle model shown in Fig. 6.17. The primary suspension system supporting the sprung mass is assumed to be a combination of linear viscous dampers (c_3, c_4) and nonlinear hysteretic springs (k_3, α_3; k_4, α_4). The equations of motion of the vehicle are:

$$m_1\ddot{X}_1 + \sum_{j=3}^{4}[c_j\dot{u}_j + \alpha_jk_ju_j + (1 - \alpha_j)k_jv_j] = 0$$

$$m_2\ddot{X}_2 + \sum_{j=3}^{4}[c_j\alpha_j\dot{u}_j + \alpha_jk_ju_j + (1 - \alpha_j)k_jv_j] = 0$$

$$m_i\ddot{X}_i - [c_i\dot{u}_i + \alpha_ik_iu_i + (1 - \alpha_i)k_iv_i] + c_{i+2}\dot{X}_i + k_{i+2}X_i$$

$$= c_{i+2}\dot{Z}_{i-2} + k_{i+2}Z_{i-2}, i = 3, 4$$

in which $u_j = X_1 + a_jX_2 - X_j$, $j = 3, 4$; k_j is the pre-yielding stiffness of hysteretic

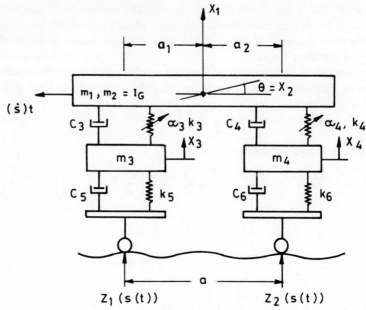

Fig. 6.17 A vehicle model.

spring and α_j is ratio of post-yielding to pre-yielding stiffness. v_i represent the hysteretic displacements and are determined by the following nonlinear differential equation (Wen, 1980):

$$\dot{v}_i = -\gamma_i |u_i| v_i |^{n_i-1} - \beta_i |u_i| v_i |^{n_i} + A_i \dot{u}_i, \; i = 3, 4$$

in which γ_i, β_i, n_i and A_i are the parameters of the hysteretic system. In the present analysis, $n_3 = n_4 = 1$.

Let $\bar{X}(t) = [X_1 X_2 X_3 X_4 v_3 v_4]^T$ represent the generalized coordinates including hysteretic displacements. The equations of motion can be expressed in the matrix form

$$\mathbf{M}\ddot{\bar{X}} + \mathbf{C}\dot{\bar{X}} + \mathbf{K}\bar{X} + \bar{g}(\dot{\bar{X}}, \bar{X}) = \bar{F}(t)$$

Using the method of equivalent linearization (Atalik and Utku, 1976) the above equations can be reduced to the linear form

$$\mathbf{M}\ddot{\bar{X}} + \mathbf{C}^*\dot{\bar{X}} + \mathbf{K}^*\bar{X} = \bar{F}(t)$$

where

$$\mathbf{C}^* = \mathbf{C} + \mathbf{C}', \text{ and } \mathbf{K}^* = \mathbf{K} + \mathbf{K}'$$

It can be shown that the non-zero elements of matrices $\mathbf{C}'$ and $\mathbf{K}'$ are (Narayanan, 1987)

$$c'_{51} = \sqrt{(2/\pi)} \, [\beta_3 \sigma_{v_3} + \gamma_3 E[\dot{u}_1 v_3]/\sigma_{\dot{u}_1}]$$

$$c'_{52} = -a_1 c'_{51}; \, c'_{53} = -c'_{51}$$

$$c'_{61} = \sqrt{(2/\pi)} \, [\beta_4 \sigma_{v_4} + \gamma_4 E[u_2 v_4]/\sigma_{\dot{u}_2}]$$

$$c'_{62} = a_2 c'_{61}; \, c'_{64} = -c'_{61}$$

$$k'_{55} = \sqrt{(2/\pi)} \, [\beta_3 E[u_1 v_3]/\sigma_{v_3} + \gamma_3 \sigma_{\dot{u}_1}$$

and

$$k'_{66} = \sqrt{(2/\pi)}\,[\beta_4\,E[u_2 v_4]/\sigma_{v_4} + \gamma_4\,\sigma_{\ddot{u}_2}]$$

Note that coefficients of equivalent linear system are functions of the cross-covariance of the response. Since we are considering nonstationary response they are time dependent.

The excitation is assumed to be generated by passing white noise through a first order filter, (6.26) and (6.30). Since the vehicle has two wheels following each other, the excitation has time variable delay. The approximation (6.127) is used to replace the delay term. The augmented equation of motion of the system is expressed in the state space by (6.119), with

$$\bar{\eta}(t) = [X_1\,X_2\,X_3\,X_4\,\dot{X}_1\,\dot{X}_2\,\dot{X}_3\,\dot{X}_4\,v_3 v_4\,Z(t)]^T$$

The zero-lag covariance matrix of $\bar{\eta}(t)$ satisfies the Lyapunov Eq. (6.124). The matrix differential Eq. (6.124) are solved by numerical integration using fourth order Runge-Kutta method based on the data in Table 6.7.

Table 6.7. Vehicle parameters

Index (i)	1	2	3	4	5	6
Generalized masses m_i (kg or kg m^2)	1000	2000	50	50	—	—
Damping Coefficients c_i (Ns/m)	—	—	2000	8000	360	360
Pre-yielding stiffness k_j (N/m)	—	—	50001	5000	18000	18000
$a = 2.5$ m; $a_1 = 1.3$ m and $a_2 = 1.2$ m						

The vehicle is assumed to move with constant acceleration. The variance of pitch and bounce are shown in Fig. 6.18 as a function of time. While integrating the matrix Eq. (6.124), it is assumed that the effect of rear wheel input comes into

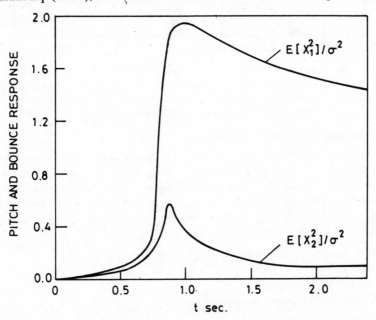

Fig. 6.18. Variance of pitch and bounce against time. $\ddot{s} = 10$, $A_3 = A_4 = 1$; $\alpha_3 = \alpha_4 = 0.25$; $\beta_3 = \beta_4 = 0.5$; $\gamma_3 = \gamma_4 = 0.5$; and $a = 0.2$ (Narayanan, 1987).

play only after it crosses the front wheel position at $s = 0$. The sudden increase in the response variance (Fig. 6.17) occurs around the time when rear wheel input becomes operative.

Figure 6.19 shows a comparison of Narayanan's approximation with the results reported by Harrison and Hammond (1986c) for a two wheel vehicle incorporating the effect of time variable delay. The results differ significantly and the method due to Narayanan (1987) yields conservative estimates. The computational effort involved in Narayanan's method is considerably less than in the method due to Harrison and Hammond (1986c).

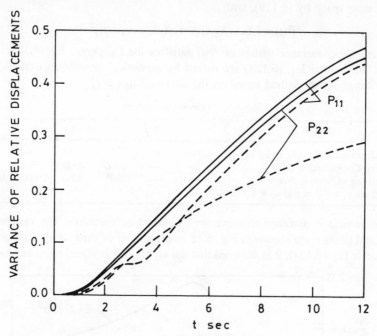

Fig. 6.19. Variance of relative displacement, against time. $\ddot{s} = 5$; $\sigma = 1$; $a = 0$, $\omega_1 = \omega_2 = 10$; $r^2 = L^2/L^2$, $\xi_1 = \xi_2 = 0.38$; and $L = 4$. (– – –) Harrison and Hammond (1986), (———) Narayanan (1987).

6.5 Design of Vehicle Systems

The considerations of economics of operation and saving in travel-time have led to higher speeds, non-stop operation over long distances, and complex vehicle systems, such as, articulated multi-wheel vehicles. This trend has made exacting demands on the design of vehicle systems—in particular, the design of suspension systems and geometric disposition of masses, linkages and supports. The major criteria for the design of vehicle systems include:

1. Ride quality—ride comfort and cargo safety.
2. Ride safety.
3. Fatigue and strength.
4. Constraints on travel of moving parts.

We have discussed the procedures for estimation of fatigue life based on random vibration analysis in Chapter 3. In Examples 6.2 and 6.3, we have illustrated the application of these procedures for establishing the fatigue life. The problem of constraints on travel of moving parts is the classical barrier-crossing problem and has been dealt with in Chapter 4. In this section, we shall discuss the criteria of ride comfort and safety from the point of view of design of vehicle systems based on random vibration analysis.

6.5.1 Ride Comfort and Cargo Safety

The evaluation of ride comfort involves assessment of human sensitivity to vibration, which depends not only on the physiological and biomechanical response of human body, but also on a number of psychological and environmental factors. The human reaction to vibration is a function of the amplitude and frequency of acceleration applied to the body, the direction (vertical, fore-aft, lateral), and the character of the motion (linear, rotational). Extensive research has been conducted on human sensitivity to vibration and its results are available in the literature (Hanes, 1970; UCTAV, 1972; ISO 2631, 1978 (E)). Basically three approaches are available for establishing comfort criteria; ride index (Carstens and Kresge, 1965), absorbed power (Lee and Pradko, 1968), power spectral density and transfer function (UTACV, 1972; ISO 2631, 1978 (E)). In the design of the vehicle systems, the comfort criteria based on psd and transfer function are most convenient. In this approach, the acceleration spectrum is divided into frequency bands (1/3-octave) and a correlation with r.m.s. acceleration, in each band, with human response to vibration is used to specify comfort criteria.

The International Standard Organisation (ISO 2631, 1978 (E)) has specified numerical values for limits of exposure to vibrations transmitted from solid surface to human body in the frequency range 1–80 Hz. These limits cover human sensitivity to vertical, lateral, and fore-aft vibration to periodic vibration over exposure time ranging from 1 min to 24 h. For broad-band excitation encountered in vehicles, the standard recommends: "In case of broad-band distributed vibration, whether random or not, occurring in more than one third—octave band, the r.m.s. value of the acceleration in each such band is to be evaluated separately with respect to the appropriate limit at the centre frequency of the band".

ISO defines three criteria corresponding to:

(a) the preservation of working efficiency (fatigue-decreased proficiency boundary);
(b) the preservation of health and safety (exposure limit); and
(c) the preservation of comfort (reduced comfort boundary).

Figure 6.20 shows the fatigue-decreased proficiency boundary for vertical vibration and gives the conversion factors for the other two criteria. Similar curves are included in the Standard for vibration in other directions. The curves in these figures are usable directly for periodic vibration. However, following the recommendation regarding broad-band excitation, these can be easily converted to psd using the following relation (Smith, 1976):

$$\log \overline{G}(\bar{f}) = (2m_s - 1) \log \bar{f} + 2b_s - \log 0.23 \qquad (6.128)$$

where $\overline{G}(\bar{f})$ is the average psd over 1/3 octave band, $\bar{f}$ is the central frequency of the band, m_s and b_s are the slope and intercept of the straight line segments in the log-log ISO plots. Fig. 6.21 shows the ISO and UTACV standards on the psd plot for vertical seat vibration. Since random vibration analysis yields psd of the response directly, ISO standard converted to psd is directly applicable for evaluating the ride comfort.

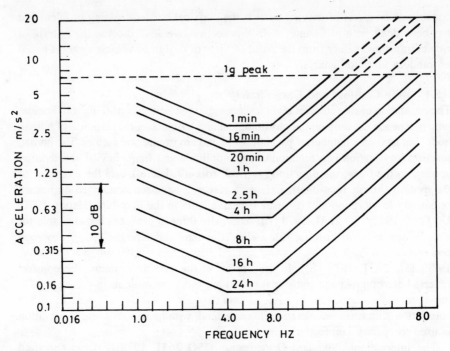

Fig. 6.20. Vertical acceleration limits as a function of exposure time and frequency: "fatigue—decreased proficiency boundary".

To obtain "exposure limits" multiply acceleration values by 2 (6 dB higher); "reduced comfort boundary" divide acceleration values by 3.15 (10 dB lower).

For broad-band excitation frequency axis represents centre frequency of third octave band.

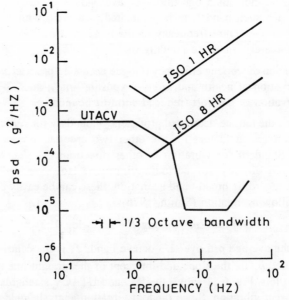

Fig. 6.21. ISO standard and UTACV specification in terms of psd for vertical direction (Smith, 1976).

ISO also provides a simple criteria based on weighted r.m.s. acceleration. In this criteria, the seat acceleration is passed through a filter before r.m.s. is computed. The following filter, which extends the range of the ISO filter from (1–80 Hz) to (0.1 to 80 Hz), has been proposed (Dahlberg, 1978):

$$
\begin{aligned}
f(\omega) &= 1.78\,\omega^{-0.69}, & 0.2\pi &< \omega < 2\pi \\
&= \sqrt{\omega/8\pi}, & 2\pi &< \omega < 8\pi \\
&= 1, & 8\pi &< \omega < 16\pi \\
&= 16\pi/\omega, & 16\pi &< \omega\,\omega < 160\pi
\end{aligned}
\tag{6.129}
$$

A plot of $f(\omega)$ is shown in Fig. 6.22.

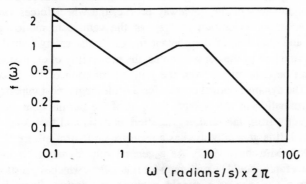

Fig. 6.22. Weight function against frequency.

Let $Q(t)$ denote the seat acceleration. Then $Q(t)$ can be related to the response through (6.83). The psd of $Q(t)$ is given by (6.84), and the weighted r.m.s. is given by

$$
\sigma_{Qw} = \left[\int_0^\infty |f(\omega)|^2\, G_{QQ}(\omega)\, d\omega \right]^{1/2}
\tag{6.130}
$$

where $G_{QQ}(\omega)$ is one-sided psd of $Q(t)$. ISO provides that for determining ride comfort limits, σ_{Qw} should be compared with the r.m.s. values in the frequency range 4–8 Hz for vertical, and 1–2 Hz for vertical and lateral directions, respectively.

The filtering of seat acceleration can also be carried out in the time-domain by a second-order shape filter (Schiehlen, 1986). Let

$$
Q_w(t) = \bar{P}^T \bar{g}(t)
\tag{6.131}
$$

and

$$
\dot{\bar{g}}(t) + \mathbf{J}\bar{g} = \bar{R}Q(t)
\tag{6.132}
$$

where $\bar{g}(t)$ is a 2×1 state vector, $\mathbf{J}$ is a 2×2 matrix, and $\bar{P}$ and $\bar{R}$ are 2×1 vectors. Schiehlen (1986) has suggested following values to match the ISO filter:

$$
\mathbf{J} = \begin{bmatrix} 0 & 13 \\ 1200 \ 1/s^2 & 500 \ 1/s \end{bmatrix};\ \bar{R} = \begin{Bmatrix} 0 \\ 1 \end{Bmatrix};\ \bar{P} = \begin{bmatrix} 500 \ 1/s \\ 50 \ 1/s \end{bmatrix}
\tag{6.133}
$$

In this approach, the state Eq. (6.132) can be adjoined to the state Eq. (6.119) of the vehicle and the weighted r.m.s. of the seat acceleration can be obtained directly (Schiehlen, 1986).

The cargo safety criteria are specified in terms of r.m.s. acceleration, or through limits between acceleration level, and frequencies (Brust, 1961; and Bekker, 1969). The significant difference in the cargo and human response to vibration is the lower sensitivity of cargo to low frequency and its higher sensitivity to impact and high frequency (Bekker, 1969).

6.5.2 Ride Safety

Ride safety of a moving vehicle depends on the lateral, longitudinal and normal forces at the wheel-guideway contact during acceleration, braking and steering. The longitudinal and lateral forces depend primarily on the wheel-guideway friction and the normal load. A fluctuation of the normal load due to guideway induced vibration, reduces the normal load and, therefore, the longitudinal and lateral forces needed for ride safety.

Due to randomness of the guideway profile, the wheel forces vary randomly with time. A random vibration analysis of the vehicle in motion provides the distribution and statistics of the contact forces. If the safety boundaries can be defined as a function of the contact forces, the probability of excursion outside the boundary can be determined from the joint distribution of the forces and their derivatives. The dynamic normal contact force is the single most important parameter for the determination of ride safety. The psd of the normal contact force can be obtained directly from the random vibration analysis and the r.m.s. value of the load estimated. It is generally used as a measure of control of directional stability (Bender, 1967; Mitschke, 1962). As a general rule, dynamic component of the normal contact force should be minimum for ride safety, which suggests minimization of its dynamic variance as a simple criterion for design. Some investigations dealing with safety of road vehicles include: (Mitschke, 1962, Quinn and Hilderbrand, 1973; Sattaripour, 1977; Dokanish and Elmadany, 1977).

In railway vehicles, derailment due to flange climbing is a major concern in ride safety. This occurs when a laterally loaded wheel flange rolls forward while being pressed against the rail. Based on the creep theory due to Kalker (1979), Gilchrist and Brickle (1967) have reexamined the criteria for derailment. The ratio of lateral to vertical force at which derailment may be incipient can be calculated using the creep data due to Kalker. The joint distribution of the lateral and vertical force ratios can be obtained from the random vibration analysis and the probability of derailment estimated based on this ratio.

6.5.3 Optimum Design

Vehicles are a complex dynamic system and are often subject to conflicting design requirements. However, for idealized models, the vehicle design problem, including optimization, can be formulated in the framework discussed in Chapter 4. The operational requirements of ride comfort and safety, and the design requirements for fatigue and strength can be incorporated as constraints in the statement of the optimum design problem. Dahlberg (1977; 1978; 1979; and 1980) has treated the optimum design of simple vehicle models using several criteria and approximations. He has shown that formal optimization leads to significant improvement in vehicle performance. In Chapter 4 we have discussed the design of a sdf system model to determine the optimum values of spring stiffness and damping constant. In Example 6.5, we discuss the optimum design of the damping elements of a five degree of freedom car model. The methods of random vibration analysis, discussed in this chapter, provide the necessary analytical tools for reliability based vehicle design.

Example 6.5:   Optimum design of a passenger vehicle for ride comfort and ride safety (Dahlberg, 1979, and 1980)

Ride Comfort

With a trend towards increase in the operating speed of passenger vehicles, the ride comfort has become an important criterion for design. We consider a five degree-of-freedom planar linear model of a passenger vehicle shown in Fig. 6.23.

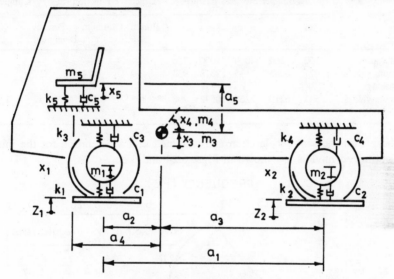

Fig. 6.23.   A vehicle model.

The seat-passenger system is modelled as sdf system with displacement X_5. The values of the vehicle parameters are shown in Table 6.8.

Table 6.8   System parameter values for vehicle in Fig. 6.22

Index	1	2	3	4	5
Generalized masses m_i (kg or kgm^2)	50	50	1000	2000	80
Spring stiffnesses k_i (KN/m)	180	180	50	50	20
Distances a_i (m)	2.5	1.3	1.2	0.3	—

The five undamped natural frequencies of the model are: $\omega_1 = 7.78$, $\omega_2 = 8.38$, $\omega_3 = 16.70$, $\omega_4 = 67.95$ and $\omega_5 = 67.99$ rad/s. The vehicle is assumed to travel with constant velocity, $V = 20$ m/s. The equations of motion of the vehicle are given by (6.77). The seat acceleration $Q(t) = X_5(t)$. The psd of seat acceleration is given by (6.84). The psd of the road profile is given by (6.2) with $S(k_0) = 120.10^{-6}$ m^3/cycle. Equation (6.24) represents the psd of the road induced excitation due to two wheels separated by a distance $l = a_1$.

The optimization problem is formulated to minimize the following two objective functions:

Case 1: Standard deviation of seat acceleration σ_Q.
Case 2: Standard deviation of weighted seat acceleration σ_{Qw}, (6.130).

Both represent measures of ride comfort (dis). The design vector $\bar{d}$ consists of five damping coefficient parameters, that is

$$\bar{d} = [c_1 c_2 c_3 c_4 c_5]^T$$

The optimization is carried out using the unconstrained minimization technique (SUMT). The initial and optimized values of the damping stiffness are shown in Table 6.9. The initial and the optimum values of the objective functions and percentage improvements are also indicated in the table.

Table 6.9.   Initial and optimum values of design variables

	Damper Stiffness (Ns/m)					Objective Function (m·s²)			Computer Time
	c_1	c_2	c_3	c_4	c_5	Initial	Optimum	Percent Decreased	
Initial values	360	360	4000	4000	750	—	—	—	—
Optimum values: (Case 1)	19627	6245	2135	13318	934	1.179	0.74	37	1
Optimum values: (Case 2)	5547	6219	2265	10572	798	0.795	0.502	37	1.75

The psd of the seat acceleration and weighted seat acceleration for the initial and optimized models are shown in Figs. 6.24 and 6.25.

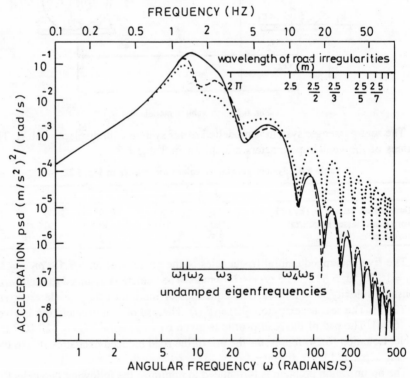

Fig. 6.24.   Seat-passenger acceleration psd: Case 1(—) initial system; (...) optimized system (Dahlberg, 1980).

The psds have three initial peaks at first three frequencies of the vehicle. Subsequent local maxima for $\omega > 100$ correspond to integral fractions of the wheel base $a_1 = 2.5$ m. The computer time for Case 2 is 1.75 times more than the time for Case 1. Dahlberg (1980) has also carried out the optimization for two design

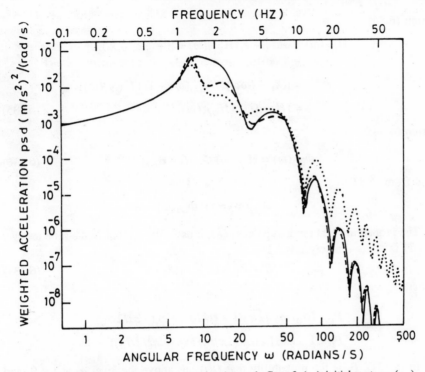

Fig. 6.25. Weighted seat-passenger acceleration psd: Case 2 (—) initial system; (....) optimized system (Dahlberg, 1980).

variables (c_3 and c_4) only. The optimum values of the objective functions for Cases 1 and 2 are 0.86 m/s^2 and 0.573 m/s^2, respectively. As expected, these values are higher than the optimum values for optimization based on five variables, but the computer time required for two variables is nearly 1/8 and 1/9.5 of the time required for five variables. Optimization was also carried out to minimize the expected maximum value of seat acceleration over exposure times 60 sec and 600 sec. Results show that optimum design is not very sensitive to the choice of objective functions and exposure time.

Ride Safety

We now consider the optimum design of the above vehicle (Fig. 6.23) for ride safety. As discussed, the ride safety of a vehicle is strongly dependent on the fluctuations in the component of the normal contact force at the wheel-guideway interface. Let $Q(t)$ represent the normal contact force at the front wheel of the vehicle. Then,

$$Q(t) = k_1(X_1 - Z_1) + C_1(\dot{X}_1 - \dot{Z}_1)$$

$$= [C_1 \mid k_1]\begin{Bmatrix} \dot{X}_1 \\ X_1 \end{Bmatrix} + [-C_1 \mid -k_1]\begin{Bmatrix} \dot{Z}_1 \\ Z_1 \end{Bmatrix} \qquad (6.134)$$

For harmonic excitation,

$$X_1 = H_{X_1 Z_1}(\omega) Z_1 + H_{X_1 Z_2}(\omega) Z_2$$

From (6.85),

$$H_{Q\bar{Z}}(\omega) = [i\omega c_1 + k_1] H_{X_1\bar{Z}}(\omega) + [(-i\omega c_1 - k_1) \ 0]$$
$$\quad\quad {\scriptstyle(1\times 2)} \quad\quad\quad {\scriptstyle(1\times 1)} \quad\quad\quad {\scriptstyle(1\times 2)} \quad\quad\quad\quad {\scriptstyle(1\times 2)}$$

$$= [(k_1 + i\omega c_1)(H_{X_1Z_1}(\omega) - 1) \mid H_{X_1Z_2}(\omega)]$$

$$= [H_{QZ_1}(\omega) \mid H_{QZ_2}(\omega)] \tag{6.135}$$

From (6.84),

$$\Phi_{QQ}(\omega) = H_{Q\bar{Z}}(\omega) \, \Phi_{\bar{Z}\bar{Z}}(\omega) \, H_{Q\bar{Z}}^{*}(\omega)^{\mathrm{T}} \tag{6.136}$$

and from (6.87),

$$\Phi_{\dot{Q}\dot{Q}}(\omega) = \omega^2 \, \Phi_{QQ}(\omega)$$

The probability that random process $Q(t)$ crosses the level q_0 in the time interval $[0, T]$ is approximately given by,

$$P = P_1 + (1 - P_1)P_2 \tag{6.137}$$

where

$$P_1 = (\sigma_Q/(q_0\sqrt{2\pi}))(1 - \sigma_Q^2/q_0^2) \exp(-q_0^2/2\sigma_Q^2)$$

$$P_2 = 1 - \exp[-\sigma_{\dot{Q}}/(2\pi\sigma_Q) \exp(-q_0^2/2\sigma_Q^2)]$$

In (6.137), P_1 is the probability that $Q(t)$ lies above the level q_0 at $t = 0$; and P_2 is the probability that $Q(t)$ crosses the level q_0 at least once in the time period $[0, T]$. σ_Q and $\sigma_{\dot{Q}}$ are obtained from the psd of $Q(t)$ and $\dot{Q}(t)$.

We now formulate the optimum design problem from the consideration of road safety. Let the design vector

$$\bar{d} = [c_3 c_4]^{\mathrm{T}}$$

The objective function is the probability P that $Q(t)$ crosses the level $q_0 = 5670 \ N$, in the time period, $T = 600$ sec. 5670 N is the static contact force, and, therefore, the objective function gives the probability of loosing contact in the time period T. We choose $\bar{d}$ to minimize P.

The optimization is carried out for the data given in Table 6.8, with $c_1 = c_2 = 360$ Ns/m, $c_5 = 750$ ns/m; $q_0 = 5670 \ N$, and $T = 600$ s, to yield (Dahlberg, 1979)

$$P_{opt} = .0269 \text{ and } c_3 = 3193 \text{ Ns/m}, c_4 = 1757 \text{ Ns/m}$$

The example illustrates the application of random vibration theory for the optimum design of vehicles.

APPENDIX 6A: **Aircraft Landing Gear**

The aircraft landing gear has a complex and strongly nonlinear dynamic behaviour. It supports the aircraft during the ground operation phase which consists of the sequence: taxi-takeoff-landing-taxi runs. Of these, the landing run, which includes landing impact followed by decelerating run, constitutes the most severe loading condition for the landing gear. The taxi and take-off phases contribute to the fatigue and other operational considerations in the landing gear design. Most modern aircraft have landing gears with oleo-pneumatic shock absorber. We shall discuss the dynamic behaviour of such landing gears, and the procedure for constructing an 'equivalent' linear model for analytical studies of aircraft response to ground induced excitation (Conway, 1958; Mcbreaty, 1948; and Milwitzky and Cook, 1953).

An oleo-pneumatic shock struct consists of two sealed telescopic tubes enclosing air and a liquid. The liquid is separated by an orifice plate with a metering pin of variable cross section which changes the effective area of the orifice during a stroke. An idealized model of the oleo-pneumatic landing gear is shown in Fig. 6A.1.

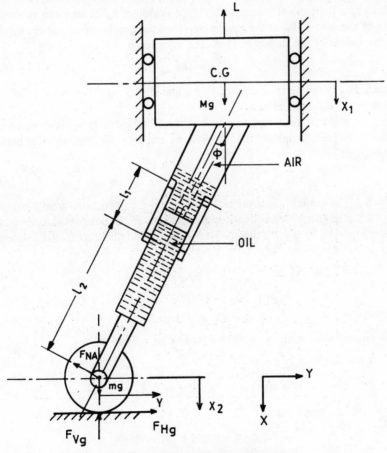

Fig. 6A.1. An idealized aircraft model.

The model is based on the following assumptions:

(i) Wing-fuselage system is treated as a rigid mass. The sprung mass M in the model represents a part of the total mass supported by the landing gear.

(ii) The effective mass of the moving parts of the shock strut, and the wheel-axle system is lumped at the wheel axle as unsprung mass m.

(iii) Shock strut is assumed to be rigid in bending.

(iv) Only vertical vibrations of the aircraft are considered.

(v) Lift and drag forces remain constant during the impact.

(vi) The tyre behaves as a nonlinear spring. The shock strut force F_s is the sum of the pneumatic spring force F_a, due to compression of air, the friction force F_f between the inner and the outer tubes, and the hydraulic force F_h due to orifice type nonlinear damping.

(vii) Spin-up drag force F_{Hg} acts at the contact point till the wheel starts rotating without skidding. The time at which the wheel stops skidding is estimated approximately from the impulse-momentum relation:

$$\int_0^{t_s} F_{Hg}\, dt = \frac{I_W V_H}{r_d^2} \tag{6A.1}$$

in which t_s is the time at which wheel stops skidding, I_W is the polar moment of inertia of the wheel, V_H is the touchdown horizontal velocity of the aircraft; and r_d is the radius of the deflected tyre. The drag force

$$F_{Hg} = \mu\, F_{Vg} \tag{6A.2}$$

in which F_{Vg} is the vertical contact force, and μ $(0.5 - 0.7)$ is the coefficient of friction for skidding condition.

(viii) The shock strut remains locked until sufficient force is developed to overcome the preload due to inflation pressure and the static friction force between the tube, that is

$$F_s = p_{ao} A_a + F_f \tag{6A.3}$$

in which p_{ao} is the initial air pressure; and A_a is the pneumatic area. During the time $(t \le t_r)$, the shock strut remains locked and the model behaves as a sdf system.

From Fig. 6A.1, it is clear that the strut stroke s, and the horizontal displacement of the axle Y are given by

$$s = \frac{X_1 - X_2}{\cos \phi} \tag{6A.4a}$$

$$Y = (X_1 - X_2) \tan \phi \tag{6A.4b}$$

The equations of motion of the model can be written as

SHOCK STRUT LOCKED: $(F_S < p_{ao} + F_f; \; t < t_r)$

$$(M + m)\ddot{X}_1 + F_{Vg} + (M + m)g(K_L - 1) = 0 \tag{6A.5}$$

with the initial conditions

$$X_1(0) = 0 \text{ and } \dot{X}_1(0) = V_v$$

in which V_v is the vertical velocity at touch down; and K_L is the lift factor (Lift force $L = K_L (M + m)g$).

SHOCK STRUT UNLOCKED: $(F_S \ge p_{ao} + F_f; \; t \ge t_r)$

$$M\ddot{X}_1 + F_s \cos \phi + F_{N_a} \sin \phi = Mg (1 - K_L) \tag{6A.6a}$$

$$M\ddot{X}_2 + F_s \cos\phi + F_{N_a} \sin\phi = F_{v_g} = mg \qquad (6A.6b)$$

where

$$F_s = F_a + F_h + F_f \qquad (6A.7)$$

$$= p_{ao} A_a \left(\frac{v_0}{v_0 - A_a^s}\right)\gamma + \frac{\dot{s}}{|\dot{s}|} \frac{\rho A_h^3}{2C_d^2(A_0 - A_p)\dot{s}^2}$$

$$+ \frac{\dot{s}}{|\dot{s}|} |F_{N_a}| \left[(\mu_1 + \mu_2)\frac{l_2 - s}{l_1 + s} + \mu_2\right] \qquad (6A.8)$$

$$F_V = k_T X_2^n \qquad (6A.9)$$

$$F_{N_a} = F_{v_g} \sin\phi - F_{Hg} \cos\phi + m\ddot{X}_1 \sin\phi - mg \sin\phi \qquad (6A.10)$$

in which γ is the polytropic constant for the air (~ 1.1); v_0 is the initial air volume; A_h is the hydraulic area; ρ is the mass density of the hydraulic fluid; A_n is the net effective area of the orifice; C_d is the discharge coefficient of the orifice; μ_1 and μ_2 are coefficients of friction at the upper and lower bearings, respectively; and k_T and n (~ 1.4) are the tyre constants. The spin-up drag force F_{Hg} becomes zero for $t \geq t_s$ (6A.1). Initial conditions for the integration of (6A.6), $X(t_r)$ and $\dot{X}(t_r)$ are obtained by integration of (6A.5).

Figure 6A.2 shows the variation of strut forces as a function of the strut stroke due to landing impact. The strut forces were computed by numerical integration of the equations of motion (6A.5) and (6A.6), using fourth order Runge-Kutta method (Goyal, 1973). The data used in the computations is given in Table 6A.1. The tyre was assumed to be linear. The strongly nonlinear behaviour of the shock strut is evident from the Fig. 6A.2. It is seen that the hydraulic force due to orifice

Table 6A.1   Values of various parameters of a oleo-pneumatic landing gear

S. No.	Items	Values
1.	Weight of uppermass, W_1	4641.0 lbs
2.	Weight of unsprung mass, W_2	119.0 lbs
3.	Polar moment of inertia of wheel, I_w	1373 lbs. in^3.
4.	Rake angle	$15°$
5.	Distance between the bearings, l_1	10.3 in.
6.	Distance between the wheel axle and lower bearing, l_2	25.5 in.
7.	Wheel size	$19'' \times 6.25'' \times 9''$
8.	Pneumatic area, A_a	11.05 in^2.
9.	Hydraulic area in compression and recoil, A_h	10.2 in^2.
10.	Orifice area in compression and recoil, A_h	0.292 in^2.
11.	Metering pin area, A_p	0
12.	Mass density of hydraulic fluid	0.03125 lbs/in^3.
13.	Initial air pressure, p_{ao}	214.7 lbs/in^2.
14.	Initial air volume, v_0	149.0 in^3.
15.	Horizontal velocity at touch down, V_H	288 ft/sec
16.	Orifice discharge coefficient, c_d	0.65
17.	Polytropic exponent	1.18
18.	Tyre stiffness (linear), k_T	4700 lbs/in.
19.	Impact velocity, V_V	120 in./sec
20.	Coefficient of ground friction	0.6

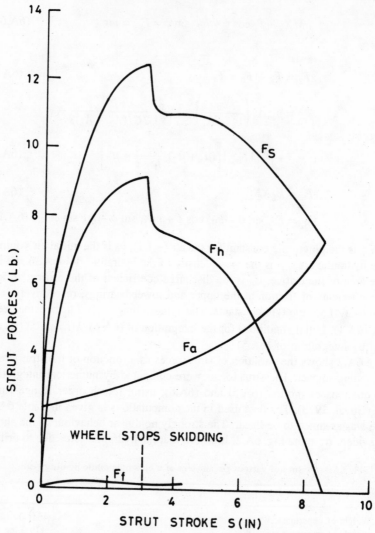

Fig. 6A.2. Load deflection curves of shock strut (Goyal, 1973).

forms the major component of the shock strut force. Towards the end of the stroke, pneumatic force becomes larger than the hydraulic force due to drop in strut velocity. The contribution of friction forces is small and remains nearly constant during the stroke. Parametric study (Goyal, 1973) showed that the response is sensitive to spin up drag force, the orifice discharge coefficient, and is not significantly affected by polytropic exponent and tyre stiffness.

6A.1 Linearized Landing Gear Model

We have seen that the dynamic behaviour of landing gear with oleo-pneumatic shock strut is strongly nonlinear during landing impact. The nonlinearities in the stiffness arise due to compression of air in the shock-strut and due to tyre deflection. The energy dissipated by the flow through the orifice and the Coulomb friction contribute to nonlinearities in the damping behaviour.

Figure 6A.3 shows a simple linearized model of the aircraft landing gear. The

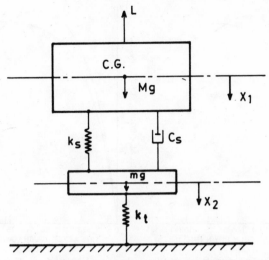

Fig. 6A.3. A linearized aircraft model.

equivalent stiffness of the pneumatic spring k_{se} and the tyre spring are obtained equating the energy stored during the maximum strut stroke and tyre displacement, respectively:

$$k_{se} = \frac{2}{s_m^2} \int_0^{s_m} F_a \, ds = \frac{2 p_{a0} A_a}{s_m^2} \int_0^{s_m} \left(\frac{v_0}{v_0 - A_{as}} \right)^{\gamma} ds \qquad (6A.11)$$

$$k_{Te} = \frac{2 k_T}{n+1} (X_m)^{n-1} \qquad (6A.12)$$

in which s_m is the maximum strut stroke and X_m is the maximum tyre displacement.

The equivalent linear damping coefficient C_{se} is determined by equating the energy dissipated during one cycle of period τ, and is given by

$$C_{se} = \frac{\displaystyle\int_0^{\tau} F_h(\dot{s}) \, \dot{s} \, dt + \int_0^{\tau} F_f(\dot{s}) \, \dot{s} \, dt}{\displaystyle\int_0^{\tau} \dot{s}^2 \, dt} \qquad (6A.13)$$

in which F_h and F_f are given by (6A.8). The integrals in (6A.13) are computed by numerical integration in conjunction with the integration of equations of motion.

Figures 6A.4 and 6A.5 show a comparison of various response parameters for nonlinear and linear models of landing gear during landing impact. The agreement is close except near the peak values, where linearized model predicts smaller values. Since the landing impact constitutes the most severe loading condition for the landing gear, it is concluded that the linearized model is adequate for analytical study of taxi, take-off and landing runs, including the landing impact.

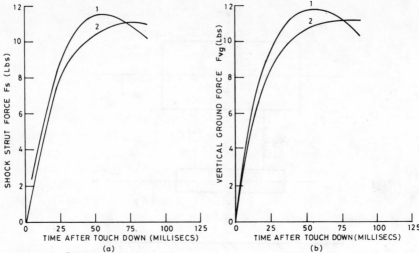

Fig. 6A.4. Time history of (a) shock strut force; (b) vertical ground force. 1—nonlinear model; 2—linear model (Goyal, 1973).

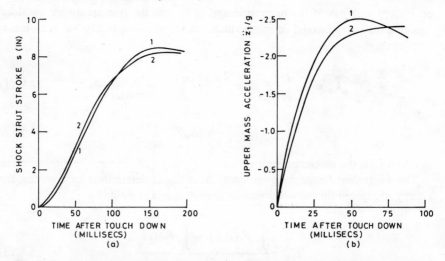

Fig. 6A.5. Time history of (a) shock strut stroke; (b) upper mass acceleration. 1—nonlinear model; 2—linear model (Goyal, 1973).

APPENDIX 6B: Active Vehicle Suspension Systems

The suspension systems in vehicles are designed:

(i) to support the vehicle body, and to isolate the vehicle system (including passengers and freight) from external disturbances, mainly induced by guideway unevenness; and

(ii) to maintain a firm contact between the guideway and the wheels, which is critical for the safety, good handling quality and the vehicle stability.

These requirements impose conflicting demands on a suspension system. In a conventional suspension consisting of a combination of passive elements, such as coil, or leaf springs; shock absorbers; and torsion bars, the designer is confronted with the problem of choosing soft spring action for good body isolation, against stiffer suspensions for a better control of both, the body and the wheel, motions. The designer is also faced with a conflict in choosing the suspension damping rates. If a high damping is provided to reduce the vehicle motion at resonance, body isolation deteriorates at high frequencies and vice versa. For better ride comfort and maneuverability, the requirement of well separated natural frequencies in bounce, pitch, and roll modes also imposes conflicting demands on suspension system design.

In recent years, considerable interest has been generated in the use of active vehicle suspensions, which can overcome some of the limitations of the passive suspension systems. These suspensions consist of: actuators (hydraulic, pneumatic, electromechanic); measuring and sensing devices (accelerometers, force transducers, potentiometers); a feedback controller to provide control commands to the actuators, and an external power source. Figure 6B.1 shows the schematic of a vehicle with active control suspension system, sensors, feedback controller and actuators. Active suspensions offer the potential of being adaptable to the quality of the guideway surface, vehicle speed, and different safety and comfort requirements.

The active suspension systems are superior in performance to passive suspensions, but their physical realization and implementation is generally complex and expensive, needing sophisticated electronically operated actuators, sensors and controllers. They involve high installation and maintenance costs and may cause reduced robustness and reliability.

The complexities involved in fully active suspension systems have led to the development of semi-active suspensions, which represent a hybrid between fully active suspensions and passive suspensions. These suspensions employ essentially passive devices, whose parameters can be varied discretely, or continuously, with ease of adaptation. Examples of semi-active suspension systems are variable spring and damper rate systems, skyhook damper devices where the suspension forces are proportional to the absolute velocity of the sprung mass, and on-off damper devices. The significant developments in the application of semi-active suspensions can be found in the works of Crosby and Karnopp (1973) Margolis et al, (1975), Kransnicki (1979, 1980), Margolis (1982a, 1982b), Horton and Crolla (1986). It has been shown that, in some cases, the performance of semi-active suspensions can be made to approach the performance of fully-active systems by a judicious choice of control strategy.

The state of the art reviews by Hedrick and Wormley (1975), Hedrick (1981), Goodall and Körtum (1983), and Sheep and Crolla (1987) cover important developments in active suspension design, including theoretical formulation and analysis, and practical aspects of hardware realization and implementation.

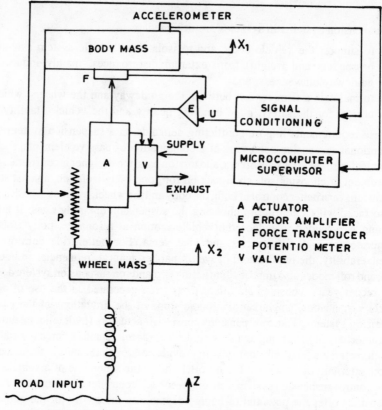

Fig. 6B.1. Schematic of an active suspension system for a road vehicle.

We present a brief review of the state-of-the-art in vehicles with active suspensions. Our discussions will be confined mainly to the control of road vehicle response to random road roughness. We shall discuss theoretical development of control strategies in the framework of stochastic optimal control theory for idealized vehicle models.

The key element, in the realization of an effective active suspension is the variation of a suitable control law relating the measured variables to the control forces. For simple systems, such as a sdf vehicle model it may be easy to derive the control law. For complex mdf vehicle models, a matrix formulation is expedient for computer implementation of the algorithms. For linear systems, a control strategy known as Linear Quadratic Regulator (LQR) theory is extensively used. It requires that: (i) input is white noise; (ii) performance index is quadratic; and (iii) all state variables are measurable (Kwakernak and Sivan, 1972).

In the active suspension problems, the performance index is typically a weighted sum of mean square values of the output variables, such as, body acceleration, wheel to body displacement, dynamic tyre load and control force expenditure. Since the road surface displacement approximates integrated white noise over the wavenumber range 0.01 to 10 cycles/m, it is necessary to augment the suspension system model with an integrator, causing the road elevation to become one of the system states. The requirement of full state information then implies that a body mounted road height sensor is needed. Determination of the optimal control law depends on a suitable state variable transformation and solution of the nonlinear matrix Riccati equation. Such a scheme has been adopted by Thompson (1976) for an automobile active suspension system represented by 2-dof quarter car model. Later, Thompson (1985) modified his formulation, by eliminating the need for a road height sensor through a suitable choice of the weighting constants in the performance index.

Alternatively, the road input can be treated as the output of a first order linear filter to white noise. Augmenting the suspension model with a filter rather than an integrator, the need for the state variable transformation can be avoided, and optimal control law includes a term dependent on the absolute road elevation, whose magnitude depends on the road roughness, or vehicle speed. Hac (1985) has used such a scheme for the active suspension design of a 2-dof quarter car model.

State variable estimation techniques have been used in conjunction with stochastic optimal control theory in active suspension design by Hac (1985) for road vehicles; by Yoshimura, Anantanarayana and Deepak (1986) for rail vehicles; and by Metz and Maddock (1986) for racing cars. Hady and Crolla (1989) developed optimal control laws for a 7-dof full vehicle model and compared the performance of competing systems, including limited state feedback. Yue, Butsen and Hedrick (1989) evaluated alternative control laws for a 2-dof vehicle model, and their results show that the linear suspension deflection retains most of the good features of full-scale feedback design.

Hac (1986) has applied the stochastic optimal control theory to a plane two ended vehicle model, including the effects of bending vibrations of the vehicle body. In this case, the rear wheel input is a time delayed version of the front wheel input. Hac neglected the correlation between the two inputs and, thus, avoided certain problems in the optimization technique. Similar work is reported by Yoshimura and Sugimoto (1990), in which an active suspension design concept is implemented in the framework of stochastic optimal control theory for a 4-dof vehicle model travelling on a flexible beam with an irregular surface.

The time delay between front and rear wheel inputs has been represented by a Pade's approximation in a full vehicle model by Frühauf, Kasper and Lückel (1985). Approximate expressions for coherence between left and right tracks are assumed, together with frequency weighting of body acceleration data in forming a discomfort parameter and applying frequency response limitations to the actuators. Full state feedback optimal control laws are derived with a Kalman filter, notionally estimating states not directly measurable. Further measurement of the road profile ahead of the front wheels is included and optimal systems are determined recognizing additional time delays between profile measurements and vehicle inputs. Louam, Wilson and Sharp (1988) also considered the time delay between front and rear wheel inputs and used a discrete time approach to treat the problem of optimal control of a half car model. The Linear Quadratic Regulator theory was applied with preview control.

The preview control is a control scheme, in which the road input is sensed before it reaches the actuator, to guide the vehicle over a rough road (Bender, 1968). The concept of track preview has been further investigated by Tomizuka (1976) and Thompson, Davis and Pearce (1980).

Elmadany (1988) used stochastic optimal control and estimation theories in the design of an active suspension system for a cab ride in a tractor-semi-trailer vehicle model. Active suspension design taking into account human tolerance criteria to vibration has been subject of investigation by Lee and Sulman (1989).

Almost all the works reviewed so far are concerned with the control of stationary response of the vehicles traversing rough roads with constant velocity. In many situations, nonstationary vehicle response arises due to:

(i) variable velocity traverse over rough roads with spatial homogeneity;

(ii) sudden encounter of a rough terrain by a vehicle moving with a constant velocity; and

(iii) vehicle traverse over nonhomogeneous terrain with constant or variable velocity.

Recently, Narayanan and Raju (1990), Raju and Narayanan (1990, 1991) have extended the concepts of active suspensions to control the nonstationary response using stochastic optimal control theory with full and limited-state feed back.

The suspension system is optimized with respect to the ride comfort, the road holding and the control force expenditure. The road profile is modelled as a spatially homogeneous random process, being the output of a second order linear spatial shaping filter to white noise. Such a model takes into account the effect of the rolling contact of the tyre with the road, (Harrison and Hammond, 1983). The second order differential equation with constant coefficients, representing the filtered profile in the space domain, is transformed into an equivalent second order differential equation with variable coefficients in the time domain. The dynamics of the vehicle is considered in combination with the equation of the road profile and expressed in state space form. The optimal values of the active suspension forces are calculated using the LQR theory to minimize the performance index, which is a weighted sum of mean square acceleration of the sprung mass, suspension working space, and control force expenditure. The corresponding nonstationary response of the vehicle is also obtained by solving numerically a matrix differential equation for the zero-lag covariance, with appropriate initial conditions. Use of estimation and optimal control theories in conjunction with Kalman-filter has been used in the case of limited state feedback. The nonstationary vehicle response, with active suspension is compared with that for a passive system, and the active system is found to give better performance in all cases.

In the sequel, we present the essentials of stochastic optimal control theory as applied to the vehicle suspension problem for a general time varying system. Two examples are given illustrating the efficacy of active suspension design in controlling the stationary and nonstationary vehicle responses.

Stochastic Optimal Control Equations

The dynamics of a linear time-varying system can be described by the matrix differential equation in state-space form

$$\dot{\bar{Y}}_1(t) = \mathbf{F}_y(t)\,\bar{Y}_1(t) + \mathbf{G}_y(t)\bar{U}(t) + \mathbf{D}_y(t)\,\bar{Z}(t), \quad \bar{Y}_1(t_0) \tag{6B.1}$$

where $\bar{Y}_1(t)$ is n-dimensional state vector, $\bar{U}(t)$ is m-dimensional control vector, $\bar{Z}(t)$ is p-dimensional disturbance vector, $\mathbf{F}_y(t)$, $\mathbf{G}_y(t)$, $\mathbf{D}_y(t)$ are respectively the system matrix; the control distribution matrix; and the excitation distribution matrix with appropriate dimensions.

In many situations, $\bar{Z}(t)$ may be considered as a vector formed by the components of independent white noise processes. For a large class of problems, $\bar{Z}(t)$ can be modelled as the output of a shaping filter to white noise excitation expressed by

$$\dot{\bar{Z}}(t) = \mathbf{F_d}(t)\,\bar{Z}(t) + \mathbf{D_d}(t)\,\bar{W}(t) \tag{6B.2}$$

where $\bar{W}(t)$ is a vector of white noises with covariance matrix

$$E[\bar{W}(t_2)\,\bar{W}^\mathrm{T}(t_1)] = \mathbf{Q}\,\delta\,(t_2 - t_1) \tag{6B.3}$$

Defining an augmented state vector $\bar{Y}^\mathrm{T}(t) = [\bar{Y}_1(t),\,\bar{Z}(t)]$, (6B.1) and (6B.2) can be combined to yield

$$\dot{\bar{Y}}(t) = \mathbf{F}(t)\,\bar{Y}(t) + \mathbf{G}(t)\,\bar{U}(t) + \mathbf{D}(t)\,\bar{W}(t) \tag{6B.4}$$

where

$$\mathbf{F}(t) = \begin{bmatrix} \mathbf{F_y}(t) & \mathbf{D_Y}(t) \\ 0 & \mathbf{F_d}(t) \end{bmatrix},\ \mathbf{G}(t) = \begin{bmatrix} \mathbf{G_y}(t) \\ 0 \end{bmatrix},\ \mathbf{D}(t) = \begin{bmatrix} 0 \\ \mathbf{D_d}(t) \end{bmatrix} \tag{6B.5}$$

If the white noise is considered as a function of space coordinate s, which, in turn is a function of time t, with $\dot{s}(t) \geq 0$, as in the vehicle problem, the basic dynamic equation can be written as

$$\dot{\bar{Y}}_1(t) = \mathbf{F}(t)\,\bar{Y}(t) + \mathbf{G}(t)\,\bar{U}(t) + \dot{s}(t)\,\mathbf{D}(t)\,\bar{W}(s(t)) \tag{6B.6}$$

with

$$E[\bar{W}\{s(t_2)\},\,\bar{W}^\mathrm{T}\{s(t_1)\}] = \mathbf{Q}\,\delta\{s(t_2) - s(t_1)\} \tag{6B.7}$$

In this form, the results of stochastic optimal control theory cannot be directly invoked as it requires the driving term to be a white noise, which is a function of t only. This problem can be overcome, as explained in Section 6.2 by considering a covariant equivalent form of $\bar{W}\{s(t)\}$, $\bar{W}_1(t)/\sqrt{\dot{s}(t)^{1/2}}$, where $\bar{W}_1(t)$ is a stationary white noise process.

Consider a quadratic integral performance criterion of the form

$$J = E\left[\bar{Y}(t_1)\,\mathbf{S}_1\,\bar{Y}^\mathrm{T}(t_1) + \int_{t_0}^{t_1} [\bar{Y}^\mathrm{T}(t)\ \bar{U}^\mathrm{T}(t)]\begin{bmatrix} \mathbf{A}(t) & \mathbf{N}(t) \\ \mathbf{N}^\mathrm{T}(t) & \mathbf{B}(t) \end{bmatrix}\begin{Bmatrix} \bar{Y}(t) \\ \bar{U}(t) \end{Bmatrix}dt\right] \tag{6B.8}$$

where $\mathbf{A}(t)$ is a symmetric positive semi-definite matrix and $\mathbf{B}(t)$ is a positive definite matrix, $\mathbf{S}_1 = \mathbf{S}(t_1)$ is symmetric final state weighting matrix, and $[t_0, t_1]$ is control period.

From stochastic optimal control theory, the control, which minimizes the performance index, is given by (Kwakernaak and Sivan 1972)

$$\bar{U}(t) = -\,\mathbf{C}(t)\,\bar{Y}(t) \tag{6B.9}$$

where
$$\mathbf{C}(t) = \mathbf{B}^{-1}\,(\mathbf{N}^\mathrm{T} + \mathbf{G}^\mathrm{T}\mathbf{S}) \tag{6B.10}$$

and $\mathbf{S}(t)$ is the solution of the matrix differential Riccati equation

$$-\,\dot{\mathbf{S}} = \mathbf{S}(t)\,\mathbf{R}(t) + \mathbf{R}^\mathrm{T}(t)\,\mathbf{S}(t) - \mathbf{SGB}^{-1}\mathbf{G}^\mathrm{T}\mathbf{S} + [\mathbf{A} - \mathbf{NB}^-\mathbf{N}^\mathrm{T}] \tag{6B.11}$$

where $\mathbf{R}(t) = \mathbf{F}(t) - \mathbf{G}(t)\mathbf{B}^{-1}(t)\,\mathbf{N}^\mathrm{T}$ and with the terminal conditions $\mathbf{S}(t_1) = \mathbf{S}_1$.

The response of the system can be described by the zero-lag covariance matrix $\mathbf{P}(t) = E[\bar{Y}(t)\,\bar{Y}^T(t)]$, which is the solution of the matrix differential Lyapunov equation (Narayanan and Raju 1990)

$$\dot{\mathbf{P}}(t) = [\mathbf{F}(t) - \mathbf{G}(t)\,\mathbf{C}(t)]\,\mathbf{P}(t) + \mathbf{P}(t)[\mathbf{F}(t) - \mathbf{G}(t)\,\mathbf{C}(t)]^T$$
$$+ \dot{s}(t)\,\mathbf{D}(t)\,\mathbf{Q}\mathbf{D}^T(t) \tag{6B.12}$$

with initial conditions $\mathbf{P}\,(t_0)$.

In the case of a time-invariant system, the matrices $\mathbf{F}$, $\mathbf{G}$ and $\mathbf{D}$ are constant, so also the control gain matrix $\mathbf{C}$ and the matrices $\mathbf{A}$, $\mathbf{N}$, and $\mathbf{B}$. Hence, the dynamics of the plant is given by

$$\dot{\bar{Y}}(t) = \mathbf{F}\bar{Y}(t) + \mathbf{G}\bar{U}(t) + \mathbf{D}\bar{W}(t) \tag{6B.13}$$

and the performance index is taken to be

$$J = \lim_{T \to \infty} \left[\frac{1}{T} \int_{t_0}^{t_1} [\bar{Y}^T(t)\,\bar{U}^T(t)] \begin{bmatrix} \mathbf{A} & \mathbf{N} \\ \mathbf{N}^T & \mathbf{B} \end{bmatrix} \begin{Bmatrix} \bar{Y}(t) \\ \bar{U}(t) \end{Bmatrix} dt \right] \tag{6B.14}$$

The optimal control strategy becomes

$$\bar{U}(t) = -\mathbf{C}\bar{Y}(t) \tag{6B.15}$$

with the control gain given by

$$\mathbf{C} = \mathbf{B}^{-1}\,(\mathbf{N}^T + \mathbf{G}^T\mathbf{S}) \tag{6B.16}$$

where $\mathbf{S}$ is a solution of the algebraic matrix Riccati equation

$$\mathbf{S}(\mathbf{F} - \mathbf{G}\mathbf{B}^{-1}\mathbf{N}^T) + \mathbf{S}(\mathbf{F} - \mathbf{G}\mathbf{B}^{-1}\mathbf{N}^T)^T$$
$$+ \mathbf{S}\mathbf{G}\mathbf{B}^{-1}\mathbf{G}^T\mathbf{S} - (\mathbf{A} - \mathbf{N}\mathbf{B}^{-1}\mathbf{N}^T) = 0 \tag{6B.17}$$

and in the steady state, the response becomes stationary, whose zero-lag covariance is given by the matrix algebraic Lyapunov equation

$$[\mathbf{F} - \mathbf{G}\mathbf{C}]\mathbf{P}(t) + \mathbf{P}(t)[\mathbf{F} - \mathbf{G}\mathbf{C}]^T + \mathbf{D}\mathbf{Q}\mathbf{D}^T = 0 \tag{6B.18}$$

Example 6B.1   Optimal active suspension of a 2-dof vehicle model (Hac, 1985)
Consider a linear 2-dof vehicle model moving with constant velocity V on a rough road as shown in Fig. 6B.2. The primary suspension supporting the sprung mass (m_1) is a combination of passive elements (k_1, c_1) and active element $(u(t))$. The equations of motion of the system are

$$m_1\ddot{X}_1 + c_1(\dot{X}_1 - \dot{X}_2) + k_1(X_1 - X_2) - u(t) = 0$$

$$m_2\ddot{X}_2 + c_1(\dot{X}_2 - \dot{X}_1) + k_1(X_2 - X_1) + k_2 X_2 + u(t) = k_2 Z(t)$$

The road excitation $Z(t)$ is given by (6.26)

$$\dot{Z} + \alpha V Z = V W(t)$$

where $W(t)$ is white noise with intensity $2\sigma^2\,\alpha V$ and α is the road roughness parameter.

Defining $\bar{Y} = \{X_1, \dot{X}_1, X_2, \dot{X}_2, Z\}^T$, the dynamics of the vehicle including the

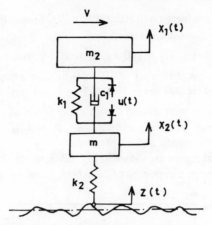

Fig. 6B.2. 2-dof vehicle model with active suspension.

roughness can be expressed as

$$\dot{\bar{Y}}(t) = \mathbf{F}\bar{Y}(t) + \mathbf{G}u(t) + \mathbf{D}W(t)$$

where

$$G = \{0,\ 1/M_1,\ 0,\ 1/M_2,\ 0\}^\mathrm{T}$$

$$D = \{0,\ 0,\ 0,\ k_2/M_2,\ 1\}^\mathrm{T}$$

$$\mathbf{F} = \begin{bmatrix} 0 & 1 & 0 & 0 & 0 \\ -\dfrac{k_1}{M_1} & -\dfrac{c_1}{M_1} & \dfrac{k_1}{M_1} & \dfrac{c_1}{M_1} & 0 \\ 0 & 1 & 0 & 1 & 0 \\ -\dfrac{k_1}{M_2} & \dfrac{c_1}{M_2} & \dfrac{-(k_1 + k_2)}{M_2} & -\dfrac{c_1}{M_2} & \dfrac{k_2}{M_2} \\ 0 & 1 & 0 & 0 & -\alpha V \end{bmatrix}$$

The suspension system is optimized with respect to ride comfort, road holding and working space of the suspension. The performance index to be minimized is assumed to be

$$J = \sum_{i=1}^{4} \rho_i J_i = E[\rho_1 \ddot{X}_1^2 + \rho_2 (X_1 - X_2)^2 + \rho_3 (X_2 - Z)^2 + \rho_4 U^2]$$

where J_i, $i = 1, 4$, are, respectively, the mean square values of acceleration of M_1, relative displacement between M_1 and M_2, relative displacement between M_2 and road surface and the control force, and ρ_i the corresponding weighting constants. The performance index can be expressed, in the standard form (6B.14) in terms of the state variables. The form of the matrices $\mathbf{A}$, $\mathbf{N}$ and $\mathbf{B}$ are given in (Hac 1985).

The results of the optimal control theory can now be directly applied with the control gain given by (6B.16) and (6B.17) and the covariance matrix of response by (6B.18).

The steps involved in solving the matrix Riccati equation, to determine the control gain matrix $\mathbf{C}$ and the matrix Lyapunov equation to determine the zero-lag covariance matrix are, enumerated in Hac (1985). We discuss the important results. Following parameter values are assumed.

$$M_1 = 500 \text{ kg}, \ M_2 = 100 \text{ kg}, \ k_2 = 200 \text{ kN/m}, \ \rho_1 = 1$$

(i) $\alpha = 0.15 \text{ m}^{-1}$, $\sigma^2 = 9 \text{ mm}^2$, $V = 10 - 50 \text{ m/s}$ for asphalt road
(ii) $\alpha = 0.45 \text{ m}^{-1}$, $\sigma^2 = 300 \text{ mm}^2$, $V = 5-30 \text{ m/s}$ for paved road

The values of k_1 and c_1 were changed with upper limits of 60 kN/m and 6 kNs/m respectively. The other weighting constants ρ_2, ρ_3 and ρ_4 were also varied and their effects on the performance index observed.

Results of the optimal design are shown in Fig. 6B.3, which shows the dependence of each performance index component (J_1, J_2, J_3, J_4), and the overall performance

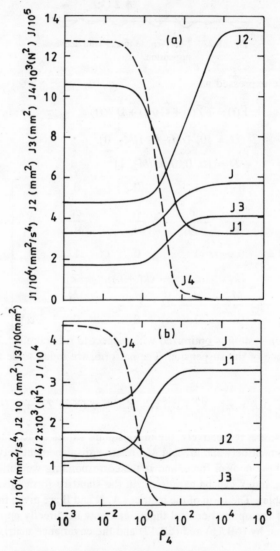

Fig. 6B.3. Variation of performance indices with weighting parameters (a) $\rho_2 = 10^4$, $\rho_3 = 10^5$); (b) $\rho_2 = 10^2$, $\rho_3 = 10^3$) (Hac, 1985).

index J upon the weighting constant (of suspension force) ρ_4 as it changes from 10^{-3} to 10^6. Taking $\rho_4 = 10^6$ gives a practically passive system (the control variable has such small values that it does not influence the dynamics of the system), while $\rho_4 = 10^{-3}$ implies arbitrarily large values of control force. The results are obtained for the vehicle model moving on an asphalt road with $V = 30$ m/s and $k_1 = 5$ kN/m and $c_1 = 1$ kNs/m, and for two combinations of: (a) $\rho_2 = 10^4$, $\rho_3 = 10^5$, and (b) $\rho_2 = 10^2$, $\rho_3 = 10^3$. The values of k_1 and c_1 are chosen on the basis of an optimization scheme of an entirely passive suspension.

From Fig. 6B.3(a), it can be seen that below a certain value of ρ_4 the variance of the control force, as well as other performance index components, become stationary with no further advantage derived by increase of control force. We also observe that use of active suspension can reduce the variance of the relative displacement (J_3) and the dynamic deflection of the suspension (J_2) considerably, but at the expense of increasing the measure of ride comfort (J_1). For a performance index with $\rho_2 = 10^2$ and $\rho_3 = 10^3$, which favours ride comfort over roadholding (Fig. 6B.3(b)), the index J_1 decreases with increase of control force as compared to the passive system, while J_2 and J_3 increase. However, in both the cases, the overall performance index reduces as compared to the passive system.

Figure 6B.4 shows the variation of J_1, J_2, J_3, J_4 and J with change in stiffness k_1 (Fig. 6B.4(a)) and the damping ratio c_1 (6B.4(b)) of the passive elements, for the vehicle moving on an asphalt road with $V = 30$ m/s $\rho_2 = 10^4$, $\rho_3 = 10^5$ and $\rho_4 = 10^{-1}$ with $c_1 = 1$ kNs/m for (a) and $k_1 = 5$ kN/m for (b). The values of J_1, J_2, J_3 and J remain almost constant with change in values of k_1 and c_1 whereas the variance of the control force J_4 changes considerably especially with change in c_1 reaching minimal values for $k_1 = 22$ kN/m (Fig. 6B.4(a)) and for $c_1 = 2.7$ kNs/m (Fig. 6B.4(b)). These values are close to optimal values obtained for a passive system under the same performance criterion with the same constant ρ_2, ρ_3 ($\rho_4 \to \infty$). Thus, it is possible to reduce the expenditure on control force by a proper choice of the passive elements of an active suspension.

Example 6B.2:   Optimal active suspension of a sdf vehicle model to control nonstationary response (Narayanan and Raju 1990)

Consider the sdf vehicle model moving on a rough road with variable velocity $\dot{s}(t)$ and with active and passive elements as shown in Fig. 6B.5, whose equation of motion is

$$\ddot{X}_1 + 2\zeta\omega_0\dot{X}_1 + \omega_0^2 X_1 = 2\zeta\omega_0\dot{Z}_1 + \omega_0^2 Z_1 + F/\sigma$$

where $X_1 = X/\sigma$, is the normalized displacement of the sprung mass, $Z_1 = Z(s)/\sigma$, is the normalized road elevation, F is the acceleration due to control force, σ^2 is the variance of the road profile.

The road elevation is considered as the ouput of second order spatial shaping filter to white noise which takes into account the rolling contact of the tyre with the road surface (Harrison and Hammond 1983), given by

$$Z''(s) + (\alpha + \beta)Z'(s) + \alpha\beta Z(s) = kW(s)$$

where α is the cut-off wave number of the road profile spectrum, β is the cut off wave number of the rolling contact filter, $k = \beta\sigma\sqrt{(2\alpha)}$ ensures that the variance

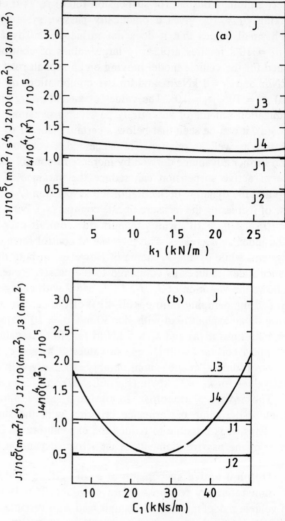

Fig. 6B.4 (a) Variation of performance indices with passive stiffness k_1, $C_1 = 1$ kNs/m; (b) Variation of performance indices with passive damping ratio C_1, $k_1 = 5$ kN/m ($V = 30$ m/s, $\rho_2 = 10^4$, $\rho_3 = 10^5$ and $\rho_4 = 10^{-1}$) (Hac, 1985).

of the true road profile is σ^2, so that the variance of $Z(s)$ is $\beta\sigma^2/(\alpha + \beta)$, $W(s)$ is spatial white noise, and the primes denote order of differentiation with recpect to s. Alternatively, in the time domain

$$\ddot{Z}_1(t) = - \dot{s}(t)(\alpha + \beta)\,\dot{Z}_1 - \dot{s}(t)\,\alpha\beta Z_1 + (k/\sigma)\,\dot{s}(t)\,W(s(t))$$

Defining the state variables $Y_1 = X_1$, $Y_2 = \dot{X}_1$, $Y_3 = Z_1$, $Y_4 = \dot{Z}_1$, and $F/\sigma = u$, an augmented state equation for the system and excitation can be expressed in the form (6B.4) where for the problem concerned

$$\mathbf{G}(t) = \{0, 1, 0, 0\}^{\mathrm{T}}, \ \bar{U}(t) = u(t), \ \mathbf{D}(t) = \{0, 0, 0, k/\sigma\}^{\mathrm{T}} \text{ and}$$

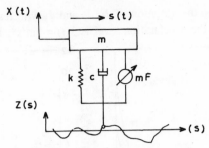

Fig. 6B.5.   sdf vehicle model with active suspension.

and

$$\mathbf{F}(t) = \begin{bmatrix} 0 & 1 & 0 & 0 \\ -\omega_0^2 & -2\zeta\omega_0 & \omega_0^2 & 2\zeta\omega_0^2\dot{s}(t) \\ 0 & 0 & 0 & \dot{s}(t) \\ 0 & 0 & -\alpha\beta\dot{s}(t) & -(\alpha+\beta)\dot{s}(t) \end{bmatrix}$$

The objective of the optimization of the suspension system is to improve vehicle response with respect to ride comfort and road holding with minimum expenditure of control energy. The overall performance index is taken to be

$$J = E\left[\bar{Y}_1(t_1) S_1 Y^T(t_1) \int_{t_0}^{t_1} (\rho_1 \ddot{X}_1^2 + \rho_2(X_1 - Z)^2 + \rho_3 u^2) \, dt \right]$$

which can be written in the standard form (6B.8) where the matrices $\mathbf{A}(t)$, $\mathbf{N}(t)$ and $\mathbf{B}(t)$ are given (Narayanan and Raju, 1990).

To solve the optimal control problem given by the matrix differential equation and Lyapunov Eqs. (6B.11) and (6B.12), fourth order Runge-Kutta numerical integration is adopted. To initiate the numerical integration scheme a set of initial conditions for the zero-lag covariance matrix coefficients $P_{ij}(0)$ are required.

The vehicle is assumed to start from rest and accelerate uniformly. The initial conditions are obtained by the following arguments. From the assumption of spatial homogeneity of the road profile the variance of $Z(s(t))$ is constant for all t, that is

$$E[Z_1^2] = \beta/(\alpha+\beta) = P_{33}(0) \text{ and } E[Z_1'^2] = \alpha\beta^2/(\alpha+\beta) = P_{44}(0)$$

At $t = 0$, no motion has taken place and so $X_1(0)$ must be equal to $Z_1(0)$. It, therefore, follows that

$$P_{11}(0) = P_{13}(0) = P_{33}(0)$$

Since no motion has taken place at $t = 0$ the vertical velocity of the mass is zero, that is, $\dot{Y}_1(0) = 0$, which leads to the following initial conditions

$$P_{12}(0) = P_{22}(0) = P_{23}(0) = P_{24}(0) = 0$$

Since $Z_1(s)$ is homogeneous, $E[Z_1(s) Z_1'(s)] = 0$. Therefore,

$$P_{14}(0) = P_{34}(0) = 0$$

The optimal control scheme is applied for the following parameter values: $\omega_0 = 10$ rad/sec, $\alpha = 0.2$ rad/m and $\beta = 2.0$ rad/m, $\rho_1 = 1$. The passive damping factor ζ and the weighting factors ρ_2 and ρ_3 are varied.

Figure 6B.6 shows the variation of J_1, J_2 and J_3 with respect to ρ_2 for $t = 20$ s with $\rho_3 = 0.1$ and the acceleration level is $\ddot{s}(t) = 5\text{m/s}^2$. It is observed that all

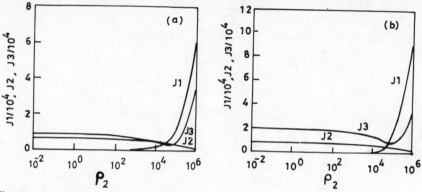

Fig. 6B.6. Influence of ρ_2 on performance indices (a) at $t = 5$ s; (b) at $t = 10$ s ($\rho_3 = 0.1$, $a_c = 5$ m/s^2, $\xi = 0.2$) (Narayanan and Raju, 1990).

three indices remain almost insensitive with respect to increase in ρ_2 upto about 10^4, beyond which road holding improves marginally, while the performances with respect to ride comfort and control force deteriorate drastically. Therefore a value of $\rho_2 = 10^4$ is chosen in further calculations.

Figure 6B.7 shows the variation of J_1, J_2 and J_3 with respect to ρ_3 keeping

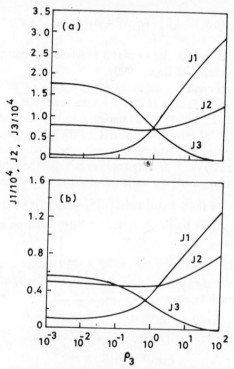

Fig. 6B.7. Influence of ρ_3 on performance indices (a) at $t = 5$ s; (b) at $t = 10$ s (Narayanan and Raju, 1990).

$\rho_2 = 10^4$. In this case, the three indices are almost constant in the range of $\rho_2 = 10^{-3}$ to 10^{-1}. Increasing ρ_3 beyond 10^{-1}, however, reduces control force effort drastically, while the performance with respect to ride comfort and road holding deteriorate drastically.

In Fig. 6B.8(a), the overall performance index is plotted as a function of non-dimensional time $\tau = \omega_0 t$, for $a = \ddot{s} = 5/\sec^2$, $\rho_2 = 10^4$, and $\rho_3 = 0.1$ with ζ as a parameter. Corresponding variations of the performance index for a passive system are also plotted alongside. From the figure, it is observed that the overall performance of the vehicle is progressively improved compared to the passive system, as the traverse time increases, the improvement being small close to the start. It is also seen from he figure that for the chosen weighting factors, the performance index for the passive system is smaller for $\zeta = 0.2$ than for $\zeta = 0.05$ and $\zeta = 0.35$, indicating

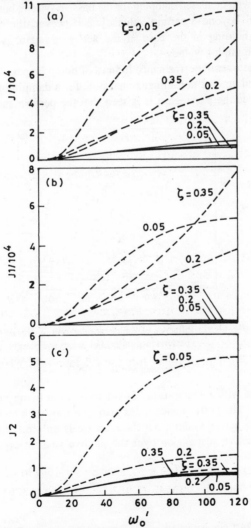

Fig. 6B.8. Effect of damping on performance indices of active and passive systems. (a) Overall index J, (b) Ride index J_1, (c) Road holding index J_2: (—) active; (– – –) passive (Narayanan and Raju, 1990).

the existence of an optimum value of passive damping factor. But, for the active system, the performance index is smaller for $\zeta = 0.05$ and increases as the damping value is increased. However, the increase in the overall performance index of active system for increased values of damping is not to significant as the reduction in the performance index of the passive system for an optimal value of damping. The performance indices corresponding to ride comfort alone are shown for both active and passive systems in Fig. 6B.8(b). They represent the mean square acceleration of the sprung mass normalized with respect to σ^2. They have the same characteristic variation as the overall index. This is, because the ride comfort criterion is weighted relatively more in the overall performance index for the weighting factors considered. The performance indices corresponding to road holding criterion alone are shown in Fig. 6B.8(c). The performance with respect to road holding improves with increased damping for the passive system, but the improvement becomes marginal beyond a certain value of ζ. However, the performance with respect to road holding in the case of the active system, is not sensitive to variations in the damping values.

In Fig. 6B.9, the overall performance indices of both, the passive and the active, systems are plotted against non-dimensional time for a damping factor of $\zeta = 0.2$, with acceleration as the parameter. It is seen that the performance of the passive

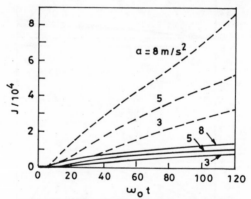

Fig. 6B.9.   Influence of vehicle acceleration or overall performance index J of active and passive systems. (———) active; (– – –) passive (Narayanan and Raju, 1990).

system deteriorates with traverse time, this deterioration being drastic for higher acceleration levels. The performance index also increases with increasing levels of acceleration for the active system, but the increase is only marginal, proving the efficiency of the active suspension over the passive suspension over a range of acceleration values.

REFERENCES

Ahlbeck, D.R. et al, 1974. Comparative Analysis of Dynamic of Freight and Passenger Rail Vehicles. Dot Rep. No. FRA-ORD & D-74-39.

Alanoly, J. and Sankar, S., 1988. Semi-active force generators for shock isolations, 126(1): J. Sound Vib. 126(1): 145–156.

Atalik, T.S. and S. Utku, 1976. Stochastic linearization of multi-degree-of freedom nonlinear systems. Earthquake Eng. Struct. Dyn. 4: 411–420.

Bekker, M.G., 1969. Introduction to Terrain-Vehicle Dynamics. Ann Arbor: The University of Michigan Press.

Bender, B.K., 1967. Optimization of random vibration characteristics of vehicle suspensions. D. Sc. Thesis, M.I.T., Cambridge.

Bender, E.K., 1968, Optimum linear preview control with application to vehicle suspension control, ASME J. Basic Engg. 90: 213–221.

Brust, J.M., 1961. Determination of fragility to meet random and sinusoidal vibration environments. SAE—430A, National Aeronautic, Space Engineering, Manufacturing Meeting, Los Angeles.

Carstens, J.P. and M.D. Kresge, 1965. Literature survey of passenger comfort limitations of high speed ground transports. United Aircraft Corporation Research Lab. Report, D-910353-1.

Caughey, T.K., 1963. Equivalent linearization techniques. J. Acoust. Soc. Amer. 35(11): 1706–1711.

Conway, H.G., 1958. Landing Gear Design. Chapman and Hall.

Corbin, J.C. and W.M. Kaufman, 1975. Classifying tracks by power spectral density. Mechanics of transportation suspension system, ADM (15): 1–20.

Crosby, M.J. and D.C. Karnopp, 1973. The active damper: A new concept for shock and vibration control, The Shock and Vib. Bull., 43(4): 119–133.

Dahlberg, T., 1977. Parametric Optimization of a 1-dof vehicle travelling on a randomly profiled road. J. Sound Vib. 55: 245–253.

Dahlberg, T., 1978. Ride comfort and road holding of 2-dof vehicle travelling on a randomly profiled road. J. Sound Vib. 58:179–187.

Dahlberg, T., 1979. Optimization criteria for vehicles travelling on randomly profiled road—A survey. Veh. Sys. Dyn. 8: 239–252.

Dahlberg, T., 1980. Comparison of ride comfort criteria for computer optimization of vehicle travelling on randomly profiled road. Veh. Sys. Dyn. 9: 291–307

Dodds, C.J. and J.D. Robson, 1973. The description of road surface roughness. J. Sound Vib. 31(2): 175–183.

Dokainish, M.A. and M.M. Elmadany, 1977. Dynamic response of tractor-semitrailer vehicle to random inputs. Proc. 5th VSD 2nd IUTAM Symp. on Dynamics of Vehicles on Roads and Tracks, Swets and Zeitlinger, 1978: 237–255.

Dokainish, M.A. and M.M. Elmadany, 1980. Random response of tractor-semitrailer system. Veh. Sys. Dyn. 9: 87–112.

Elmadany, M.M., M.A. Dokainish and A.B. Allan, 1979. Ride dynamics of articulated vehicles— a literature survey. Veh. Sys. Dyn. 8: 287–316.

Elmadany, M.M., M.A. Dokainish, and A.B. Allan, 1980. Road vehicle train response to random road surface undulations. ASME, Winter Annual Meeting, Dynamic System and Control Div.

Elmadany M.M., 1988. Stochastic optimal control of highway tractors with active suspensions. Veh. Sys. Dyn., 17: 193–210.

Frühauf, F., R. Kasper and J. Luckel, 1985. Design of an active suspension for a passenger vehicle model using input processes with time delays. Proc. Dynamics of Vehicles on Roads and on Railway Tracks. O. Nordstorm (ed.). 9th IASVD Symposium, Linkoping: 126–138.

Garivaltis, D.S., V.K. Garg, and A.F. D'souza, 1980a. Dynamic response of a six-axle locomotive to random track inputs. Veh. Sys. Dyn. 9: 117–147.

Garivaltis, D.S., V.K. Garg and A.F. D'souza, 1980b. Fatigue damage of the locomotive suspension elements under random loading, ASME Winter Annual Meeting, Chicago, Paper No. 80-WA/DE-7.

Gelb, A., and R.S. Warren, 1972. Direct statistical analysis of nonlinear ystems—CADET. AIAA Conference on Guidance and Control, Stanford, Paper No. 72–875.

Goldman, D.E. and H.E. von Gierke, 1960. The effects of shock and vibration on man. Naval Medical Research Institute, Lecture and Review Series No. 60-3.

Goodall, R.M. and W. Kortüm, 1983. Active controls in ground transportation: A review of the state of the art and future potential. Veh. Sys. Dyn., 12: 225–257.

Goyal, S.K., 1973. Response of aircraft to ground environment. M. Tech. Thesis, Department of Aeronautical Engineering, I.I.T., Kanpur.

Hac, A., 1985, Suspension optimization of a 2-DOF vehicle model using stochastic optimal control technique. J. Sound Vib. 100(3) : 343–357.

Hac, A., 1986. Stochastic optimal control of vehicles with elastic bodies and active suspension. ASME J. of Dyn. Sys. Meas. and Control, 108: 106–110.

Hady, M.B.A. and D.A. Crolla, 1989. Theoretical analysis of active suspension performance using a four-wheel vehicle model. Proc. Ins. Mech. Engg. J. of Automobile Engineering (D), 203: 125–135.

Hanes, R.M., 1970. Sensitivity to whole body vibration in urban transportation systems: a literature review. Applied Physical Laboratory, DOT No. APL/JHU-TPR004.

Harrison, R.F. and J.K. Hammond, 1981. The response of vehicles on rough ground. Proc. Institute of Acoustics, Spring Conference, New Castle upon Tyne, England.

Harrison, R.F. and J.K. Hammond, 1986a. Evolutionary (frequency/time) spectral analysis of response of vehicles moving on rough ground by using "covariance equivalent" modelling. J. Sound Vib. 107(1): 29–38.

Harrison, R.F. and J.K. Hammond, 1986b. Approximate, time domain nonstationary analysis of stochastically excited nonlinear systems with particular reference to motion of vehicles on rough ground. J. Sound Vib. 105(3) : 361–371.

Harrison, R.F. and J.K. Hammond, 1986c. Analysis of the nonstationary response of vehicles with multiple wheels. ASME J. of Dyn. Sys. Mmt. Control, 108 : 69–73.

Haug, E.J. (ed.), 1984. Computer Aided Analysis and Optimization of Mechanical System Dynamics. Berlin, Springer Verlag.

Hedrick, J.K. Rail system inputs. Department of Mechanical Engineering, M.I.T., Cambridge.

Hedrick, J.K., 1981. Railway vehicle active suspensions. Veh. Sys. Dyn. 10: 267–283.

Hedrick, J.K. (ed.), 1984. The Dynamics of Vehicles. Lesse, Swets and Zeitlinger.

Hedrick, J.K. and D.N. Wormley, 1975. Active suspensions for ground transport vehicles—A state of the art review in mechanics of transportation suspension systems, B. Paul, et al (eds.), ASME AMD 15: 21–40.

Hooks, A.E.W. and T.G. Pearce, 1974. The lateral dynamics of the linear induction motor test vehicle. ASME paper No. 74-Aut-0.

Horton, H.N.L. and D.A. Crolla. 1986. Theoretical analysis of a semi-active suspension fitted to an off-road vehicle. Veh. Sys. Dyn. 15: 351–372.

International Standards Organization 1972. Proposal for generalized road inputs to vehicles. Poc. No. ISO/TC 108/WG9 (Secretariat-2) 5.

International Standards Organisation, 1978. Guide for the evaluation of human response to whole body Vibration. ISO 2631.

Iwan, W.D. and A.B. Mason, Equivalent linearization for systems subjected to nonstationary random excitation. Int. J. of Nonlinear Mech. 15: 71–82.

Iwan, W.D. and I.M. Yang, 1972. Application of statistical linearization techniques to nonlinear multi-degree-of-freedom systems. J. App. Mech. 39: 545–550.

Jaiswal, O.R. and R.N. Iyengar, 1991. Dynamic analysis and random process modelling of railway tracks. Report No. 1-1991, Department of Civil Engineering, I.I.Sc., Bangalore.

Kalker, J.J., 1967. On the rolling contact of two elastic bodies in the presence of dry friction. Ph.D. Thesis, Delft University.

Kalker, J.J., 1979. Survey of wheel-rail rolling contact theory. Veh. Sys. Dyn. 8: 317–358.

Kamash, K.M.A. and J.D. Robson, 1977. Implications of isotropy in road-surface modelling. J. Sound Vib. 57: 89–100.

Kirk, C.L. and P.J. Perry, 1971. Analysis of taxing induced vibration in aircraft by power spectral density method. Aeronautical Journal 75: 182–194.

Krasnicki, E.J. 1979. Comparison of analytical and experimental results for a semi-active vibration isolator in Proc. 50th Shock and Vibration Symposium, Colarado Springs.

Kwakernaak, H., and Sivan, R., 1978. Linear Optimal Control Systems. John Wiley & Sons Inc.

La Barre, R.P. and R.T. Forkes, 1969. The measurement and analysis of road surface roughness. MIRA Report No. 1970/5.

Lee, A.Y. and M.A. Salman, 1989. On the design of active suspension incorporating human sensitivity to vibration. Opt. Control. App. and Methods. 10: 189–195.

Louam, N., D.A. Wilson and R.S. Sharp. 1988. Optimal control of a vehicle suspension incorporating the time delay between front and rear wheel inputs. Veh. Sys. Dyn. 17: 317–336.

Margolis, D.L., 1982a. Semi-active heave and pitch control for ground vehicles. Veh. Sys. Dyn., 11: 31–42.

Margolis, D.L. 1982b. The response of active and semi-active suspensions in realistic feed back signals. Veh. Sys. Dyn., 11: 267–282.

Margolis, D.L., J.L. Tylee and D. Hrovat. 1975. Heave mode dynamics of a tracked air cushion vehicle with semi-active air bay secondary suspension ASME J. Dyn. Sys. Meas. and Control., 97: 399–407.

Mcbreaty, J.F., 1948. A critical study of aircraft landing gears. J. Aero. Sci. 15: 263.

Metz, D. and J. Maddock. 1986. Optimal ride height control for championship race cars. Automatica. 22: 509–520.

Milwitzky, B. and F.E. Cook, 1953. Analysis of landing gear behaviour. N.A.C.A. Report No. 1154.

Mitschke, E.M., 1962. Influence of road and vehicle dimensions and amplitudes of body motions and dynamic wheel loads (theoretical and experimental evaluation investigations) SAE, 70: 434–496.

Narayanan, S., 1987. Nonlinear and nonstationary random vibration of hysteretic systems with application to vehicle dynamics in Proc. Nonlinear Stochastic Dynamic Engineering Systems. F. Ziegler and G.I. Schueller (eds.), IUTAM Symp., Innsbruck, Austria: Springer Verlag.

Narayanan, S. and G.V. Raju. 1990. Stochastic optimal control of nonstationary response of a single degree of freedom vehicle model. J. Sound. Vib. 140(3).

Newland, D.E., 1986a. The effect of a footprint on perceived surface roughness. Proc. Roy. Soc. (London), A. 405: 303–327.

Newland, D.E., 1986b. General theory of vehicle response to random road roughness in Random Vibration—Status and Recent Developments. I. Elishakoff and R.H. Lyon (eds.), Amsterdam : Elsevier.

Nigam, N.C., 1983. Introduction to Random Vibrations, Cambridge: The MIT Press.

Nigam, N.C. and D. Yadav, 1974. Dynamic response of accelerating vehicles to ground roughness. Proc. Noise, Shock and Vibration Conference, Monash University, Melbourne: 280–285.

Packeja, H.B. (ed.), 1976. The Dynamics of Vehicles, Amesterdam: Swets & Zeitlinger.

Parkhilovskii, I.G. 1968. Investigations of the Probability characteristics of the surfaces of distributed types of roads. Avtom Prom. 8: 18–22.

Quinn, B.E. and S.E. Hildebrand, 1973. Effect of road roughness on vehicle steering. Highway Research Record, 471: 62–75.

Raju, G.V. and S. Narayanan, 1990. Active Control of nonstationary response of a 2-dof vehicle model. in Proc. Int. Cong. on Recent Developments in Air-and-Structure-Borne Sound and Vibration. Crocker, M.J. and P.K. Raju (eds.) Auburn: 687–694.

Raju, G.V. and S. Narayanan, 1991. Optimal estimation and control of nonstationary response of a two-degree of freedom vehicle model. J. Sound. and Vib. 149: 413–428.

Rakheja, S. and S. Sankar 1985. Vibration and Shock isolation performance of semi-active "on-off" damper. ASME J. of Vib. Acous. Stress and Reliability in Design. 107: 383–403.

Robson, J.D. 1978. The role of Parkhilovskii model in road description Veh. Sys. Dyn. 7: 153–162.

Robson, J.D. and C.J. Dodds, 1970. The responses of vehicle components to random road surface indulations. Proc. 13 FISITA Congress, Paper 17-2D, Brussels.

Robson, J.D. and C.J. Dodds, 1975/76. Stochastic road inputs and vehicle response. Veh. Sys. Dyn. 5: 1–13.

Sankar, T.S. and M. Samaha, 1986. Research in rail vehicle dynamics state-of-the-art. The Shock and Vibration Digest. 18(2) : 9–18.

Sattaripour, A., 1977. The effect of road roughness on vehicle behaviour. Proc. 5th VSD—2nd IUTAM Symp., Swets & Zeitlinger, 229–235.

Schiehlen, W.O., 1982a. Modelling of complex vehicle systems in Wickens, A.H. (ed.) The Dynamics of Vehicles, Lisse: Swets & Zeitlinger.

Schiehlen, W.O., 1982b. Nonstationary random vibrations in K. Hennig (ed.). Random Vibrations and Reliability. Proc. IUTAM Symp., Frankfurt. (ORDER): Akademic Verlag.

Schiehlen, W.O., 1986. Probabilistic analysis of vehicle vibrations. Probabilistic Engineering Mechanics, 1(2): 99–104.

Sharp, R.S. and D.A. Crolla. 1987. Road Vehicle suspension system design—A review. Veh. Sys. Dyn. 16: 167–192.

Slibar, A. and H. Springer (eds.), 1978. The Dynamics of Vehicles, Amsterdam: Swets & Zeitlinger.

Thompson, A.G. 1976. An active suspension with optimal linear state feedback, Veh. Sys. Dyn. 5: 187–203.

Thompson, A.G., 1984. Optimal and sub-optimal linear active suspensions for road vehicles. Veh. Sys. Dyn. 13: 61–72.

Thompson, A.G., B.R. Davis and C.E.M. Peance, 1980. Opimal linear active suspension with finite road preview. SAE paper 800520.

Tomizuka, M., 1976. Optimum linear preview control with application to vehicle suspension— Revisited. ASME. J. of Dyn. Sys. Meas. Control 98: 309–315.

UCTAV, 1972. Performance specifications and engineering design requirements for the UCTAV. DOT specification.

Wen, Y.-k., 1976. Method of random vibration of hysteretic systems, J. Eng. Mech. Div. Proc. ASCE 102 (EM) : 389–401.

Wen, Y-k, 1980. Equivalent linearization of hysteretic systems subjected to random excitation. J. of App. Mech. Trans. ASME 47 : 150–154.

Wickens, A.H. and A.O. Gilchrist, 1977. Railway vehicle dynamics—the emergence of practical theory. MacRobert Award Lecture. Council of Engineering Institutions, U.K.

Wickens, A.H. (ed.), 1982. The Dynamics of Vehicles. Lisse: Swets & Zeitlinger.

Willermeit, H.P. (ed.), 1980. The Dynamics of Vehicles. Lisse: Swets & Zeitlinger.

Wilson, D.A., R.S. Sharp and S.A. Hassan. 1986. Application of linear optimal control theory to the design of automobile suspensions. Veh. Sys. Dyn. 15: 103–118.

Yadav, D., 1971. Response of moving vehicles to ground induced excitation Ph.D. Thesis, Department of Aeronautical Engineering, I.I.T., Kanpur.

Yadav, D. and N.C. Nigam, 1978. Ground induced nonstationary responses of vehicles. J. Sound Vib. 61(1): 117–126.

Yadav, D. and N.C. Nigam, 1984. Reliability analysis of landing gear flatigue life. IUTAM Symp. on Probabilistic Methods in Mechanics of Solids and Structures, Stockholm.

Yoshimura, T., N. Ananthanarayana and D. Deepak. 1986. An active vertical suspension for track vehicle systems. J. Sound. Vib. 106(2): 217–225.

Yoshimura, T. and M. Sugimoto, 1990. An active isolation suspension for a vehicle for travelling on flexible track vehicle systems. J. Sound. Vib. 138(3): 433–445.

Yue, C., T. Butsen and J.K. Hedrick, 1989. Alternative control laws for automative active suspensions. ASME. J. of Dyn. Sys. Meas. and Control. 111: 286–291.

7. Response of Structures to Earthquakes

7.1 Introduction

Earthquakes are a source of critical loading condition for structures located in the seismically active regions of the earth. A significant feature of the earthquake loading is a large measure of uncertainty associated with the earthquake phenomena. To establish the seismic loading condition for a structure, it is necessary to anticipate the number, size and location of future earthquakes in the region surrounding the site during the service life of the structure. Further, this assessment should be coupled with the prediction of structural response, and damage due to random vibration induced by ground motion of a given intensity. Both these steps involve uncertainty at several stages. The problem is compounded by the fact that potentially damaging strong motion earthquakes occur after long intervals of time and the available data for such events is often statistically insufficient. A probabilistic treatment of earthquake engineering problems, which involves assessment of seismic risk and random vibration analysis, therefore, provides a rational and consistent basis for aseismic design.

In this chapter, we shall discuss the probabilistic analysis and design of structures subjected to earthquake loading. The characteristics of ground motion during earthquakes, its stochastic models and simulation are discussed first. The random vibration analysis of structures due to earthquake excitation is discussed in detail. Finally, we discuss the probabilistic design of structures in seismic regions, including the effect of seismic risk.

In Appendix 7A, we provide a brief review of seismic risk analysis. In Appendix 7B, we discuss the concept of response spectrum and its relationship to power spectral density. The emphasis is primarily on the application of random vibration theory to aseismic design of structures modelled as sdf and mdf systems. Both stationary and nonstationary models of ground acceleration are considered, and the distributions and moments of peak values of the response are derived. Response of structures to the simultaneous action of more than one component is treated, and a brief discussion is included on the behaviour of discrete inelastic systems. Effort is made to present approximate results which are of practical value in reliability based aseismic design.

Reliability based design requires estimation of total risk. In earthquake engineering, it is convenient to treat the risk in two parts:

(i) Risk associated with the response of the structure to random ground motion for given seismic inputs. This provides the conditional distribution of the response and is treated in Section 7.5.

(ii) Risk associated with the seismicity of the region surrounding the site. This is treated in Appendix 7A.

The total risk is computed by convolution of the conditional distribution of response with the distribution of seismic risk. In Section 7.6, we discuss the concept of 'damage', and design based on total risk. A few examples are included to illustrate the applications of the concepts.

7.2 Characteristics of Earthquake Ground Motion

Most earthquakes of engineering significance are of tectonic origin and are caused by slip along geological faults. While specific source mechanisms leading to a slip vary in different regions of the earth, and are not always fully understood, four basic types of faulting can be identified with strong-motion earthquakes (Housner, 1977): (i) low-angle, compressive under-thrust faulting caused by compressive forces generated due to the movement of the sea-floor crustal plate against continental plate; (ii) compressive over-thrust faulting due to shear failure on an inclined fault with the upper portion of the rock moving upward under the action of compressive forces; (iii) extensional faulting on inclined faults due to extensional strains in the earths' crust causing rocks overlying the fault to move downwards; and (iv) strike-slip faulting which consists of a relative horizontal displacement of the two sides of a fault across an essentially vertical plane. In most earthquakes the actual slip mechanism is a combination of two, or more, of the above types of faulting. Often slip occurs on an irregular surface and on more than one fault. The characteristics of the ground motion during an earthquake in the vicinity of the causative fault (near-field) are strongly dependent on the type of faulting and the time-history motion of fault displacement. As we move away from the fault (far-field), the nature of ground motion is primarily determined by the travel-path geology. The nature of ground motion at a point on the earth surface is also influenced by the local site conditions, that is, soil properties and topography. Characteristics of the source mechanisms, travel-path geology and local site conditions, therefore, determine the nature of ground motion due to an earthquake. The basic characteristics of the seismic waves depend primarily on: the stress drop during the slip; total fault displacement; size of slipped area; roughness of the slipping process; fault shape; and the proximity of the slipped area to the a ground surface. As the waves radiate from the fault, they undergo geometric spreading and attenuation due to loss of energy in the rocks. Since the interior of the earth consists of heterogeneous formations, the waves undergo multiple reflections, refractions, dispersion and attenuation as they travel. The seismic waves arriving at a site on the surface of the earth are a result of complex superposition giving rise to irregular motion, which may be modelled as a random vector varying randomly in space and time.

A simple physical model of the slip on a fault can be constructed by visualising that a fault is formed by the superposition of a large number of incremental shear dislocations (Housner, 1955). As the stresses build-up in the interior of the earth, the stress level at one, or more, of the dislocations exceeds the limiting stress, the dislocations snap and stress waves are generated. By analysing the sudden release of stresses on the face of a penny-shaped crack in an elastic solid, Housner (1955) showed that the ground motion generated is similar to the recorded motions during a slip over small faults. Since major earthquakes are generated by a slip over large areas with nonuniform prestress, the ground motion during such earthquakes is, in effect, a result of superposition of simple pulses arriving randomly in time. Several important characteristics of strong-motion earthquakes can be explained on the basis of this simple physical model.

A formulation of earthquake engineering analysis, or design, problem may proceed either from the source, or from certain postulates regarding the nature of ground motion at the site based on a set of ground motions recorded under comparable conditions. It is generally convenient to start with the latter, and eventually embed the analysis into the total problem by incorporating the effects of seismicity. We shall adopt this approach. In Appendix 7A, we briefly review the relevant aspects of seismology and seismic risk. Since we are primarily concerned with the applications of random vibration theory to earthquake engineering, the treatment is limited to providing a link between seismicity and the dynamic response of ground-based systems to a given ground motion intensity.

An earthquake causes both translation and rotation at a point on the surface of the earth. For most problems the rotational component can be disregarded, and ground motion treated as a random vector with three orthogonal translational components—two horizontal and one vertical. Each component can be expressed either by an acceleration, velocity or displacement function of time. Although the three forms contain equivalent information and can be derived from each other by differentiation, or integration, it is generally convenient to represent and record earthquake ground motion as acceleration, and derive velocity and displacement through integration, if required. Based on its characteristics, a ground acceleration time-history due to earthquakes may be classified into four broad groups (Newmark and Rosenblueth, 1971):

(i) time-history containing essentially a single shock. Such motions occur at short distances on firm ground during moderate to shallow focus earthquakes. The records exhibit a strong unidirectional character and represent predominantly short period oscillatory motion;

(ii) time-history containing moderately long and extremely irregular motion. Such motions occur on firm ground at moderate distances from the focus of moderate to large earthquakes. They contain a wide range of frequencies (0.1–30 Hz), and are generally of comparable severity in the three directions;

(iii) time-history of long duration containing a dominant frequency of vibration. Such motions result from the filtering of the second type of ground motion through layers of soft soil and from successive wave reflections in the mantle; and

(iv) motions consisting of large-scale, permanent deformation of ground, such as, slides or soil liquefaction.

The actual ground motion during an earthquake may contain the characteristics of two, or more, of the type of motions described above. The first type of motions can be adequately treated as a deterministic vibration problem due to its simplicity. The fourth type of motions are not amenable to formal analytical treatment. The second and the third type of earthquake ground motions can be modelled as random processes and, therefore, cause random vibration of ground-based systems. We shall confine subsequent discussion to these types of ground motions. The ground acceleration, velocity and displacement time-history of a typical ground motion of the second type (NS: EL—Centro, May 18, 1940) are shown in Fig. 7.1.

The dynamic behaviour of structures during an earthquake is determined primarily by the amplitude, frequency and duration of ground motion. The amplitude of strong-motion earthquake acceleration records generally exhibit: (i) a rapid build-up at the beginning of the motion; (ii) a nearly constant value during the strong-

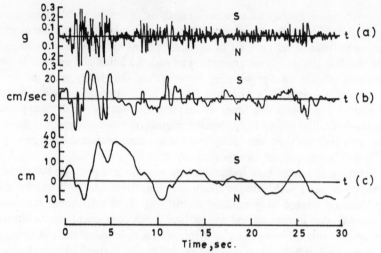

Fig. 7.1. Ground motion time histories of El-Centro, May 18, 1940, NS—component earthquake record: (a) acceleratioon; (b) velocity; and (c) displacement (Blume et al, 1961).

motion shaking; and (iii) an exponentially decaying tail. The frequency characteristics are reflected by the Fourier amplitude spectra; power spectral density (psd); response spectra (Appendix 7B); or response envelope spectra (Trifunac, 1971). The second type of earthquakes possess broad-band characteristics and their acceleration time-history can be adequately modelled as a uniformly modulated stationary random process. The high frequency components of ground motion attenuate with distance faster than the low frequency components, and this strongly influences the spectral characteristics of ground motion as a function of the distance. The important parameters defining the gross characteristics of earthquake motion at a site are: the peak values of ground acceleration (A_g), velocity (V_g), displacement (D_g); the r.m.s. value of ground acceleration (σ_g); the response spectra (SA, SV, SD); the spectrum intensity (SI); a site intensity such as NMI (I); and the duration (T'). The relationship of these parameters to the magnitude (M), and focal distance (R), of the earthquakes are given in Appendix 7A.

7.3 Stochastic Models of Earthquake Ground Motions

Housner (1947) was the first to suggest that "acceleration records of earthquake exhibit characteristics of randomness". If a large number of ground motion records were available for a particular site, the parameters of a stochastic process model could be determined directly by statistical analysis. However, this approach is not possible at the present time in any part of the world, due to availability of only a few strong-motion records. It, therefore, becomes necessary to use considerable judgement in constructing, and validating stochastic models of ground motion on the basis of a few available records at a site, or at comparable locations, coupled with seismological and geological data, and local site conditions.

Let $Z_i(t)$, $i = 1, 2, 3$, represent the three components of ground displacement at a point due to an earthquake. Consistent with general characteristics of strong-motion earthquakes discussed in the preceding section, each component of the ground acceleration can be expressed as

$$\ddot{Z}_i(t) = A_i(t)\,\ddot{Y}_i(t); \quad 0 \leq t \leq T' \tag{7.1}$$

$$= 0; \qquad \text{otherwise}$$

where $i = 1, 2, 3$; $A_i(t)$ are slowly varying random functions of time called the envelope; $\ddot{Y}_i(t)$ are a segment of stationary random processes; and T' is the duration of the motion. Thus each component of ground acceleration is a realisation of a nonstationary random process. The stochastic model represented by (7.1) may be described to varying degree of the completeness by the joint distribution functions, the joint moment or spectral functions and other properties such as level crossings, peaks etc.

We shall first discuss the properties and models of individual components. We have discussed the general theory of stochastic modelling in Chapter 2. We shall, therefore, confine our discussion to specific aspects relevant to earthquake ground motion. Three types of basic models, and their minor variations reflecting increasing level of complexity, have been used to model earthquake ground acceleration:

  (i) white noise (Housner, 1947; Byrcroft, 1960; Hudson, Housner and Caughey, 1960);
 (ii) stationary process (Tajimi, 1960; Housner and Jennings, 1964); and
(iii) nonstationary process (Bolotin, 1960; Cornell, 1964; Shinozuka and Sato, 1967; Amin and Ang, 1968; Boore, 1983).

If the short duration of initial rise and exponentially decaying tail are disregarded, the central strong-motion portion of acceleration records on firm ground can be treated as a segment of stationary random process. Byrcroft (1960) has shown that properly scaled segment of white noise have response spectra similar to the response spectra of real earthquakes and can, therefore, be used to model the strong-motion portion of ground acceleration.

The stationary process model is an improvement over the white noise as it can be shaped to represent the frequency characteristics of the actual ground motions. The model makes it possible to incorporate the effect of local soil conditions explicitly. A segment of stationary process is, therefore, a more realistic model of the strong-motion portion of ground acceleration records. Following analytical expression due to Tajimi (1960) based on the work of Kanai (1957), called Kanai-Tajimi spectra, has been extensively used to represent the psd of ground acceleration:

$$\Phi(\omega) = \Phi_0 \frac{1 + 4\zeta_g^2(\omega/\omega_g)^2}{[1 - (\omega/\omega_g)^2]^2 + 4\zeta_g^2(\omega/\omega_g)^2} \tag{7.2}$$

A filtered white noise with psd given by (7.2) can be generated by the following second order filter

$$\ddot{U} + 2\zeta_g\omega_g\dot{U} + \omega_g^2 U = W(t) \tag{7.3}$$

where $W(t)$ is white noise with psd Φ_0 and $\ddot{Z}(t) = \ddot{U}(t) + W(t)$. From an analysis of actual records, it is found that $\zeta_g = 0.6$ and $\omega_g = 5\pi$ correspond closely to the spectral properties for firm ground. For a specific site, the parameters ζ_g and ω_g should be chosen suitably to represent local site conditions.

The variance and other spectral parameters for the Kanai-Tajimi spectrum are given by

$$\sigma_g^2 = \frac{\pi \Phi_0 \omega_g}{2 \zeta_g} (1 + 4 \zeta_g^2) \qquad (7.4)$$

$$\Omega_g = \left(\frac{\lambda_2}{\lambda_0} \right)^{1/2} = 2.1 \, \omega_g \qquad (7.5)$$

$$\delta_g = \left(1 + \frac{\lambda_1^2}{\lambda_0 \lambda_2} \right)^{1/2} = 0.67 \qquad (7.6)$$

where λ_i ($i = 0, 1, 2, \ldots$) are the spectral moments of the random process $\ddot{Z}(t)$ (4.131*). In (7.4) and (7.5), cut-off frequency $\omega_0 = 4 \, \omega_g$ and $\zeta_g = 0.6$ (Vanmarcke, 1977). Bolotin (1960) proposed the following expression to represent a comparable auto-correlation function for stationary ground acceleration:

$$R(\tau) = \sigma_g^2 \, e^{-\alpha |\tau|} \cos \beta \tau$$

The nonstationarity in the ground acceleration records arises primarily through envelope functions $A_i(t)$ in (7.1). These are slowly varying random functions of time. A simple and adequate nonstationary model of ground acceleration can be constructed by assuming $A_i(t) = A(t)$ to be a deterministic function and $\ddot{Y}_i(t)$ a set of stationary random processes with specified psd. The model may be constructed either by multiplying a segment of filtered white noise by a modulating function (MFWN) or alternatively by multiplying white noise by a modulating function first and then filtering the product process (FMWN). If the filter characteristics and the modulating functions are identical for the two models, the difference in their characteristics depends on the smoothness of $A(t)$. For earthquakes with long quasi-stationary motion, both models yield similar characteristics.

Several functions have been used to model the envelope of the ground acceleration records. Following expressions are commonly used:

Iyengar and Iyengar (1969):

$$A(t) = (A_0 + A_1 t) \, e^{-at} \, H(t) \qquad a > 0 \qquad (7.7)$$

Shinozuka and Sato (1967):

$$A(t) = A_0 (e^{-at} - e^{-bt}) \, H(t) \qquad b > a > 0 \qquad (7.8)$$

Jennings et al (1969):

$$
\begin{aligned}
A(t) &= A_0 \left(\frac{t}{t_1} \right)^a H(t) & 0 \le t \le t_1 \\
&= A_0 H(t - t_1) & t_1 \le t \le t_2 \\
&= A_0 \exp \left[- b(t - t_2) \right] H(t - t_2) & t_2 \le t \le t_3 \\
&= A_0 [.05c + d(T' - t)^2] H(t - t_3) & t \ge t_3
\end{aligned}
\qquad (7.9)
$$

where $H(t)$ is heaviside unit step function.

The envelope functions represented by (7.9) cover significant features of a complete range of recorded earthquake ground accelerations. Jennings et al (1969) have suggested suitable values for the parameters of the envelope function to model four different types of earthquake motions (A, B, C and D) which are considered to be significant for engineering structures. The values of the parameters are given in Table 7.1.

Table 7.1   Parameters of $A(t)$ in (7.9) (Jennings et al, 1969)

Type	Target Magnitude (Richter)	Target Spectrum intensity	Total Duration (sec)	Duration (sec)			a	b	c	d
				t_1	t_2	t_3				
A	≥ 8	1.5 (El-Centro, 1940)	120	4	35	80	2	0.0357	1	0.938×10^{-4}
B	6–8	El-Centro, 1940	50	4	15	30	2	0.0992	1	0.5×10^{-2}
C	5.5–6	Golden Gate, 1957	12	2	4	—	2	0.268	—	—
D	4.5 – 5.5	Parkfield, 1966	10	2	2.5	3.5	3	1.606	2	0.237×10^{-2}

The envelope functions defined by (7.7) and (7.8) are analytically simpler and are used extensively in theoretical treatments. The nonstationary nature of ground motions has a significant influence on the tail value of the probability estimates, nonlinear response and soil behaviour. Further, the stationary models are inadequate for small near-field earthquake ground motions.

In Chapter 2 we have seen that filtered Poisson process models can be used to represent both stationary and nonstationary random processes. The early models of earthquake ground motion (Housner, 1947, 1955) were based on superposition of pulses arriving randomly in time. Such models have the advantage that the response of system to such a model is also a filtered Poisson process with a modified pulse shape and, therefore, the moment functions of the response can be readily determined by the Campbell's Theorem. Bogdanoff et al (1961) and Goldberg et al (1964) proposed random functions to model and simulate earthquake acceleration, thus expressing member functions of the random process directly. Hsu and Bernard (1978) have proposed a simple model of random functions of the form:

$$\ddot{Z}(t) = A(t)\,\ddot{Y}(t), \qquad t > 0 \qquad\qquad (7.10)$$

$$= 0, \qquad\qquad t \leq 0$$

where

$$A(t) = bt \exp(-\alpha t)$$

$$\ddot{Y}(t) = (2/n)^{1/2} \sum_{j=1}^{n} \cos(f_j t) + \theta_j)$$

in which $b = a(n/2)^{1/2}$, $A(t)$ is a deterministic function specified by parameters α, a and n; $\ddot{Y}(t)$ is a stationary random process expressed explicitly as a sum of harmonic functions with random frequency parameters. The random frequencies f_j are identically distributed random variables with a distribution function established from the data of the past records, and θ_j are independent random variables distributed uniformly over the interval $[0, 2\pi]$. Hsu and Bernard (1978) have discussed, in detail, the procedures for estimation of parameters, testing of the model and simulation of ground acceleration records.

By invoking the central limit theorem, it can be argued that since the ground motion at a site is a result of irregular slips on a fault, or faults, and multiple reflections and refractions through heterogeneous formations, its distribution should be normal. A statistical analysis of available records shows that ground acceleration can be assumed to be normal (Saragoni et al, 1980).

Time series models (AR, MA, ARMA, and ARIMA) provide an alternative approach to the construction and simulation of both stationary and nonstationary models of earthquake ground motion (Polhemus and Cakmak, 1981; Nau et al, 1982; Gersh and Kitagawa, 1985). In Chapter 2 we have discussed the general theory of time-series models and an example of ARMA model to simulate earthquake ground motion due to Polhemus and Cakmak (1981).

We have discussed several stochastic models to represent strong motion ground acceleration time histories during strong motion earthquakes. The choice of a particular model should depend on its adequacy for specific application. As a rule, the model should be as simple as possible, consistent with the required level of accuracy, and should incorporate the characteristics that are important to the problem at hand. The uniformly modulated random process model, incorporates the significant features of the ground acceleration and is analytically tractable. However, it does not reproduce the time-varying frequency content of actual ground acceleration records. Generally, these records show a shift from higher frequencies to lower frequencies towards the end of the record. This feature is perhaps insignificant for many structures, but may be significant for inelastic behaviour of structures (Amin et al, 1969). Trifunac (1971) proposed a method for synthesizing strong motion accelerograms based on Amplitude Envelope Spectrum (AES) which incorporates the nonstationarity in the frequency content.

7.3.1 Multi-component Ground Motion

We have so far discussed the stochastic models of the individual components of ground acceleration during earthquakes. Since the three components of earthquake ground motion act simultaneously on a structure, it is necessary to construct stochastic models which incorporate their joint properties. Also, since the choice of axes along which the components of an earthquake are recorded is arbitrary, it is necessary to establish relations for the transformation of their properties due to the rotation of the coordinate systems.

Consider the three components of ground acceleration defined by (7.1) along the axes (123). Let $A_i(t) = A(t)$ be a deterministic function, and $\ddot{Y}_i(t)$ zero-mean, stationary random processes. Then the elements of the covariance matrix of $\ddot{Z}_i(t)$ are given by

$$K_{\ddot{Z}_i \ddot{Z}_j}(t, \tau) = E[\ddot{Z}_i(t)\,\ddot{Z}_j(t + \tau)] = A(t)\,A(t + \tau)\,E[\ddot{Y}_i(t)\,\ddot{Y}_j(t + \tau)]$$

$$= A(t)\,A(t + \tau)(R_{\ddot{Y}_i \ddot{Y}_j}(\tau));\ i, j = 1, 2, 3 \qquad (7.11)$$

Since the correlation time of the accelerograms is usually very small, the effect of changing the coordinate directions on the covariance functions can be investigated by considering (7.11) for $\tau = 0$, that is

$$E[\ddot{Z}_i(t)\,\ddot{Z}_j(t)] = A^2(t)\,E[\ddot{Y}_i(t)\,\ddot{Y}_j(t)] \qquad (7.12)$$

which can be expressed in the matrix form

$$\mathbf{R}_{\ddot{Z}\ddot{Z}}(t) = A^2(t)\,\mathbf{R}_{\ddot{Y}\ddot{Y}} \qquad (7.13)$$

Since $\ddot{Y}(t)$ are stationary, the elements of variances matrix $\mathbf{R}_{\ddot{Y}\ddot{Y}}$ are constants. Also $\mathbf{R}_{\ddot{Y}\ddot{Y}}$ is a real symmetric and positive definite matrix. It is, therefore, possible to find a canonical transformation matrix $\mathbf{P}$, such that

$$\ddot{Y}_P(t) = \mathbf{P}\,\ddot{Y}(t) \qquad (7.14)$$

and

$$\mathbf{R}_{\ddot{Y}_P\ddot{Y}_P} = \mathbf{P}^T\mathbf{R}_{\ddot{Y}\ddot{Y}}\mathbf{P} = \mathrm{diag.}\,(R_{11}, R_{22}, R_{33}) \qquad (7.15)$$

where $R_{11} \ge R_{22} \ge R_{33}$ are the principal variances of the matrix $\mathbf{R}_{\ddot{Y}\ddot{Y}}$, and the columns of $\mathbf{P}$ define the three orthogonal principal directions (Penzien and Watabe, 1975). Further, from (7.13)

$$\mathbf{R}_{\ddot{Z}\ddot{Z}}(t) = A^2(t)\,\mathrm{diag}\,(R_{11}, R_{22}, R_{33}) \qquad (7.16)$$

Hence, the principal variances of $\ddot{Z}$ are given by $A^2(t)R_{ii}$ (no sum), $i = 1, 2, 3$. Note that variances are functions of time but the principal directions are time-independent. If different modulation functions, $A_i(t)$, are used for each component, the principal directions will be time-dependent. Under the assumptions stated above, it follows that the components of ground motion are correlated, but it is always possible to find at least one set of three principal directions along which the components are uncorrelated. The variances along these three principal directions represent maximum, minimum and intermediate values.

By analysing six different strong-motion records, Penzien and Watabe (1975) concluded that:

(i) the ratios of the minor and intermediate principal variances to major principal variance are of the order of 1/2 and 3/4 respectively;

(ii) the principal values of the cross-correlation coefficients ($\rho_{ij} = (R_{ii} - R_{ij})/(R_{ii} + R_{ij})$) ρ_{12}, ρ_{23} and ρ_{13} are approximately 0.14, 0.2 and 0.33 respectively.

(iii) the major principal axis is generally in the direction of the epicentre and the minor principal axis is vertical; and

(iv) the directions of principal axes are reasonably stable over successive time intervals. It is, therefore, reasonable to assume the same envelope function, $A(t)$, in the three directions.

A significant conclusion of the above discussion is that the three components of ground acceleration can be modelled as mutually uncorrelated random processes, provided they are directed along a set of principal axes with major principal axis directed towards the expected epicentre, and the minor principal axis directed vertically. Hadjian (1981) has analysed the correlation properties of the horizontal components of recorded ground motion to obtain the probability density function of the correlation coefficient. Recognising that the uncertainties in the value of the correlation coefficient arise due to two distinct sources—geometric and seismological, Hadjian has synthesised a probability density function shown in Fig. 7.2. He has suggested that an 'equivalent' rectangular distribution, may be used for design codes and simulation purposes.

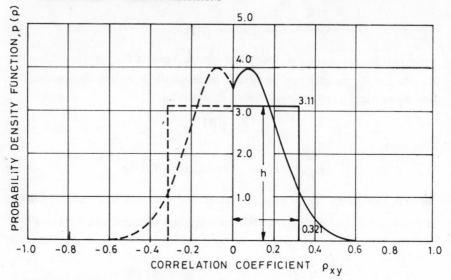

Fig. 7.2. Probability distribution of correlation coefficient and an 'equivalent' rectangular distribution (Hadjian, 1981).

7.3.2 Estimation of Model Parameters

Stochastic models of earthquake ground acceleration are used for two purposes: (i) random vibration analysis of system response to earthquake excitation; and (ii) simulation of an ensemble of ground acceleration records for Monte-Carlo type studies. In either case, it is essential to determine the parameters of the model and to validate it. This is done by using both the available seismological information, and the past earthquake records at the site, and at comparable locations. The peak values of ground motion parameters, duration of motion etc. can be estimated from attenuation laws, if potential sources can be identified (Appendix 7A). Local site conditions can be incorporated in modelling the frequency content. The available ground motion records can be grouped into large, medium and small earthquakes (Jennings et al, 1969) and an ensemble of earthquake records can be compiled for each group, after amplitude and time scaling, if required. The parameters of the model can be estimated on the basis of such information. The limitation of data is a serious problem in earthquake engineering and, therefore, considerable ingenuity and judgement is required in the estimation of model parameters.

Stochastic models of earthquake ground acceleration involve two major components—the modulating function reflecting the nonstationary amplitude characteristics, and the associated stationary process reflecting the frequency content. The estimation of parameters and validation of the model is carried out by using a combination of the following characteristics (Housner and Jennings, 1964; Iyengar and Iyengar, 1969; Levy et al, 1971, and Hsu and Bernard, 1978):

  (i) peak ground acceleration (A_g); the time (t_m) at which the peak occurs; ratio (t_m/T),
 (ii) time-dependent variance; intensity moments (Nigam, 1982),
(iii) rate of zero crossing (N_0), rate of maxima (N_m), ratio (N_0/N_m),
(iv) covariance function; p.s.d.; spectral moments (Vanmarcke, 1972),

(v) average response spectra; response envelope spectra; time response spectra
     (Schiff and Bogdanoff, 1966), and

(vi) spectrum intensity.

In time-series models, parameters are estimated on the basis of comparison of
estimated statistics of the generated sequence and the statistics of actual records.
We have discussed the estimation of parameters and validation of an ARMA
model in Chapter 2 using San Fernando earthquake record (Polhemus and Cakmak,
1981). Digital processing technique has also been used to estimate the parameters
of FMWN model relying on a single record (Nau et al, 1982). Parameters of
stochastic models can also be estimated from theoretical considerations (Trifunanc,
1971a; Boore, 1983).

7.4   Simulation of Earthquake Accelerograms

A complete characterisation of a random process requires an ensemble containing
infinite number of sample functions. We have noted that due to availability of only
few comparable strong-motion records, a probabilistic formulation of earthquake
engineering problem is singularly ill-conditioned in the statistical sense. For linear
systems, random vibration theory provides the analytical framework for determining
the exact, or approximate, estimates of response statistics, and reliability measures,
if a stochastic model of the excitation is available. For non-linear systems, and
some problems in linear systems, such as, low-cycle fatigue, first-passage across
a barrier, analytical solutions are generally not available. Such problems can be
treated by Monte-Carlo methods which require a reasonably large ensemble of
sample functions. The ensemble can be constructed by simulation of earthquake
ground motion on an analog, or a digital computer to supplement the available
records, if any.

In Chapter 2 we have discussed the basic theory and methods of stochastic
modelling and simulation of random processes. In this section we shall discuss
some aspects of simulation specific to earthquake ground motion.

For the simulation of earthquake ground motion at a site two broad approaches
are available:

(i) construction of a stochastic model of the ground motion on the basis of
available records at the site, or at comparable locations, and generation of sample
functions to match the characteristics of the model; and

(ii) construction of the physical models of the source, travel-path, local-site
conditions, and generation of ground motion at the site starting with an assumed
mechanism of energy release at the source and following its travel to the site.

In the preceding section we have discussed the stationary and nonstationary
models of earthquake ground acceleration. We shall first consider the simulation
of a single component.

7.4.1   Single Component Excitation

The simulation problem can be divided in two parts:

(i) choice of envelope function, $A(t)$, and estimation of its parameters; and

(ii) generation of stationary random process, $\ddot{Y}(t)$, with a specified psd or response
spectrum.

In Appendix 7A, we discuss the relationship between site intensity parameters, such as, peak ground acceleration, duration, the magnitude, and the focal distance. For a given site, seismicity and geometry of the potential sources can be used to establish the distribution of these parameters. The parameters of the envelope functions can be estimated for a confidence level, using these distributions.

The sample functions of stationary random process with a specified psd can be generated either by filtering white noise or by direct simulation through (2.107). We give below a sequence of steps which may be used to simulate an ensemble of uniformly modulated stationary random process to represent earthquake accelerograms:

1. Generate a segment of white noise, $W(t)$, with unit psd and a duration T' using the procedure described in Appendix 2A. The sampling interval should be chosen to approximate white noise within a given tolerance corresponding to the smallest period of interest.

2. Assume the white noise process to be a straight line segment on each time interval.

3. Pass each sample function through a second order filter defined by (7.3) to yield absolute acceleration response, $\ddot{Y}(t) = \ddot{U}(t) + W(t)$, having the psd given by (7.2). Since the excitation is assumed to be made up by a series of straight line segments, integration can be performed in a closed form and evaluated arithmetically (Nigam and Jennings, 1969).

4. Multiply the random process, $\ddot{Y}(t)$, by the envelope function $A(t)$, given by Eqs. (7.7–7.9) to impart the nonstationary characteristics.

5. Apply the parabolic base-line correction to filter out the long period components (Berg and Housner, 1961). Apply a linear correction to the first second, or half-second, of the accelerogram to remove the offset introduced by the parabolic correction at the beginning.

6. Pass each sample function through a second order filter having the transfer function

$$T\left(\frac{T}{T_f}\right) = \frac{1}{\left(1 + \left(\frac{T}{T_f}\right)^4\right)^{1/2}} \tag{7.17}$$

where T_f is the cut-off period. The filter (7.17) acts as a low-pass filter to match the characteristics of real earthquakes. Jennings et al (1969) have suggested $T_f = 7$ secs for A and B type, and $T_f = 2$ secs for C and D type, earthquakes.

7. Scale the sample functions by multiplying them with ratio of target and computed 20% spectrum intensity, r.m.s. value, or peak ground acceleration.

The above procedure gives modulated filtered white noise (MFWN) process. To simulate FMWN process, step 4 should precede step 3.

Response Spectrum Consistent Accelerograms

The response spectrum is widely used to represent the spectral properties of earthquake ground motion. Aseismic design codes in most countries require analysis and design of structures based on design spectra. Simulation of earthquake ground motion to fit a target response spectrum is, therefore, of considerable interest. A comprehensive review of methods for simulation of spectrum consistent accelerograms is available in the papers of Ahmadi (1979), Spanos (1983), and Permount (1984). A discussion on mathematical aspects of the problem, such as, existence and uniqueness is contained in the paper by Levy and Wilkinson (1976).

In Appendix 7B, we discuss the approximate and iterative methods to determine psd consistent with a target spectrum in a certain probabilistic sense. Knowing the psd, consistent with median or mean response spectra, or spectra with a specified confidence level, the corresponding random process can be simulated by using the methods discussed in the preceding section. The accelerograms consistent with a deterministic spectrum can be synthesized directly as (Scanlan and Sachs, 1974)

$$\ddot{Z}(t) = A(t) \sum_{i=1}^{N} Y_i \cos\left(\frac{2\pi i t}{T'} + \phi_i\right) \tag{7.18}$$

where $A(t)$ is a deterministic modulating function (7.7–7.9); T' is the preselected duration; and ϕ_i are random phase, uniformly distributed in the interval $[0, 2\pi]$. The amplitudes, Y_i, are initially selected as the ordinates of target spectrum for zero damping. Final values of amplitudes are selected through an iterative procedure which involves scaling of Y_i by the ratio of the target over synthesized spectrum for each $\omega = \omega_i$. Procedures for synthesis of accelerograms which are simultaneously compatible with a design spectrum and a peak acceleration value (Kast et al, 1978); or with spectra corresponding to two different damping ratios (Iyengar and Rao, 1979) have also been proposed.

Phase-difference Method

The nonstationary character of the ground acceleration can also be incorporated in the simulation by recognizing the relationship between the moments of the intensity function $(I(t) = A^2(t))$, and the probability density function of the phase derivative of the Fourier transform of a uniformly modulated white noise process (Nigam, 1982, 1984; Ohsaki, 1979; Ohsaki et al, 1979). Let $V(\omega)$ be the phase of the Fourier transform of the uniformly modulated random process ((7.1) with $A_i(t) = A(t)$—a deterministic function). It can be shown (Nigam, 1982) that the probability density function of $V'(\omega) = dV/d\omega$ is given by

$$p(v') = \frac{\rho^2/\gamma_0}{2[(v' - t_1)^2 + \rho^2/\gamma_0]^{3/2}}, \quad -\infty < v' < \infty \tag{7.19}$$

where

$$\rho = \gamma_2 - \frac{\gamma_1^2}{\gamma_0} \tag{7.20}$$

$$t_1 = \gamma_1/\gamma_0$$

$$\gamma_i = \frac{\phi_0}{4\pi} \int_0^{T'} t^i I(t)\, dt, \quad i = 0, 1, 2, \ldots \tag{7.21}$$

γ_i are called the intensity moments and depend on the shape of the envelope function $A(t)$, thus reflecting the gross nonstationary properties of a random process.

Since the earthquake ground acceleration can be adequately modelled by a modulated white noise process (Nigam, 1984), the distribution of phase derivative can be used to impart the nonstationary character to a simulated random process (Ohsaki, 1979; Ohsaki et al, 1979). Ohsaki has suggested a procedure called phase-differences method in which phase angles in (7.18) are generated from the relation

$$\phi_{i+1} = \phi_i + \Delta\phi_i; \ i = 0, 1, 2, \ldots$$

where $\Delta\phi_i$ are random variables with distribution given by (7.19). This eliminates the need to multiply the simulated stationary random process by the envelope function $A(t)$. Ohsaki (1979) has shown that 'phase-difference method' gives faster and uniform convergence as compared to conventional methods.

Simulation of Critical Excitation

In Appendix 4B, we have discussed the concept of design based on critical excitation. Due to a large measure of uncertainty inherent in earthquake ground motion at a site, this concept is particularly suited for aseismic design (Drenick, 1970, 1973, and 1977). The procedure developed by Iyengar and Manohar (1985, 1987) yields the psd of the critical excitation consistent with the general characteristics of the ground motion (such as, variance, duration, envelope) that may be reasonably expected at a site. An ensemble of earthquake ground motions can be generated consistent with the critical psd by any one of the methods discussed in this section. We shall illustrate the simulation of critical excitation by an example.

Example 7.1:   Critical excitation for aseismic design at an earth dam
(Iyengar and Manohar, 1987)

Consider an earth dam of triangular cross-section shown in Fig. 7.3. The material properties at the dam are:
shear modulus $G = 1.92 \times 10^5$ kN/m^2; viscous damping coefficient $\zeta = 0.2$; and density, $\rho = 2.04 \times 10^4$ kN/m^3. The natural frequencies and mode shapes of dam based on a shear beam model are:

$$\omega_i = \left(\frac{Z_i}{l}\right)\left(\frac{G}{\rho}\right)^{1/2}$$

and

$$u_i(s) = \frac{2J_0(Z_i s/l)}{Z_i J_1(Z_i)} \tag{7.22}$$

where J_0 and J_1 are Bessel's functions of the first kind, and Z_i are the zeros of J_0.

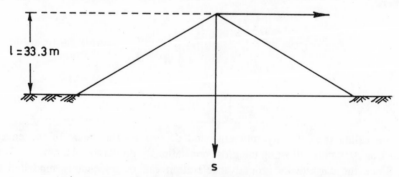

$l = 33.3$ m

s

Fig. 7.3. An earth dam.

The variance of the top displacement relative to the ground is

$$\sigma_x^2(0, t) = \int_0^\infty G(\omega) H(0, \omega, t) \, d\omega \tag{7.23}$$

with

$$H(s, \omega, t) = \int_0^t \int_0^t \sum_i \sum_j u_i(s) u_j(s) A(\tau_1) A(\tau_2) h_i(t - \tau_1) h_j(t - \tau_2)$$
$$\times \cos \omega (\tau_2 - \tau_1) \, d\tau_1 d\tau_2 \tag{7.24}$$

In (7.24), $A(t)$ is the envelope function given by

$$A(t) = \exp(-0.13t) - \exp(-0.45t)$$

The duration of the excitation, $T = 30$ sec; and the average r.m.s. of the ground acceleration is specified to be 0.1 g as in example 4B.1. The critical psd is determined to maximize the variance (7.23), subject to the constraint (4B.32). Seven exponential functions (4B.33) with $p = 0.032$ are used in the series expansion (4B.24). The response variance is scanned over the 0–30 sec interval to determine the highest value, which occurs at $t_c = 4$ sec and is equal to 8.8 cm². The psd function corresponding to the critical excitation is shown in Fig. 7.4a and a sample function consistent with this psd is shown in Fig. 7.4b. It may be remarked that the sample function has the appearance of a typical strong motion earthquake ground acceleration on a firm ground.

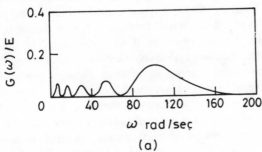

(a)

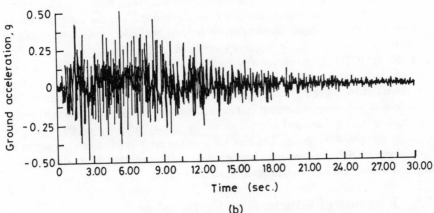

(b)

Fig. 7.4. (a) Power spectral density, and (b) a sample function of the critical excitation.

The critical excitations developed by the above procedure are by definition dependent on the system and the choice of response quantity to be maximized. Thus, at the same site this procedure leads to different critical excitations for different systems. Similarly, for the same system, different critical excitations are obtained depending on the choice of the response quantity to be maximized. A better alternative is to determine the critical excitation that is site-dependent, but is independent of the system or the response quantity. This may be done by constructing a suitable soil deposit model for the site, and then determine the critical excitation that must be applied at the base rock level to maximize the site response at the surface level (Iyengar and Manohar, 1987).

7.4.2   Multi-component Excitations

We have so far discussed the simulation of a single component of ground acceleration. In the preceding section, we have shown that ground acceleration during an earthquake can be represented by three uncorrelated components along principal directions having maximum, intermediate and minimum variance respectively. The three components along principal directions can be generated as independent sample functions consistent with specified response spectra or psd. Watabe and Tohodo (1978) have proposed the simulation of the three components using the 'phase-wave' of recorded earthquakes along principal directions.

7.4.3   Simulation Based on Physical Models of Sources, Travel-path and Site

The simulation of earthquake ground motion based on physical models of sources, travel path and local site conditions starts with an assumed source mechanism. Housner (1955) was the first to suggest a simple source mechanism in the form of a stress dislocation, and simulate earthquake ground motion by superposition of randomly arriving double sine-pulses. Rascon and Cornell (1968, 1969) developed a stochastic model based on physical arguments to simulate strong-motion earthquake records on firm ground at moderate to large distances from a fault. The model combines the available information about known physical features with analytical and statistical analysis of the parameters of the source and recorded ground motions. The simulation starts with a set of double-couple source mechanisms along a line and incorporates the effects of spherical spreading, multiple reflection and refraction and attenuation of P and S waves. The P and S waves arriving from each elementary source are modelled as a nonstationary Poisson process. The resulting process is non-homogeneous-multiple-filtered Poisson process and incorporates time variation of both the intensity and the frequency contents of the motion.

Trifunac (1971) has developed a method for synthesis of accelerograms based on the knowledge of temporal and spatial properties of source mechanism, incorporating the effects of local geology through group-velocity dispersion data at a given site. The simulation procedure assumes that most of the near-field strong-motion is caused by shallow earthquakes and is represented by energy propagating through low-velocity surface-wave guide. The method is based on a detailed interpretation of several recorded strong-motion and incorporates the properties of the local geological site and source mechanism whenever these are known.

7.5   Response of Structures to Earthquakes

All structures resting on the ground, or connected to ground-based structures, are subjected to random excitation during earthquakes. The response of such structures to earthquakes is, therefore, a random process. A reliability based design, in the framework of random vibration analysis, provides a rational and consistent basis for aseismic design of structures. To implement such a design procedure, it is necessary to establish the seismic risk at the site, construct a random process model of the ground motion, and carry out a random vibration analysis of the structure. Further, it is necessary to identify the response parameters of interest, the constraints on these parameters to maintain the integrity of the system, and the distribution of 'failure-events' or damage to the structure. In this section, we shall discuss the random vibration analysis of structures for elastic and inelastic behaviour for a specified ground motion intensity.

We have noted that the ground acceleration at a point on the earth's surface during an earthquake is a random vector, which may be modelled by three non-stationary random processes representing the ground motion along two orthogonal horizontal directions and a vertical direction. We have also noted that along principal directions, these random processes are uncorrelated. A structure resting on the ground is, therefore, subjected to simultaneous action of three components at each point of contact. In earthquake engineering, it is generally assumed that same motion takes place at all points of contact, implying that foundation soil is rigid and the excitation is completely defined by a single random vector. If the base dimensions of the structure are small as compared to the wavelength of the ground motion, this assumption does not result in significant error. However, for long structures, such as, dams, bridges, pipelines, it is necessary to assume multiple support excitations (Clough and Penzien, 1975). We shall confine our discussion to single support excitation.

We assume that the nonstationary nature of each component of ground acceleration is adequately modelled by a uniformly modulated stationary random process (7.1). A segment of stationary random process is a special case of such a model with a box-car function as the modulating function. We assume that the ground acceleration is Gaussian and each component is partially described by one, or more, of the parameters, such as, peak ground acceleration, r.m.s. value, psd, response spectra.

The civil engineering structures, such as, buildings, chimneys, nuclear power plants, dams, bridges etc. cover a wide range of shapes and forms. Some structures, such as, dams, and water-tanks involve special effects, such as, fluid-structure interaction. Almost all structures resting on ground involve soil-structure interaction, which causes the free-field ground motion to be modified. In all cases, an analytical treatment of random vibration requires an idealized model of the structure and interacting media. We shall confine our discussion to discrete models exhibiting linear and nonlinear behaviour. Extension of the results to continuous models is generally straight forward, if normal mode method can be used.

The general theory of random vibration of discrete and continuous systems, both for linear and nonlinear behaviour, is covered in the texts on random vibration (Lin, 1967; Nigam, 1983). We shall confine our discussion to special aspects associated with earthquake excitations. Our primary objective will be to determine the design values of response parameters of interest for specified reliability and ground motion intensity.

7.5.1 Discrete linear elastic systems

A single-degree-of-freedom (sdf) system is the simplest, and the most important of discrete linear systems. A large class of structures can be adequately modelled by a sdf system, and it is well known that analysis of mdf and continuous systems can, in most cases, be reduced to the analysis of a set of sdf systems. We shall first consider the response of a sdf system to a single component of ground acceleration. The equation of motion can be expressed as

$$\ddot{X} + 2\zeta\omega_0\dot{X} + \omega_0^2 X = -m\ddot{Z}(t) \tag{7.25}$$

where $X(t)$ is the relative displacement, and $\ddot{Z}(t)$ the ground acceleration during an earthquake. Since the system is at rest when the earthquake occurs, $X(0) = \dot{X}(0) = 0$.

The displacement of a sdf system to a typical earthquake is shown in Fig. 7.5. From the figure, and the general theory of the response of sdf systems to random excitation (Nigam, 1983), we conclude that for the range of damping values of interest ($0.1 \leq \zeta \leq 0.1$), the response is a narrow-band, nonstationary random process. The nonstationary character is due to sudden start, and also due to nonstationary nature of ground acceleration. However, if the ground acceleration is modelled as a segment of stationary random process, the response tends to become stationary for large times, $t \gg 1/\zeta\omega_0$ (Caughey and Stumpf, 1961). For example, if $\zeta = 0.1$ the response becomes stationary within 2–3 natural periods ($T_0 = 2\pi/\omega_0$), whereas for $\zeta = 0.025$, stationarity is attained after about 10 natural periods. The envelope of the response is a slowly varying function of time, and the nonstationary spectral characteristics can be closely modelled by the evolutionary psd.

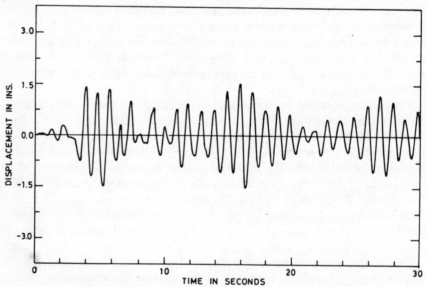

Fig. 7.5. Displacement response of a linear sdf system to earthquake excitation (Nigam, 1970).

The second-order statistics and other properties of the response, such as, spectral moments, level crossings, peaks and the maximum value in a specified period, for the sdf system can be obtained from random vibration theory (Nigam, 1983). We shall use these statistics to determine the distributions of the maximum values of the response parameters, such as, the displacement SD (ζ, ω_0), the velocity SV (ζ, ω_0) and the absolute acceleration SA(ζ, ω_0) called the *response spectra*. In Appendix 7B, we define the concept of response spectra and discuss the relationship among them, and ground motion intensity parameters. We show that the maximum forces acting on a structure, and therefore, the member forces can be derived in terms of response spectra.

Since the ground motion is treated as a random process, the response spectra, SD, SV and SA, are random variables and we are interested in their values for specified reliability levels. This is a classical problem in random process theory and is closely related to D-type first-passage problem. The exact solution of this problem is not available, but several approximate solutions have been attempted which yield results of practical interest (Crandall, 1970; Lutes, Chen and Tzuang, 1980; Crandall and Zhu, 1983).

Let the displacement spectrum

$$SD(\zeta, \omega_0) = X_m(T) = \max_{0 \leq t \leq T} (|X(t)|) \tag{7.26}$$

The distribution function of SD can be expressed as:

$$F_{SD}(\alpha) = P[SD \leq \alpha] = P[T_f > T] = p, \tag{7.27}$$

where T_f is the first-passage time of D-type barrier $|X(t)| = \alpha$; p is the probability that the level will not be exceeded, and $T = T' + T_0/2$ (Appendix 7B).

Stationary Ground Acceleration
The ground acceleration during an earthquake is nonstationary. However, stationary models, which are analytically much simpler to treat, give acceptable results for linear systems under fairly general conditions (Byrcroft, 1960; Rosenblueth and Bustamente, 1962). Further, the results for empirical models can be used to take into account the nonstationary effects by empirically modifying a few parameters (Vanmarcke, 1977; Kiureghian, 1980).

Stationary Response
Let $\ddot{Z}(t)$, $0 \leq t \leq T$ be a segment of stationary random process with psd, $\Phi_{\ddot{z}\ddot{z}}(\omega)$. The response of sdf system to $\ddot{Z}(t)$ will be initially nonstationary, but will tend to become stationary for $t \gg 1/\zeta\omega_0$. We first consider the case when the response $X(t)$ is stationary. For this case, (7.27) can be expressed as (Crandall, 1970).

$$F_{SD}(\alpha) = L_{T_f}(T) = L_0 e^{-\nu(\alpha)T} = p \tag{7.28}$$

where $\nu(\alpha)$ is called the limiting decay rate, and L_0 is the probability of survival at $t = 0$. Since the system starts from rest, $L_0 = 1$. Let $\eta = \alpha/\sigma_x$, and

$$SD_{T;p} = \eta_{T;p} \sigma_x \tag{7.29}$$

$$\overline{SD} = E[SD] = \overline{\eta} \, \sigma_x \tag{7.30}$$

$$\widetilde{SD} = \text{Var}[SD] = \tilde{\eta} \, \sigma_x^2 \tag{7.31}$$

where $SD_{T;p}$ is obtained by solving (7.28) for α and a specified p. $\eta_{T;p}$ is the peak factor for the response spectra, and $\bar{\eta}$ and $\tilde{\eta}$ are the peak factors for its mean and variance. σ_x is the standard deviation of the displacement. For stationary response (8.28*)

$$\sigma_X^2 = \pi \Phi_{\ddot{z}\ddot{z}}(\omega_0)/2\zeta\omega_0^3 \tag{7.32}$$

The distribution of displacement spectra given by (7.28) depends on the choice of limiting decay rate v. We shall consider two cases corresponding to independent X-crossing, and two-state Markov crossing assumptions. Expressions for peak factor $\eta_{T;p}$; mean and variance of the peak factor represented respectively by $\bar{\eta}, \tilde{\eta}$, are given below for the two cases:

(i) Independent X-crossings (Section 6.7.1*)

$$v = 2N_0 \exp\left(- \eta^2/2\right) \tag{7.33}$$

$$F_{SD}(\eta) = \exp\left(- 2N_0 \, T \exp\left(- \eta^2/2\right)\right) \tag{7.34}$$

$$\eta_{T;p} = 2[\ln\left(- 2N_0T/\ln\left(p\right)\right)]^{1/2} \tag{7.35}$$

$$= C_1 - \frac{\ln\left(- \ln\left(p\right)\right)}{C_1} \tag{7.36}$$

$$\bar{\eta} = C \tag{7.37}$$

$$\tilde{\eta} = \frac{\pi^2}{6C_1^2} \tag{7.38}$$

where

$$N_0 = \frac{\Omega}{2\pi} = \frac{1}{2\pi}\left(\frac{\lambda_2}{\lambda_0}\right)^{1/2} = \frac{\omega_0}{2\pi} \tag{7.39}$$

$$\lambda_i = \int_0^\infty \omega^i G_{XX}(\omega)d, \quad i = 0, 1, 2, \ldots \tag{7.40}$$

$$C_1 = [\ln\left(2N_0T\right)]^{1/2} \tag{7.41}$$

$$C = C_1 + \frac{0.5772}{C_1} \tag{7.42}$$

(ii) Two-state Markov crossings (Vanmarcke, 1969; 1977)

$$v = 2N_0 \frac{1 - \exp\left(\pi/2\right)^{1/2} \delta_e \eta)}{\exp\left(\eta^2/2\right) - 1} \tag{7.43}$$

where

$$\delta = (1 - \lambda_1^2/\lambda_0\lambda_2)^{1/2} = k/\sqrt{2\pi} = \frac{2}{\sqrt{\pi}} \zeta^{1/2} \tag{7.44}$$

and

$$\delta_e = \delta^{1+b}; b \simeq 0.2 \tag{7.45}$$

$$F_{SD}(\eta) = \exp\left[- 2N_0T \frac{1 - \exp\left(-\frac{\pi}{2} \delta_e \eta\right)}{\exp\left(\eta^2/2\right) - 1}\right] \tag{7.46}$$

$$\eta_{T;p} = [2\ln\{-2N_0T/\ln p\,[1 - \exp(1 - \delta_e(\pi\ln(-2N_0T/\ln p)^{1/2}))]\}]^{1/2} \quad (7.47)$$

$$\bar{\eta} \approx (2\ln(2\text{Noe }T))^{1/2} + \frac{0.5772}{(2\ln(2\text{Noe }T))^{1/2}}, \quad \text{(Kiureghian, 1980; Smeby and}$$

$$\text{Kiureghian, 1985)} \quad (7.48)$$

$$\tilde{\eta} = \left[\frac{1.2}{(2\ln 2\text{Noe }T)^{1/2}} - \frac{5.4}{13 + (2\ln 2\text{Noe }T)^{3/2}}\right]^2 \quad (7.49)$$

$$\text{Noe} = \begin{cases} \max(1.05,\, 2\delta N_0 T); & 0.0 < \delta \le 0.1 \\ (1.63\,\delta^{0.45} - 0.38)\,N_0 T; & 0.1 < \delta \le 0.69 \\ N_0 T; & 0.69 < \delta < 1.00 \end{cases} \quad (7.50)$$

Note that $\bar{\eta}\sigma_x$ gives the average response spectra and $\eta_{T;0.5}\,\sigma_x$ its median value.

A detailed discussion of the independent X-crossings, and two-state Markov crossing assumptions is given in Vanmarcke (1969) and Nigam (1983). The distribution of response spectra based on independent X-crossings, (7.34), are asymptotically exact for high reliability (high α-level), but are overly conservative for intermediate and low reliability. Two-state Markov crossing assumption (7.46), which takes into account the effect of clumping, gives relatively accurate results at all levels. Several other approximations are available in the literature (Crandall, 1970; Iwan and Mason, 1980), including some empirical expressions (Shinozuka and Yang, 1971; Lutes, Chen and Tzuang, 1980).

Nonstationary Response
The response of a sdf system to a segment of stationary random process is initially nonstationary. In the preceding treatment we neglected the nonstationary behaviour of initial response. Corotis, Vanmarcke and Cornell (1972) have shown that this assumption may lead to overly conservative estimates of design values for intermediate and low reliability levels. The nonstationary behaviour can be easily incorporated in the analysis by taking into account the time-dependent nature of response statistics and a time-dependent limiting decay rate $v(t)$. The distribution function of the displacement spectra can be expressed as (6.183*)

$$F_{\text{SD}}(\alpha, T) = L_{T_f}(T) = L_0 \exp\left(\int_0^T v(t)\, dt\right) \quad (7.51)$$

Since the system starts from rest, $L_0 = 1$.

The limiting decay rate, $v(t)$, is given by the following relations for the independent X-crossings, and two-state Markov crossing assumptions:

(i) Independent X-crossings

$$v(t) = 2N_0(t) \exp\left[-\frac{1}{2}\left(\frac{\alpha}{\sigma_x(t)}\right)^2\right]^2 \quad (7.52)$$

(ii) Two-state Markov crossings

$$v(t) = 2N_0(t) \frac{1 - \exp(-(-\pi/2)^{1/2}\delta_e\alpha/\sigma_x(t))}{\exp[1/2\,(\alpha/\sigma_x(t))^2] - 1} \quad (7.53)$$

where $\sigma_x(t)$ is time-dependent standard deviation of the displacement given by (8.80*).

$$\sigma_x^2(t) = \frac{\pi \Phi_{\ddot{Z}\ddot{Z}}(\omega_0)}{2\zeta \omega_0^3} \left\{ 1 - \frac{e^{-2\zeta\omega_0 t}}{\omega_d^2} [\omega_d^2 + 2(\zeta\omega_0 \sin \omega_d t)^2 + \zeta\omega_0\omega_d \sin 2\omega_d t] \right\}$$

(7.54)

$$\simeq \frac{\pi \Phi_{\ddot{Z}\ddot{Z}}(\omega_0)}{\zeta\omega_0^3} (1 - e^{2\zeta\omega_0 t}) = \frac{\pi \Phi_{\ddot{Z}\ddot{Z}}(\omega_0)}{2\zeta_t \omega_0^3}, \text{ if } \zeta \ll 1$$

(7.55)

where

$$\zeta_t = \zeta (1 - e^{-2\zeta\omega_0 t})^{-1}$$

(7.56)

$$N_0(t) = \frac{1}{2\pi} \left(\frac{\lambda_2(t)}{\lambda_0(t)} \right)^{1/2} = \frac{\Omega(t)}{2\pi}$$

(7.57)

$$\delta(t) = \left(1 - \frac{\lambda_1^2(t)}{\lambda_0(t) \lambda_2(t)} \right)^{1/2}$$

(7.58)

and $\lambda_j(t)$, $j = 0, 1, 2$ and time-dependent spectral moments are given by

$$\lambda_j(t) = \int_0^\infty \omega^j G_{XX}(\omega, t) \, d\omega$$

(7.59)

where

$$G_{XX}(\omega, t) = |\tilde{H}(\omega, t)|^2 G_{\ddot{Z}\ddot{Z}}(\omega)$$

(7.60)

is the evolutionary spectral density of the displacement. Exact and approximate ($\zeta \ll 1$) expressions for $\lambda_j(t)$, $j = 0, 1, 2$ are available in Corotis, Vanmarcke, Cornell (1972), and (8.81*-8.87*).

The displacement spectrum for reliability level p-$SD_{T;p}$ can be obtained by solving for α in the equation

$$F_{SD}(\alpha, T) = \exp \left[-\int_0^T v(t) \, dt \right] = p$$

(7.61)

where $v(t)$ is given by (7.52) or (7.53).

A solution of (7.61) to determine $SD_{T;p}$ requires a numerical procedure. Vanmarcke (1977) has proposed an approximate procedure based on 'equivalent stationary response' which permits a closed form solution of (7.61) to a reasonable accuracy. Since the system starts from rest, $\sigma_x(t)$ increases with time from 0 to $\sigma_x(T)$, and the decay rate $v(t)$ increases more rapidly. The major contribution to the integral in (7.61), therefore, comes from values of time t close to T. Hence, we can define an 'equivalent stationary response duration' T_e such that

$$\int_0^T v(t) \, dt = T_e \, v(T)$$

(7.62)

where $T_e \leq T$. The ratio T_e/T can be obtained from the relation (Vanmarcke, 1977)

$$\frac{T_e}{T} = \exp\left[-2\left(\frac{\sigma_x^2(T)}{\sigma_x^2(T/2)}\right) - 1\right] \qquad (7.63)$$

which is obtained by approximating the area under $v(t)$ from 0 to T, in terms of $v(T)$ and $v(T/2)$.

The distribution of displacement spectra, the peak factor $\eta_{T;\,p}$, its mean and variance can now be obtained from (7.33) to (7.50) by replacing T by T_e and σ_x by $\sigma_x(T)$. Simulation studies (Vanmarcke, 1977) show that this procedure yields sufficiently accurate response spectra.

A Markov process formulation provides an alternative method for approximate determination of the distribution of response spectra (Rosenblueth, Bustamante, 1962; Caughey and Gray, 1963; Roberts, 1976). The response of a sdf system to white noise is a second-order Markov process. While the problem of determining the distribution of the maximum response can be formulated, its solution is very difficult (Crandall, 1970, Nigam, 1983). However, if we define a random process

$$R(t) = [(\omega_d X(t))^2 + (\zeta\omega_0 X(t) + \dot{X}(t))^2]^{1/2} \qquad (7.64)$$

it can be shown that under conditions approximately satisfied by earthquakes, $R(t)$ is a first-order Markov process. The distribution of $R(t)$ can be obtained (Rosenblueth and Bustamante, 1962; Gray, 1966; Nigam, 1983). Rosenblueth and Bustamante (1962) have presented the results which can be readily used for reliability based design.

From (7.64), it is clear that, if $\zeta \ll 1$, $\dot{X}(t)$ will be very small when $X(t)$ attains a maximum value. Hence

$$R(\zeta, \omega_0) = \max_{0 \le t \le T} [R(t)] = \omega_0 \max [|X(t)|]$$

$$= \omega_0 \mathrm{SD}(\zeta, \omega_0) = \mathrm{SV}(\zeta, \omega_0) \qquad (7.65)$$

Thus, distribution of $R(t)$, can be used to represent distribution of response spectra.

Nonstationary Ground Acceleration
The nonstationary character of earthquake ground acceleration can be modelled by a random process with evolutionary spectra. Let the evolutionary psd of the ground acceleration be expressed as

$$\Phi_{\ddot{Z}\ddot{Z}}(\omega, t) = |A(\omega, t)|^2 \, \Phi(\omega) \qquad (7.66)$$

where $A(\omega, t)$ is the modulating function and $\Phi(\omega)$ is the psd of an associated broad-band random process. A uniformly modulated, or a separable, random process, with $A(\omega, t) = A(t)$, is a special case of the evolutionary process.

The response of a sdf system to an excitation having evolutionary psd $\Phi_{\ddot{Z}\ddot{Z}}(\omega, t)$ given by (7.66), is also a nonstationary random process with evolutionary psd.

$$\Phi_{XX}(\omega, t) = |M(\omega, t)|^2 \, \Phi(\omega) \qquad (7.67)$$

where

$$M(\omega, t) = \int_0^t h(t - \tau) A(\tau, \omega) e^{-i\omega(t-\tau)} d\tau \tag{7.68}$$

and $h(t)$ is the impulse response function of the sdf system.

The psd of the response (7.67) to nonstationary excitation is of the same form as the psd of the nonstationary response to stationary excitation treated earlier (7.60). The distribution of the response spectra and associated response statistics can therefore be obtained as before. Markov property of the response of a sdf system to evolutionary excitation (Spanos, 1980; Spanos and Lutes, 1980) provides an alternative method to obtain the distribution of the response spectra.

Multi-degree-of-freedom (mdf) Systems

A large class of structural/mechanical systems can be modelled as mdf systems. The equations of motion of mdf systems excited by the three components of ground motion during an earthquake can be expressed as:

$$\mathbf{M} \ddot{\bar{X}} + \mathbf{C} \dot{\bar{X}} + \mathbf{K} \bar{X} = \bar{F}(t) = -\mathbf{M} \mathbf{I} \ddot{\bar{Z}} \tag{7.69}$$

where $\mathbf{M}$, $\mathbf{C}$ and $\mathbf{K}$ are mass, damping and stiffness matrices respectively; $\bar{X}(t)$ is the relative displacement vector; $\bar{Z} = (Z_1 Z_2 Z_3)^T$ is the vector of ground displacement components, in which Z_1 and Z_2 are horizontal and Z_3 is the vertical component; and $\mathbf{I}$ is the influence matrix such that its kth column I_k couples the degrees of freedom of the structure to ground motion component $Z_k(t)$.

We first consider the case for which $\mathbf{M}$, $\mathbf{C}$ and $\mathbf{K}$ satisfy the conditions for the existence of classical normal modes (Nigam, 1983). Let $\mathbf{U} = [\bar{u}^1, \bar{u}^2, ..., \bar{u}^n]^T$ be the modal column matrix and the vector of normal coordinates. Setting

$$\bar{X} = \mathbf{U} \bar{Y} \tag{7.70}$$

the jth uncoupled model equation is given by

$$\ddot{Y}_j + 2\zeta_j \omega_i \dot{Y}_j + \omega_j^2 Y_j = \bar{\Gamma}_j^T \ddot{\bar{Z}} \tag{7.71}$$

where ζ_j and ω_j are the damping and natural frequency in the jth mode; $\bar{\Gamma}_j = [\Gamma_j^{(1)} \Gamma^{(2)} \Gamma^{(3)}]^T$ is the vector of participation factors for mode j with elements

$$\Gamma_j^{(k)} = \frac{(\bar{u}^j)^T M \bar{I}_k}{(\bar{u}^j)^T M \bar{u}^j}, \quad k = 1, 2, 3 \tag{7.72}$$

The design of a system involves response quantities such as displacement, member force, stresses, which can be expressed as a linear combination of nodal displacements. Let $Q(t)$ be a response quantity of interest, and let

$$Q(t) = \bar{C}^T \bar{X} = \sum_{i=1}^n C_i X_i \tag{7.73}$$

where $\bar{C}^T$ is a constant vector. Then following the standard procedure (Nigam, 1983), the evolutionary psd of $Q(t)$ can be expressed as

$$G_{QQ}(\omega, t) = \sum_k \sum_l \sum_i \sum_j L_i^{(k)} L_j^{(l)} \tilde{H}_i(\omega, t) \tilde{H}_j(\omega, t) G_{\ddot{Z}_k \ddot{Z}_l}(\omega, t) \tag{7.74}$$

where $L_i^{(k)} = \Gamma_i^{(k)} \bar{C}^T \bar{u}^i$, is the effective participation factor associated with mode i; $\tilde{H}_i(\omega, t)$ is the frequency response function for mode i; and $G_{\ddot{Z}_k \ddot{Z}_l}(\omega, t)$ is k, l element of the psd matrix, $\mathbf{G}_{\ddot{Z}\ddot{Z}}(\omega, t)$. Note that ground motion components are modelled as evolutionary random processes. Under the assumptions stated below, (7.74) reduces to:

(a) $\ddot{Z}(t)$ is modelled as a stationary random vector

$$G_{QQ}(\omega, t) = \sum_k \sum_l \sum_i \sum_j L_i^{(k)} L_j^{(l)} \tilde{H}_i(\omega, t) \tilde{H}_j^*(\omega, t) G_{\ddot{Z}_k \ddot{Z}_l}(\omega) \qquad (7.75)$$

As $t \to \infty$, $Q(t)$ becomes stationary, and

$$G_{QQ}(\omega) = \sum_k \sum_l \sum_i \sum_j L_i^{(k)} L_j^{(l)} \tilde{H}_i(\omega) \tilde{H}_j^*(\omega) G_{\ddot{Z}_k \ddot{Z}_l}(\omega) \qquad (7.76)$$

(b) $Z(t)$ are uncorrelated (principal directions)

$$G_{QQ}(\omega) = \sum_k \sum_i \sum_j L_i^{(k)} L_j^{(k)} \tilde{H}_i(\omega, t) \tilde{H}_j^*(\omega, t) G_{\ddot{Z}_k \ddot{Z}_k}(\omega, t) \qquad (7.77)$$

Response to Single Earthquake Component
For a single earthquake component, $Z(t)$, (7.74) reduces to

$$G_{QQ}(\omega, t) = G_{\ddot{Z}\ddot{Z}}(\omega, t) \sum_i \sum_j L_i L_j \tilde{H}_i(\omega, t) \tilde{H}_j^*(\omega, t) \qquad (7.78)$$

If $\ddot{Z}(t)$ is modelled as a stationary random process, then

$$G_{QQ}(\omega, t) = G_{\ddot{Z}\ddot{Z}}(\omega) \sum_i \sum_j L_i L_j \tilde{H}_i(\omega, t) \tilde{H}_j^*(\omega, t) \qquad (7.79)$$

and the spectral moments of $Q(t)$ are

$$\lambda_{m,Q}(t) = \int_0^\infty \omega^m G_{QQ}(\omega, t) \, d\omega \qquad (7.80)$$

$$= \sum_i \sum_j L_i L_j \int_0^\infty \omega^m \tilde{H}_i(\omega, t) \tilde{H}_j^*(\omega, t) G_{\ddot{Z}\ddot{Z}}(\omega) \, d\omega \qquad (7.81)$$

Vanmarcke (1972) has shown that significant contributions to the spectral moments in (7.81), usually come from the terms for which $i = j$, particularly when modal frequencies are well separated and damping is low. Following his treatment, (7.81) can be expressed as

$$G_{QQ}(\omega, t) \simeq G_{\ddot{Z}\ddot{Z}}(\omega) \sum_i |\tilde{H}_i(\omega, t)|^2 \left(L_i^2 + \sum_{i \neq j} L_i L_j A_{ijt}\right) \qquad (7.82)$$

in which A_{ijt} is a function of frequency ratio $r = \omega_j/\omega_i$ and the equivalent damping values ζ_{it}, ζ_{jt}, and is given by

$$A_{ijt} = \frac{8r \zeta_{it}(\zeta_{jt} + \zeta_{it})[(1 - r^2)^2 - 4r(\zeta_{it} - \zeta_{jt} r)(\zeta_{jt} - \zeta_{it} r)]}{8r^2[(\zeta_{it}^2 + \zeta_{jt}^2)(1 - r^2)^2 - 2(\zeta_{jt}^2 - \zeta_{it}^2 r^2)(\zeta_{it}^2 - \zeta_{jt}^2 r^2)] + (1 - r^2)^4}$$

$$\qquad (7.83)$$

A_{ijt} is plotted in Fig. 7.6 as a function of r for different pairs of ζ_{jt} and ζ_{it}. It is

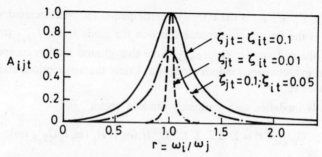

Fig. 7.6. Factors A_{ijt} showing interaction between mode i and j of a mdf system (Vanmarcke, 1977).

seen that A_{ijt} vanishes when r is either very small or very large, and is sharply peaked for small values of damping at $r = 1$.

Substitution of (7.82) in (7.80) and integration over all frequencies gives

$$\lambda_{m,Q}(t) = \sum_{i=1}^{n} \alpha_{it} L_i^2 \lambda_{m,ii}(t) \tag{7.84}$$

where

$$\alpha_{it} = 1 + \sum_{i \neq j} (L_j/L_i) A_{ijt} \tag{7.85}$$

and $\lambda_{m,ii}(t)$ is the mth spectral moment of modal coordinate $Y_i(t)$. For $m = 0$,

$$\sigma_Q^2(t) = \lambda_{0,Q}(t) = \sum_i \alpha_{it} L_i^2 \sigma_{y_i}^2(t) \tag{7.86}$$

Clearly, the contribution of cross-terms ($i \neq j$) in (7.84) and (7.86) will be insignificant if $\alpha_{it} \simeq 1$, which happens when: (i) modal frequencies are well separated; (ii) damping is small; and (iii) time t is sufficiently large. Under these conditions:

$$\lambda_{m,Q}(t) \simeq \sum_i L_i^2 \lambda_{m,ii}(t) \tag{7.87}$$

The spectral parameters of $Q(t)$, that is, $\Omega_Q(t)$ and $\delta_Q(t)$ required for determining the characteristics, such as, level-crossing, peak, first-passage time, peak-response factor can be determined as follows. Equation (7.82) can be written as

$$G_{QQ}(\omega, t) = \sum G_{Q_iQ_i}(\omega, t) = \sum_i G_{\ddot{Z}\ddot{Z}}(\omega) \, |\tilde{H}_i(\omega, t)|^2 L_i^2 \alpha_{it} \tag{7.88}$$

Let

$$p_i = \frac{\displaystyle\int_{-\infty}^{\infty} G_{Q_iQ_i}(\omega, T) \, d\omega}{\displaystyle\int_{-\infty}^{\infty} G_{QQ}(\omega, T) \, d\omega} = \frac{\alpha_i T L_i^2 \sigma_{y_i}^2(T)}{\sigma_Q^2(T)} \tag{7.89}$$

and let Ω_i and $\delta_i(t)$ be the spectral parameters of $Q_i(t)$. Clearly, $\sum p_i = 1$, and it can be shown that (Vanmarcke, 1977)

$$\Omega_Q(t) = (\sum_i p_i \, \Omega_i^2)^{1/2} \tag{7.90}$$

$$\delta_Q(t) = [1 - \{\sum_i p_i (\Omega_i/\Omega_Q)(1 - \delta_i^2)^{1/2}\}^2]^{1/2} \tag{7.91}$$

The maximum value of $|Q(t)|$, $0 \le t \le T$, and associated probability p; its expected value and variance can be expressed as

$$Q_{m;T;p} = \eta_{T;p}\, \sigma_Q(T) \qquad (7.92)$$

$$E[Q_m] = \bar{\eta}\, \sigma_Q(T) \qquad (7.93)$$

$$\mathrm{Var}\,[Q_m] = \tilde{\eta}\, \sigma_Q^2(T) \qquad (7.94)$$

where $\eta_{T;p}$, $\bar{\eta}$ and $\tilde{\eta}$ are given by (7.35) to (7.50) for X-crossing and Markov crossing assumptions respectively.

Stationary Response

If the response is assumed to be stationary, the relations for the nonstationary response presented above get modified by setting

$$\tilde{H}(\omega, t) = H(\omega)\, e^{i\omega t}$$

$$\zeta_t = \zeta$$

$$G_{QQ}(\omega, t) = G_{QQ}(\omega)$$

$$\lambda_{m,Q}(t) = \lambda_{m,Q}, \text{ in particular } \sigma_Q(T) = \sigma_Q \qquad (7.95)$$

Response to Multi-component Ground Motion

In Section 7.3, we have seen that for the proposed ground motion model (Penzien and Watabe, 1975), the three components are uncorrelated in the principal directions during the strong-motion phase. Denoting the components in the principal directions by the vector $\bar{Z}(t)$, the ground motion components $\bar{Z}'(t)$ along arbitrary orthogonal directions can be expressed as

$$\bar{Z}' = A\bar{Z}$$

where A is a 3×3 transformation matrix. The psd of the ground acceleration processes for the two systems of axes are related through

$$G_{\ddot{\bar{Z}}'\ddot{\bar{Z}}'}(\omega) = A G_{\ddot{\bar{Z}}\ddot{\bar{Z}}}(\omega)\, A^T \qquad (7.96)$$

Note that $G_{\ddot{\bar{Z}}\ddot{\bar{Z}}}$ is diagonal. In most structures, the vertical axis is a principal axis and coincides with the minor principal axis of the ground motion. Hence,

$$A = \begin{vmatrix} \cos\theta & \sin\theta & 0 \\ -\sin\theta & \cos\theta & 0 \\ 0 & 0 & 1 \end{vmatrix} \qquad (7.97)$$

where θ is the angle of rotation between the two sets of horizontal axes (Fig. 7.7).

The instantaneous correlation coefficient between the horizontal components of ground motion for this model can be shown to be (Smeby, 1982)

$$\rho_{\ddot{z}_1 \ddot{z}_2} = \frac{-(1-\alpha)\sin 2\theta}{[(1+\alpha)^2 - (1-\alpha)^2 \cos 2\theta]^{1/2}} \qquad (7.98)$$

where α is the ratio of the mean-square intensities of the horizontal components

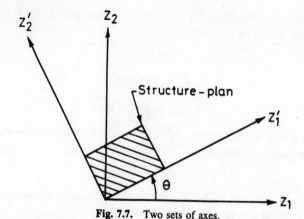

Fig. 7.7. Two sets of axes.

along the principal directions. It is seen that for $\alpha = 1$, correlation is zero for all values of θ. It is maximum for $\theta = \pi/4$ and increases with increasing difference between intensities of the principal components.

Substituting (7.96) and (7.97) in (7.74), the psd of the response quantity $Q(t)$ can be expressed as:

$$G_{QQ}(\omega) = \sum_k \sum_i \sum_j L_i^{(k)} L_j^{(k)} H_i(\omega) H_j^*(\omega) G_{\ddot{Z}_k \ddot{Z}_k}(\omega) - \sum_i \sum_k [L_i^{(1)} L_j^{(1)} - L_i^{(2)} L_j^{(2)}]$$

$$H_i(\omega) H_j^*(\omega) [G_{\ddot{Z}_1 \ddot{Z}_1}(\omega) - G_{\ddot{Z}_2 \ddot{Z}_2}(\omega)] \sin^2 \theta - \sum_i \sum_j [L_i^{(1)} L_j^{(2)} - L_i^{(2)} L_j^{(1)}]$$

$$H_i(\omega) H_j^*(\omega) [G_{\ddot{Z}_1 \ddot{Z}_1}(\omega) - G_{\ddot{Z}_2 \ddot{Z}_2}(\omega)] \sin \theta \cos \theta \tag{7.99}$$

The spectral moments of $Q(t)$

$$\lambda_m = \int_0^\infty \omega^m G_{QQ}(\omega) \, d\omega, \quad m = 0, 1, 2, \dots \tag{7.100}$$

can be expressed as

$$\lambda_m = \sum_k \sum_i \sum_j L_i^{(k)} L_i^{(k)} \lambda_{m,ij}^{(k)} - \sum_i \sum_j [L_i^{(1)} L_j^{(1)} - L_i^{(2)} L_j^{(2)}][\lambda_{m,ij}^{(1)} - \lambda_{m,ij}^{(2)}]$$

$$\times \sin^2 \theta - 2 \sum_i \sum_j L_i^{(1)} L_j^{(2)} [\lambda_{m,ij}^{(1)} - \lambda_{m,ij}^{(2)}] \sin \theta \cos \theta \tag{7.101}$$

where

$$\lambda_{m,ij}^{(k)} = \mathrm{Re}\left[\int_0^\infty \omega^m H_i(\omega) H_j^*(\omega) G_{\ddot{Z}_k \ddot{Z}_k}(\omega) \, d\omega \right] \tag{7.102}$$

The distribution of the maximum value and other statistical measures of response can now be obtained in terms of λ_0, λ_1 and λ_2, in the same way as for single component excitation. For broad-band excitation, generally applicable to earthquake ground motion, the expression for λ_m in (7.101) can be simplified by introducing the following approximations. Let

$$\lambda_{m,ij}^{(k)} = \rho_{m,ij}^{(k)} w_{m,ij}^{(k)} (\lambda_{0,ii}^{(k)} \lambda_{0,jj}^{(k)})^{1/2} \tag{7.103}$$

$$\rho_{m,ij}^{(k)} = \frac{\lambda_{m,ij}^{(k)}}{(\lambda_{m,ii}^{(k)} \lambda_{m,jj}^{(k)})^{1/2}} \qquad (7.104)$$

$$w_{m,ij}^{(k)} = \frac{(\lambda_{m,ii}^{(k)} \lambda_{m,jj}^{(k)})^{1/2}}{(\lambda_{0,ii}^{(k)} \lambda_{0,jj}^{(k)})^{1/2}} \qquad (7.105)$$

Following approximate expressions can be introduced in (7.101) for $\rho_{m,ij}$ and $w_{m,ij}$, $m = 0, 1, 2$. These expressions are based on response to white noise and are sufficiently accurate when significant modes of vibration are within the dominant frequencies of the broad-band input. Let

$$\omega_a = \frac{\omega_i + \omega_j}{2}; \quad \omega_b = \omega_i - \omega_j \qquad (7.106)$$

$$\zeta_a = \frac{\zeta_i + \zeta_j}{2}; \quad \zeta_b = \zeta_i - \zeta_j \qquad (7.107)$$

Assuming broad-band process in each principal direction, it can be shown that independent of principal directions (Kiureghian, 1980),

$$\rho_{0,ij} \simeq R_{ij}\left(\zeta_a + \frac{\zeta_b \omega_b}{4\omega_a}\right); \quad \rho_{1,ij} \simeq R_{ij}\left(\zeta_a - \frac{\omega_b^2}{2\pi\omega_a^2}\right)$$

$$\rho_{2,ij} \simeq R_{ij}\left(\zeta_a - \frac{\zeta_b \omega_b}{4\omega_a}\right)$$

$$w_{0,ij} \simeq 1; w_{1,ij} \simeq \left[\omega_i \omega_j\left(1 - \frac{2\zeta_i}{\pi}\right)\left(1 - \frac{2\zeta_j}{\pi}\right)\right]^{1/2}; w_{2,ij} \simeq \omega_i \omega_j \quad (7.108)$$

where $R_{ij} = 4\omega_a^2 \sqrt{(\zeta_i \zeta_j)/(\omega_b^2 + 4\omega_a^2 \zeta_a^2)}$. With these approximations (7.101) can be recast as

$$\lambda_m = \sum_k \sum_i \sum_j \rho_{m,ij} w_{m,ij} L_i^{(k)} L_j^{(k)} (\lambda_{0,ii}^{(k)} \lambda_{0,jj}^{(k)})^{1/2} - \sum_i \sum_j \rho_{m,ij} w_{m,ij}$$

$$- [L_i^{(1)} L_j^{(1)} - L_i^{(2)} L_j^{(2)}] [(\lambda_{0,ii}^{(1)} \lambda_{0,jj}^{(1)})^{1/2} - (\lambda_{0,ii}^{(2)} \lambda_{0,jj}^{(2)})]^{1/2} \sin^2 \theta$$

$$- 2 \sum_i \sum_j \rho_{m,ij} w_{m,ij} L_i^{(1)} L_j^{(2)} [(\lambda_{0,ii}^{(1)} L_{0,jj}^{(2)})^{1/2}$$

$$- (\lambda_{0,ii}^{(2)} \lambda_{0,jj}^{(2)})]^{1/2} \sin \theta \cos \theta \qquad (7.109)$$

The spectral moments λ_0, λ_1 and λ_2 of the response quantity, $Q(t)$, can therefore be obtained from the variance $\lambda_{0,ii}(t)$, of each modal response for the two earthquake components.

It is reasonable to assume that the two horizontal components of ground acceleration along principal directions have identical psd shapes. Let the ratio of the standard deviation of the weak to the strong component be denoted by

$$\gamma = \left(\frac{\lambda_{0,ii}^{(2)}}{\lambda_{0,ii}^{(1)}}\right)^{1/2} \qquad (7.110)$$

(7.109) can now be rewritten as

$$\lambda_m = \sum_k \sum_i \sum_j \rho_{m,ij}\, w_{m,ij}\, L_{(i)}^{(k)} L_j^{(k)} (\lambda_{0,ii}^{(k)} \lambda_{0,jj}^{(k)})^{1/2} - (1 - \gamma^2) \sum_i \sum_j \rho_{m,ij}\, w_{m,ij}$$

$$\times [L_i^{(1)} L^{(1)} - L_i^{(2)} L_j^{(2)}] (\lambda_{0,ii}\, \lambda_{0,jj})^{1/2} \sin^2 \theta$$

$$- 2(1 - \gamma^2) \sum_i \sum_j \rho_{m,ij}\, w_{m,ij}\, L_i^{(1)} L_j^{(2)} (\lambda_{0,ii}^{(1)} \lambda_{0,jj}^{(1)})^{1/2} \sin \theta \cos \theta \qquad (7.111)$$

In (7.111), the second and third terms vanish, if $\theta = 0$, that is the principal axes of the structure and the ground motion coincide, or if $\gamma = 1$, that is the ground motions in the two horizontal directions are of equal intensity. In the general case, when $\theta \neq 0$ and $\gamma < 1$, the second term represents the effect of rotation of the axes, and the third term represents the effect of correlation between the input components along the structure axes. It has been shown (Smeby, 1982) that contribution of these terms is usually small, except when γ is small. In practice, $\gamma = 1$ and, therefore, the second and third terms can be neglected. Further, if $L_i^{(1)} L_j^{(1)} + L_i^{(2)} L_j^{(2)} = 0$, for $i \neq j$, in the first term of (7.111), the cross-product can be dropped, irrespective of the closeness of the frequencies. This occurs when the model vectors $\bar{u}^i$ and $\bar{u}^j$ are situated in perpendicular planes (Smeby and Kiureghian, 1985).

We have seen that the response of a structure is a function of θ, the angle of rotation. In many design situations, the direction of principal axes may not be known and, therefore, value of θ is uncertain. In such cases, one of the two procedures may be adopted: (i) θ may be assumed to be random variable with a prescribed distribution; or (ii) the most critical value of θ may be determined for the response quantity of interest.

Let $p(\theta)$, $0 \leq \theta \leq 2\pi$ be the probability density function of θ, and $F_{Q_m|\theta}(\alpha | \theta)$ be the conditional distribution function of $Q_m = \max | Q(t) |$, $0 \leq t \leq T$. Then, the unconditional distribution of Q_m is given by

$$F_{Q_m}(\alpha) = \int_0^{2\pi} F_{Q_m|\theta}(\alpha | \theta)\, p(\theta)\, d\theta \qquad (7.112)$$

The critical value of θ will be different for each response quantity. For example, the critical value of θ for the standard deviation of $Q(t)$ can be obtained from (7.101), $(m = 0)$, by setting $d\lambda_0/d\theta = 0$. Hence,

$$\theta_{cr} = -\frac{1}{2} \tan^{-1} \frac{2 \sum_i \sum_j \rho_{0,ij}\, L_i^{(2)} L_j^{(2)} (\lambda_{0,ii}^{(1)} \lambda_{0,jj}^{(2)})^{1/2}}{\sum_i \sum_j \rho_{0,ij} [L_j^{(1)} L_j^{(1)} - L_i^{(2)} L_j^{(2)}] (\lambda_{0,ii}^{(1)} \lambda_{0,jj}^{(2)})^{1/2}} \qquad (7.113)$$

Note that $\omega_{0,ij} = 1$ (7.108) and θ_{cr} are independent of γ. For an indepth discussion of the effect of θ on the response to multi-component excitation, see Smeby and Kiureghian (1985).

Example 7.2: Space frame subjected to multi-component ground motion (Matsushima, 1974)

Consider a simple three-dimensional structure shown in Fig. 7.8 and subjected to three components of stationary ground acceleration. The expressions for the psd matrix are given in Table 7.2 and plotted in Fig. 7.9.

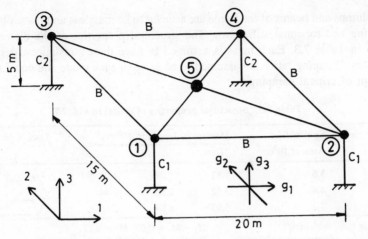

Fig. 7.8.   Three dimensional structural model (Matsushima, 1974).

Table 7.2   Mathematical expressions of spectra for strong motion accelerograms

Expressions of Spectra	Parameters
$G_{ij}(\omega) = \varepsilon_{ij} \sqrt{G_{ii}(\omega) \cdot G_{jj}(\omega)}$ $G_{ii}(\omega) = \dfrac{R_{ii}(0) \cdot \rho_i}{2\pi} \left\{ \dfrac{1}{(\omega - \omega_i)^2 + \rho_i^2} \right.$ $\left. + \dfrac{1}{(\omega + \omega_i)^2 + \rho_i^2} \right\}$ for $i, j = 1, 2, 3$	$\varepsilon_{ij} = \begin{cases} 1.0 & \text{for} \quad i = j \\ 0.6 & \text{for} \quad i \neq j \end{cases}$ $R_{ii}(0) = \begin{cases} 1.2 \times 10^4 \text{ gal}^2 & \text{for} \quad i = 1, 2 \\ 0.6 \times 10^4 \text{ gal}^2 & \text{for} \quad i = 3 \end{cases}$ $\omega_i = \begin{cases} 4\pi \text{ sec}^{-1} & \text{for} \quad i = 1, 2 \\ 8\pi \text{ sec}^{-1} & \text{for} \quad i = 3 \end{cases}$ $\rho_i = \begin{cases} 10 \text{ sec}^{-1} & \text{for} \quad i = 1, 2 \\ 15 \text{ sec}^{-1} & \text{for} \quad i = 3 \end{cases}$

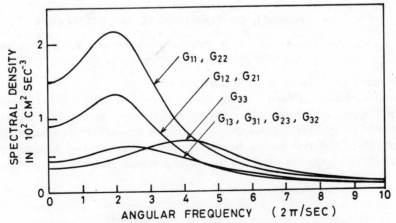

Fig. 7.9.   Components of ground acceleration spectral density matrix (Matsushima, 1974).

Columns and beams of the frame are asumed to be massless and posses bending, shearing and torsional stiffnesses. The structural properties of the elements are given in Table 7.3. Each mass is assumed to have three translational degrees of freedom. Damping ratios are assumed to be same in each mode and equal to three percent of critical damping.

Table 7.3   Structural properties of model in Fig. 7.7

Member	Sectional Area in $10^3 cm^2$	Moment of Inertia in 10^6 cm⁴			Local Coordinates
		I_x	I_y	I_z	
C_1	3.6	1.82	1.08	1.08	
C_2	4.8	3.12	2.56	1.44	
B	7.2	5.93	8.64	2.16	

Mass ton · sec² · cm⁻¹ $M_1 \sim M_4 = 0.03, M_5 = 0.12$
Modulus of elasticity ton · cm⁻²   Bending $E = 210$, Shear $G = 90$

The r.m.s. values of the absolute acceleration at A, B, C and D (directions shown in the inset) are plotted in Fig. 7.10 for one, two and three components acting simultaneously. It is seen that for a structure of the type considered, simultaneous action of the three components may have a significant effect on the response.

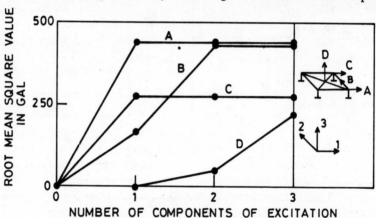

Fig. 7.10.   Response–r.m.s. value of absolute acceleration at A, B, C and D in the directions indicated (Matsushima, 1974).

Response Spectrum Method for mdf Systems
A representation of design ground motion through smooth response spectra is a standard practice in most countries. It is, therefore, important to express the response quantities in terms of response spectra. The results derived from the random vibration analysis of mdf systems can be readily expressed in terms of response spectral ordinates of respective modes, if the classical normal modes exist, by noting that:

$$\rho_{0,ii} = \left[\frac{\overline{SD}_i^{(k)}}{\bar{\eta}_i} \right]^2 \tag{7.114}$$

where $\overline{SD}_i^{(k)}$ is the mean response spectral value for the ith mode, corresponding to kth earthquake component, and $\bar{\eta}_i$ is the expected value of the peak factor for mode $- i$. Let $\bar{\eta}_Q$ be the mean peak factor for the response process $Q(t)$, then the mean peak value of the response $Q(t)$ can be expressed as ((7.86), (7.93))

$$E[Q_m^{(k)}] = \bar{\eta}_Q \left[\sum_i \alpha_{iT} L_i^2 \left(\frac{\overline{SD}_i^{(k)}}{\bar{\eta}_i} \right)^2 \right]^{1/2}, \; k\text{-fixed} \qquad (7.115)$$

which is an improved rule for combining modal maxima. Further, if it is assumed that $\bar{\eta}_Q / \bar{\eta}_i \approx 1$, that is, the peak factors of the total response and modal response are the same, (7.115) reduces to

$$E[Q_m^{(k)}] = [\sum_i \alpha_{iT} L_i^2 (\overline{SD}_i^{(k)})^2]^{1/2}, \; k\text{-fixed} \qquad (7.116)$$

If the cross-product terms in α_{iT} are neglected, which is justified for well-separated modal frequencies and small damping, (7.116) reduces to well known SRSS rule (Rosenblueth, 1951)

$$E[Q_m^{(k)}] = [\sum_i L_i^2 (\overline{SD}_i^{(k)}]^2, \; k\text{-fixed}$$

For multi-component excitation, (7.111) can be used to express the mean value of peak response,

$$E[Q_m] = \{ \sum_k \sum_i \sum_j \rho_{0,ij} L_i^{(k)} L_j^{(k)} \overline{SD}_i^{(k)} \overline{SD}_j^{(k)}$$

$$- (1 - \gamma^2) \sum_i \sum_j \rho_{0,ij} [L_i^{(1)} - L_j^{(1)} - L_i^{(2)} L_j^{(2)}] \overline{SD}_i^{(1)} \overline{SD}_j^{(1)} \sin^2 \theta \qquad (7.117)$$

$$- 2(1 - \gamma^2) \sum_i \sum_j \rho_{0,ij} L_i^{(1)} L_j^{(2)} \overline{SD}_i^{(1)} \overline{SD}_j^{(1)} \sin \theta \cos \theta \}^{1/2}$$

where we have assumed $\bar{\eta}/\tilde{\eta}_i = 1$ and $\gamma = (\overline{SD}_i^{(2)}/\overline{SD}_i^{(1)})$.

Equation (7.117) reduces to extended SRSS rule for multi-component excitation under conditions for which $\rho_{0,ii} = 1$ and $\rho_{0,ii} = 0$ if $i \neq j$. Smeby and Kiureghian (1985) have shown that for multi-component excitation SRSS rule gives good results even when modal frequencies are closely spaced, if: (i) input components are of equal intensity, and (ii) the modal vectors corresponding to closely spaced frequencies lie in perpendicular planes.

Example 7.3: Random vibration analysis of multi-storey flexible base buildings to earthquake excitation (Asthana and Datta, 1990)

We have discussed the methods for determining the peak response of linear mdf systems to random seismic excitations represented either by psd, or by mean response spectra. In this example, we shall compare the two methods with the 'exact' time domain analysis based on an ensemble of spectrum (psd) compatible simulated earthquake records. The influence of base flexibility on the response is also investigated.

Consider two multi-storey buildings with symmetric and antisymmetric floor-plans shown in Fig. 7.11. The buildings are modelled as a three-dimensional lumped mass structure with three degrees of freedom at each storey level and resting on the surface of an elastic half-space. The base of the buildings is assumed

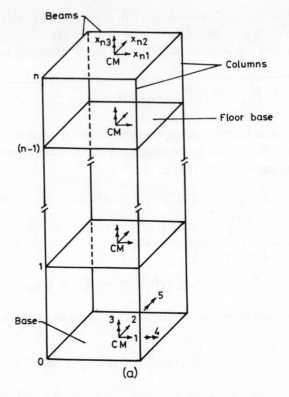

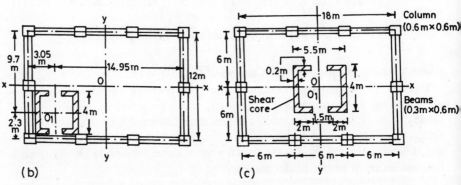

Fig. 7.11. (a) Multi-storey building with: (b) antisymmetric floor-plan; and (c) symmetric floor plan.

to be an equivalent rigid circular disc footing of negligible thickness. Five base degrees of freedom, namely, two horizontal translations, and three rotations are used to model the base flexibility. The soil stiffness and the soil damping are represented through frequency independent impedance functions for a rigid circular footing (Veletsos and Wei, 1971). Due to the inclusion of base flexibility, the above mdf model has non-classical damping. An energy approach is used to define an equivalent modal damping in each mode of vibration of the flexible base structure, making it amenable to normal mode analysis.

The analysis is performed for buildings having 16 storeys, each with a uniform storey height of 3 m. The lumped masses at the centre of mass of each floor are:

translational mass = 143.568 kNs^2/m and rotational mass = 7217.7 kNs^2 m (for symmetric building), and rotational mass = 7831.17 kNs^2 m (for asymmetric building). The modulus of elasticity = 21 GN/m^2; the Poisson's ratio = 0.15; the ratio of modulus of elasticity to shear modulus = 2.3; and mass density 2.4 kNs^2/m^4. The responses of buildings are obtained for fixed base and shear wave velocity of 200 m/s. The radius of the footing (r_0) is taken as that of a circle having the same area as the plan of the building (r_0 = 8.5 m). The modal damping for the building is taken as 5% of critical damping.

The buildings are assumed to be excited by a single ground acceleration component acting in the y-direction, and modelled as a broadband, stationary Gaussian process with the mean acceleration spectrum specified in IS 1893–1975 (Fig. 7.12). The psd of the ground acceleration is obtained from the response spectrum using the method proposed by Kaul (1978). The time histories of ground accelerations are generated by Monte Carlo simulation from the derived psd. An ensemble of twenty samples of time histories are generated and used for averaging the responses.

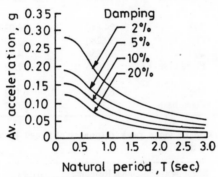

Fig. 7.12. Mean acceleration spectrum (IS 1893–1975).

The mean peak responses of the two buildings are determined by the following methods:

(A) Response spectrum method (7.115).
(B) Frequency domain method based on psd.
(C) Time domain method based on simulation.

Tables 7.4 and 7.5 show a comparison of the mean peak responses obtained by the three methods for the two buildings. The variations in the responses obtained by the three methods lie between 3 and 6%. The base flexibility has a significant influence on the response causing a significant increase in the displacement and a reduction in the forces.

Example 7.4: Response of a mdf structure to multi-component excitation
                (Smeby and Kiureghian, 1985)

Consider the mdf structure shown in Fig. 7.13a. The first four modal frequencies of the structure are: ω_1 = 17.7, ω_2 = 19.4, ω_3 = 27.9 and ω_4 = 47.2 (all in rad/sec). The displacements of the top of the structure are represented by X_1 and X_2 under the simultaneous action of the two horizontal components of earthquake ground motion.

Table 7.4   Comparison of mean peak responses obtained by different methods (asymmetric building)

S. No.	Response Quantities	Shear Wave Velocity (m/s)	(A): Response Spectrum Method	(B): Frequency Domain Method	(C): Time Domain Method
1	*Top-floor displacements*				
	X_{n1} (cm)	200	3.030	2.860	3.190
		Fixed base	2.988	2.750	3.08
	X_{n2} (cm)	200	5.290	4.872	5.048
		Fixed base	4.678	4.260	4.408
	X_{n3} ($\times 10^{-2}$ rad)	200	0.5003	0.4724	0.5033
		Fixed base	0.4868	0.4561	0.4900
2	*Base shears*				
	V_{01} (kN)	200	605.1	603.4	631.8
		Fixed base	618.0	613.0	652.0
	V_{02} (kN)	200	966.3	966.8	875.1
		Fixed base	1032.0	1025.0	921.6
3	*Base moments*				
	M_{01} (kNm)	200	24,840	26,032	23,350
		Fixed base	30,286	32,721	29,048
	M_{02} (kNm)	200	15,858	14,785	16,205
		Fixed base	19,006	17,723	20,628
4	*Base torque*				
	T_{03} (kNm)	200	6,563	6,172	6,688
		Fixed base	7,365	7,092	7.448

Table 7.5   Comparison of mean peak responses obtained by different methods (symmetric building)

S. No.	Response Quantities	Shear Wave Velocity (m/s)	(A): Response Spectrum Method	(B): Frequency Domain Method	(C): Time Domain Method
1	*Top-floor displacements*				
	X_{n2} (cm)	200	8.750	8.432	8.756
		Fixed base	7.751	7.350	7.753
2	*Base shear*				
	V_{02} (kN)	200	1273	1303	1267
		Fixed base	1413	1408	1407
3	*Base moment*				
	M_{01} (kNm)	200	36,600	37,230	36,532
		Fixed base	41,047	42,345	40,658

Note: $X_{n1} = X_{n3} = V_{01} = M_{01} = T_{03} = 0$.

The mean peak response of the structure is given by (7.117). Two important parameters determining the response are θ and γ. Figure 7.13b, c show the peak mean displacement response of X_1 and X_2 plotted against θ for $\gamma = 0.25$, 0.50 and 0.87. The broken lines represent the response for uncertain θ, which is modelled as a uniformly distributed random variable over 0 to 2π. It is seen that the response is sensitive to both θ and γ, and the critical values of θ are independent of γ.

Percent contribution of the third term in (7.117) is plotted in Fig. 7.14a as a function of θ. This term represents the correlation between the input components to the response X_1. It is seen that contribution of the third term is only significant

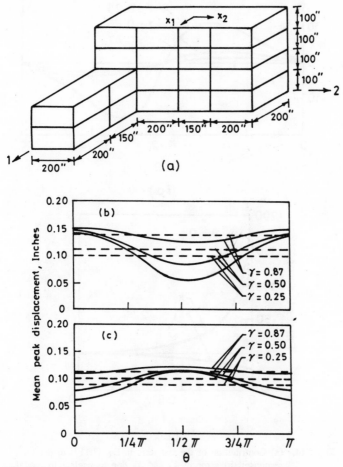

Fig. 7.13. (a) A mdf structure. Mean peak response of the structure: (b) displacement, X_1; and (c) displacement, X_2; (- - -) θ uniformly distributed over 0 to 2π (Smeby and Kiureghian, 1985).

when γ is small, that is, when the intensities of the two components are significantly different. In practice, γ is near unity and, therefore, the contribution of this term is expected to be relatively small.

Figure 7.14b shows the percent error in the responses X_1 and X_2, due to the neglect of the effect of modal correlation, as function of γ for $\theta = 0$. For $\gamma = 1$, that is, for equal intensity components the error is almost zero. This is so, because the modal vectors corresponding to the closely spaced frequencies of the structure (ω_1 and ω_2) lie in two nearly perpendicular planes along Z_1' and Z_2', and the cross-modal terms resulting from the two input components cancel out. As γ approaches zero, that is, as the input approaches a single-component excitation, the error due to neglect of modal correlations increases, and the SRSS rule tends to underestimate the response in the direction of the strong component (X_1), and overestimate the response in weak component (X_2). Smeby and Kiureghian (1985) have shown that for a structure in which closely spaced frequencies are associated with modal

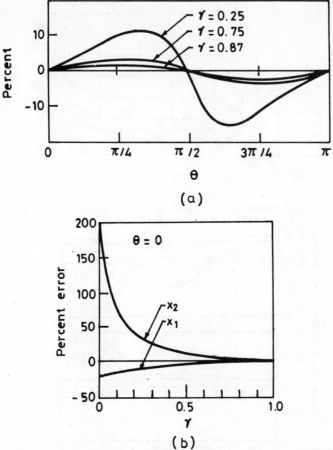

Fig. 7.14. (a) Contribution of the last term in Eq. (7.117) to the total response; (b) error in X_1 and X_2 due to neglect to modal correlation (Smeby and Kiureghian, 1985).

vectors situated in parallel planes, the error is consistently large for all values of γ. In such cases, modal correlations are important even for input components of equal intensity.

7.5.2 Discrete Inelastic System

So far we have considered systems modelled by linear, elastic force-deformation relationship. It is well known that both steel and concrete structures have a large reserve of strength beyond the elastic range (Housner, 1956, 1959), (Nigam, 1970), (Sues et al, 1983). In the design of structures to resist earthquakes, inelastic behaviour is of special interest due to reduction in the intensity of shaking caused by dissipation of energy through hysteresis. A structure can be designed to remain elastic during frequent small earthquakes, and allowed to undergo inelastic deformation without collapse, during infrequent large earthquakes. This basic approach to aseismic design is incorporated in most codes. The random vibration analysis of inelastic systems is, therefore, of considerable interest.

The force-deformation relationships for inelastic behaviour are both nonlinear and hysteretic. Commonly used models of inelastic behaviour are shown in Fig. 7.15.

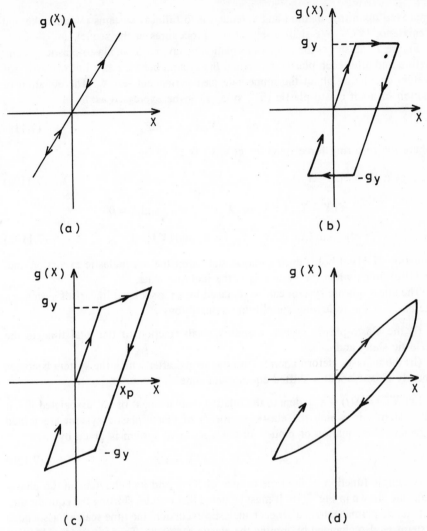

Fig. 7.15. Idealised models of inelastic force-displacement behaviour: (a) elastic; (b) elastoplastic; (c) bilinear hysteretic; and (d) general yielding.

Wen (1976, 1980), Baker and Wen (1981) have proposed models of inelastic behaviour, which, besides the nonlinear hysteretic behaviour, incorporate degradation of stiffness and/or strength. The mathematical difficulties in a nonlinear random vibration analysis of inelastic behaviour are considerable (Nigam, 1983). Approximate methods, such as, equivalent linearization (Caughey, 1960, 1963); Gaussian closure technique (Iyengar and Dash, 1978), and Markov vector formulation (Wen, 1976) have been used to obtain the response statistics of such systems. Due to analytical difficulties numerical integration of the equations of motion to real, or simulated, ground has received considerable attention for the study of inelastic behaviour, and has formed the basis of validating approximate methods.

Karnopp and Scharton (1965) developed a simple procedure for random vibration analysis of elasto-plastic sdf systems to determine statistics of accumulated plastic deformation and its application to low cycle fatigue. Vanmarcke (1969, 1977) improved upon this method and extended it to bilinear systems (Vanmarcke and Veniziano, 1973). We shall discuss these procedures in the sequel.

Figure 7.15b shows the force-deformation behaviour of an elasto-plastic system. Before and after each plastic excursion, the system behaves as a linear elastic sdf system oscillating about the immediate past permanent-set, X_p. The equation of motion of a sdf elasto-plastic (EP) system can be expressed as:

$$m\ddot{X} + c\dot{X} + g(X, \dot{X}) = -m\ddot{Z} \tag{7.118}$$

where the restoring force function $g(X, \dot{X})$ is given by

$$g(X, \dot{X}) = k(X - X_{pn}), \tag{7.119a}$$

$$\text{if } |X - X_{pn}| < X_y \text{ or } |X - X_n| = X_y \text{ and } \dot{X} = 0$$

$$= g_y \, \text{sgn}\,(\dot{X}), \text{ if } |X - X_{pn}| = X_y \text{ and } |\dot{X}| > 0 \tag{7.119b}$$

Equation (7.119a) holds during elastic state after the nth inelastic excursion and, (7.119b) holds when the system is is the inelastic state.

The elasto-plastic system can be replaced by an associated linear sdf system on the basis of the following simplifying assumptions:

(i) the elasto-plastic system spends a small fraction of the total time in the inelastic state; and

(ii) the inelastic deformation is 'instantaneous' after which the system becomes linearly elastic about a shifted equilibrium state.

Let $X'(t) = X(t) - X_{pn}$, denote the relative displacement of an 'associated linear sdf system's in which the elastic segments of elasto-plastic system are joined together. The equation of motion of the associated system is given by

$$m\ddot{X}' + C\dot{X}' + kX' = -m\ddot{Z} \tag{7.120}$$

A sample function of the time history of $X'(t)$; and its behaviour in the phase-plane are shown in Fig. 7.16. It must be noted that besides shifting the equilibrium-state to the permanent-set after each inelastic excursion, the time scale of associated systems is also changed by joining the elastic segments. Thus if, T is the duration of $X(t)$, the duration of $X'(t)$,

$$T' = T - \sum_{i=1}^{n} \Delta t_{pi} \tag{7.121}$$

where n is the number of inelastic excursions and Δt_{pi} is the time spent during the ith excursion.

The subsequent treatment is based on the plausible assumption that the response statistics of the elastic behaviour of EP system and associated linear elastic system are not significantly different, and the system response during inelastic excursions can be determined on the basis of the physical argument that during an inelastic excursion, the kinetic energy of the mass at the beginning of the excursion is dissipated through plastic work, and the system returns to elastic-state with the initial conditions $X'(t_0) = \pm X_y$ and $\dot{X}'(t_0) = 0$. Thus

$$g_y \, |\delta| = \frac{m}{2} (\dot{X}')^2, \quad |X(t)| = X_y \tag{7.122}$$

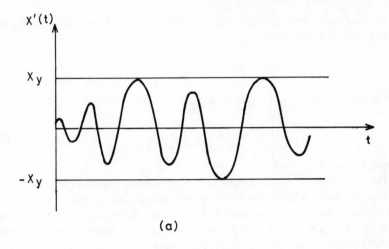

(a)

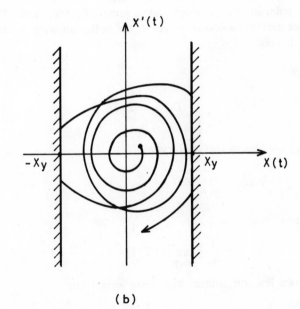

(b)

Fig. 7.16. A sample function of $X'(t)$. (a) time history; (b) behaviour in phase-plane.

where δ is the plastic displacement during an inelastic excursion. Setting $g_y = kX_y$

$$|\delta| = \frac{1}{2\omega_0^2 X_y}(\dot{X}')^2, \quad |X'(t)| = X_y \qquad (7.123)$$

For random excitation, δ is random and its expected value is given by

$$E[|\delta|] = \frac{1}{2\omega_0^2 X_y}\int_{-\infty}^{\infty} \dot{x}'^2 p(\dot{x}'|x' = X_y)\, d\dot{x}' \qquad (7.124)$$

For practical aseismic design, it is realistic to assume that inelastic excursions will be infrequent. If so, the response can be assumed to be stationary at the entry to the next excursion, and, therefore, $E[X'\dot{X}'] = 0$. The conditions probability in (7.124) becomes the probability of $\dot{X}'$ alone, and

$$E[|\delta|] = \frac{1}{2\omega_0^2 X_y} \sigma_{\dot{X}'}^2.$$ (7.125)

Assuming $\sigma_{\dot{X}'} = \sigma_{\dot{X}}$ and $\sigma_{\dot{X}} = \sigma_X \omega_0$,

$$E[|\delta|] = \frac{\sigma_X^2}{2X_y} = \frac{\sigma_X}{2\eta_y}$$ (7.126)

where $\eta_Y = X_y/\sigma_X$.

The response of a sdf system is a narrow-band random process for which excursions above a level tend to occur in 'clumps', particularly if the level is not too high and the damping is small. In each clump, there may be several successive inelastic excursions during which energy is dissipated, followed by relatively long elastic regime. As an approximation the cluster of excursions in a clump may be treated as a point in time at which plastic jumps, D_i, take place. The number of clumps can be treated as a Poisson process to define the following response parameters of practical interest.

Plastic Drift

$$D(T) = \sum_{i=1}^{N(T)} D_i$$ (7.127)

Total Energy Dissipated Due to Yielding

$$DA(T) = \sum_{i=1}^{N(T)} g_y |D_i|$$ (7.128)

Ductility Ratio

$$r = \frac{1}{X_y} [\max_{0 \leq t \leq T} |D(T)|] + 1 = \frac{X_{pm}}{X_y} + 1$$ (7.129)

where $N(t)$ is a Poisson process with mean arrival rate,

$$v_c = \frac{N_x^+(X_y)}{E[M_1]}$$

and $E[M_1]$ is the expected clump-size (6.81*). The tendency for clumping is expected to be less in elasto-plastic systems, as compared to corresponding elastic systems due to energy dissipation during each inelastic excursion. Further, for design levels of practical interest clumping is expected to be limited and expected clump size may be assumed to be unity (Vanmarcke, 1969).

Total Energy Dissipated

Let the excitation be a Gaussian random process. We assume that the response of the elasto-plastic system is also Gaussian. Under this assumption, $\dot{X}/\sigma_{\dot{X}}$ is a standard Gaussian random variable, and therefore $\dot{X}^2/\sigma_{\dot{X}}^2$ has χ^2-distribution with one-degree-of-freedom, with mean and variance equal to 1 and 2 respectively. In (7.123), $|\delta|$ and, therefore, $|D_i|$ are proportional to $\dot{X}^2/\sigma_{\dot{X}}^2$. Hence

$$E[|D_i|] = \frac{\sigma_{\dot{X}}^2}{2\omega_0^2 X_y} E\left[\frac{(\dot{X}')^2}{\sigma_{\dot{X}}^2}\right] = \frac{\sigma_{\dot{X}}^2}{2X_y} E\left[\frac{\dot{X}^2}{\sigma_{\dot{X}}^2}\right] = \frac{\sigma_{\dot{X}}^2}{2X_y} \qquad (7.131)$$

$$\text{Var}[|D_i|] = \left(\frac{\sigma_{\dot{X}}^2}{2X_y}\right)^2 \text{Var}\left[\frac{\dot{X}^2}{\sigma_{\dot{X}}^2}\right] = 2\left(\frac{\sigma_{\dot{X}}^2}{2X_y}\right)^2 \qquad (7.132)$$

Assuming $|D_i|$ in (7.128) to be mutually independent, identically distributed and also independent of $N(t)$, we have (Parzen, 1962)

$$E[DA(T)] = g_y E[N(T)] E[|D_i|]$$

$$= g_y v_c T \frac{\sigma_{\dot{X}}^2}{2X_y} = g_y 2N_x(X_y) T \frac{\sigma_{\dot{X}}^2}{2X_y}, E[M_1] = 1$$

$$= g_y \frac{\omega_0}{2\pi} T \frac{\sigma_x}{\eta} e^{-(\eta_y^2/2)} \qquad (7.133)$$

$$\text{Var}[DA(T)] = \{E[N(T)] \text{Var}[|D_i|] + \text{Var}[N(T)] E^2[|D_i|]\} g_y^2$$

$$= \frac{3}{2} \frac{\omega_0}{2\pi} T \frac{\sigma_{\dot{X}}^2}{\eta_y^2} g_y^2 e^{-(\eta_y^2/2)} \qquad (7.134)$$

Total Plastic Drift

The total plastic deformation $D(T)$ defined by (7.127) has a zero-mean, as each inelastic excursion D_i is equally likely to be positive or negative with mean and variance equal to 0 and $2 (E[|\delta|])^2$ respectively. Hence,

$$E[D(T)] = v_c TE[D_i] = 0 \qquad (7.135)$$

$$\text{Var}[D(T)] = v_c T \text{Var}[D_i] + E^2[D_i] = 2 (v_c T) (E[|\delta|])^2$$

$$= \frac{\omega_0 T}{4} \left(\frac{\sigma_y}{\eta_y}\right)^2 e^{-(\eta_y^2/2)} \qquad (7.136)$$

Ductility Ratio

To establish the distribution of ductility ratio, r, in (7.129), it is necessary to establish the distribution of peak inelastic deformation X_{pm}, the maximum value of $|D(t)|$ in the interval $0 \le t \le T$. Vanmarcke (1977) has shown that the distribution function of X_{pm} can be approximated as

$$F_{X_{pm}}(\alpha) = \exp[-(\exp(v_c T) - 1) \exp(-\alpha/\delta)], \alpha \ge 0 \qquad (7.137)$$

The distribution of ductility ratio follows from (7.129).

In the above discussion we have treated the random vibration of elasto-plastic sdf system for infrequent excursions. Karnopp and Scharton (1965), have also given approximate results for frequent excursions taking into account the effect of clumping and nonstationary behaviour in between excursions.

Gazetas (1976), Ziegler (1987) have extended the procedure due to Karnopp and Scharton (1965) to mdf systems, including the effects of nonstationary excitation.

7.6 Probabilistic Design for Earthquakes

Aseismic design encompasses choice of design parameters to avoid, or limit, 'damage' to a system due to earthquakes during its service-life. The term 'damage' is used in generic sense to denote any, 'unfavourable behaviour' such as, loss of life, structural collapse, cost of repair or interruption of services, discomfort to occupants etc. Due to a large measure of uncertainty contributed by several sources, a probabilistic formulation provides a consistent and realistic basis for aseismic design. In aseismic design, it is convenient to partition the total risk into two steps:

(i) determination of the conditional distribution of damage to a system due to an ensemble of ground-motions representing design earthquake(s) in terms of one, or more, *given* site-intensity parameters; and

(ii) determination of the distribution of site-intensity parameters (seismic inputs) at the site during the service-life of the system.

The total risk is the combined effect of the different ways in which the system may respond to a given event, and all possible seismic events that may occur during its service-life. It is computed by a convolution of the conditional distribution of system response for given seismic input, with the distribution of seismic input. A general expression for the total risk can be written as (Whitman and Cornell, 1977).

$$P[D_i] = \sum_{\text{all } j} P[D_i \mid Y_j] P[Y_j] \qquad (7.138)$$

where D_i denotes the event that the damage state of the system is i and Y_j denotes the seismic input at 'level' j; and $P[D_i \mid Y_j]$ is the conditional probability that the damage state of the system is D_i, *given* that seismic input Y_j has occurred. If the damage and seismic input represent continuous variables, (7.138) can be expressed as

$$F_D(d) = P[D \leq d] = \int F_{D|Y}(d \mid y)\, p_Y(y)\, dy \qquad (7.139)$$

If the seismic input is specified in terms of more than one site-intensity parameters, say by a m-dimensional vector $\bar{Y}$, total risk can be written as

$$F_D(d) = \underbrace{\int \ldots \int}_{m} F_{D|\bar{Y}}(d \mid \bar{Y})\, p_{\bar{Y}}(\bar{y})\, d\bar{y} \qquad (7.140)$$

In Section 7.3, we have derived the distributions of structural response for elastic and inelastic behaviours. These are conditional distributions for *given* seismic input. The damage to a structure during a seismic event, or a sequence of seismic events, can be expressed as a stochastic, and under certain conditions and approximations, as a deterministic function of response parameters. Its conditional distribution can be determined from the distribution of response parameters using the theory of random functions. The distribution of site-intensity parameters can be established through a determination of seismic risk treated in Appendix 7A. We, therefore, have the necessary framework for evaluating total risk; or choosing design parameters, or design intensities, for a specified total risk, if the functional relationship between damage and response parameters is established.

Damage to a Structure Due to Earthquakes
The damage to a structure due to earthquakes may be treated under two broad classes:

(i) instantaneous damage, which results when at an instant of time one, or more, response parameters cross the safe operating domain for the first time, which, is the classical first-passage problem; and

(ii) cumulative damage, which accumulates during one or over several seismic events, and the 'failure' occurs when damage function exceeds a specified limit.

As mentioned earlier, the damage is generally a stochastic function of the response. However, under certain conditions, it may be treated as a deterministic function of the response. We shall consider the latter case first.

Deterministic Damage-response Relationship
We have seen that response of a structure due to an earthquake depends on the given value of site-intensity. If the damage is deterministically related to the response, it will still be a random function of the site-intensity. Let the damage measure, D, be a monotonic function of the site-intensity, Y, that is

$$D = f(Y); \frac{df}{dy} > 0 \text{ for } y, \text{ and } D \geq 0 \qquad (7.141)$$

Inverting (7.141),

$$Y = g(D) \qquad (7.142)$$

Let $D_{max}^{(t)}$ be the maximum value of D in time interval t. Then the distribution of $D_{max}^{(t)}$

$$F_{D_{max}^{(t)}}(d, t) = P[D_{max} \leq d] = P[Y_{max} \leq g(d)]$$

$$= F_{Y_{max}}(g(d), t), \qquad (7.143)$$

From (7.A.21)

$$F_{D_{max}^{(t)}}(d, t) = \exp\left[-C\hat{v}G(g(d))^{-\beta/b_2}t\right] \qquad (7.144)$$

and

$$P[D_{max}^{(t)} > d] = 1 - \exp\left[-C\hat{v}G(g(d))^{-\beta/b_2}t\right]$$

for the range of probabilities of practical interest in design.

If the damage is proportional to Y, that is $D = aY$, where a is constant, as in the case of expected peak response of high frequency sdf systems (Cornell, 1971),

$$P[D_{max}^{(t)} > d] = C\hat{v}G\left(\frac{d}{a}\right)^{-\beta/b_2}t \qquad (7.145)$$

It must be noted that, in the above derivation, it has been assumed that there has been no deterioration due to previous lower intensity loads and the events are independent. If this condition remains true, that is either previous damage does not influence the response significantly, or the system is restored through repair, the cumulative total damage, TD, can be expressed as

$$TD = \sum_{j=1}^{N(t)} D_j = \sum_{j=1}^{N(t)} f(Y_j) \qquad (7.146)$$

where $N(t)$ is a Poisson process. For $D = aY$,

$$\mu_{TD} = E[TD] = \hat{v} t a \mu_y \tag{7.147}$$

and

$$\sigma_{TD} = \text{Var}\,[TD] = \hat{v} t a^2 E[Y^2] \tag{7.148}$$

The characteristic function of TD is given by

$$M_{TD}(\theta) = \exp\left[\hat{v}t\{M_D(\theta) - 1\}\right] \tag{7.149}$$

where $M_D(\theta)$ is characteristic function of D_j. If $\hat{v}t$ is not small, TD will be approximately normally distributed. A Markov model can be used, if the incremental damage is stochastically dependent on the total damage todate (Vanmarcke, 1969).

A simple example, where damage is directly related to peak ground motion occurs in nuclear power plants which sustain economic loss whenever accelerometer scramming control shuts down the plant, if peak ground acceleration exceeds a threshold level. If damage due to each shut down is d, the total expected damage in time t is

$$E[TD] = d\,\hat{v}t\,CG\,y^{-\beta/b_2} \tag{7.150}$$

from (7A.21).

Stochastic Damage-Response Relationship
Due to random nature of earthquake ground motion, the response is a random process and the damage is, therefore, a stochastic function of system response. The conditional distribution of system response is obtained through random vibration analysis. The damage can be generally related to the peak response of the system. In section 7.5, we have derived the approximate expressions for the distribution, and second order statistics, of the peak response of sdf and mdf systems. These are conditional distributions for given seismic input. The total risk, or design based on specified total risk, can be obtained by substituting these in (7.139) or (7.140).

Example 7.5 Reliability based aseismic design of a sdf system
Let the ground acceleration be modelled as a segment of stationary random process with psd $\Phi(\omega)$ and variance σ_g^2. Let Y denote the expected value of peak ground acceleration. Assuming independent crossings, it follows from application of (7.30), (7.37) and (7.40–7.42) to ground acceleration process, that

$$Y = \bar{\eta}_g \sigma_g \tag{7.154}$$

where

$$\bar{\eta}_g = C_1 + \frac{0.5772}{C_1} \tag{7.155}$$

and

$$C_1 = \left[2 \ln\left(\frac{\Omega_g T'}{\pi}\right)\right]^{1/2} \tag{7.156}$$

T' is the duration of the record and Ω_g is the central frequency.

Consider the stationary response of a sdf system to the ground acceleration described above. For independent crossing assumption, the conditional distribution of peak displacement, X_m, defined by (7.26), is given by (7.34)

$$F_{X_m|Y}(x \mid Y = y) = \exp\left[-\frac{T\omega_0}{\pi} \exp\left(\frac{x^2}{2\sigma_{X|Y}^2}\right)\right] \tag{7.157}$$

$$E[X_m \mid Y = y] = \bar{\eta}_X \sigma_{X|Y}^2 \tag{7.158}$$

and

$$\text{Var}\,[X_m \mid Y = y] = \tilde{\eta}_X \sigma_{X|Y}^2 \tag{7.159}$$

where

$$\sigma_{X|Y}^2 = \frac{\pi\Phi(\omega_0)}{2\zeta\omega_0^3} = \frac{\pi\sigma_g^2\,\Phi^*(\omega_0)}{2\zeta\omega_0^3} \tag{7.160}$$

$\bar{\eta}_X$ and $\tilde{\eta}_X$ are given by (7.37) and (7.38) respectively, and $\Phi^*(\omega)$ is the normalized psd of the ground acceleration.

From (7.154) and (7.160), we have

$$\sigma_{X|Y} = \frac{Y}{\bar{\eta}}\left(\frac{\pi\Phi^*(\omega_0)}{2\zeta\omega_0^3}\right)^{1/2} \tag{7.161}$$

and (7.157), (7.158) and (7.159), can be rewritten as

$$F_{X_m(Y(x)|Y=y)} = \exp\left[-\frac{T\omega_0}{\pi} \exp\left(-x^2(2\zeta\omega_0^3)^{-2}\,\eta_g/y^2\,(\pi\Phi^*(\omega_0)\right)\right] \tag{7.162}$$

$$E[X_m \mid Y = y] = \frac{\bar{\eta}_x}{\bar{\eta}_g}\left(\frac{\pi\Phi^*(\omega_0)}{2\zeta\omega_0^3}\right)^{1/2} y \tag{7.163}$$

$$\text{Var}\,[X_m \mid Y = y] = \frac{\tilde{\eta}_x}{\bar{\eta}_g}\left(\frac{\pi\Phi^*(\omega_0)}{2\zeta\omega_0^3}\right) y^2 \tag{7.164}$$

The above expressions give the conditional distribution and moments of peak displacement response of a sdf system, for a given value of peak ground acceleration. The unconditional distributions and moments are given by

$$F_{X_m}(x) = \int_0^\infty F_{X_m|Y}(x \mid Y = y)\, p_Y(y)\, dy \tag{7.165}$$

$$E[X_m] = \int_0^\infty E[X_m \mid Y = y]\, p_Y(y)\, dy = \frac{\bar{\eta}_x}{\bar{\eta}_g}\left(\frac{\pi\Phi(\omega_0)}{2\zeta\omega_0^3}\right)^{1/2} E[Y] \tag{7.166}$$

$$\text{Var}\,[X_m] = \int_0^\infty \text{Var}\,[X_m \mid Y = y]\, p_Y(y)\, dy = \frac{\tilde{\eta}_x}{\bar{\eta}_g}\left(\frac{\pi\Phi(\omega_0)}{2\zeta\omega_0^3}\right) E[Y^2] \tag{7.167}$$

where $p_Y(y) = dF_Y(y)/dy$ is the probability density function of peak ground acceleration obtained from the seismic risk analysis (7A.9).

Special Case
If the coefficient of variation of X_m, given Y, is small compared to that of Y itself, (7.165) reduces approximately to

$$F_{X_m}(x) = P[X_m \leq x] = P[Y \leq f^{-1}(x)] = F_Y(f^{-1}(x)) \tag{7.168}$$

where

$$f(y) = \frac{\bar{\eta}_x}{\bar{\eta}_g} \left(\frac{\pi \Phi(\omega_0)}{2\zeta\omega_0^3} \right)^{1/2} y = E[X_m \mid Y = y] \tag{7.169}$$

From (7A.9)

$$F_{X_m}(x) = P[X_m \leq x] = P\left[Y \leq x \frac{\bar{\eta}_g}{\bar{\eta}_x} \left(\frac{2\zeta\omega_0^3}{\pi \Phi^*(\omega_0)} \right)^{1/2} \right] \tag{7.170}$$

$$= 1 - CG\left\{ x \frac{\bar{\eta}_g}{\bar{\eta}_x} \left(\frac{2\zeta\omega_0^3}{\pi \Phi^*(\omega_0)} \right)^{1/2} \right\}^{-\beta/b_2} \tag{7.171}$$

Let $X_{max}^{(t)}$ denote the maximum value of X_m over a t-year period. It follows from (7A.15)

$$F_{X_{max}^{(t)}}(x) = P[X_{max}^{(t)} \leq x] = \exp[-(1 - F_{X_m}(x)\,\hat{v}t]$$

$$= \exp\left[- CG\hat{v}t \left\{ x \frac{\bar{\eta}_g}{\bar{\eta}_x} \left(\frac{2\zeta\omega_0^3}{\pi \Phi(\omega_0)} \right)^{1/2} \right\}^{-\beta/b_2} \right] \tag{7.172}$$

The assumption made in the above derivation that the coefficient of variation of X_m given Y, is small (< 0.2) as compared to the coefficient of variation of Y, is generally valid, as the uncertainty due to seismicity is usually much more than the random vibration uncertainty (Cornell, 1971).

Cumulative Damage
If the damage due to each event, i, is proportional to X_m, the total damage can be expressed as

$$TD = \sum_{i=1}^{N} CX_{mi} \tag{7.173}$$

Assuming Poisson events

$$E[TD] = \hat{v}t\,CE[X_m] \tag{7.174}$$

$$\text{Var}[TD] = \hat{v}t\,C^2 E[X_m^2] = \hat{v}t\,C^2\{E^2[X_m] + \text{Var}[X_m]\} \tag{7.175}$$

where $E[X_m]$ and $\text{Var}[X_m]$ are given by (7.166) and (7.167).

In Section 7.5.2 we have derived the distributions and moment functions of the response quantities, such as, plastic drift, energy dissipated, and ductility ratio of discrete inelastic systems. These are conditional distributions and moments for given site-intensity. The unconditional distributions and moments can be obtained, as before, by convolution with the distributions of seismic intensity through (7.139) or (7.140). Unlike the simple cases treated above, where moment functions of damage could be expressed as a polynomial function of site intensity Y, the expressions for moment functions are complex for the inelastic system. However, these can be expressed as polynomials of Y using the Taylor series expansion about $E[Y]$ (Cornell,, 1971), and unconditional moment functions of damage can be expressed in terms of moments of Y.

Example 7.6: **Optimum design of the support-structure of a water tank to earthquake loading** (Narayanan and Nigam, 1983)

We shall now consider the minimum weight design of a planar-truss structure supporting an elevated water tank, Fig. 7.17, subjected to random base excitation due to earthquakes (Narayanan and Nigam, 1983). The optimization problem is formulated and solved in the reliability framework discussed in Chapter 4.

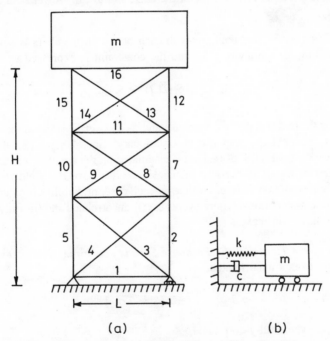

Fig. 7.17.   (a) An elevated water tank; and (b) a sdf model.

DESIGN VECTOR: The design vector $\bar{D}$ consists of 12 variables representing the areas of cross-sections and length of the truss members:

$$\bar{D} = \{d_1 d_2, \dots d_{12}\}^{\mathrm{T}}$$

$$= \{A_1, A_2, A_3, A_6, A_7, A_8, A_{11}, A_{12}, A_{13}, A_{16}, L_2, L_7\}^{\mathrm{T}}$$

where A_i and L_i are the area of cross-section, and the length of the ith member respectively. From symmetry $A_2 = A_5$, $A_3 = A_4$, $A_7 = A_{10}$, $A_8 = A_9$, $A_{12} = A_{15}$, $A_{13} = A_{14}$; $L_2 = L_5$; $L_3 = L_4$; $L_7 = L_{10}$; $L_8 = L_9$; $L_{12} = L_{15}$; and $L_{13} = L_{14}$. Also $L_1 = L_6 = L_{11} = L_{16} = L$ and $L_{12} = H - (L_2 + L_7)$, where L is the specified base width, and H is the given total height of the truss.

OBJECTIVE FUNCTION: The objective function is the total volume of the truss structure:

$$W(\bar{D}) = \sum_{i=1}^{16} A_i L_i \tag{7.176}$$

CONSTRAINTS: The optimization problem is formulated with constraints on stresses in the members; absolute acceleration of the mass; natural frequency of the water tank treated as a sdf system; and the non-negativity constraints on design variables.

FREQUENCY CONSTRAINT: The first three sloshing frequencies of partially filled tank are constrained to remain well below the natural frequency of the tank to avoid large scale oscillations. It is expressed as

$$\omega_0(\bar{D}) \geq \omega^L \tag{7.177}$$

where ω_0 is the natural frequency of the tank; and ω^L, the lower bound on natural frequency.

STRESS CONSTRAINT: The axial stress in each of the 16 members is constrained to remain below the yield stress S_y, and the constraint is expressed as

$$P[\underset{\substack{i=1 \\ 0 \leq t \leq T}}{\overset{16}{\cup}} \{|S_i(\bar{D}, t)| \geq S_y\}] \geq p_1 \tag{7.178}$$

where T is the duration of the earthquake and $S_i(\bar{D}, t)$ is the stress in the ith member.

   We assume that the response is a stationary, Gaussian random process. The probability on the L.H.S. of (7.178) is calculated on the basis of two assumptions: (i) fraction of time the stress in members spends above the yield stress S_y; and (ii) the first excursion failure. Further, in each case, probability is evaluated under two extreme assumptions: failure events are (i) uncorrelated and (ii) fully correlated. Thus, we have four cases:

Fraction of the Time that the Response Spends above the Yield Stress Level S_y
(a) Failure events uncorrelated,

$$P[\cdot] = \sum_{i=1}^{16} \text{erfc}\,[S_y/\sqrt{2}\,\sigma_{Si}] \leq p_1 \tag{7.179a}$$

(b) Failure events fully correlated,

$$P[\cdot] = \max\,\{\text{erfc}\,[S_y/\sqrt{2}\,\sigma_{Si}] \leq p_1 \tag{7.179b}$$

Failure Due to First Excursion above the Stress Level S_y
(c) Failure events uncorrelated,

$$P[\cdot] = \sum_{i=4}^{16} \left\{1 - \exp\left[-\frac{\omega_0 T}{\pi} \exp\left[-S_y^2/2\sigma_{Si}^2\right]\right]\right\} \leq p_1 \tag{7.179c}$$

(d) Failure events fully correlated,

$$P[\cdot] = \max\left\{1 - \exp\left[-\frac{\omega_0 T}{\pi} \exp\left(-S_y^2/2\sigma_{Si}^2\right)\right]\right\} \leq p_1 \tag{7.179d}$$

ACCELERATION CONSTRAINT: The absolute acceleration of the tank is constrained to remain less than a specified value a, and the constraint is expressed as

$$P[\ |\ A(t)\ |\ \geq a] \leq p_2 \tag{7.180}$$

where $A(t) = \ddot{X} + \ddot{Z}$ is absolute acceleration. Assuming the acceleration to be a stationary, Gaussian random process, and evaluating the probability in terms of fractional occupation time (7.180) can be expressed as

$$\text{erfc}\,(a/\sqrt{2}\,\sigma_A) \leq p_2 \tag{7.181}$$

NON-NEGATIVITY CONSTRAINT: The non-negativity of design variables is expressed as

$$d_i \geq 0, \ i = 1, 2, \ldots 12 \tag{7.182}$$

The optimization problem is solved using the sequential unconstrained minimization technique discussed in Chapter 4. The design ground acceleration due to earthquakes is assumed to be the segment of a stationary, Gaussian random process with psd given by

$$\Phi(\omega) = \frac{31.72\,(1 + \omega^2/147.8)}{(1 - \omega^2/242)^2 + \omega^2/147.8} \tag{7.183}$$

to correspond to El Centro 1940 earthquake. Following design data is assumed:

$m = 382.8$ kg/cm^2	$H = 27.45$ m	$L = 9.14$ m
$E = 2.53 \times 10^8$ kg/cm^2	$S_y = 253.0$ kg/cm^3	$\omega^L = 5$ rad/sec
$p_1 = 10^{-3}$	$p_2 = 10^{-3}$	$\zeta = 0.2$
$a = 914.4$ cm/sec^2	$T = 25$ sec	

The liquid sloshing frequencies assuming that the tank has a square cross-section 6.1 m $\times$ 6.1 m, and the height 6.1 m, are:

Half-full: $\Omega_1 = 1.28$; $\Omega_2 = 2.16$ and $\Omega_3 = 2.74$ rad/sec,
Three Fourth-full: $\Omega_1 = 1.45$, $\Omega_2 = 2.44$ and $\Omega_3 = 2.76$ rad/sec.

The initial and optimal values of the design parameters for the four cases are given in Table 7.6. The sequence of minima converging to the optimum design are shown in Fig. 7.18.

It is seen that the optimum design is sensitive to the assumption under which probability for the stress constraint is evaluated. Comparing cases (a) and (c), it is seen that optimum volume is increased by 58 per cent when failure is expressed in terms of first excursion in comparison to the fractional occupation time. Further, optimum design is sensitive to the degree of correlation between member stresses. As expected, optimum volume is higher when stresses are assumed to be uncorrelated.

In Chapter 8, we shall discuss the optimum design of the above system to wind induced excitation.

Table 7.6. Truss structure—aseismic design

		d_1	d_2	d_3	d_4	d_5	d_6	d_7	d_8	d_9	d_{10}	d_{11}	d_{12}	$W(D) \times 10^6$
(a)	Initial	254	254	254	254	254	254	254	254	254	254	914.4	914.4	109.0
	Optimal	97.3	118.1	142.2	360.7	81.5	115.6	134.1	54.4	72.1	52.3	972.8	906.8	48.7
(b)	Initial	355.6	355.6	547.2	457.2	355.6	355.6	381.0	431.8	304.8	304.8	914.4	914.4	161.3
	Optimal	143.8	149.6	178.8	345.4	127.0	138.9	197.4	83.8	91.4	125.7	848.4	919.5	62.1
(c)	Initial	457.2	457.2	457.2	457.2	355.6	406.4	406.4	406.4	355.6	355.6	914.4	914.4	175.7
	Optimal	237.2	167.4	233.7	309.9	153.2	198.4	225.3	93.2	127.2	168.9	795.0	970.3	77.3
(d)	Initial	457.2	304.8	304.8	381.0	304.8	254.0	381.0	301.8	254	355.6	914.4	914.4	132.4
	Optimal	365.8	218.5	274.3	342.9	191.8	212.3	297.2	126.5	165.9	271.8	332.8	850.9	96.6

(a) Member stresses correlated ⎫ Stress constraint in terms of mean fractional occupation of time.
(b) Member stresses uncorrelated ⎭

(c) Member stresses correlated ⎫ Stress constraint in terms of first excursion failure.
(d) Member stresses uncorrelated ⎭

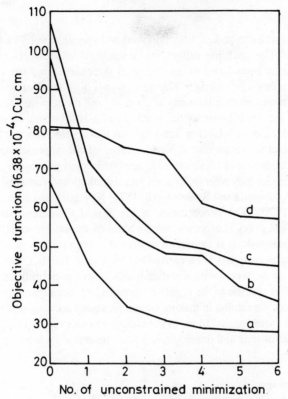

Fig. 7.18.   Sequential convergence to optimum design
"(a, b, c and d) refer to Table 7.6 (Narayanan
and Nigam, 1983).

Appendix 7A: Seismology and Seismic Risk

The occurrence of earthquakes, both temporal and spatial, involves a large measure of uncertainty. The multiple reflections, refractions and dispersions of seismic waves at irregular boundaries in the course of their travel to a site compound the level of uncertainty even further. This results in a high level of variability in the estimates of ground motion intensity at a site during the service life of a structure and is the major contributor to the total seismic risk, which is determined by embedding the random vibration analysis into the seismic risk analysis.

In this appendix, we present a broad outline of the approach to seismic risk analysis and the principal results needed for our limited purpose. For an in-depth treatment, the reader may refer to the texts on seismology and earthquake engineering (Richter, 1958; Lomnitz and Rosenblueth, 1976; Newmark and Rosenblueth, 1971; and Wiegel, 1970). The proceedings of the World Conference on Earthquake Engineering held every four years contain state-of-the-art information on aspects related to seismic risk. It is important to note that the research in seismology and earthquake engineering has been carried out by a large number of investigators in different parts of the world with significant differences in definitions, assumptions, and data which is specific to the region covered by the investigation. It is, therefore, necessary to exercise caution in making comparisons and drawing general conclusions on the basis of these results. Idriss (1978) has given an excellent review of various approaches, definitions and developments from time to time relating to earthquake ground motion.

7A.1   Seismology

It is convenient to divide the study of seismicity under two broad headings: (i) *local seismicity*, which deals with the probabilities of earthquakes occurring in a given portion of earths crust; and (ii) *regional seismicity*, which deals with the probabilities of earthquakes of given intensities shaking a given region of earth's surface (Newmark and Rosenblueth, 1971). The study of local seismicity basically involves knowledge of the geotectonic features and seismic history at micro- and macro-levels. Regional seismicity studies combine this information with the transmission of seismic waves, local geological information and statistical data on intensities of past earthquakes to establish estimates of seismic risk. The historical data for a site is seldom sufficient to permit direct statistical estimate of seismic risk. Seismic risk is, therefore, assessed through deductive probabilistic models using the available data. The Bayesian approach, which makes it possible to systematically supplement quantitative data with qualitative information is ideally suited for seismicity investigations (Esteva, 1969).

Some Definitions and Formulae

Magnitude (*M*) is a measure of the size of the earthquake in terms of the energy released. There are several definitions of magnitude (Idriss, 1978). The original definition due to Richter (1958) defines magnitude as the common logarithm of the trace amplitude, in microns, of a standard seismograph located 100 km. from the epicentre. *Intensity* (*I*) is a measure of earthquake's local destructiveness. It is measured on a subjective scale, such as, the modified Mercalli (*MM*) scale. *Focus* or *Hypocentre* is the point in the earths crust where the first seismic waves originates. *Epicentre* is the vertical projection of the focus on the earth surface. The location of a site in relation to an earthquake may be specified by the *focal distance* (R); or by *epicentral distance* (X) and *focal depth* (H). Clearly,

$$R = (X^2 + H^2)^{1/2} \tag{7A.1}$$

Peak ground acceleration (A_g), *peak ground velocity* (V_g), and peak ground displacement (D_g), are respectively the peak values of the absolute value of the ordinates in the ground acceleration velocity and displacement records at a site. An empirical relationship between the peak ground motion parameters $(A_g, V_g$ or $D_g)$ and the magnitude and focal distance, called *attenuation relations*, can be expressed in the following general form (Esteva and Rosenblueth, 1964; Idriss, 1978)

$$Y = b_1 e^{b_2 M} R^{-b_3} \tag{7A.2}$$

where Y may denote, A_g, V_g or D_g; and b_1, b_2 and b_3 are constants which depend on the quantity represented by Y. Esteva and Rosenblueth suggest that, for southern California, the average values of constants (b_1, b_2, b_3) may be chosen as (2000, 0.8, 2), (16, 1.0, 1.7) and (7, 1.2, 1.6), if Y denotes A_g, V_g or D_g in units of centimeters and seconds, and R is measured in km. It may be remarked that available data shows a large scatter around the mean curve represented by (7A.2) (Esteva, 1970). We shall discuss this aspect later.

Duration (T_g) is the length in time of significant shaking at a site during an earthquake. Several definitions have been proposed for T_g (Idriss, 1978). The definitions based on Husid plot (Husid, 1969) are being adopted increasingly. We shall express the relation between the duration, magnitude and focal distance in the form

$$T_g = C_1 e^{C_2 M} + C_3 R \tag{7A.3}$$

where T_g is in seconds, R is in km, and the average values of the constants (C_1, C_2, C_3) are (0.02, 0.74, 0.3) (Esteva and Rosenblueth, 1964).

The *r.m.s value* (σ_g) of the ground acceleration can be expressed in terms of the peak median value of $\hat{A}_g$ in the following form (Vanmarcke, 1977),

$$\sigma_g = \hat{A}_g \left(2 \ln \left(\frac{2.8\,\Omega_g}{2\pi} \right) \right)^{-1/2} \tag{7A.4}$$

where $\hat{A}_g$ is the median value of A_g, and $\Omega_g \simeq 2.1\,\omega_g$ for the Kanai-Tajimi psd with $\omega \le 4\omega_g$ and $\zeta_g = 0.6$. Through (7A.2), (7A.4) σ_g gets related to M and R.

The *half slipped length* (S) of fault can be expressed in terms of magnitude in the following form (Housner, 1970)

$$S = 1/2 \exp [a_1 (M - m_0)^{a_2}] \tag{7A.5}$$

where m_0 is the lower limit on earthquakes of engineering interest, and a_1 and a_2 are constants.

7A.2 Seismic Risk Analysis

The seismic risk at a site is usually expressed in terms of the probability of site 'intensity' exceeding a certain value in a given period of time. The term 'intensity' is used here in a generic sense to denote any one, or a function of several, ground motion parameters (Kiureghian, 1981). To determine the seismic risk at a site, it is necessary to construct the stochastic models of the source parameters, both temporal and spatial, and the travel path. A seismic source is characterised by its geometry, temporal characteristics and size of the seismic events. The geometric shape of the source on the surface of the earth can be idealized as a point, a line, or an area based on the knowledge of the spatial distribution of past earthquakes and known geotectonic features. The spatial distribution of earthquakes can generally be assumed to be homogeneous in a source, and, if necessary, a source may be subdivided into homogeneous sources. The focal depth of earthquakes is usually assumed to be constant. However, if focal depth data is available, a distribution can be fitted (Basu, 1977; Basu and Nigam, 1977). The occurrence of earthquakes, in time, at a source is generally assumed to be Poisson, so that

$$P_N(n, t) = \frac{(vt)^n \exp(-vt)}{n!} \tag{7A.6}$$

where $P_N(n, t)$ denotes the probability of occurrence n earthquakes of magnitude greater than say m_0, during the time interval t, and $v = v(m_0)$ is the expected rate of occurrence of such earthquakes. A Poisson model assumes stationarity and independence of successive events. Both assumptions are not fully substantiated by data and evolutionary models of earthquake occurrence, specially if fore- and after-shocks are included. However, the model is considered adequate for the service life of most structures (Knopoff, 1964; Lomnitz 1966).

The probability density function of the magnitude of earthquakes may be expressed as

$$P_M(m) = \beta \exp[-\beta(m - m_0)], \, m \geq m_0 \tag{7A.7}$$

where m_0 is the threshold magnitude below which the events are not of engineering interest and $\beta(1.5\text{-}2.3)$ is a constant for the source.

It is clear that the variables M and R in (7A.2) are random variables, and therefore, the site intensity, Y, is also a random variable. The distribution of M is given by (7A.7), and the distribution of R can be derived for a given source-geometry, assumed spatial distribution of epicentres and focal depth. For example, for a line-source of length l, uniform distribution of epicentres along the line and constant focal depth, it can be shown (Cornell, 1968) that

$$p_R(r) = \frac{2r}{l(r^2 - d^2)^{1/2}}, \quad d \leq r \leq r_0 \tag{7A.8}$$

where d and r_0 are indicated in Fig. 7A.1. Assuming that R and M are independent random variables, it can be shown from the theory of functions of random variables that (Cornell, 1968)

$$P_y = P[Y \geq y] = 1 - F_Y(y) = \frac{C}{l} Gy^{-\beta/b_2}, \, y \geq y' \tag{7A.9}$$

where

$$C = \exp(\beta m_0)(b_1)^{\beta/b_2} \tag{7A.10}$$

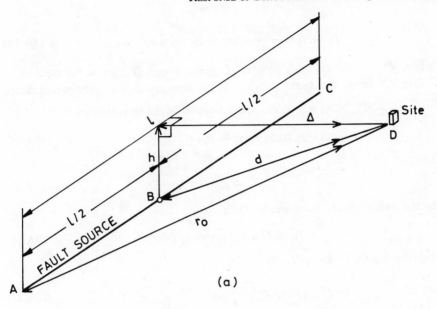

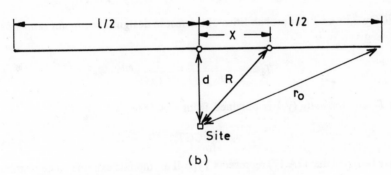

Fig. 7A.1 Line source (a) perspective; (b) plan (Cornell, 1967).

$$G = \frac{2}{d^{\alpha}} \int_{0}^{\sec^{-1}(r_0/d)} (\cos u)^{\alpha-1} \, du \qquad (7A.11)$$

$$\alpha = \beta \frac{b_3}{b_2} - 1 \qquad (7A.12)$$

and

$$y' = b_1 \exp(b_2 m_0) \, d^{-b_3} \qquad (7A.13)$$

Note that parameter G depends on the geometry of the source and can be similarly derived for the area source. For point source $G = (r)^{-(1+\alpha)}$.

Equation (7A.9) gives the probability that the site intensity Y, will exceed the value y, given that an event of magnitude $M \geq m_0$ has occurred on the source. Since the occurrence of such events at a source is assumed to be Poisson with arrival rate v, it is clear that the events with site-intensity $y \geq y'$ are 'special events' with arrival rate $P_y \, v$, and

$$P_N^*(n, t) = \frac{(P_y \, vt)^n}{n!} \exp(-P_Y \, vt), n = 0, 1, 2, \ldots \qquad (7A.14)$$

where $N^*(t)$ is the counting process of special events $(Y \geq y)$, at the site.

Let $Y_m^{(t)}$ denote the maximum value of the intensity over a t-year period. Since

$$P[Y_m^{(t)} \leq y] = P \text{ (Zero special-events in the time interval } t)$$

$$F_{Y_m^{(t)}}(y) = P_N^*(0, t) = \exp[-P_y \, vt] \qquad (7A.15)$$

Let

$$Y_m = Y_m^{(t)} \mid t = 1$$

be the *annual maximum intensity*. Then

$$P_N^*(0, 1) = F_{Y_m}(y) = \exp[-P_y \, v] \qquad (7A.16)$$

Substituting for P_y from (7A.9) in (7A.16)

$$F_{Y_m}(y) = \exp\left[-\frac{v}{l} CGy^{-\beta/b_2}\right], \quad y \geq y' \qquad (7A.17)$$

$$\simeq 1 - \hat{v}CGy^{-\beta/b_2}, \quad y \geq y', \, \hat{v} = v/l$$

if $\hat{v}CGy^{-\beta/b_2} \ll 1$, for probabilities of interest in design. The *annual return period* (T_y) is given by

$$T_y = \frac{1}{1 - F_{Y_m}(y)} = \frac{1}{\hat{v}CG} y^{\beta/b_2} \qquad (7A.18)$$

The *T-year intensity* (y_T) is obtained from (7A.18)

$$y_T = (\hat{v}CGT_y)^{-\beta/b_2} \qquad (7A.19)$$

It may be noted that (7A.17) represents Type-II asymptotic extreme value distribution of largest values.

The above results have been derived for a single source. If a site may experience shaking by more than one source, the above results can be easily extended, if the sources are assumed to be independent. Consider m such sources. It is clear that $P[Y_m^{(t)} \leq y]$ is the probability that the maximum value from *each* source is less than or equal to y, that is

$$F_{Y_m}^{(t)}(y, t) = \prod_{j=1}^{m} F_{Y_{mj}}^{(t)}(y, t)$$

$$= \exp\left[-\sum_{j=1}^{m} \hat{v}_j \, C_j \, G_j y^{-\beta/2b_{2j}} \, t\right], \, y \geq y' \qquad (7A.20)$$

where $F_{Y_{mj}}^{(t)}(y, t)$ is the distribution of the maximum, in time t, for jth source, and y' is the largest y_j'. If constants β, b_i, $i = 1, 2, 3$ are the same for each source, (7A.20) reduces to

$$F_{Y_m}^{(t)}(y, t) = \exp[-C\hat{v}Gy^{-\beta/b_2} \, t] \qquad (7A.21)$$

where

$$\hat{v}G = \sum_{j=1}^{m} \hat{v}_j G_j \qquad (7A.22)$$

Equations (7A.21), (7A.22) indicate that each source contributes approximately in an additive way to the risk. The design intensity for a specified probability ($y_{t;p}$), the return period for a specified intensity, and T-year intensity can be obtained obtained from (7A.20) or (7A.21) for multiple source situation.

7A.3 Some Extensions and Comments
The discussion in the preceding sections covered the basic approach to engineering seismic risk analysis, which can be integrated with random vibration analysis to compute the total risk. In this section, we shall discuss some extensions and comment on the possible improvements.

Limited Magnitude Distribution
In the preceding treatment, the earthquake magnitude could take any value above threshold value m_0. On physical grounds it can be argued that there should be a limit on the size of the earthquakes (Housner, 1970). Such a limit can be established on the basis of historical or geophysical data for a region or a source. Let m_1 be the upper limit on the magnitude of the earthquake. The probability density function for M can be expressed as

$$p_M(m) = k_{m_1} \beta \exp\left[-\beta(m - m_0)\right], m_0 \leq m \leq m_1 \qquad (7A.23)$$

where

$$k_{m_1} = \left[1 - \exp\left[\beta(m_1 - m_0)\right]\right]^{-1} \qquad (7A.24)$$

is the normalising factor. The restriction on magnitude to an upper limit m_1, defines a focal distance r_y beyond which no earthquake will cause ground motion in excess of y. This distance is

$$r_y = \left(\frac{y}{b_1}\right)^{-(1/b_3)} \exp\left(\frac{b_2 m_1}{b_3}\right) \qquad (7A.25)$$

and the distribution for $Y_m^{(t)}$ can be expressed as (Cornell and Vanmarcke, 1969),

$$F_{Y_m^{(t)}}(y, t) = \exp\left[-\hat{v}_y t(1 - k_{m_1})\right] \exp\left[-k_{m_1} C \hat{v}_y G_y y^{-\beta/b_2}\right], \quad y \geq y' \qquad (7A.26)$$

where $\hat{v}_y G_y$ is given by (7A.22). The subscript y on $\hat{v}$ and G implies that sources, or part thereof, are considered within the radius r_y. Computations of seismic risk show that major contribution to risk comes from more frequent smaller earthquakes close to the site, and the risk is not significantly decreased by the upper-limit on magnitude (Cornell and Vanmarcke, 1969).

Scatter in Attenuation Relations
The attenuation relations (7A.2) provide only a crude correlation with the large scatter of observed data (Esteva, 1970). To incorporate the effect of scatter on seismic risk estimates, the attenuation relations may be defined as

$$Y = (b_1 e^{b_2 M_R - b}) \varepsilon \qquad (7A.27)$$

where ε is an assumed random error, defined as the ratio between observed and measured ground intensities. Esteva (1970) has shown that in ε is approximately normally distributed with zero mean and standard deviation σ, lying in the range 0.20 to 1.10. Integrating over the error term in the seismic risk analysis, it can be shown that for fixed R, that is for a point source

$$P_Y = P(Y > y) = (1 - k_{m1}) \, \Phi^*(Z/\sigma)$$

$$+ k_{m1} \, \Phi^*(Z/\sigma - \beta\sigma/b_2) \, e^{\beta^2 \sigma^2/2b_2^2} \, e^{\beta m_0}$$

$$\times R^{-\beta b_3/b_2} \, (y/b_1)^{-\beta/b_2} \qquad (7A.28)$$

where $\Phi^*(\cdot)$ is the complimentary cumulative distribution function of the standardized Gaussian distribution, and

$$Z = \ln(y) - \ln(b_1 e^{b_2 m_1} R^{-b_3}) \qquad (7A.29)$$

The above result for the point source can be used to compute the exceedance probability for arbitrary sources numerically. The effect of including the attenuation law uncertainty can result in almost doubling the computed risk.

Cornell and Vanmarcke (1969) have examined in detail the sensitivity of seismic risk estimates to various parameters. They have found that risk may be sensitive to the focal depth H, to the attenuation constant, b_3, and to the ratio β/b_2. Esteva (1967) has suggested replacement of focal-depth H, by 'effective focal-depth' $(H^2 + 20^2)^{1/2}$, to improve the fit in the near-source zone. Basu (1977) has fitted uniform and truncated lognormal distributions to the focal depth data for Indian subcontinent and incorporated these in the seismic risk analysis.

Line-source Model Incorporating Length of Fault Slip

In the seismic risk analysis described above, the distance considered is the focal distance from the site. This implies that the total energy in an earthquake is concentrated at the focus, and in this sense the analysis is based on a point-source model. The energy released during a strong-motion earthquake is derived from the total area of fracture in earth's crust (Housner, 1970). In the point-source model, no consideration is given to the length of fault-slip and the shortest distance from the site to the slip. A line-source model takes into consideration the relation between the earthquake magnitude and the length of the fault slip, and also the closest distance of the site to the fault slip (Kiureghian and Ang, 1977). The length of fault-slip is computed from (7A.5) and the attenuation rules (7A.2) are used in the risk analysis with R replaced by the shortest distance from site to the slip. The distribution of the maximum value of the site intensity can be derived on this basis. The results obtained for the point-source and line-source models show that the seismic risk estimates based on the former tend to be unsafe, particularly for higher levels of ground motion intensity.

Multivariate Seismic Risk Analysis (Kiureghian, 1981)

The seismic risk analysis described above was based on a single ground motion intensity, implying that a single ground motion intensity descriptor is sufficient to describe the performance or safety of the structural systems. Such an analysis may be called a *univariate risk analysis*. Generally, joint behaviour of several ground motion intensity descriptors may be needed to adequately describe the performance or safety of a structural system, and such an analysis is called *multivariate risk analysis*.

Let the performance of a structure subjected to earthquakes be described by a function $g(\bar{Y})$, where Y_i, $i = 1, 2, 3, \ldots n$, are random variables describing the ground motion at a site due to a random earthquake. Let $\bar{y}$ be a specific ground motion, and let $g(\bar{y}) \leq 0$ imply failure and $g(\bar{y}) \geq 0$ imply survival. The boundary between the two regions is the *failure surface* in an n-dimensional space and is given by

$$g(\bar{y}) = g(y_1, y_2, \ldots y_n) = 0 \tag{7A.30}$$

Let the attenuation relation for each Y_i be expressed as

$$Y_i = \varepsilon_i f_i(M, R), i = 1, 2, 3, \ldots n \tag{7A.31}$$

where ε_i is random error. The performance function can now be expressed as

$$g(\varepsilon_1 f_1(M, R), \ldots \varepsilon_n f_n(M, R)) = G(M, R, \bar{\varepsilon}) \tag{7A.32}$$

and the failure surface as

$$G(m, r, \bar{\varepsilon}) = 0 \tag{7A.33}$$

The probability of failure for a random earthquake is given by $P[G(M, R, \bar{\varepsilon}) \leq 0]$, and assuming Poisson arrivals,

$$P\{G(M, R, \bar{\varepsilon}) \leq 0 \text{ in } (0, t)\} = 1 - \exp\{-vt\, P[G(M, R, \bar{\varepsilon}) \leq 0]\}$$

$$\simeq vt\, P[G(M, R, \bar{\varepsilon}) \leq 0] \tag{7A.34}$$

where

$$P[G(M, R, \bar{\varepsilon}) \leq 0] = \sum_j P[G(M, R, \bar{\varepsilon}) \leq 0 \mid E_j] P(E_j) \tag{7A.35}$$

in which E_j = event that the earthquake occurs in source j, and

$$P[G(M, R, \bar{\varepsilon}) \mid E_j] = \int_m^{m_j} \int_{\bar{\varepsilon}} P[G(m, R, \bar{\varepsilon}) \leq 0 \mid E_j]\, p_\varepsilon(\bar{\varepsilon})\, p_M(m)\, d\bar{\varepsilon}\, dm \tag{7A.36}$$

It may be reasonable to replace $\bar{\varepsilon}$ by ε, reducing the n-fold integral on ε to a single integral. For an upper-limit on M, a failure distance can be computed, which circumscribes the potential sources that may cause failure (Kiureghian, 1981). The return period can be determined from (7A.34) as before.

APPENDIX 7B: **Response Spectra**

The response spectra of a component of an earthquake represents the maximum value of the response of a single-degree of freedom (sdf) system as a function of the fraction of critical damping ζ, and the natural period $T_0 = 2\pi/\omega_0$. Let $\ddot{Z}(t)$ represent a single component of ground acceleration during an earthquake and let $X(t)$ be the relative displacement of a sdf system. Then (Housner, 1941)

$$SD(\zeta, T_0) = \max_{0 \leq t \leq T} [|X(t)|] \tag{7B.1}$$

$$SV(\zeta, T_0) = \max_{0 \leq t \leq T} [|\dot{X}(t)|] \tag{7B.2}$$

$$SV(\zeta, T_0) = \max_{0 \leq t \leq T} [|\ddot{X}(t) + \ddot{Z}(t))|]$$

where SD, SV and SA are called the displacement, velocity and absolute acceleration spectra respectively of $\ddot{Z}(t)$. Since the maximum value may occur upto half-cycle beyond the duration of earthquake, $T = T' + T_0/2$, where T' is the duration of significant ground motion. Fig. 7B.1 shows the velocity spectrum of the N-S component of El Centro earthquake, May 18, 1940.

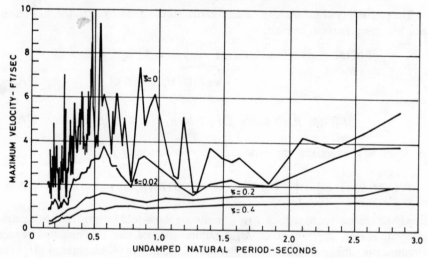

Fig. 7B.1. Velocity spectrum: El Centro, May 18, 1940; NS component.

It is convenient to define the pseudo-spectra of a component of an earthquake,

$$PSV(\zeta, T_0) = \omega_0 SD(\zeta, T_0) = \frac{2\pi}{T_0} SD(\zeta, T_0) \tag{7B.4}$$

$$PSA(\zeta, T_0) = \omega_0 PSV(\zeta, T_0) = \omega_0^2 SD(\zeta, T_0) \tag{7B.5}$$

where PSV and PSA denote the pseudo-relative velocity and pseudo-absolute acceleration spectra respectively. The relations (7B.4) and (7B.5) between SD, PSV and PSA, make it possible to represent the spectral properties of a component of an earthquake on a tripartite log-log plot. Fig. 7B.2 shows the pseudo-spectra of NS component of El Centro, 1940 earthquake on a log-log plot.

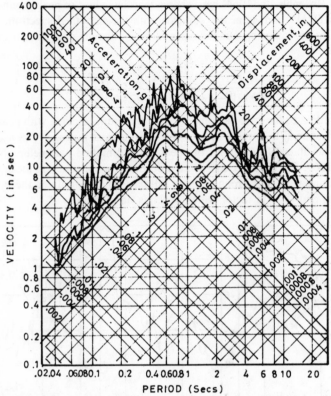

Fig. 7B.2.   EI Centro, May 18, 1940; NS component spectra on log-log plot ζ = 0, 0.02, 0.5, 0.10 and 0.20.

The *spectrum intensity* of a component of an earthquake is defined (Housner, 1952) as

$$SI(\zeta) = \int_{0.1}^{2.5} SV(\zeta, T_0)\, dt \qquad (7B.6)$$

It represents a gross measure of the severity of an earthquake component as regard to its effect on elastic structures.

The Fourier spectra of a component of an earthquake represents the final energy of an undamped sdf system as a function of its period. The undamped velocity spectra, SV $(0, T_0)$, on the other hand, represents the maximum value of the energy of a sdf system excited by the earthquake component. The shape of SV $(0, T_0)$ is similar to the Fourier spectra but its amplitudes are somewhat higher (Hudson, 1963).

The spectral ordinates of a component of an earthquake can be normalized with respect to one of the ground motion parameters, such as, peak ground acceleration A_g, r.m.s value σ_g or spectrum intensity SI. Such spectra are called normalized spectra. Housner (1959) constructed a set of smooth response spectra by normalizing the eight horizontal components of four major California earthquakes. These spectra reflect, in an average sense, the response to be expected in a strong earthquake and have been recommended for use as *design-spectrum* curves after scaling for a site.

Esteva and Rosenblueth (1964) have shown that the values of peak ground motion parameters D_g, V_g and A_g nearly coincide with the smoothed spectrum for 25% damping. The smoothed spectra for other values of damping can be constructed by the relation

$$\frac{Av\,[S(\zeta,\,T)]}{Av\,[S(0.25\ T)]} = \frac{(1 + 0.6\ \zeta\omega_0 T)^{0.45}}{(1 + 0.15\zeta\omega_0 T)^{0.45}} \qquad (7B.7)$$

Since the ordinates for 25% damping average spectrum coincide with peak ground motion, equation (7B.7) makes it possible to construct average response spectra for different values of damping from peak ground motion parameters. Newmark and Hall (1969, 1973) have suggested a procedure for constructing mean response spectra starting with peak ground motion estimates and multiplying these, for each value of ω_0 and ζ, by a factor which depends on damping.

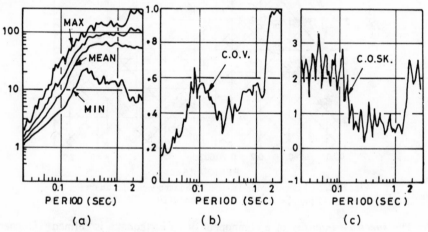

Fig. 7B.3. Statistics of scaled response spectra for 0.5% damping (a) maximum, mean, mean + 1σ, and minimum value; (b) coefficient of variation; and (c) coefficient of skewness. (Vanmarcke, 1977).

The response spectra of an ensemble of earthquake records scaled to a common maximum show a wide variation due to random nature of ground motion. Fig. 7B.3 shows the maximum, minimum and the mean spectra; the coefficient of variation; and coefficient of skewness of 39 historical strong-motion records scaled to common maximum acceleration (Vanmarcke, 1977). The ordinates of the response spectra are random variables, and their distribution can be established through random vibration analysis (7.28) or (7.51). The response spectra corresponding to a specified exceedance probability, the mean spectra and the variance of spectral ordinates can be determined from the distribution.

7B.1 Response Spectra and Power Spectral Density of Ground Acceleration

The power spectral density (psd) of the excitation finds extensive application in the frequency domain random vibration analysis of systems, and in the simulation of random processes. In earthquake engineering, however, spectral characteristics of ground motion are commonly incorporated in the analysis through response spectra. It is, therefore, important to establish an explicit relationship between psd and response spectra of ground motion. If psd is given, response spectra can be determined directly through random vibration analysis (7.29–7.31). The inverse problem of determining psd for a specified response spectra involves iterative solution of an integral equation through a numerical procedure. However, approximate explicit expressions for psd can be derived for a given response spectra.

Rosenblueth and Bustamante (1962) (see also the discussion by Caughey and Gray, 1963) established a simple relationship between the maximum energy of a sdf system and the constant psd of white noise excitation. Housner and Jennings (1964) extended the relationship to express the one-sided psd of ground acceleration in terms of associated average velocity spectra

$$G_{\ddot{Z}\ddot{Z}}(\omega) = a_1 \left[\frac{2\zeta\omega_0}{a_2\pi} (1 - e^{-2\zeta_0\omega_0 T/a_2}) \{E[SV(\zeta, \omega_0)]\}^2 \right] \qquad (7B.8)$$

where a_1 is a multiplicative constant chosen to fit the intensity of ground motion, and $a_2 = 2.5 \pm 10\%$.

The relationship between the response spectra $SD_{T;p}$ and the standard deviation of the response σ_X is given by (7.29) as

$$SD_{T;p} = \eta_{T;p}\,\sigma_X \qquad (7B.9)$$

where $\eta_{T;p}$ is the peak response factor given by (7.35) or (7.47) depending on the assumption regarding the limiting decay rate. Now

$$\sigma_X^2 = \lambda_0 = \frac{\pi G_{\ddot{Z}\ddot{Z}}(\omega_0)}{4\zeta\omega_0^3} \qquad (7B.10)$$

and

$$SD_{T;p} = PSA_{T;p}/\omega_0^2 \qquad (7B.11)$$

Combining (7B.9), (7B.10) and (7B.11),

$$G_{\ddot{Z}\ddot{Z}}(\omega_0) = \frac{4\zeta}{\pi\omega_0} \left(\frac{PSA_{T;p}}{\eta_{T;p}} \right)^2 \qquad (7B.12)$$

Kaul (1978) has derived the above result (with $\eta_{T;p}$ given by (7.35), starting with the distribution function of peaks due to Cartwright and Longuet-Higgins (1956) and stating the approximations at each stage. He has also solved the inverse problem exactly using a numerical integration procedure, and has shown that the approximate relation (7B.12) gives good agreement with exact results.

Figures 7B.4 and 7B.5 show respectively the response spectra, and derived psd, obtained from the exact and approximate procedures. Equation (7B.12) is based on the assumption that $\ddot{Z}(t)$ is a stationary random process, and therefore of infinite duration. Rosenblueth and Elorduy (1969) have proposed a simple correction for transient nature of response and have proposed that damping in (7B.12) is replaced by

$$\zeta_e = \zeta + \frac{2}{\omega_0 T} \qquad (7B.13)$$

Vanmarcke (1977) has proposed a simple iterative procedure to determine the psd for a specified acceleration spectra. He has shown that

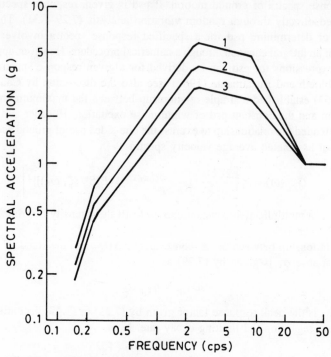

Fig. 7B.4. NRC design response spectrum, 1—ζ = .005, 2—ζ = .02 and 3—ζ = 0.05 (Kaul, 1978).

$$\frac{SA_{T;p}}{\eta_{T;p}} = \sigma_A^2(T) = G_{\ddot{Z}\ddot{Z}}(\omega_0)\,\omega_0\left(\frac{\pi}{4\zeta_T} - 1\right) + \int_0^\omega G_{\ddot{Z}\ddot{Z}}(\omega)\,d\omega \qquad (7B.14)$$

The second term in the above equation makes-up for low and intermediate frequencies. Let $\omega_0^{(1)} < \omega_0^{(2)} < \ldots < \omega_0^{(i)} < \ldots$ represent a sequence of frequencies, for which psd is proposed to be determined. $\omega_0^{(1)}$ is determined first from (7B.14) by neglecting the second term. At an arbitrary frequency $\omega_0^{(i+1)}$, $i = 1, 2, \ldots$, the integral of $G_{\ddot{Z}\ddot{Z}}(\omega)$ is evaluated numerically upto $G_{\ddot{Z}\ddot{Z}}(\omega_0^{(i)})$, and $G_{\ddot{Z}\ddot{Z}}(\omega_0^{(i+1)})$ is computed from (7B.14).

Response Envelope Spectrum
Although response spectra are widely used in earthquake engineering, it must be recognised that they provide an incomplete description of ground motion. A number of ground motion records with wide variation in duration and intensity can match a given design spectrum. The application of response spectrum to mdf systems introduces significant errors, as the response spectra do not indicate the time at which the peaks occur.

To remove both these limitations Trifunac (1971) defined response envelope spectrum, $A(\zeta, T_0, t)$ as the envelope of the response of a sdf system. The response envelope spectrum contains standard response spectrum, since

$$SD(\zeta, \omega_0) = \max_{0 \le t \le T} [A(\zeta, \omega_0, t)] \qquad (7B.15)$$

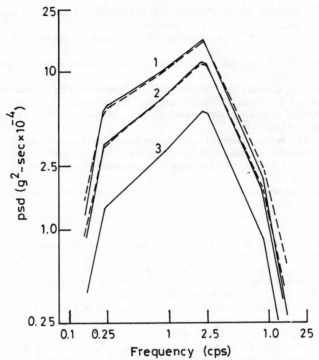

Fig. 7B.5. Response spectrum consistent acceleration psd: (—) exact; (– – –) approximate. 1—$\zeta = 0.05$; 2—$\zeta = 0.02$; and 3—$\zeta = 0.05$ (Kaul, 1978).

Response envelope spectrum of an earthquake allows a more precise modal combination for mdf systems. Also, for a zero-mean, broad-band Gaussian model, $E[A^2 (\zeta, \omega_0, t)]$ is closely related to the evolutionary psd. The practical value of response envelope spectrum is, however, limited by the difficulty in plotting $A(\zeta, \omega_0, t)$. For a fixed ζ, it represents a surface in (ω, t) plane. Also, it cannot be simply related to response statistics and, therefore, its applicability in random vibration analysis is limited.

REFERENCES

Ahmadi, G., 1979. Generation of artificial time-histories compatible with given response spectra—A review. *Solid Mech. Arch.*, 4(3): 207–239.

Amin, M. and A. H-S. Ang, 1968. Nonstationary stochastic model of earthquake motions. *J. Engg. Mech. Div., Proc. ASCE*, 94(EM2): 559–583.

Amin, M., H.S., Ts'AO, and A.H-S. Ang. 1969. Significance of nonstationarity of earthquake motions, *Proc. 4th WCEE*, Santigo, Chile.

Ang, A.H-S., 1973. Structural risk analysis and reliability-based design. *J. Str. Div., Proc. ASCE*, 99(ST9): 1891–1910.

Asthana, A.K. and T.K. Datta, 1990. A simplified response spectrum method for random vibration analysis of flexible base buildings. *J. of Eng. Struct.*, 12: 185–194.

Baber, T.T. and Y.K. Wen, 1981. Random vibration of hysteretic degrading systems. *J. Engg. Mech. Div., Proc. ASCE*, 107 (EMG):

Baber, T.T. and Y.K. Wen, 1982. Stochastic response of multistorey yielding frames. *Earth. Engg. Str. Dyn.*, 10: 403–416.

Basu, S., 1977. Statistical analysis of seismic data and seismic risk analysis of Indian Peninsula. Ph.D. thesis, Dept. of Civ. Engg., I.I.T., Kanpur.

Basu, S. and N.C. Nigam, 1977. Seismic risk analysis of Indian peninsula. *Proc. 5th WCEE*: 2, New Delhi: 425–433.

Berg, G.V. and G.W. Housner, 1961. Integrated velocity and displacement of strong ground motions. *Bull. Seis. Soc. Am.*, 51:

Bogdanoff, J.L., J.E. Goldberg and M.C. Bernard, 1961. Response of a simple structure to a random earthquake type disturbance. *Bull. Seism. Soc. Am.*, 51: 293–310.

Bolotin, V.V., 1960. Statistical theory of aseismic design of structures. *Proc. 2WCEE, Tokyo and Kyoto*, Vol. II: 1365–1374.

Boore, D.M., 1983. Stochastic simulation of high-frequency ground motions based on seismological models of the radiated spectra. *Bull. Seism. Soc. Am.*, 73(6): 1865–1894.

Byrcroft, G.N., 1960. White noise representation of earthquakes. *J. Engg. Mech. Div., Proc. ASCE*, 86(EM2): 1–16.

Cartwright, D.E. and M.S. Longuet-Higgins, 1956. The statistical distribution of maxima of a random function. *Proc. Roy. Soc. London*. A. 237: 212–232.

Caughey, T.K., 1960. Random excitation of a system with bilinear hysteresis. *J. App. Mech.*, 27: 649–652.

Caughey, T.K., 1963. Equivalent linearisation techniques. *J. Acoust. Soc. Amer.*, 35(11): 1706–1711.

Caughey, T.K. and H.J. Stumpf, 1961. Transient response of a dynamic system under random excitation. *J. Appl. Mech., Proc. ASME*, 28: 563–566.

Caughey, T.K. and A.H. Gray, Jr., 1963. Discussion of distribution of structural response to earthquakes. *J. Engg. Mech. Div., Proc., ASCE*, 89(EM2): 159–168.

Clough, R.W. and J. Penzien, 1975. *Dynamics of Structures*. New York: McGraw-Hill.

Cornell, C.A., 1964. Stochastic process models in structural engineering. Tech. Report No. 34. Stanford University, Deptt. of Civil Engg.

Cornell, C.A., 1968. Engineering seismic risk analysis. *Bull. Seism. Soc. Am.*, 58(5): 1583–1606.

Cornell, C.A., 1969. Bayesian statistical decision theory and reliability-based design. *Intern. Conf. Struct. Safety and Reliability of Engg. Struct.*, Washington.

Cornell, C.A. and E.H. Vanmarcke, 1969. The major influences on seismic risk. *Proc. 4th WCEE*, Santigo, Chile.

Cornell, C.A., 1971. Probabilistic analysis of damage to structures under seismic loads. *Dynamic Waves in Civil Engineering*. D.A. Howell, ed., Wiley Interscience.

Corotis, R.B., E.H. Vanmarcke, and C.A. Cornell, 1972. First passage of nonstationary random processes. *J. Engg. Mech. Div., Proc. ASCE*, 98(EM2): 401–414.

Crandall, S.H., 1970. First crossing probabilities of linear oscillator. *J. Sound and Vib.* 12(3): 285–289.

Crandall, S.H. and W.Q. Zhu, 1983. Random vibration: a survey of recent developments. *J. App. Mech.*, 50: 953–962.

Drenick, R.F., 1970. Model-free design of aseismic structures. *J. Engg. Mech. Div., Proc. ASCE*, 96(EM4): 483–493.

Drenick, R.F., 1973. Aseismic design by way of critical excitation. *J. Engg. Mech. Div., Proc. ASCE*, 99(EM4): 649–667.

Drenick, R.F., 1977. On the class of non-robust problems in stochastic dynamics. in *Stochastic Problems in Dynamics*. B.L. Clarkson, eds., London: Pitman.

Elishakoff, I., 1983. *Probabilistic Methods in the Theory of Structures*. John Wiley, New York.

Esteva, L., 1969. Seismicity prediction: a Bayesian approach. *Proc. 4th WCEE*, Santiago, Chile.

Esteva, L., 1970. Seismic risk and seismic design decisions. *In. Seismic Design of Nuclear Power Plants*. R.J. Hanson, (ed.) Cambridge, Mass, MIT Press.

Esteva, L. and E. Rosenblueth, 1964. Espectros de Temblores a Distancias Moderadas Y. Grandes. *Bol. Soc. Mex. Ing. Sism.*, 2(1): 1–18.

Gardner, J.K. and L. Knopoff, 1974. Is the sequence of earthquakes in Southen California, with aftershocks removed, Poisson. *Bull. Seism. Soc. Am.*, 64(5): 1363–1368.

Gazetas, G., 1976. Vibration analysis of inelastic multi-degree-of-freedom systems subjected to Earthquake ground motion. M.I.T. Report.

Gersh, W. and G. Kitagawa, 1985. A time varying AR coefficient model for modelling and simulating of earthquake ground motion. *Earthquake Eng. Struct. Dyn.*, 13: 243–254.

Goldberg, J.E., J.L. Bogdanoff and D.R. Sharpe, 1964. The response of simple nonlinear systems to random disturbance of earthquake type. *Bull. Seism Soc. Am.*, 54: 263–276.

Gray, A.H., Jr., 1966. First-passage time in random vibrational system. *J. Appl. Mech.* 33: 189–191.

Hadjian, A.H., 1981. Correlation of the components of strong ground motion. *Bull. Seism. Soc. Am.*, 71(4): 1323–1331.

Housner, G.W., 1941. Calculating the response of an oscillator to arbitrary ground motion. *Bull. Seism. Soc. of Am.* 68(1): 143–149.

Housner, G.W., 1947. Characteristics of strong motion earthquakes. *Bull. Seism. Soc. Am.*, 37(1): 19–37.

Housner, G.W., 1955. Properties of strong ground motion. *Bull. Seism. Soc. Am.*, 45(3): 197–218.

Housner, G.W., 1956. Limit design of structures to resist earthquakes. *Proc. Ist WCEE*, San Francisco.

Housner, G.W., 1959. Behaviour of Structures during earthquakes. *J. Engg. Mech. Div., ASCE,* 85 (EM4): 109–129.

Housner, G.W., 1970. Strong ground motion. In *Earthquake Engineering*, R.L. Wiegel (ed.), Prentice-Hall, New Jersey.

Housner, G.W., 1977. Important properties of earthquake ground motion. *Proc. 5th WCEE.*, New Delhi.

Housner, G.W. and P.C. Jennings, 1964. Generation of artificial earthquakes. *J. Engg. Mech. Div., Proc. ASCE*, 90 (EM1): 113–150.

Hsu, T.-I and M.C. Bernard, 1978. A random process for earthquake simulation. *Earthquake Engg. Str. Dyn.*, 347–362.

Hudson, D.E., 1963. Some problems in the application of spectrum techniques to strong-motion earthquake analysis. *Bull. Seism. Soc. Am.*, 53(2).

Hudson, D.E., G.W. Housner and T.K. Caughey, 1960. Discussion-White noise representation of earthquakes. *J. Engg. Mech. Div., Proc. ASCE.* 86 (EM4): 193–195.

Husid, R.L., 1969. Analysis de terremotos: Analysis general Revista del IDIEM, Santiago, Chile, 8(1): 21–42.

Idriss, I.M., 1978. Characteristics of earthquake ground motion. State-of-the-art report. *Proc. Speciality Conference on Earthquake Engineering and Soil Dynamics, ASCE, III:* 1151–1266.

Irschik, H. and F. Ziegler, 1985. Nonstationary random vibrations of yielding frames. *Nuclear Engg. and Design*, 90: 357–364.

Iwan, W.D. and A.B., Mason, Jr., 1980. Equivalent linearization of systems subjected to nonstationary random excitation. *Int. J. Nonlinear Mech.*, 15: 71–82.

Iyengar, R.N. and K.T.S.R. Iyengar, 1969. A nonstationary random process model for earthquakes and its application. Structures Research Report I/69. Deptt. of Civil Engg., I.I.Sc. Bangalore.

Iyengar, R.N. and C.S. Manohar, 1985. System dependent critical stochastic seismic excitations, M15/6, 8 *Int. Conf. SMIRT*, Brussels.

Iyengar, R.N. and C.S. Manohar, 1987. Nonstationary random critical seismic excitations. *J. Engg. Mech. Div., Proc. ASCE*, 113 (EM4): 529–541.

Iyengar, R.N. and P.N. Rao, 1979. Generation of spectrum compatible accelerograms. *Earth. Engg. Str. Dy.*, 7: 253–263.

Iyengar, R.N. and P.K. Dash, 1978. Study of random vibration of nonlinear systems by Gaussian closure technique. *J. Appl. Mech.*, 45: 393–399.

Jennings, P.C., G.W. Housner, and N.C. Tsai, 1969. Simulated earthquake motions for design purposes. *Proc. 4th WCEE*, Chile: 145–160.

Kanai, K., 1957. Semi-empirical formula for seismic characteristics of the Ground. *Bull. Earthq. Res. Inst.*, Univ. of Tokyo, 35: 309–325.

Karnopp. D. and T.D. Scharton, 1965. Plastic deformation in random vibration. *J. Acoust. Soc. Amer.*, 39(6): 1154–1161.

Kast, G., T. Tellkamp, H. Kamil, A. Gantyat and F. Weber, 1978. Automated generation of spectrum compatible time histories. *Nucl. Engg. Des.*, 45: 243–249.

Kaul, M.K., 1978. Stochastic characterization of earthquakes through their response spectrum. *Earth. Engg. Str. Dyn.*, 6: 493–509.

Kiureghian, A.D., 1980. Structural response to stationary excitation. *J. Engg. Mech. Div., Proc. ASCE*, 106(EM6): 1195–1213.

Kiureghian, A.D., 1981. Seismic risk analysis of structural systems. *J. Engg. Mech. Div., Proc. ASCE*, 107(EM6): 1133–1153.

Kiureghian, A.D. and A.H.-S. Ang, 1977. A fault-rupture model for seismic risk analysis. *Bull. Seism. Soc. Am.*, 67(4): 1173–1194.

Kiureghian, A.D. and A.H.-S. Ang, 1977. A fault-rupture model for eismic risk analysis. *Bull. Seism. Soc. Am.*, 54: 1871–1873.

Levy, R.F. Kozin and R.B.B. Moorman, 1971. Random processes for earthquake simulation. *J. Engg. Mech. Div. Proc. ASCE*, 97: 495–517.

Levy, S. and J.P.D. Wilkinson, 1976. Generation of artificial time-histories, rich in all frequencies from given response spectra. *Nucl. Engg. Res.*, 38: 241–251.

Lin, Y.K., 1967. *Probabilistic Theory of Structural Dynamics*, New York: McGraw Hill.

Lomnitz, C., 1966. Statistical prediction of earthquakes. *Reviews of Geophysics*, A(3): 337–393.

Lomnitz, C., 1973. Poisson processes in earthquake studies. *Bull. Seism. Soc. Am.*, 63(2): 735.

Lomnitz, C. and E. Rosenblueth (ed.), 1977. *SeismicRisk and Engineering Decisions.* Elsevier, New York

Lutes, L.D., Y.-T.T. Chen and S.-H. Tzung, 1980. First-passage approximations for simple oscillators. *J. Engg. Mech. Div. Proc. ASCE,* 106 (EM6): 1111–1124.

Matsushima, Y., 1974. Stochastic response of structure due to three-dimensional earthquake excitations. *Building Research Institute, Japan. Research Paper No.* 58.

McGuire, R.K., 1976. Methodology for incorporating parameter uncertainties into seismic hazard analysis for low risk design intensities. *Int. Symp. on Earthquake. Str. Engg.*, St. Louis Missouri: 1007–1021.

Nau, R.F. et al, 1982. Simulating and analysing artificial nonstationary earthquake ground motions. *Bull. Seism. Soc. Am.*, 72(2): 615–636.

Newmark, N.M. and E. Rosenblueth, 1971. *Fundamentals of Earthquake Engineering*. Englewood Cliffs, N.J., Prentice-Hall.

Newmark, N.M. and W.J. Hall, 1969. Seismic design criteria for nuclear reactor facilities. *Proc. 4th WCEE*, Santiago.

Newmark, N.M. and W.J. Hall, 1973. A rational approach to seismic design standards for structures. *Proc. 5th WCEE*, Rome.

Nigam, N.C., 1970. Yielding in framed structures under dynamic loads. *J. Engg. Mech. Div., Proc. ASCE.*, EM(5): 387–409.

Nigam, N.C., 1982. Phase properties of a class of random processes. *Earthq. Engg. Str. Dyn.,* 10(5): 711–717.

Nigam, N.C., 1983. *Introduction to Random Vibrations*. Cambridge; Mass: The MIT, Press.

Nigam, N.C., 1984. Phase properties of earthquake ground acceleration records. *Proc. 8th WCEE.* San-Fransisco.

Ohsaki, Y., 1979. On the significance of phase content in earthquake ground motion. *Earthq. Engg. Str. Dyn.,* 7: 427–439.

Ohsaki, Y., R. Iwasaki, I. Ohakawa, and T. Masao, 1979. Phase characteristics of earthquake accelerogram and its application Trans—*5th Int. Conf. Str. Mech. in Reactor Tech.*, K(a): 1–8.

Parzen, E., 1962. *Stochastic Processes*, Holden-day, San Fransisco.

Penzien, J. and M. Watabe, 1975. Characteristics of 3-dimensional earthquake ground motions. *Earthq. Engg. Str. Dyn.,* 3: 365–373.

Permount, A., 1984. The generation of spectrum compatible accelerogram for the design of nuclear power plants. *Earth. Egg. Str. Dyn.*, 12: 481–497.

Polhemus, N.W. and A.S. Cakmak, 1981. Simulation of earthquake ground motions using ARMA models. *Earthq. Engg. Str. Dyn.*, 9(4): 343–354.

Richter, C.F., 1958. *Elementary Seismology*. W.H. Freeman, San Francisco.

Roberts, J.B., 1976. First-passage time for the envelope of a randomly excited linear oscillator. *J. Sound Vib.* 46(1): 1–14.

Rosenblueth, E., 1951. *A basis for aseismic design*, Ph.D. Thesis, Univ. of Illinois, Urbana.

Rosenblueth, E. and J.I. Bustamante, 1962. Distribution of structural responses to earthquakes. *J. Engg. Mech. Div. Proc. ASCE*, 86 (EM2): 1–16.

Rosenblueth, E. and J. Elorduy, 1969. Response of linear systems to certain transient disturbances. *Proc. 4th WCEE*, Santiago, Chile: 185–196.

Saragony, G.R., R. Alarcon and J. Crempin, 1980. Gaussian properties of earthquake ground motion. *Proc. 7th WCEE*, 2: 491–498.

Scanlan, R.H. and K. Sachs, 1974. Earthquake time-histories and response spectra. *J. Engg. Mech. Div., Proc. ASCE.*, 100 (EM4): 635–655.

Schiff, A. and J.L. Bogdanoff, 1966. *Some Problems in Engineering Seismology*. Centre for applied stochastics, Purdue University.

Shinozuka, M., 1970. Maximum structural response to seismic excitations. *J. of Engg. Mech. Divn.*, *Proc. ASCE*, 96: 729–738.

Shinozuka, M. and Y. Sato, 1967. Simulation of nonstationary random process. *J. Engg. Mech. Div.*, *Proc. ASCE*, 93(EM4): 11–40.

Shinozuka, M. and J.N. Yang, 1971. Peak structural response to nonstationary random excitations. *J. Sound Vib.* 14(4): 505–517.

Smeby, W., 1982. Stochastic analysis of response of structures and multiple supported secondary systems to multidirectional ground motion. Ph.D. Thesis, Univ. of Calif., Berkeley.

Smeby, W. and A.D. Kiureghian, 1982. Stochastic analysis of response of structures and multiple supported secondary systems to multidirectional ground motion. Report No. 82-0510, Det. Norske Veritas, Oslo, Norway.

Spanos, P.T.D., 1980. Probabilistic earthquake energy spectra equations. *J. Engg. Mech. Div., Proc. ASCE*, 106 (EM4): 147–159.

Spanos, P-T.D., 1983. Digital synthesis of response-design spectrum compatible earthquake records for dynamic analysis. *The Shock and Vibration Digest*, 15(3): 21–29.

Spanos, P-T.D. and L.D. Ludes, 1980. Probability of response to evolutionary process. *J. Engg. Mech. Div., Proc. ASCE*, 106(EM2): 213–224.

Sues, R.H., Y.K. Wen and A.H-S. Ang., 1983. Stochastic seismic performance evaluation of buildings. Civil Engineering studies, SRS No. 56, University of Illinois at Urbana, Illinois.

Tazimi, H., 1960. A statistical method of determining the maximum response of a building structure during an earthquake. *Proc. 2nd WCEE*, Vol. II, Tokyo and Kyoto, Japan.

Trifunac, M.D., 1971. Response envelope spectrum and interpretation of earthquake ground motion. *Bull. Seism. Soc. Am.*, 61(2): 343–356.

Trifunac, M.D., 1971a. A method for synthesizing realistic strong ground motion. *Bull. Seism. Soc. Am.*, 61(6): 1739–1753.

Vanmarcke, E.H., 1969. First passage and other failure criteria in narrow band random vibration. A discrete state approach. Ph.D. dissertation, Department of Civil Engineering, MIT.

Vanmarcke, E.H., 1972. Properties of spectral moments with applications to random vibration. *J. Engg. Mech. Div. Proc. ASCE*, 98 (EM2): 425–446.

Vanmarcke, E.H., 1977. Structural response to earthquakes. In *Seismic Risk and Engineering Decisions*. C. Lomnitz and E. Rosenblueth, ed., Amsterdam; Elsevier: 287–337.

Vanmarcke, E.H. and D. Veneziano, 1973. Seismic damage at inelastic systems. A random vibration approach. M.I.T. Deptt. of Civil Engg. Rep. R73-5.

Veletsos, A.J. and Y.T. Wei, 1971. Lateral and rocking vibrations of footings. *J. Soil Mech. and Foundation Engg. ASCE*, 97: 1227–1248.

Watabe, M. and M. Tohado, 1978. Research on simulation of three-dimensional earthquake ground motions—Part I and II. *Proc. 5th Japan Earthq. Engg. Symp*: 105–120.

Whitman, R.V. and C.A. Cornell, 1977. *Seismic Risk and Engineering Decisions*. (ed.) C. Lomnitz and E. Rosenblueth, Elsevier: 339–380.

Wiegel, R.L. (Ed.), 1970. *Earthquake Engineering*, Prentice Hall, Englewood Cliffs, New Jersey.

Wen, Y.K., 1976. Method for random vibration of hysteretic systems. *J. Engg. Mech. Div. Proc. ASCE*, 102 (EM2): 249–263.

Wen, Y.K. 1980. Equivalent linearization for hysteretic systems under random excitations. *J. Appl. Mech.*, 47(1): 150–154.

Ziegler, F., 1987. Random vibrations: A spectral method for linear and nonlinear structures. *J. Prob. Engg. Mech.*, (2): 92–99.

8.   Response of Structures to Wind

8.1   Introduction

Wind effects on structures have many special features. Unlike earthquakes, which occur in seismically active regions of the world, wind is everywhere. Severe wind conditions, which may cause extreme loading on structures, occur primarily in certain coastal regions, but freak extreme winds, such as, thunderstorms, tornadoes can occur anywhere. The wind characteristics at a site are determined by general weather pattern over many years—called the *wind climate*; and local effects, such as, terrain, proximity to a large body of water—called the *wind structure*. Both, the wind climate and the wind structure are major sources of uncertainty in predicting the response of structures to wind. The wind induced loads are a result of complex interactions between the approaching wind and the structure. The spatial and temporal random fluctuations in the wind velocity; and the shape, size, permeability, surface finish and the dynamic properties of structure determine the characteristics of the wind loads. In flexible structures, such as, tall buildings, cables, masts, long-span bridges, dynamic effects are often very significant and may cause strong random vibrations.

A schematic of the wind loading chain is shown in Fig. 8.1. The wind induced load W can be expressed as (NBC, Canada, 1975)

$$W = qC_eC_gC_p \tag{8.1}$$

where q is the reference wind pressure; C_e is the exposure factor determined by the terrain; C_g is the gust factor; and C_p is the pressure coefficient. In (8.1), q reflects the wind climate; and C_e and C_g the wind structure. The uncertainty associated with each of these factors may be assessed by determining the variance of the wind load, which can be expressed as

$$V_w^2 = V_q^2 + V_{c_e}^2 + V_{c_g}^2 + V_{c_p}^2 \tag{8.2}$$

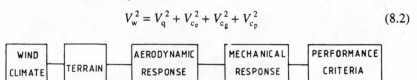

Fig. 8.1.   The wind loading chain.

Table 8.1 gives approximate estimates of the variances in (8.2) based on code and wind tunnel tests (Surry and Davenport, 1979). It is clear from relative values of the variance that wind climate (q) is the dominant source of uncertainty in determining the wind load. The uncertainty due to wind structure (C_e and C_g) is also significant, but it can be reduced by wind tunnel testing.

The random dynamic forces experienced by a structure due to wind are caused primarily by the turbulence in the approaching flow, trail of vortices shed in the wake, and motion of the structure. The trend towards reduction in stiffness. mass

Table 8.1   Approximate estimates of variance in (8.2)

Type of Wind	V	V_{c_e}	V_{c_g}	V_{c_p}	V_w	Remarks
Hurricane	0.35				0.44	
		0.20	0.10	0.15		Code
Extratropical	0.25				0.37	
Hurricane	0.35				0.36	
		0.05	0.05	0.05		Wind Tunnel
Extratropical	0.25				0.26	

and damping in modern light weight monolithic structural systems have contributed to increased importance of dynamic effects. Due to random nature of wind-induced loads, a probabilistic approach based on random vibration analysis, provides a rational and consistent basis for analysis and design of structures against wind. The pioneering work of Davenport and his group at the University of Western Ontario, has provided a strong base for a world wide research activity in this area (Davenport, 1961a; Sachs, 1972; Simiu and Scanlan, 1978; Gould and Abu-sitta, 1980; and Proceedings: Wind Effects on Buildings and Structures).

In this chapter we shall discuss the random vibration analysis and design of structures subjected to wind loading. The basic characteristics of wind flow in atmospheric boundary layer will be briefly discussed. Stochastic models of wind climate, wind structure, and methods for computing extreme wind velocities from past data will be discussed in detail. We will then consider various aspects of wind-structure interaction and the stochastic properties of wind induced loads. Response of structures to wind loads will be discussed in the framework of linear random vibration theory. Finally, some special aspects of probabilistic design of structures against wind, including the effect of wind direction will be discussed.

Due to complex nature of interaction between wind flows and structures, an entirely theoretical approach for determining wind loads and response is not feasible. A combined treatment involving meteorological information, wind tunnel testing and response analysis is, therefore, commonly used. In this context, a wind tunnel may be looked upon as a analog computer to determine the response of a structure to wind. From the measurements made on aeroelastic and pressure models exposed to turbulent boundary layer flows of varied speed, direction and structure, wind loads and response can be directly established. A large body of literature is now available on wind tunnel testing for determination of loads and response. We will not cover this aspect. Reader may refer to texts and publications in this area (Sachs, 1972; Simiu and Scanlan, 1978; and Rao, 1985).

8.2   Characteristics of Wind

Wind is the motion of air relative to the surface of the earth due to differential solar heating and rotation of the earth. The temperature differences due to solar radiation, and the radiation from the earth, produce large scale pressure systems, which drive the air masses generating the wind. The characteristics of the wind are strongly influenced by the geographical location, topographic features, and proximity of large bodies of water. In tropical and semi-tropical regions, the wind environment is relatively calm. Close to the equator ($< 10°$ latitude), severe storms are rare.

However, in many coastal regions of the tropics, strong winds are caused by tropical cyclones, which nucleate and intensify over a body of warm water. As these cyclones hit a coastline and travel inland, they degenerate rapidly. The severe wind conditions in such storms are, therefore, confined to narrow coastal belts. Outside the tropics, winds are primarily associated with heat transfer from equator to the poles through large scale pressure systems, or cyclones, travelling from east to west due to the rotation of the earth. In some temperate regions, extreme winds are also associated with infrequent small scale disturbances, such as, thunderstorms, tornadoes and squalls.

The extreme winds caused by tropical cyclones, thunderstorms, tornadoes are called *non-well-behaved winds*. Due to infrequent occurrence, the data-base for such storms is limited. The modelling and simulation of wind environment in regions prone to non-well-behaved winds, therefore, require special techniques, such as, Monte Carlo method (Russel, 1971; Tryggvason et al, 1976). Over large geographical areas, however, the winds are relatively *well-behaved*. Our discussion will be confined primarily to such winds.

8.2.1   The Boundary Layer

The earth's atmosphere extends to several kilometers above its surface. The shear action of the surface roughness retards the wind velocity to near zero at the surface. It increases gradually with height and attains a nearly constant value at the gradient height. This region varies from 300 to 500 m and is called the earth's boundary layer. The forces determining the flow pattern in the lower part of earth's atmosphere are: (i) the forces due to pressure systems; (ii) Coriolis forces due to rotation of the earth; (iii) the frictional resistance caused by the rough boundary of the earth; (iv) buoyancy forces due to vertical temperature gradients; and (v) the viscous forces. At gradient height, the wind speed and direction are determined by the form, intensity and geographic location of the pressure system. In the boundary layer, the flow in dominated by the influence of ground roughness and surface friction.

8.2.2   Long-term Spectra

Figure 8.2 shows a typical power spectrum of a long-term wind velocity record in earth's boundary layer. The spectrum covers periods ranging from one cycle/year to one cycle/second, and shows characteristic periodicities of the order of one

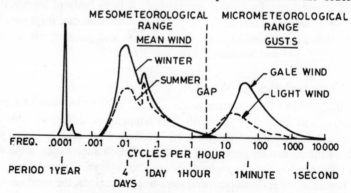

Fig. 8.2. Long-term wind speed spectrum

year, several days, one day, and a few minutes. Some of the distinctive features of the spectrum are:

(i) The energy is distributed over two distinct frequency ranges—mesometeorological and micrometeorological—separated by a low energy region called *spectral gap* (1 cycle/hour).

(ii) In mesometeorological range, the peak at one year period arises from annual variation in temperature gradient between polar and equatorial regions. The dominant peak at 4 day period corresponds to large scale pressure systems, and the smaller peak at 1 day period is due to diurnal heating and cooling effects.

(iii) In micrometeorological range, the high frequency energy is associated with boundary layer turbulence. Large scale eddies having dimensions comparable to boundary layer thickness are generated due to instabilities in the flow near the ground. The Reynold stresses break down the large scale motions into progressively small scale motions, which are finally dissipated through viscosity. A peak occurs at about one cycle/minute, and, therefore, there is significant energy in the range of natural frequencies of typical slender structures.

(iv)   The energy in the high-frequency range increases approximately as the square of the average wind speed. Thus, for strong winds, the spectrum level is higher than for light winds.

In modelling the wind environment, it is convenient to distinguish between wind climate and wind structure. We shall define the wind climate to include "those characteristics of wind determined over many years by the general weather patterns, as distinct from those wind characteristics dependent on local environment" (Surry and Davenport, 1979). The wind structure shall include wind fluctuations having periods less than twenty minutes, which are associated with mechanically or convectively generated turbulence, including characteristics due to non-homogeneous local effects. As noted earlier, both the wind climate and the wind structure are major sources of uncertainty in predicting the response of structures to wind.

8.3   Stochastic Models of Wind

In the preceding section, we have noted that the structure and the climate represent distinctive features of wind, and both contribute to the uncertainty inherent in the wind environment. In this section, we shall discuss the stochastic models of wind structure and wind climate.

8.3.1   Wind Structure

Wind velocity records in earth's boundary layer exhibit both spatial and temporal randomness, as shown in Fig. 8.3. Figure 8.3a shows typical time-dependent fluctuations in wind velocity and direction during an extra-tropical cyclone. Figure 8.3b shows a short segment of wind velocity record; and Fig. 8.3c the variation of wind velocity with height. From these records, it can be concluded that:

(i) both magnitude and direction of wind velocity exhibit rapid fluctuations as a function of time. The intensity of fluctuations increases with the average velocity defined by the trace centre-line;

(ii) the average velocity is a slowly varying function of time, and over small time intervals of the order of 20 min-1 hr, it may be assumed to be constant; and

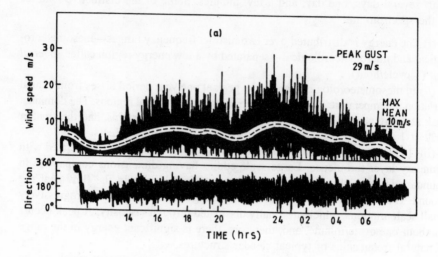

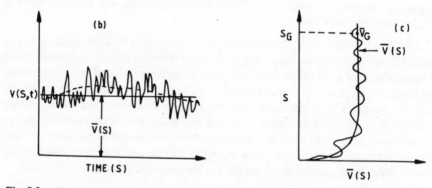

Fig. 8.3. (a) Anemograph showing fluctuations in the wind velocity and the direction during an extra-tropical cyclone; (b) a short segment of wind velocity record; and (c) variation of wind velocity with height in atmospheric boundary layer.

(iii)   the wind velocity is nearly zero at the earth's surface and it increases gradually with height to a nearly constant value at gradient height s_G. It exhibits random fluctuations with height.

Consistent with the wind characteristics described above, it is convenient to express the instantaneous wind velocity $V(s, t)$ in the form

$$V(s, t) = \bar{V}(s) + V'(s, t) \tag{8.3}$$

where s is the height above the surface, and $\bar{V}(s)$ is the *mean wind velocity* given by

$$\bar{V}(s) = \frac{1}{T_0} \int_{t - T_0/2}^{t + T_0/2} V(s, t)\, dt \tag{8.4}$$

In (8.4), T_0 is the averaging interval, which may range from 10 minutes to 1 hour corresponding to spectral gap in the long-range wind spectrum. $V'(s, t)$ in (8.3) represents random fluctuations about the mean wind and is called the *gust component*. Over short intervals of time (~ T_o), it can be treated as a locally stationary random process with zero mean and variance

$$\sigma_{V'}^2(s) = \sigma_V^2(s) = \frac{1}{T_o} \int_{t-T_o/2}^{t+T_o/2} (V(s, t) - \bar{V}(s))^2 \, dt$$

$$= \frac{1}{T_o} \int_{t-T_o/2}^{t+T_o/2} V'^2(s, t) \, dt \tag{8.5}$$

Actual observations of the gust components at different heights show that $\sigma_V(\approx 0.1\bar{V}_G)$ tends to decrease slightly with height, and increase slightly with terrain roughness. The relative gustiness, or the intensity of turbulence, in the wind is denoted by I and defined as:

$$I = \frac{\sigma_V}{\bar{V}_{10}} \tag{8.6}$$

where $\bar{V}_{10}$ is the mean wind velocity 10 m above ground level, and $\bar{V}_G$ is the gradient mean wind velocity.

A statistical analysis of measured wind records shows that the distribution of wind velocity in a specified direction is approximately normal and can be expressed as

$$p_V(v) = \frac{1}{(2\pi)^{1/2}\sigma_V} \exp\left[-\frac{1}{2}\left(\frac{v - \bar{V}}{\sigma_V}\right)^2\right] \tag{8.7a}$$

$$p_{V'}(v') = \frac{1}{(2\pi)^{1/2}\sigma_V} \exp\left[-\frac{1}{2}\left(\frac{v'}{\sigma_V}\right)^2\right] \tag{8.7b}$$

Thus, $\bar{V}$ and σ_V completely describe the statistical variation of wind velocity over a time interval T. Let V_m represent the maximum value of the instantaneous wind velocity in the time interval T. Then,

$$V_m(s) = \max_{0 \le t \le T} [V(s, t)] = \bar{V}(s) + \max_{0 \le t \le T} [V'(s, t)] \tag{8.8}$$

Since $V(s, t)$ is $N(\bar{V}, \sigma_V)$, the distribution function and the expected value of V_m are given by (Nigam, 1983)

$$F_{V_m}(v) = \exp\left[-N_{\bar{V}}^+ T \exp\left\{-\frac{1}{2}\left(\frac{v - \bar{V}}{\sigma_V}\right)^2\right\}\right] \tag{8.9}$$

and

$$E[V_m] = \bar{V} + \eta\sigma_V$$

where the peak factor

$$\eta = [2 \ln (N_{\bar{V}}^+ T)]^{1/2} + \frac{0.5772}{[2 \ln (N_{\bar{V}}^+ T)]^{1/2}}$$

$N_{\bar{V}}^{+}$ is the expected rate of crossing of the level $\bar{V}$ with a positive slope—a measure of the average frequency of gusts.

Mean wind Velocity Profile

We have seen that the surface wind consists of a mean flow determined by large scale pressure system and superimposed gust fluctuations caused by surface roughness. For strong winds, which are of primary interest in engineering design, the mechanical processes in the atmosphere dominate over thermal processes and the stability is neutral. In the free atmosphere away from the ground, the pressure gradient is balanced by the inertial effects. Near the ground, however, this balance is disturbed by the effect of drag forces induced by the flow around and over obstacles. This results in retardation of air flow near the ground, changes in direction away from isobars, and turbulence. The rougher the surface, the more pronounced these effects are.

The shear action of the surface roughness retards the mean wind velocity to near zero at the surface, which increases gradually with height and attains a nearly constant value $\bar{V}_G$ at the gradient height s_G (Fig. 8.3c). The mean velocity profile may be expressed by the power law (Davenport, 1967)

$$\bar{V}(s) = \bar{V}_G \left(\frac{s}{s_G}\right)^\alpha \tag{8.10}$$

where $\alpha > 0$ is called the *power law exponent*. A more realistic description of mean wind velocity profile is given by the logarithmic law (Simiu and Lozier, 1975)

$$\bar{V}(s) = \frac{1}{k} V^* \ln\left(\frac{s - s_d}{s_o}\right) \tag{8.11}$$

where k is the von Karman coefficient (approximately 0.4); V^* is the friction wind velocity corresponding to a reference height; s_d is the zero plane displacement; and s_o is the roughness length. The parameters, α, s_G, and s_o, depend on the nature of the terrain. Typical values are given in Table 8.2.

Table 8.2 Values of gradient height s_G; power law exponent α; surface drag coefficient κ; and roughness length s_o (Davenport, 1977)

Terrain	Gradient Height $s_{G\ (m)}$	Roughness Length $s_o(\ m)$	Power Law Exponent α	Surface Drag Coeff. κ
Rough sea	250	0.005–0.01	0.12	0.001
Open grassland	300	0.01 –0.1	0.16	0.005
Forest and suburban areas	400	0.3 – 1.0	0.28	0.015
City centres	500	1.0 – 5.0	0.40	0.050

Structure of Gust

The gust components of the wind velocity vector represent broad-band random fluctuations about the mean wind. If V_i', $i = 1, 2, 3$, represent the horizontal, transverse and vertical components respectively,

$$\sigma_v^2 = \sigma_{v_{1'}}^2 + \sigma_{v_{2'}}^2 + \sigma_{v_{3'}}^2 \qquad (8.12)$$

Near the earth's surface, $\sigma_{v_{1'}}$ is much larger than $\sigma_{v_{2'}}$ and $\sigma_{v_{3'}}$ and therefore, $\sigma_{v'} \simeq \sigma_{v_{1'}}$.

The spectral distribution of energy in the gust fluctuations is represented by the psd. Let $G_{VV}(f)$ represent the one-sided psd. Then

$$\sigma_{v'}^2 = \int_0^\infty G_{VV}(f)\,df = \int_0^\infty f G_{VV}(f)\,d(\log_e f) \qquad (8.13)$$

where $f G_{VV}(f)$ is called the *logarithmic power spectrum*.

The shape of the psd is specific to a location and wind condition. Davenport (1961a) suggested the following expression for psd, which is representative of several sites and wind conditions over a wide frequency range,

$$\frac{f G_{VV}(f)}{\kappa \bar{V}_{10}^2} = \frac{200 \bar{f}^2}{(1 + \bar{f}^2)^{4/3}} \qquad (8.14)$$

in which κ is the surface drag coefficient (Table 8.2); and $\bar{f}$ is the reduced frequency given by

$$\bar{f} = \frac{\bar{f} L_0}{\bar{V}_{10}} \qquad (8.15)$$

where L_0 is a representative scale length ($\simeq 1200$ m). Figure 8.4 shows a typical plot of the psd, given by (8.14), together with some field observations.

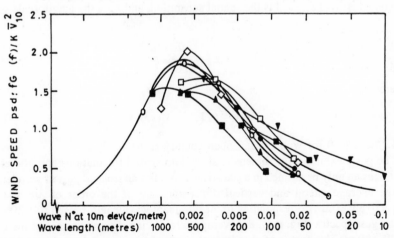

Fig. 8.4. Spectrum of horizontal wind speed for strong winds. ($\square$, $\lozenge$, $\blacksquare$, $\blacktriangledown$,-field observations) (Davenport, 1977).

It is seen from the spectrum that the energy in atmospheric turbulence is associated primarily with the wavelengths in the range 200–2000 m, with a peak at about 600 m. There is very little energy at wavelengths less than 10 m, and more than 5000 m. At wind speeds of interest in terms of wind loading, say 40 m/s, energy is primarily in the period range 5–50 sec, which covers large flexible structures.

Simiu (1974) proposed the following two expressions for the psd in terms of friction wind velocity V^*:

$$\frac{fG_{VV}(f, s)}{(V^*)^2} = \frac{4\hat{f}^2}{(1 + \hat{f}^2)^{4/3}} \tag{8.16}$$

and

$$\frac{fG_{VV}(f, s)}{(V^*)^2} = \frac{200\hat{f}}{(1 + 50\hat{f})^{5/3}} \tag{8.17}$$

in which

$$\hat{f} = \frac{fs}{\tilde{V}(s)} \tag{8.18}$$

Equation (8.17) applies to a wide frequency range, while (8.16) is applicable only in the higher range.

Several other expressions have been proposed for psd to fit the data (Harris, 1972; Merchant, 1964). Of these, the expression for psd due to Merchant, (2.25), is of special interest for the construction of filtered Poisson model as discussed in Chapter 2.

The spatial random fluctuations of the wind velocity can be modelled by the coherence function

$$\phi(s^1, s^2; f) = \exp\left(-\frac{|\Delta s|}{L_c}\right) \tag{8.19}$$

in which $|\Delta s| = |s^1 - s^2|$ is the spatial separation, and L_c is the length scale given by

$$L_c = \frac{\tilde{V}}{f\overline{C}} \tag{8.20}$$

where

$$\tilde{V} = \frac{1}{2}(V(s^1) + V(s^2)) \tag{8.21}$$

and $\overline{C} = 6$ to 9 for wooded and open country respectively.

The gusts exhibit three dimensional random spatial fluctuations and, therefore, *gust size* can only be defined in a statistical sense. For an isolated gust of wavelength, λ, the average lateral and vertical dimensions are of the order of $\lambda/8$ and $\lambda/5$ respectively. Thus, corresponding to the dominant wavelength-range (200–2000 m) in the gust spectra (Fig. 8.4), the average across-wind dimensions are of the order of 40–400 m. This fact is of significance as it reduces the total wind load on large structures.

8.3.2   Wind Climate

The wind climate represents the absolute strength of the wind environment at a site determined, over many years, by the general weather pattern. It is best described by the gradient wind speed outside the range of frictional influence of the ground, and nonhomogeneous local effects (Surry and Davenport, 1979). A large measure of uncertainty is associated with the wind climate and, therefore, it may be described statistically by the parent process distributions representing the common events, and the extreme value distributions representing the rare events.

Parent Process distributions
The gradient wind velocity at a location varies slowly as a function of time and direction. If the variations due to direction are disregarded, the Weibull distribution is found to fit the data well and the distribution function of the wind velocity can be expressed as

$$F_{\bar{V}}(v) = P(\bar{V} \leq v) = 1 - \exp\left\{-\left(\frac{v}{C}\right)^k\right\} \tag{8.22}$$

Note that, if $k = 2$, (8.22) reduces to a Rayleigh distribution. Taking double logarithm, (8.22) can be expressed as

$$\log_e\left(-\log_e\left(1 - F_{\bar{V}}(v)\right)\right) = k \log_e v - k \log_e C \tag{8.23}$$

which represents a straight line if $\log_e\left(-\log_e\left(1 - F_V - (v)\right)\right)$ is plotted against $\log_e v$. The Weibull parameters k and C can be determined from such plots (Davenport, 1977).

At many locations, wind climate displays strong directional characteristics corresponding to prevailing wind. The response of structures may also exhibit directional sensitivity, as some structures are particularly weak in certain directions called the critical directions. If a critical direction coincides with the prevailing wind direction, the structure is subjected to the extreme loading condition. To take into account the directional effects, it is necessary to construct joint distributions of the wind velocity and direction. Two distributions given below have been found to fit the data well. The first distribution is

$$F_{\bar{V},\theta}(v, \theta) = 1 - A(\theta) \exp\left\{-\left(\frac{v}{C(\theta)}\right)^{k(\theta)}\right\} \tag{8.24}$$

where θ is the wind direction, and $A(\theta)$ satisfies the condition

$$\int_0^{2\pi} A(\theta)\, d\theta = 1 \tag{8.25}$$

Note that (8.24) is an extension of Weibull distribution, (8.22), to include directional effects.

The second distribution—a modified bivariate Gaussian distribution, which takes into account the variation of surface exposure with direction, can be expressed in the form

$$p_{\bar{V},\theta}(v, \theta) = v \exp\left\{-\left(A(\theta)v^2 + B(\theta)v + C(\theta)\right)\right\} \tag{8.26}$$

in which

$$A(\theta) = \sum_{j=0}^{\infty}\left(A_j \cos 2\pi j\theta + A_j' \sin 2\pi j\theta\right) \tag{8.27}$$

with similar expressions for $B(\theta)$ and $C(\theta)$.

The values of direction dependent parameters in the above distributions are determined from observational data usually collected over 16°, 22.5° sectors. An example of directional distribution obtained by fitting (8.24) is shown in Fig. 8.5.

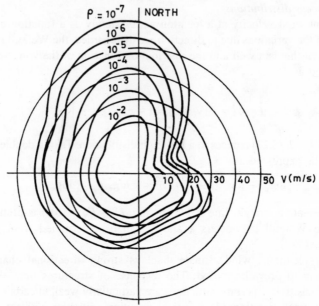

Fig. 8.5. Directional distribution of wind speed. Probability of exceeding V in 22.5° sector.

Extreme Value distributions

In Chapter 7, we have seen that in risk analysis distributions of the extreme values play a dominant role. For wind loading, extreme value distributions are generally determined on the basis of maximum annual hourly mean velocity data obtained from long term records. Alternatively, parent process distribution can be used to derive the extreme value distribution following the procedure discussed in Appendix 7A. If the parent distribution from which the extremes are drawn is of exponential type, such as the Weibull distribution (8.24), the extreme value distribution is of Type I (Fisher-Tippett) and can be expressed as

$$P_1 = L_{\bar{V}_E}(v) = P(\bar{V}_E > v) = 1 - \exp\left[-\exp\left\{-a(v - \tilde{U})\right\}\right] \qquad (8.28)$$

in which $\bar{V}_E$ is the annual extreme wind velocity, $\tilde{U}$ is the modal wind velocity (the mode), and $1/a$ is a scale wind velocity (the dispersion).

Through a double logarithmic transformation, (8.28) can be expressed as

$$v = \tilde{U} - \frac{1}{a}\left[\log_e\left(-\log_e\left(1 - P_1\right)\right)\right] \qquad (8.29)$$

which represents a straight line, if v is plotted against $\log_e(\log_e(1 - P_1))$, and can be used to determine the parameters $\tilde{U}$ and $1/a$ at a given site from the observed data.

The annual probability of exceeding the velocity v is related to the return period

$$T_v = \frac{1}{P_1} \qquad (8.30)$$

Substituting (8.30) in (8.29), and noting that

$$\log_e \left(1 - \frac{1}{T_v}\right) \simeq -\frac{1}{T_v}$$

and if T_v is large (> 5), it follows that

$$v \simeq \tilde{U} + \frac{1}{a} \log_e (T_v) \tag{8.31}$$

The probability that the design wind speed is exceeded at least once in N consecutive years s given by

$$P_N = L_{N, \bar{V}_E}(v) = P[\bar{V}_E > v \mid N \text{ consecutive years}]$$

$$= 1 - \exp(-P_1 N) = 1 - \exp\left(-\frac{N}{T_v}\right) \tag{8.32a}$$

$$\simeq \frac{N}{T_v}, \ N \ll T_v \tag{8.32b}$$

The design wind speed for a specified value P_N can be obtained by substituting for P_1 from (8.28) in (8.32a) and solving for v.

The choice of acceptable risk is governed by codes and standards which regulate public safety. The ANSI standard specifies risk directly in terms of return period T_v as given in Table 8.3.

Table 8.3   Risk in terms of return period (ANSI)

Condition	T_v (years)
(a) Structures having no human occupants or negligible risk to human life	25
(b) All permanent structures except those presenting a high degree of hazard to life and property in case of failure	50
(c) Exceptions to (b)	100

The preceding treatment is based on Type I extreme value distribution. Based on data in United States, Thom (1960, 1968) proposed a Type II (Frechet) distribution in the following form:

$$P_1 = L_{\bar{V}_E}(v) = 1 - \exp\left\{-\left(\frac{v - \tilde{\mu}}{\sigma}\right)^{-\gamma}\right\}, \quad v > 0 \tag{8.33}$$

in which $\tilde{\mu}$ is the location parameter; σ is the scale parameter, and γ is the tail length parameter. The parameters $\tilde{\mu}$, σ and γ can be determined from the straight line plot between $\log_e (-\log_e (P_1))$ and $\log_e V$. Keeping in view the inherent errors in the estimation of parameters associated with Type I and Type II distributions, either distribution can be used for all practical purposes. However, Type I distribution has an edge over Type II distribution, since parent distributions of wind speed tend to be of exponential type. Further, as $\gamma \to \infty$, Type II distribution reduces to Type I distribution, which is found to adequately model well-behaved wind climate and has been adopted in most countries (Simiu and Filliben, 1978; Ito and Fujino, 1977; and Dorman, 1983).

The Type I (8.28) and the Type II (8.33) distributions have also been used to model the non-well behaved winds in regions visited by infrequent cyclones and tornadoes. Thom (1967) proposed a mixed distribution to model such events

$$F_{\overline{V}_E}(v) = p_T \exp\left[-\left(\frac{v}{\sigma}\right)^{-4.5}\right] + (1 - p_T) \exp\left[-\left(\frac{v}{\sigma}\right)^{-9}\right] \qquad (8.34)$$

where the first and the second terms respectively represent the probabilities that winds associated with tropical cyclones and extratropical storms will not exceed the value v in any year, and p_T represents the probability of occurrence of tropical cyclone.

Figure 8.6 shows the histogram and the fit of the Type I and the Type II distributions to the cyclonic data for the east coast of India during the period 1891–1982 (Venkateswarlu et al, 1985). It includes cyclones with wind velocities greater than 62 knots. The mean and the standard deviation of the sample are 86.0 knots and 18.98 knots, respectively. The parameters of the Type I and the Type II distributions are estimated as:

Type I: $a = 0.06$; $\tilde{U} = 77.03$

Type II: $\sigma = 76.03$; $\tilde{\mu} = 0.0$; $\gamma = 7.2$

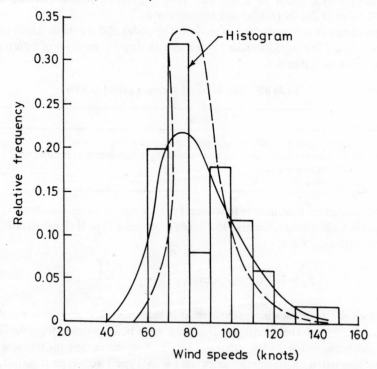

Fig. 8.6. Histogram of cyclonic wind data for east coast of India and fitted Type I (—) and Type II (– – –) models (Venkateswarlu et al, 1985).

Kolomogrov-Smirnov (K-S) goodness-of-fit test shows that both distributions are statistically admissible at 5% significance level, but the Type I gives a better fit. Based on the Type I distribution, the design wind velocity for a 50 year service life and risk levels 0.63 and 0.1 are:

$$\bar{V}_D = \bar{V}_{E,\,50;\,0.63} = 141 \text{ knots}$$

and

$$\bar{V}_D = \bar{V}_{E,\,50;\,0.1} = 179 \text{ knots}$$

Extreme Value and Parent Process Distributions

In Appendix 7A, we have derived the extreme value distribution from the distribution of the parent process for earthquakes. Following the same approach, we shall derive the extreme value distribution for wind using the parent process distribution (8.22). Let $N^+(v)T$ be the expected number of times the speed v is crossed with a positive slope in the time period T. For large values of v, crossings may be assumed to be independent and Poisson, so that

$$F_{\bar{V}_E}(v) = \exp\left[-N^+(v)T\right] \tag{8.35}$$

where

$$N^+(v) = \int_0^\infty \dot{v}\, p_{\bar{V}\dot{\bar{V}}}(v, \dot{v})\, d\dot{v}$$

Assuming that $\bar{V}(t)$ is stationary, $p_{\bar{V}\dot{\bar{V}}}(v, \dot{v}) = p_{\bar{V}}(v)\, p_{\dot{\bar{V}}}(\dot{v})$, and

$$N^+(v) = 2\pi C n \sigma_{\bar{V}} p_{\bar{V}}(v) \tag{8.36}$$

where

$$C = \frac{\displaystyle\int_0^\infty \dot{v} p_{\dot{\bar{V}}}(\dot{v})\, d\dot{v}}{\sigma_{\dot{\bar{V}}}}$$

and

$$n = \frac{\sigma_{\dot{\bar{V}}}}{\sigma_{\bar{V}}} = \left\{ \frac{\displaystyle\int_0^\infty f^2 G_{\bar{V}\bar{V}}(f)\, df}{\displaystyle\int_0^\infty G_{\bar{V}\bar{V}}(f)\, df} \right\}^{1/2} \tag{8.37}$$

where $G_{\bar{V}\bar{V}}(f)$ is the long term spectrum of $\bar{V}$. The constant C depends on the distribution of $\dot{\bar{V}}$, and for normal distribution it equals $\sqrt{2\pi}$ ($= 2.51$). Based on direct measurements, Gomes and Vickery (1974) have suggested $C = 2.26$. The value of n exhibits a large variation in the range 500–1000 cycles per annum. Hence, we may assume

$$N^+(v) \simeq 1500\, \sigma_{\bar{V}}\, p_{\bar{V}}(v)\, \text{year}^{-1} \tag{8.38}$$

Assuming a Weibull distribution (8.22) for $\bar{V}$, it can be shown, that for normal values of k, $\sigma_{\bar{V}} \simeq 0.93\, C/k$, and

$$\sigma_{\bar{V}} p_{\bar{V}}(v) = 0.93 \left(\frac{v}{C}\right)^{k-1} \exp\left\{-\left(\frac{v}{C}\right)^k\right\} \tag{8.39}$$

Substituting (8.39) in (8.38), we have

$$N^+(v) \simeq 1300 \left(\frac{v}{C}\right)^{k-1} \exp\left\{-\left(\frac{v}{C}\right)^k\right\}$$

and from (8.35)

$$F_{\tilde{V}_E}(v) = \exp\left[-1300T\left(\frac{v}{C}\right)^{k-1} \exp\left\{-\left(\frac{v}{C}\right)^k\right\}\right] \qquad (8.40)$$

For annual extremes, $T = 1$, and

$$F_{\tilde{V}_E}(v) = \exp\left[-1300\left(\frac{v}{C}\right)^{k-1} \exp\left\{-\left(\frac{v}{C}\right)^k\right\}\right] \qquad (8.41)$$

The distribution (8.41) can be approximated by Type I distribution (8.28), with modal value $\tilde{U}$ obtained by solving the equation

$$-1300 \left(\frac{\tilde{U}}{C}\right)^{k-1} \exp\left\{-\left(\frac{\tilde{U}}{C}\right)^k\right\} = 1 \qquad (8.42)$$

and dispersion factor a is given by

$$a = \frac{d}{dv}\left[-1300 \left(\frac{v}{C}\right)^{k-1} \exp\left\{-\left(\frac{v}{C}\right)^k\right\}\right] \qquad (8.43)$$

for $v = \tilde{U}$.
Davenport (1977) has suggested following approximate values:

$$\tilde{U} \simeq C7^{1/k}$$

and

$$\tilde{U}a \simeq 7k \qquad (8.44)$$

8.4 Response of Structures to Wind

8.4.1 Forces Due to Flow Past a Body
We have discussed the stochastic nature of wind environment, which consists of mean flows, superimposed by turbulent flows. A body obstructing the flow alters the flow pattern and gives rise to steady and fluctuating pressure distributions on the body surfaces. The nature of flow around the body, and consequential pressure distributions, depend upon the characteristics of the approaching wind flow; and the shape, size, surface finish, permeability and the motion, if any, of the body. The integration of pressure field over the body gives rise to forces *along* (called *drag*, denoted by the subscript D), and *across* (called *lift*, denoted by the subscript L) the direction of flow, and in some cases to *moments* (denoted by the subscript M).

Figure 8.7 shows the significant characteristics of the mean and fluctuating flows around a bluff body representative of the cross-section of a structure. The flow consists of three distinct regions (Davenport, 1977):

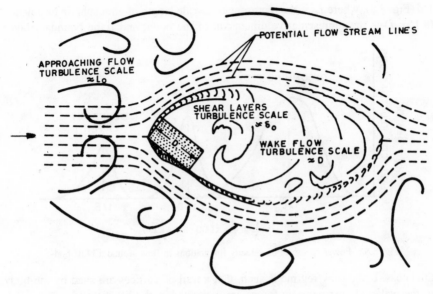

Fig. 8.7.   Flow of fluid around a bluff body in turbulent flow.

(i) the approaching flow due to wind;
(ii) the boundary layers attached to body surface and extending down stream in the form of free shear layers; and
(iii) the wake behind the body, which may include trail of vortices shed by the body.

Mean Flow Characteristics
The approaching flow follows the profile of the body, detaches from it and, after curving around the body re-attaches behind it. A thin boundary layer separates this region from the upstream face and preserves the 'no slip' condition at the surface of the body. Behind the separation points, the boundary layer becomes free shear layer and separates the approaching flow region from the wake. In the approaching flow region, the interaction between the mean flow and turbulence is insignificant with little dissipation of energy. The mean flow streamlines closely follow the theoretical potential flow lines and the mean pressure may be determined from Bernoulli's equation. In the wake region, the energy loss is significant and the mean pressures computed from potential flow using Bernoulli's equation may significantly overestimate the actual pressures.

Fluctuating Flow Characteristics
The fluctuating pressures on the body are due to the cumulative effect of the eddy motions in the three regions. Let λ represent the wave length corresponding to mean flow velocity $\bar{V}(s)$ and the frequency f. Then,

$$\lambda = \frac{\bar{V}(s)}{f} \tag{8.45}$$

The dominant frequency of the velocity fluctuations in the approaching flow, the wake and the boundary layer regions correspond, respectively to $1/L_0$, $1/D$ and

$1/\delta$ (Fig. 8.8), where L_0 is the representative scale length of atmospheric turbulence (8.15); D is the representative dimension of the body; and δ the boundary layer

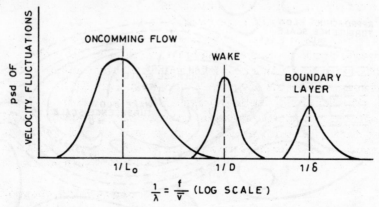

Fig. 8.8. Power spectrum of velocity fluctuations in flow around a bluff body.

thickness. If the flow regime is such, that a trail of vortices are shed by the body in the wake, almost periodic forces are generated in the lift direction. Figure 8.9 shows the nature of vortices for uniform flow, turbulet flow, and turbulent flow coupled with body in transverse motion. The vortex shedding frequency is determined by the Strouhal number $N_s = fD/\bar{V}$. Figure 8.9b shows that turbulent flow not only induces direct pressure fluctuations, but also modulates the vortex shedding frequency. Turbulence may also influence the mean forces and re-attachment of the approaching flow. The transverse motion of the body (Fig. 8.9c) induces well correlated fluid responses, and secondary flows interact with other forces in the wake causing mixing of frequencies.

(a)

(b)

(c)

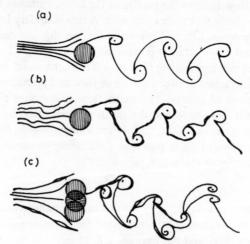

Fig. 8.9. Vortex shedding by bluff body (a) smooth flow; (b) turbulent flow; and (c) turbulent flow with body in motion.

The aerodynamic excitation of a body, due to the flow regimes described above consists of three types of dynamic forces induced by:

(i) turbulent fluctuations in the approaching flow causing both quasi-static and resonant responses in the along-wind and across-wind directions;

(ii) vortices shed in the wake of the body, causing primarily across-wind resonant responses; and

(iii) motion of the body, causing aerodynamic damping forces which determine the resonant response amplitude. The aerodynamic damping may add to, or subtract, from structural damping, and in the latter case may cause large amplitude oscillations in the across-wind direction.

The response of the structure is influenced by the combined action of these forces. It consists of an irregular, slowly varying component, called *background response* in the wind direction, or/and oscillatory components in the both along and across-wind directions with a dominant frequency (or frequencies) called the *resonant response* (Fig. 8.10).

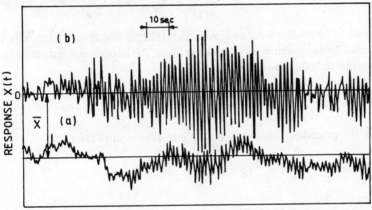

Fig. 8.10.   (a) Along-wind, and (b) across-wind measured response of a 100 m chimney (Van Koten, 1968).

Size Effect

The net load on a structure due to turbulence depends on the spatial gust size, λ, in relation to the size of the structure, $\sqrt{A}$, that is on the ratio

$$\frac{\lambda}{\sqrt{A}} = \frac{\bar{V}}{f\sqrt{A}} \tag{8.46}$$

where A is the area normal to the flow.

For high frequency components, $\lambda/\sqrt{A} \ll 1$, and pressures are correlated over small areas only. The total effect is, therefore, small due to simultaneous increase and decrease in pressure over different parts of the surface. For low frequency components, $\lambda/\sqrt{A} \gg 1$, and the influence of gust is correlated over the whole, or a large part of the total surface, resulting in large total effect.

We first consider a structure with area A, placed normally in an approaching flow in which the dominant wave length is much greater than the typical dimensions of the structure, that is, $\lambda/\sqrt{A} \gg 1$. We call such a structure, *point-like*, and we may regard the flow past it as quasi-steady. The drag force on the structure may be expressed as:

$$F_D(t) = \frac{1}{2} C_D \rho A V^2(t) = \frac{1}{2} C_D \rho A[\bar{V} + V'(t)]^2 \tag{8.47}$$

where C_D is the drag coefficient (determined from wind-tunnel test); and ρ is the density of the air. Ignoring the terms of the order $(V')^2$ in comparison with $\bar{V}^2$, the mean drag force

$$\bar{F}_D = \frac{1}{2} C_D A \rho \bar{V}^2 \tag{8.48}$$

and the fluctuating drag force due to gust

$$F_D'(t) = C_D A \rho \bar{V} V'(t) \tag{8.49}$$

The psd of $F_D'(t)$ is related to the psd of the wind velocity by the relation

$$G_{F_D F_D}(f) = (C_D \rho A \bar{V})^2 G_{VV}(f) \tag{8.50a}$$

$$= \frac{4(\bar{F}_D)^2}{(\bar{V})^2} G_{VV}(f) \tag{8.50b}$$

If the size of the structure is not small as compared to the dominant wavelength, adjustment must be made for the reduced spatial correlation of forces. This is conveniently introduced through an *aerodynamic admittance function*

$$\left| \chi \left(\frac{f\sqrt{A}}{\bar{V}} \right) \right|$$

From the preceding discussion on the size effects, it is clear that

$$|\chi| \to 1 \text{ as } \frac{f\sqrt{A}}{\bar{V}} \to 0$$

and

$$|\chi| \to 0 \text{ as } \frac{f\sqrt{A}}{\bar{V}} \to \infty$$

Figure 8.11 shows the empirical data for flat plates and prisms, and an empirically fitted relationship

$$\chi(f) = \frac{1}{1 + \left[\dfrac{2f\sqrt{A}}{\bar{V}} \right]^{4/3}}$$

Incorporating the admittance function, the psd of the drag force can be expressed as

$$G_{F_D F_D}(f) = \frac{4(\bar{F}_D)^2}{(\bar{V})^2} \chi^2 \left(\frac{f\sqrt{A}}{\bar{V}} \right) G_{VV}(f) \tag{8.51}$$

Lift Forces on a body
The lift force on a body due to turbulent fluctuations may be expressed by eqs. (8.47) through (8.50), by replacing the drag coefficient C_D with the lift coefficient C_L.

The lift force on a body may also be caused by the vortex shedding in the wake. The vortex shedding excitation has been found to be approximately Gaussian, and the psd of the induced lift force may be expressed as (Vickery, 1979)

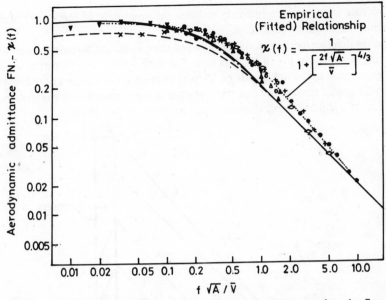

Fig. 8.11. Aerodynamic admittance for flat plates and prisms normal to the flow ($N_R \approx 2 \times 10^4$). (o, Δ, $\Diamond$, $\times$, $\blacktriangle$, $\bullet$, —experimental observations) (Vickery and Davenport, 1967).

$$\frac{f G_{F_L F_L}(f)}{F_L^2} \, \alpha \, \frac{\gamma}{\sqrt{2\pi} I} \exp\left[-\frac{1}{2}(\gamma - .1)^2/I^2\right] \qquad (8.52)$$

where

$$\gamma = \frac{f D}{\bar{V} N_s} = \frac{f}{f_e}; \qquad f_e = \frac{\bar{V} N_s}{D}$$

and I is the intensity of turbulence. The spectrum is peaked at the critical reduced frequency and its bandwidth is proportional to the turbulence intensity (Fig. 8.12).

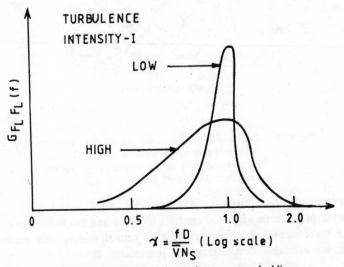

Fig. 8.12. Lift force spectra due to vortex shedding.

An increase in turbulence intensity reduces the amplitude of force spectrum and broadens the range of excitation.

The tip displacement response of a guyed mast to lateral turbulence and vortex shedding is shown in Fig. 8.13. It is seen that the response due to vortex shedding is large, even at low wind speeds (< 15 m/s) in modes 3–6. Studies of the response of similar guyed mast show that the response is quite sensitive to the distribution and size of cylindrical exteriors, and the location of the shroud.

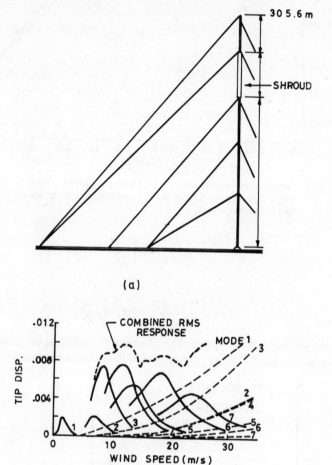

Fig. 8.13.   Response of a cylindrical guyed mast to lateral turbulence and vortex shedding (a) guyed mast; (b) tip response: (—) vortex shedding; (– – –) lateral turbulence (Davenport, 1985).

8.4.2   Response of Structures to Wind Forces

The response of structures to wind consists of mean and fluctuating components. Following three approaches, involving wind tunnel testing and mathematical modelling, are available to determine the response:

(i) *Aeroelastic modelling*:   All significant dynamic and aerodynamic properties of the structure and the wind are simulated in a wind tunnel. The response of full scale structure is inferred from the measured response of the model. This requires an understanding of scaling laws, but only a broad understanding of aerodynamic behaviour.

(ii) *Force measurements*:   Direct Measurements of static and dynamic forces induced on a rigid model placed in correctly scaled flow carried out in a wind tunnel. Responses are determined using a mathematical model of the structure.

(iii) *Analytical model*:   Responses are inferred from the mean and turbulence properties of wind and the aerodynamic characteristics of the structure.

We shall follow the second approach. Consider a sdf system subjected to force $F(t) = \bar{F} + F'(t)$, with psd given by (8.51). Let the displacement of the system be represented by $X(t) = \bar{X} + X'(t)$. Then, the psd of the displacement is given by

$$G_{XX}(f) = \frac{1}{k^2} |H(f)|^2 G_{FF}(f) \tag{8.53}$$

where

$$|H(f)|^2 = \frac{1}{[1 - (f/f_0)^2]^2 + 4\zeta^2 (f/f_0)^2} \tag{8.54}$$

and k is the stiffness; f_0 is the natural frequency; and ζ is the fraction of critical damping of the sdf system. Substituting for $G_{FF}(f)$ from (8.51) in (8.53),

$$G_{XX}(f) = \frac{4(\bar{F})^2}{k^2(\bar{V})^2} \chi^2 \left(\frac{f\sqrt{A}}{\bar{V}} \right) |H(f)|^2 G_{VV}(f) \tag{8.55}$$

and the variance of the displacement is given by

$$\sigma_X^2 = \frac{4(\bar{F})^2}{k^2(\bar{V})^2} \int_0^\infty \chi^2 \left(\frac{f\sqrt{A}}{\bar{V}} \right) |H(f)|^2 G_{VV}(f) \, df \tag{8.56}$$

The mean value of the displacement is given by

$$\bar{X} = \frac{\bar{F}}{k}. \tag{8.57}$$

Combining (8.56) and (8.57), it can be shown that (Davenport, 1977)

$$\frac{\sigma_X^2}{(\bar{X})^2} \simeq \frac{4\sigma_V^2}{(\bar{V})^2} \left[B + \chi_0^2 \frac{E}{\zeta} \right] = (2I)^2 \left[B + \chi_0^2 \frac{E}{\zeta} \right] \tag{8.58}$$

where I is the intensity of turbulence (8.4), $\chi_0 = \chi (f_0 \sqrt{A}/\bar{V})$ and

$$B \simeq \frac{1}{\sigma_V^2} \int_0^\infty \chi^2 \left(\frac{f\sqrt{A}}{\bar{V}} \right) G_{VV}(f) \, df \tag{8.59a}$$

and

$$E \simeq \frac{G_{VV}(f_0)\zeta}{\sigma_V^2} \int_0^\infty |H(f)|^2 \, df = \frac{\pi}{4} \frac{f_0 G_{VV}(f_0)}{\sigma_V^2} \tag{8.59b}$$

are called the *background* (low frequency gust) *excitation factor*, and the *gust energy factor*, representing quasi-static and resonant components, respectively. Let X_B and X_R represent the background and resonant responses. Then

$$\sigma_X = (\sigma_{X_B}^2 + \sigma_{X_R}^2)^{1/2} \tag{8.60}$$

where

$$\sigma_{X_B} = \bar{X}(2I)\sqrt{B} \text{ and } \sigma_{X_R} = \bar{X}(2I)\chi_0\left(\frac{E}{\zeta}\right)^{1/2}$$

Let

$$X_m = \max_{0 \le t \le T} [X(t)]$$

then, the expected value of X_m

$$\bar{X}_m = E[X_m] = \bar{X} + \bar{X}_m' = \bar{X} + \bar{\eta}\sigma_X \tag{8.61}$$

where $\bar{\eta}$ is the *average peak factor* (7.42), and can be expressed as

$$\bar{\eta} = (2 \log_e N_0 T)^{1/2} + \frac{0.5772}{(2 \log_e N_0 T)^{1/2}} \tag{8.62}$$

In (8.62), N_0 is the expected rate of zero-crossing by the dynamic response $X'(t)$. In most cases, $\bar{\eta}$ has a value between 3 and 4, and a value of 3.5 is a fair approximation. Combining (8.58) and (8.61), the *gust response factor*

$$\frac{\bar{X}_m}{\bar{X}} = 1 + \bar{\eta}\left(\frac{2\sigma_v}{\bar{V}}\right)\left(B + \frac{\chi_0^2 E}{\zeta}\right)^{1/2} = G \tag{8.63}$$

Velozzi and Cohen (1968) have derived a similar expression for the gust response factor, including the effect of nonuniformity in the mean flow.

Response of One-dimensional Structures to Wind

The response of one-dimensional (line-like) structures, for which the smallest significant wavelength is large as compared to the breadth b, may be treated under the assumption of quasi-steady aerodynamics (Davenport, 1962). The approach is valid for structures such as: slender chimneys, towers; suspension cables; and some bridge structures.

We consider a slender structure shown in Fig. 8.14. The along-wind dynamic response of the structure can be expressed by the partial differential equation

$$m(s)\frac{\partial^2 X(s, t)}{\partial t^2} + c(s)\frac{\partial X(s, t)}{\partial t} + L(s)[X(s, t)] = F(s, t) \tag{8.64}$$

where $m(s)$ is the mass per unit length of the structure; $c(s)$ is the damping coefficient per unit length; and $L(s)$ is the structural operator. $F(s, t)$ is the force per unit length and can be expressed as

$$F(s, t) = \bar{F}(s) + F'(s, t) \tag{8.65}$$

where the dynamic component (8.49)

$$F'(s, t) = C_D b\rho \bar{V} V'(s, t) \tag{8.66}$$

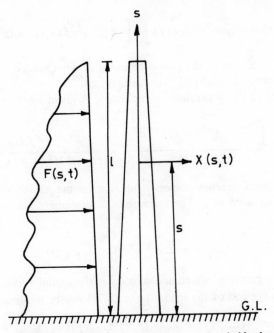

Fig. 8.14. Response of a slender structure to wind load.

Let $u_j(s)$ denote the jth mode shape satisfying the orthogonality conditions

$$\int_0^l m(s)u_j(s)u_p(s)\,ds = M_j, \quad \text{if } j = p$$
$$= 0, \quad \text{if } j \neq p \tag{8.67a}$$

and

$$\int_0^l c(s)u_j(s)u_p(s)\,ds = 2M_j\zeta_j\omega_j \quad \text{if } j = p$$
$$= 0, \quad \text{if } j \neq p \tag{8.67b}$$

Further, let the dynamic component of the displacement be

$$X'(s, t) = \sum_{j=1}^{\infty} u_j(s)\,q_j(t) \tag{8.68}$$

where $q_j(t)$ are the generalised coordinates. Using the orthogonality properties (8.67), the equation of motion (8.64) can be reduced to the canonical form (Nigam, 1983)

$$\ddot{q}_j(t) + 2\zeta_j\omega_j\dot{q}_j(t) + \omega_j^2 q_j(t) = \frac{Q_j(t)}{M_j}, \quad j = 1, 2 \ldots \tag{8.69}$$

where the modal force

$$Q_j(t) = \int_0^l F'(s, t)\,u_j(s)\,ds = \int_0^l C_D b\rho \bar{V}(s)V'(s, t)\,u_j(s)\,ds \tag{8.70}$$

Assuming that C_D, b, ρ, and $\bar{V}$ are independent of s, the psd of the modal force is given by

$$G_{Q_j Q_j}(f) = (C_D b \rho \bar{V})^2 G_{VV}(f) \int_0^l \int_0^l \phi(s^1, s^2; f) u_j(s^1) u_j(s^2) \, ds^1 ds^2 \quad (8.71)$$

where $G_{VV}(f)$ is the psd of the gust component of wind velocity; and $\phi(s^1, s^2; f)$ is spatial coherence function (8.19).

The variance of the generalised coordinate q is given by

$$\sigma_{q_j}^2 = \frac{\int_0^\infty G_{Q_j Q_j}(f) |H_j(f)|^2 \, df}{(2\pi f_j)^4 M_j^2} \simeq \frac{\pi}{4(2\pi f_j)^4 M_j^2} \left(\frac{f_j G_{Q_j Q_j}(f_j)}{\zeta_j} \right) \quad (8.72)$$

where $H_j(f)$ is the frequency response function for the jth mode and is same as (8.54), with f_0 replaced by f_j. The variance of displacement can now be expressed as

$$\sigma_X^2(s) = \sum_{j=1}^\infty \sigma_{q_j}^2 u_j^2(s) = \sum_{j=1}^\infty \sigma_{X_j}^2(s) \quad (8.73a)$$

Let $\gamma(s)$ be a function, which accommodates the spatial variations in C_D, b, ρ and $\bar{V}(s)$, then the psd of the modal force (8.71) can be expressed as

$$G_{Q_j Q_j}(f) = (C_{D_o} b_o l \rho_o \bar{V})^2 G_{VV}(f) |J(f)|^2 \quad (8.73b)$$

where subscript o denotes nominal values, and

$$|J(f)|^2 = \int_0^l \int_0^l \phi(\xi^1, \xi^2; f) \gamma(\xi^1) \gamma(\xi^2) u_j(\xi^1) u_j(\xi^2) \, d\xi^1 d\xi^2 \quad (8.74)$$

In (8.74), $\xi = s/l$ is a normalized coordinate; and $|J(f)|^2$ the *Joint Acceptance Function* (J.A.F.). It incorporates the sensitivity of the interaction between the turbulence characteristics and structural modes of vibration, and depends on the relative size of the 'span' to the scale of turbulence in the spanwise direction. Davenport (1977) has discussed in detail the nature of typical Joint Acceptance Functions for several structures.

Assuming spanwise correlation of force to be the same as that for the transverse correlation of the longitudinal wind component (8.19), the coherence function in (8.74) can be expressed as

$$\phi(\xi^1, \xi^2; f) = \exp(-\beta |\xi^1 - \xi^2|), \quad 0 \le |\xi^1 - \xi^2| \le 1 \quad (8.75)$$

where (8.20)

$$\beta = \frac{f \bar{C} l}{\bar{V}} = \frac{l}{L_c}$$

Two asymptotic values of β are of interest:

(i) $\beta \to 0$, that is, the span of the structure is small as compared to the significant wavelengths, so that, gust is more or less correlated along the entire span. For this case $\exp(-\beta |\xi^1 - \xi^2|) \to 1$ regardless of separation, and

$$|J(f)|^2 \to \left[\int_0^1 \gamma(\xi) u(\xi) \, d\xi \right]^2 \quad (8.76)$$

For $\gamma(\xi) = 1$, and antisymmetric mode shapes, $|J(f)|^2 = 0$.

(ii) $\beta \to \infty$, that is, the significant wavelengths are small compared to the span of the structure, so that, gust is correlated only over small parts of the span. For this case, coherence function is non-zero only when $|\xi^1 - \xi^2| \to 0$, and

$$|J(f)|^2 \to \frac{2}{\beta} \int_0^1 \gamma^2(\xi) u^2(\xi)\, d\xi \qquad (8.77)$$

Assuming,

$$\gamma(\xi) = \xi^\alpha$$

and

$$u(\xi) = \xi^\delta$$

Davenport (1977) has shown that $|J(f)|^2$ may be approximated by

$$|J(f)|^2 \simeq \frac{1}{(\alpha + \delta + 1)^2 + \left(\alpha + \delta + \dfrac{1}{2}\right)} = \frac{1}{A + B\beta} \qquad (8.78)$$

which provides a simple estimate for J.A.F. of one-dimensional structures with mode shapes of one sign, as is the case with fundamental mode of a variety of structures. Figure 8.15 shows general characteristics of J.A.F. of one-dimensiqnal structures.

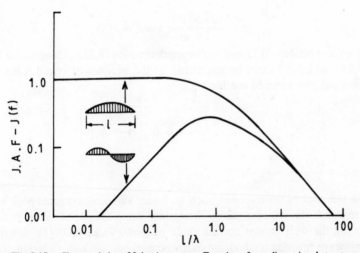

Fig. 8.15.   Characteristics of Joint Acceptance Function of one-dimensional structure.

We have noted that J.A.F. (8.74) combines the size effect reflected by aerodynamic admittance function, and the influence of the mode shape. For two ànd three-dimensional structures, J.A.F. can be estimated under the simplifying assumption that the frontal area is replaced by a mesh of slender elements. Under this assumption, the psd of modal force $Q_j(t)$ on an area A, can be expressed as

$$G_{Q_jQ_j}(f) = \iiiint\limits_A C_{pp}\,(s_1^1,\, s_2^1,\, s_1^2,\, s_2^2,\, f)\, u(s_1^1,\, s_2^1)\, u(s_1^2,\, s_2^2)$$

$$\times\, \gamma(s_1^1,\, s_2^1)\, \gamma(s_1^2,\, s_2^2)\, ds_1^1 ds_2^1 ds_1^2 ds_2^2 \qquad (8.79)$$

where C_{pp} is the cross-spectrum of pressure at points (s_1^1, s_2^1) and (s_1^2, s_2^2) on the structure; $u(s_1, s_2)$ is the mode shape; and $\gamma(s_1, s_2)$ is the adjustment factor for spatial variations in force coefficient, mean velocity, and psd of the gust component. Let b_1 and b_2 be the dimensions of the structure, and let $\xi_1 = s_1/b_1$; and $\xi_2 = s_2/b_2$. Equation (8.79) can be expressed as

$$G_{Q_jQ_j}(f) = (C_{D_o}A\rho_o\bar{V})^2 \, G_{VV}(f) \, | \, J(f) \, |^2 \tag{8.80}$$

where

$$| \, J(f) \, | = \iiiint_A \phi(\xi_1^1, \xi_2^1, \xi_1^2, \xi_2^2, f) \, u(\xi_1^1, \xi_2^1) \, u(\xi_1^2, \xi_2^2)$$

$$\gamma(\xi_1^1, \xi_2^1) \, \gamma(\xi_1^2, \xi_2^2) \, d\xi_1^1 d\xi_2^1 d\xi_1^2 d\xi_2^2 \tag{8.81}$$

The coherence function ϕ can be approximated by (Davenport, 1977)

$$\phi(\xi_1^1, \xi_2^1, \xi_1^2, \xi_2^2, f) = \exp\left[-\alpha \left\{ \bar{C}_1 \frac{fb_1}{\bar{V}} | \xi_1^1 - \xi_1^2 | \right.\right.$$

$$\left.\left. + \bar{C}_2 \frac{fb_2}{\bar{V}} | \xi_2^1 - \xi_2^2 | \right\}\right] \tag{8.82}$$

where

$$\alpha \simeq \frac{(1 + r^2)^{1/2}}{(1 + r)}; \quad r = \bar{C}_1 b_2 / \bar{C}_1 b_2$$

α has a value between 0.71 and 1. The approximation (8.82) uncouples the integrals in (8.81), and J.A.F. can be expressed as the product of two J.A.F.s of one-dimensional structures of the form

$$| \, J(f) \, |^2 = \left[\frac{1}{A_1 + B_1\beta}\right]\left[\frac{1}{A_2 + B_2\beta}\right] \tag{8.83}$$

where $\beta = \dfrac{f\bar{C}_1 b_1}{\bar{V}}$.

The statistics of the response, such as, mean, variance, expected peak values or gust response factors, etc., can be determined using the psd of modal force (8.80) following the well known normal mode method (Vickery, 1971). Alternatively, the influence function method; approximate methods, such as, Galerkin, Rayleigh-Ritz methods; or transfer matrix method (Yang and Lin, 1981) may be used (Nigam, 1983). The total peak response may be expressed as

$$X_m = \bar{X} + \bar{\eta} \, [\sigma_{X_B}^2 + \sum_{j=1}^{\infty} \sigma_{X_{R_j}}^2]^{1/2} \tag{8.84}$$

where $\bar{X}$ is the mean response; $\bar{\eta}$ the peak factor ($\simeq 3.0$); X_B is the background response, and X_{R_j} the resonant response in the jth mode due to atmospheric turbulence and/or excitation due to vortex shedding as shown in Fig. 8.16.

Aerodynamic Damping

The movement of structure influences the time-dependent forces induced by the

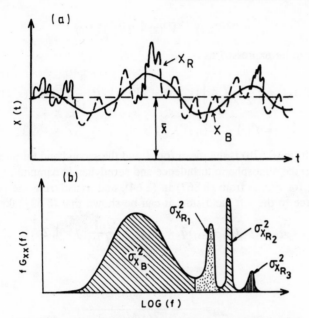

Fig. 8.16. Characteristics of the dynamic response to wind excitation (a) response history, and (b) response spectrum.

wind. In the preceding treatment, we have ignored this effect. The forces induced due to the movement of structure may be proportional to the acceleration, the velocity and the displacement. The forces proportional to the acceleration are of the order 1% of the inertia forces of a massive structure, such as a building, and can generally be ignored. However, these forces may be significant for very light, but large structures. The forces proportional to displacement are analogous to spring forces and can also be ignored. However, the rotational displacement of the deck of a bridge may have a significant influence on the lift, drag and torsion on the section.

The effect of forces proportional to the velocity is analogous to structural damping forces, and is called the *aerodynamic damping effect*. These forces are small, as compared to inertia and stiffness forces on a structure. However, they may be comparable to the structural damping forces and may have a significant influence on the wind induced dynamic response of structures.

Let $X(s, t)$ denote the displacement of a structure in the direction of the wind. The relative velocity between the structure and wind is given by $[V(s, t) - \dot{X}(s, t)]$, and the wind induced force can be expressed as:

$$F(s, t) = \left(\frac{1}{2} \rho_a CA \right) [V(s, t) - \dot{X}(s, t)]^2$$

$$= \left(\frac{1}{2} \rho_a CA \right) [V^2(s, t) - 2V(s, t)\dot{X}(s, t)], \text{ if } \dot{X} \ll V(s, t) \quad (8.85)$$

where ρ_a is the density of the air; and C is the force coefficient. Assuming the gust component $V'(t) \ll \bar{V}(s)$

$$V^2(s, t) \simeq \bar{V}^2(s) + 2\bar{V}(s)V'(t) \tag{8.86}$$

and (8.85) can be expressed as

$$F(s, t) = \left(\frac{1}{2} \rho_a CA\right) [\bar{V}^2(s) + 2\bar{V}(s)V'(t) - 2\bar{V}(s)\dot{X}(s, t)] \tag{8.87a}$$

$$= \bar{F}(s) + F'_T(s, t) + F'_A(s, t) \tag{8.87b}$$

The first term in (8.87b) is the mean force, and the second and third terms are the forces due to the atmospheric turbulence and aerodynamic damping, respectively. Substituting for $F(s, t)$ from (8.867) in (8.64), and transferring the aerodynamic damping force to the left hand side, it can be shown that (8.69) takes the form

$$\ddot{q}_j + 2\omega_j(\zeta_j + \zeta_{aj})\dot{q}_j + \omega_j q_j^2 = \frac{Q_j(t)}{M_j(t)}, \ j = 1, 2 \ldots \tag{8.88}$$

where

$$\zeta_{aj} = \frac{\int_0^1 \rho_a CA\bar{V}(s)u_j^2(s)\, ds}{2\omega_j M_j} \tag{8.89}$$

is the aerodynamic damping in the jth mode.

From (8.89), it is seen that ζ_a is significant for low natural frequencies, that is, for tall flexible structures. If ζ_a is positive, it augments the available structural damping and reduces the response. However, if it is negative, it reduces the structural damping, and can lead to instability, if its value exceeds the value of structural damping. The aerodynamic damping can be generally expressed in the following form

$$\zeta_a = (\rho_a/\rho_s) \, C_a(V^*) \tag{8.90}$$

where C_a is a function of the reduced velocity, $V^* = V/(fD)$ and ρ_s is average density of the structural envelope. It exhibits following characteristics (Davenport, 1985):

(i) At high reduced velocities ζ_a approaches quasi-steady values in both drag and lift.

$$C_a(V^*) = V^* C_D/4\pi \quad \text{(drag)} \tag{8.91}$$

$$= V^*\left(\frac{\partial C_L}{\partial \alpha} + C_D\right)\bigg/8\pi \quad \text{(lift)} \tag{8.92}$$

The above results indicate linearity with reduced wind speed. Lift damping is negative, if

$$\left(\frac{\partial C_L}{\partial \alpha} + C_D\right) < 0$$

Similar expressions do not exist for torsion.

(ii) Bluff structures subject to vortex shedding indicate strong tendency to negative aerodynamic damping in lift at wind speeds just above critical vortex shedding wind speed.

(iii) Aerodynamic damping is inversely proportional to ρ_s.

The effect of aerodynamic damping on the response of a bridge deck is shown in Fig. 8.17. The lift shows resonance due to vortex shedding excitation, and torsion shows instability setting in at the critical wind speed.

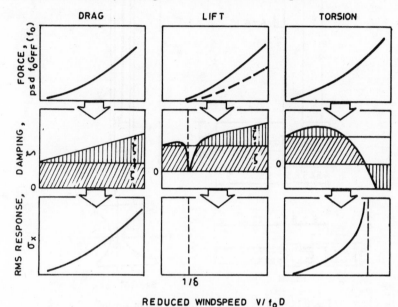

Fig. 8.17.   Response of suspension bridge deck to turbulent wind—effect of aerodynamic damping (Devenport, 1985).

In vertical structures, the variations in mean velocity, the cross-section and the turbulence intensity with height, significantly influence the response. "Changes in mean wind speed and diameter frequently run counter to each other as in tapered chimneys. This diffuses the peak excitation over a range of frequencies, lowers the spectral amplitude at any given frequency and it also reduces the effective (negative) aerodynamic damping averaged over the height of the structure (Davenport, 1985)".

Example 8.1: Dynamic analysis of a pylon (Nigam et al, 1981)

Figure 8.18 shows the three views and the cross-section at A-A of a pylon located in the Mahatma Gandhi Institute in Mauritius-an island in the Indian Ocean. Several high intensity cyclones have visited the region in the past causing severe wind damage. The unsymmetrical cross-section and a V-notch at the tip of the pylon makes it specially susceptible to dynamic wind action. A 1/50 scale, rigid model of the pylon was tested in the closed circuit low speed wind tunnel at IIT Kanpur, India. The plot of the force and the torsional moment coefficients along the height of the pylon are shown in Fig. 8.19, for the worst wind directions.

A random vibration analysis of the pylon was carried out to determine the static and dynamic forces and moments at different stations along the height. The design mean wind speed was assumed to be 120 miles/hr. and the gust component was modelled as a stationary, Gaussian random process with Davenport spectrum (8.14). The pylon was modelled as an Euler beam and the variance of the response was

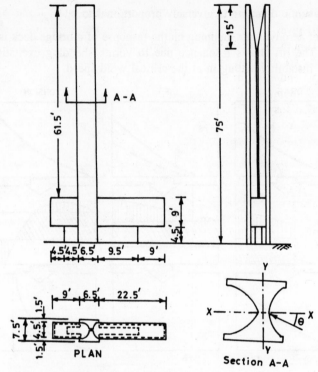

Fig. 8.18. Three views and cross-section of a pylon.

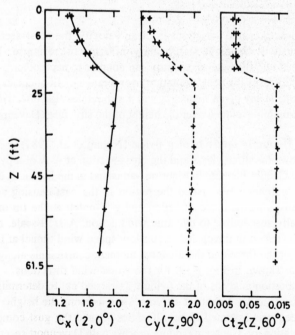

Fig. 8.19. The force and torsional moment coefficients as a function of height for worst wind direction.

determined using the fundamental mode. The maximum value of the response is assumed to be given by $X_{max} = \bar{X} + 3\sigma_x$. The static, the maximum dynamic and the total maximum values of the shear force and bending moment in X-X and Y-Y directions are given in Table 8.4. It is seen that the Gust Response Factors (8.63) are 6.35 and 6.92 for the shear force and bending moments, respectively, near the top of the pylon in the X-X direction. Such high values of Gust Response Factor are due to unsymmetrical plan-form of the pylon and the V-notch at the top. The design forces based on random vibration analysis are as much as 90% more than the estimates based on BS-Code.

Table 8.4   Maximum shear force and bending moments and gust response factor

Station (Distance) from Top, ft	Shear force (kips), X-X			Gust Factor	Bending Moment X-X (ft kips)			Gust Factor
	Static	Dynamic	Total Maximum		Static	Dynamic	Total Maximum	
6	1.56	8.31	9.87	6.35	4.26	24.99	29.25	6.92
12	4.02	16.02	20.04	5.0	20.53	98.49	119.02	5.80
15	5.66	19.38	25.04	4.4	35.01	151.68	186.69	5.32
25	12.28	26.07	38.35	3.12	125.33	367.68	493.01	3.92
35	18.31	33.27	51.58	2.82	278.80	686.55	965.35	3.36
45	23.65	39.33	62.98	1.66	489.02	942.78	1431.80	2.92
55	28.31	42.15	70.46	2.49	749.46	1378.50	2127.96	2.84
61.5	31.0	42.27	73.27	2.36	942.10	1574.88	2516.98	2.57
	Shear force (kips), Y-Y				Bending moment Y-Y, (ft kips)			
6	4.85	7.44	12.29	2.54	14.40	22.98	37.38	2.60
12	10.28	13.32	23.60	2.30	59.57	86.10	145.67	2.45
15	13.34	15.75	29.09	2.18	95.00	129.75	224.75	2.36
25	18.94	21.30	40.24	2.12	256.690	304.5	561.19	2.28
35	24.17	28.35	52.52	2.18	472.54	571.2	1043.74	2.20
45	28.94	31.41	60.35	2.08	738.427	677.10	1415.53	1.92
55	33.15	33.96	67.11	1.98	1049.363	1178.40	2227.76	2.12
61.5	35.53	34.11	69.64	2.05	1272.723	1338.72	2611.44	2.05

Filtered Poisson Models for Wind Gust and Response

In Chapter 2 (Example 2.2), we have discussed the filtered Poisson model for the gust component of wind velocity. The model was constructed by superposition of rectangular pulses arriving randomly as Poisson events (2.18). The response of a structure to such a process is also a filtered Poisson process, the pulse shape being the response of the structure to a rectangular pulse. For an sdf system, it can be expressed as (Racicot, 1969)

$$X'(t) = \sum_{n=1}^{N(t)} w(t - \tau_n, Y_n, D_n)$$

where

$$w(s, y, d) = Ay\left[1 - e^{-\zeta \omega_0 s}\left(\frac{\zeta}{\sqrt{1 - \zeta^2}} \sin \omega_D s + \cos \omega_D s\right)\right], s \le d$$

$$= Aye^{-\zeta \omega_0 s}[f_1(d) \sin \omega_D s + f_2(d) \cos \omega_D s], s > d$$

and

$$s = t - \tau$$

$$A = \rho \bar{V} C_D A / k$$

$$\omega_D = \omega_0 \sqrt{1 - \zeta^2}$$

$$f_1(d) = e^{-\zeta \omega_0 d} \left(\frac{\zeta}{\sqrt{1 - \zeta^2}} \cos \omega_D d + \sin \omega_D d \right) - \frac{\zeta}{\sqrt{1 - \zeta^2}}$$

$$f_2(d) = e^{-\zeta \omega_0 d} \left[\cos \omega_D d - \frac{\zeta}{\sqrt{1 - \zeta^2}} \sin \omega_D d \right] - 1$$

The autocovariance function of the response can be determined directly from Campbells' theorem (5.36*). The psd of the response can be determined as the Fourier transform of autocovariance function. The special advantage of filtered Poisson process models is the ease of simulation, and analytical determination of moment functions from Campbells' theorem. Racicot (1969) has demonstrated the use of this model to determine the response statistics, including univariate and bivariate distributions, crossing rates, and first-excursion failure.

8.5   Design for Wind Loads

Design of structures against wind induced loads involves both significant dynamic effects and uncertainty inherent in the wind environment due to random fluctuations in (i) mean wind velocity; (ii) atmospheric turbulence; (iii) wind direction; and (iv) wake behind the structure.

The stochastic models of wind climate of the region, and wind structure specific to a site, take into account the first three effects, whereas the uncertainty in the wake is usually accounted for through wind tunnel testing. We have seen that wind environment may be adequately modelled by segments of stationary, Gaussian random processes in space and time. The response of linear systems may, therefore, be treated as segments of stationary, Gaussian random processes completely determined by the second order statistics. The response may be expressed in terms of displacements; internal forces, such as, bending moments, shear forces; or the state of stresses. The design criteria generally include constraints on: strength, flexibility, serviceability and cumulative damage due to fatigue during the service life of the structure. For design in a probabilistic framework, the constraints are expressed in terms of limits on probability of exceeding a specified level during an extreme event such as a storm; and/or in terms of cumulative damage due to fatigue. The design, including optimization, can be implemented within the probabilistic frame-work as discussed in Chapter 4.

Simple design procedures generally adopted in the codes   (BIS: 875, 1975; NBC, Canada, 1975; BSI, CP3, 1972; ANSI A58-1, 1972; Sachs, 1978) involve static analysis based on "design wind velocities or pressures", determined for a specified exceedance probability over the service life of the structures through return-period zoning maps, or wind velocity data. An improvement over this procedure is a design based on "peak response", determined by multiplying the "mean response" by a suitable "gust response factor" which takes into account the dynamic effects.

More elaborate reliability based design procedures involve random vibration analysis and a probabilistic formulation of the design problem. In wind engineering, this may involve response statistics in terms of:

(i) parent process: $X(t) = \bar{X} + X'(t)$

(ii) modular envelope process: $A(t) = \bar{A} + A'(t)$

(iii) storm peak envelope process: $\hat{X}(t) = \bar{X} + \hat{X}'(t)$

(iv) annual maximum: $\hat{X}^* = \max(\hat{X}(t)); \quad 0 \le t \le 1 \text{ yr}$

as shown in Fig. 8.20.

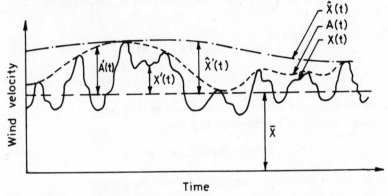

Fig. 8.20. Response Processes—parent process, $X(t)$; modular envelope process, $(A(t)$; and storm peak envelope process, $\hat{X}(t)$.

Each of the above random processes may be modelled by stringing together segments of stationary random processes. For each stationary segment of duration ΔT_j the distribution function of peak response may be expressed as

$$F_{\hat{X}}(\alpha) = \exp(-v(\alpha)\,\Delta T_j) \tag{8.93}$$

where $v(\alpha)$, the limiting decay rate, may be replaced by $N_X^+(\alpha)$ or $N_A^+(\alpha)$ the rates of up-crossings of the level α by the random processes $X(t)$ or $A(t)$, respectively. The up-crossing rate for a stationary random process $X(t)$ can be expressed as (8.35)

$$N_X^+(\alpha) = 2\pi C n_x \sigma_x p_x(\alpha) \tag{8.94}$$

where $\quad C = \dfrac{\displaystyle\int_0^\infty \dot{x} p_{\dot{x}}(\dot{x})\, d\dot{x}}{\sigma_{\dot{x}}} \quad$ and $\quad n_x = \dfrac{\sigma_{\dot{x}}}{\sigma_{\dot{x}}}$

If $p_{\hat{X}}(\dot{x})$ is gaussian, $C = 1/\sqrt{2\pi}$, and (8.94) reduces to

$$N_x^+(\alpha) = \sqrt{2\pi}\, n_x \sigma_x p_x(\alpha) \tag{8.95}$$

The parameter C in (8.94) is comparatively insensitive to distribution, and, therefore, (8.95) may be adopted without any significant error. Equation (8.93) implies independent crossings. Due to clumping, this may not be true. It is possible to incorporate the effect of clumping in the limiting decay rate (Nigam, 1983).

The random vibration analysis discussed in the preceding section is valid for a specified value of mean wind velocity $\bar{V}$, and a fixed wind direction. The response so determined is, therefore, conditional on the fixed values of $\bar{V}$ and θ. We have seen that both $\bar{V}$ and θ are random in nature and their joint distribution is given by $p_{\bar{V}\theta}$, $(\bar{v}, \theta)$, (8.24, 8.26). The unconditional response statistics can, therefore, be obtained by combining the conditional response with the joint distribution as shown below.

Up-crossing of Level α by the Parent Process $X(t)$

$$N_x^+(\alpha) = \int_0^{2\pi} \int_0^\infty N_{x'}^+ \{(\alpha - \bar{X}(\bar{v}, \theta)) \mid \bar{v}, \theta\} \, p_{\bar{v}, \theta} \, (\bar{v}, \theta) \, d\bar{v} \, d\theta$$

$$= \int_0^{2\pi} \int_0^\infty \sqrt{2\pi} \eta_{X'}(\bar{v}, \theta) \, \sigma_{X'}(\bar{v}, \theta)$$

$$\times p_{X'} \{(\alpha - \bar{X}(\bar{v}, \theta)) \mid \bar{v}, \theta\} \, p_{\bar{v}, \theta} \, (\bar{v}, \theta) \, d\bar{v} \, d\theta \qquad (8.96)$$

If a complete year consists of m stationary segments of duration ΔT_j ($j = 1, 2, m$), the expected number of up-crossings in a year

$$N_x^+(\alpha; 1\,\text{yr.}) = \sum_{j=1}^m {}_jN_x^+(\alpha) \, \Delta T_j \qquad (8.97)$$

where ${}_jN_x^+(\alpha)$ is rate of up crossing of jth stationary segment.

Up-crossing of Level α by Modular Envelope Process $A(t)$

$$N_A^+(\alpha) = \int_0^{2\pi} \int_0^\infty \sqrt{2\pi} \eta_{A'}(\bar{v}, \theta) \, \sigma_{A'}(\bar{v}, \theta)$$

$$\times p_{A'} \{(\alpha - \bar{X}(\bar{v}, \theta)) \mid \bar{v}, \theta\} \, p_{\bar{v}, \theta} \, (\bar{v}, \theta) \, d\bar{v} \, d\theta$$

where

$$P_A(a) = \frac{a}{\sigma_A^2} \exp\left(-\frac{a^2}{2\sigma_A^2}\right), \, 0 \le a \le \infty \qquad (8.98)$$

For the whole year

$$N_A^+(\alpha; 1\,\text{yr}) = \sum_{j=1}^m {}_jN_A^+(\alpha) \, \Delta T_j \qquad (8.99)$$

Probability Distribution of Peak Storm Envelope $\hat{X}(t)$

$$F_{\hat{X}}(\alpha) = \int_0^{2\pi} \int_0^\infty F_{\hat{X}}^+ \{(\alpha - \bar{X}(\bar{v}, \theta)) \mid \bar{v}, \theta\} \, p_{\bar{v}, \theta} \, (\bar{v}, \theta) \, d\bar{v} \, d\theta \qquad (8.100)$$

and the up-crossing rate of $\hat{X}(t)$,

$$N_{\hat{X}}^+(\alpha) = \sqrt{2\pi} \, n_{\hat{X}}, \, \sigma_{\hat{X}}, \, p_{\hat{X}}(\alpha) \qquad (8.101)$$

where $p_{\hat{X}}(\alpha) = \dfrac{dF_{\hat{X}}(\alpha)}{d\alpha}$.

In (8.101), $n_{\hat{x}}$, is approximately equal to the cycling rate of mean wind-$n_{\bar{v}} \approx 0.1$ cycles per hour (Davenport, 1971).

Probability Distribution of Annual maximum response, $\hat{X}^*$

The maximum response for the year may be computed as a compound event based on the responses for the segments during the year which are assumed to be stationary and independent. Let ΔT_j denote the duration of jth segment. Then

$$F_{\hat{X}^*}(\alpha) = \prod_{j=1}^{m} \exp\left(-_j N_{\hat{X}}^+(\alpha)\, \Delta T_j\right)$$

$$= \exp\left(-\sum_{j=1}^{m} {}_j N_{\hat{X}}^+(\alpha)\, \Delta T_j\right) \tag{8.102}$$

The above relations can be used to determine the probability associated with the peak response. Figure 8.21 shows the sequence of steps involved and a graphical representation of peak response probabilities.

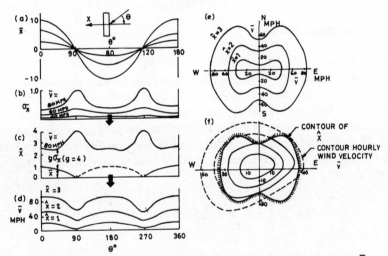

Fig. 8.21.   Steps for determination of peak response, (a) mean response $\bar{X}$, (b) rms response σ_X, (c) peak response ($\hat{X} = \bar{X} + g\sigma_x$), (d) and (e) contours of peak response, and (f) probability of peak

response $P(> \hat{X}) = \oint \oint p(v_1, v_2)\, dv_1\, dv_2$ (Davenport, 1971).

Storm Duration

To determine the probability of failure during a storm, it is necessary to specify the duration of the storm. The mean wind velocity during a storm rises slowly to a peak value and then subsides. Konishi et al (1968) have suggested the following equation to represent a strong storm of duration T_s:

$$\bar{V}(t; \bar{V}(s), T_s) = \bar{V}(s)\left(1 - \frac{2\,|t\,|}{T_s}\right)^{1/2}, \, -T_s \leq t \leq T_s \tag{8.103}$$

where $\bar{V}(s)$ is maximum fastest mile velocity at height s, and $\bar{V}$ is the quasi-stationary mean wind velocity at height s. Figure 8.22 shows the rise and fall of $\bar{V}$ as given

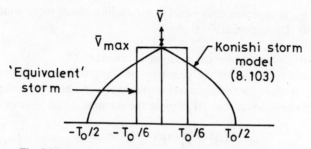

Fig. 8.22 Variation of mean wind velocity occurring storms.

by (8.103). To simplify the computational work, the storm profile may be replaced by an 'equivalent' profile of duration $T_s/3$ as shown in Fig. 8.22.

The duration of storm T_s exhibits a large variation. Konishi, et al (1968) have proposed the following simple relation

$$T_s = \bar{V}_{max}^2/5.0 \text{ sec} \tag{8.104}$$

where $\bar{V}_{max}$ is given in ft/sec.

Fatigue Damage

The cumulative damage due to fatigue can be computed for each stationary segment using the methods discussed in Chapter 3. The total damage over a period can be determined by first incorporating the effect of fluctuations in the mean wind velocity and wind direction, and then summing up the damage of each segment. Let $E[_jD \mid \bar{v}, \theta]$ represent the expected value of the cumulative fatigue damage during jth segment. Then, the expected value of total damage in m segments can be expressed as

$$E[D] = \sum_{j=1}^{m} \int_0^{2\pi} \int_0^\infty E[_jD \mid \bar{v}, \theta] \, p_{\bar{V}, \theta}(\bar{v}, \theta) \, d\bar{v} \, d\theta \tag{8.105}$$

Example 8.2: Reliability based design of a sdf system to wind loading
(Racicot, 1968)

We consider the design of a sdf system to wind loading for a prescribed failure probability under the following assumptions:

(i) The distribution of maximum yearly mean wind velocity is given by (8.33) with $\tilde{\mu} = 0$; $\gamma = 9.0$; and $\sigma = 49.2$. Accordingly, the reference mean design wind velocity with exceedance probability 0.00029 over a one year period is give by

$$\bar{V}_D = \bar{V}_{E, 1; 0.00029} = 161 \text{ ft/sec}$$

(ii) The duration of the storm is given by (8.104).

(iii) The gust component is modelled as (a) a Gaussian process; and (b) a filtered Poisson Process (F.P.P.).

(iv) The system is assumed to 'fail', if its displacement crosses a threshold level α, for the first time. The threshold level is prescribed in terms of a factor of safety n, as

$$\alpha = n \left(\frac{1}{2} \frac{\rho C_D A}{k} \bar{V}_D^2 \right)$$

where k is the stiffness of the sdf system.

The total displacement of the system consists of the static and the dynamic components, that is

$$X(t) = \bar{X} + X'(t)$$

$$= \frac{1}{2} \frac{\rho C_D A}{k} \bar{V}_E^2 + X'(t)$$

If the static response $\bar{X} > \alpha$, the first passage probability is 1. If $\bar{X} > \alpha$, failure may occur due to dynamic fluctuations about the mean, if the gust response crosses the level

$$\alpha' = \alpha - \bar{X}$$

$$= \frac{1}{2} \frac{\rho C_D A}{k} (n\bar{V}_D^2 - \bar{V}_E^2)$$

The first passage probability for crossing the level α' by the random process X' (t), over the 'equivalent' storm duration $T_s/3$, can be computed. Two cases are considered (Nigam, 1983): (i) independent crossings; and (ii) independent clumps.

Combined with the two distributions—Gaussian and F.P.P.—we have four different estimates of failure probability: (a) Gaussian-independent crossings; (b) Gaussian-independent clumps; (c) F.P.P.-independent crossings; and (d) F.P.P.-independent clumps.

Let $P(\alpha \mid \bar{v}, t_s)$ denote the first passage probability of crossing the level α, for mean wind velocity $\bar{V}_E$ and storm duration T_s. The total probability of failure is then given by

$$P_f = \int_0^\infty P(\alpha \mid \bar{v}, T_s) \, p_{\bar{v}_E}(\bar{v}) \, d\bar{v}$$

where $p_{\bar{v}_E}(\bar{v}) = \frac{d}{d\bar{v}} (F_{\bar{v}_E}(\bar{v}))^N$, and N is the service life of the structure.

Figure 8.23 shows the probability of first excursion as a function of the mean wind velocity for the four cases. It is seen that failure is nearly certain, if the mean wind velocity exceeds 110 ft/sec, and approaches zero if the wind velocity is less than 90 ft/sec. In the range (90–110) ft/sec, the first excursion probability differs in the four cases—being highest for F.P.P.—with independent crossings.

Figure 8.24 shows the total failure probability as a function of factor of safety, n. As expected, the probability of failure decreases with increase in the factor of safety, and for a prescribed failure probability, the model—F.P.P. with independent crossing—requires highest value of factor of safety. Table 8.5 shows the required factor of safety to yield failure probability of 5×10^{-3} in the four cases.

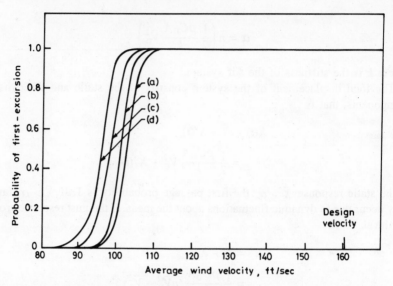

Fig. 8.23. First-excursion probability as a function of average wind velocity ($n = 1.0$; $f_o = 5.0$ cps; $\zeta = 0.015$) (Racicot, 1968).

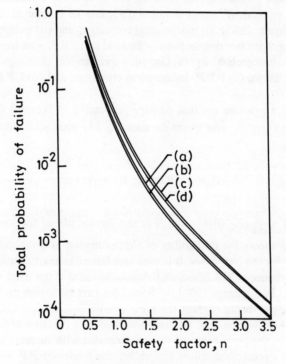

Fig. 8.24 Total probability of failure as a function of safety factor ($f_0 = 5.0$ cps; $\zeta = 0.015$) (Racicot, 1968).

Table 8.5 **Safety fact or required for a prescribed failure probability**
(f_o = 5.0 sec, ζ = 0.015)

Case	Required Safety Factor (n) to Yield a $P_f = 5 \times 10^{-3}$
(a) Gaussian–independent crossings	2.13
(b) Gaussian–independent clumps	2.08
(c) F.P.P.–independent crossings	2.22
(d) F.P.P.–independent clumps	2.16

Example 8.3   Optimum design of the support-structure of a water tank to wind loading (Narayanan and Nigam, 1983)

In Chapter 7, we have discussed the minimum weight design of a planar truss structure supporting an elevated water tank to earthquake excitation. We shall now consider the minimum weight design of the same structure to wind induced excitation. The formulation of optimum design problem remains the same as in Example 7.6. We will solve the optimization problem under the following assumptions:

(i) mean wind profile is given by (8.10) with α = 0.16 and $\bar{V}_G$ = 42.7 m/sec;

(ii) the psd of the gust component is given by (8.14) with κ = 0.005 and L_0 = 1200 m; and

(iii) the failure probability in the stress constraint (7.183) is evaluated as one-sided first excursion problem, assuming full correlation between member stresses.

Design data is same as in Example 7.6 except as given below:

$$p_1 = 10^{-5} \qquad\qquad p_2 = 10^{-4} \qquad\qquad T = 3600 \text{ sec}$$
$$\rho_a = 1,237 \text{ kg/m}^3 \qquad C_D = 2.0 \qquad\qquad \zeta = 0.02$$

The projected area of the tank is assumed to be equal to 37.16 m². Each member is assumed to have the minimum area $d_i \geq 2$ cm², i = 1, 2, ... 10.

The optimization is carried out using the sequential unconstrained minimization technique discussed is Chapter 4. The initial and optimal values of design parameters are given in Table 8.6(a). Comparison with the optimum design for earthquake in Table 7.4 shows that assumed design earthquake provides much more severe loading condition than the assumed design wind. This is due to the fact that the structure has been designed for a severe earthquake: El-Centro 1940. The optimum design for an earthquake, approximately one-third the size of El-Centro 1940 earthquake, and failure probability $p_1 = 10^{-4}$ is shown as case (b) in Table 8.6. The optimum aseismic design for this case is comparable to the optimum design (a) for wind.

Table 8.6 Truss structure—design for (a) wind and (b) earthquake loads

	d_1	d_2	d_3	d_4	d_5	d_6	d_7	d_8	d_9	d_{10}	d_{11}	d_{12}	$W(\bar{D}) \times 10^5$
(a) Initial	101.6	101.6	101.6	101.6	101.6	101.6	101.6	101.6	101.6	101.6	914.4	914.4	43.60
Optimal	31.0	111	38.1	93.0	48.0	39.1	11.7	23.6	35.3	5.08	965.2	978.0	17.58
(b) Initial	127.0	152.4	127.0	101.6	127.0	101.6	127.0	127.0	101.6	101.6	914.4	914.4	51.20
Optimal	26.7	122.7	37.1	54.9	49.3	43.2	51.3	24.6	35.0	21.6	942.3	950.0	19.90

(b) Design Earthquake (1/3rd El-Centro, 1940); $p_1 = 10^{-4}$, $p_2 = 10^{-3}$.

REFERENCES

American National Standards Institute (ANSI) A58-1, 1972. *Building Code Requirements for Minimum Design Loads in Buildings and other Structures.* Chapter 6: Wind Loads.

British Standards Institution (BSI), CP3, 1972. *Code of Practice, Basic Data for the Design of Buildings.* Chapter V, Part 2: Wind Loads.

Bureau of Indian Standards (BIS), IS-875, 1975. *Code of Practice for Structural Safety of Buildings*: Loading Standards.

Davenport, A.G., 1961a. The spectrum of horizontal gustiness near the ground in high winds. *Quart. J. Roy. Met. Soc.*, 87: 194–211.

Davenport, A.G., 1961b. The application of statistical concepts to the wind loading of structures. *Proc. Instt. Civ. Engrs.*, (19): 449–473.

Davenport, A.G., 1962. The response of slender líne-line structures to gusty wind. *J. Instt. Civ. Engrs.*, (23): 449–472.

Davenport, A.G., 1967. Gust loading factors. *J. Str. Div., Proc. ASCE*, 93(ST3): 11–34.

Davenport, A.G., 1971. On the statistical prediction of structural performance in the wind environment. ASCE, National Structure Engineering Meeting, Baltimore.

Davenport, A.G., 1977. Wind structure and wind climate. Proc. Int. Res. Seminar. Safety of Structures Under Dynamic Loading. Norwegian Institute of Technology, Trondheim: 209–283.

Davenport, A.G., 1982. The dynamic response of structures to wind turbulence. Seminar on Structural Aerodynamics, St—Remy—les—Chevreuse.

Davenport, A.G., 1985. The representation of the dynamic effect of turbulent wind by equivalent static loads. Int. Course on Design of Civ. Engr. Str. for Wind Loads, Civ. Engg. Deptt., University of Roorkee.

Gould, P.L. and S.H. Abu-Sitta, 1980. *Dynamic response of Structures to Wind and Earthquakes.* Pentech Press.

Harris, R.I. 1972. Measurements of wind structure. Proc. Symp. Wind Loading on Structures, Bristol.

Ito, M. and Y. Fujino, 1977. Some probabilistic considerations in wind resistant design. Deptt. of Civ. Engg., Univ. of Tokyo, Rep. No. 7910.

Konishi, I., M. Shinozuka and H. Itagaki, 1968. Safety analysis of suspension bridges. 8th Congress, Int. Assoc. Bridge and Struct. Engr., New York.

Merchant, D.H., 1964. A stochastic model of wind gust. Deptt. of Civ. Engg., Stanford University, Tech. Rep. No. 83 FM6.

National Building Code (NBC) of Canada, 1975. Part A, Design. national Research Council.

Narayanan, S. and N.C. Nigam, 1983. Optimal design of a truss structure supporting a water tank. *Int. J. of Str.* 1(4): 129–139.

Nigam, N.C., 1983. *Introduction to Random Vibrations.* Cambridge, MA: The MIT Press.

Nigam, N.C., A.K. Gupta and S.P. Mathur, 1981. Wind tunnel study and probabilistic dynamic analysis of pylon for wind loading. *J. Institution of Engineers (India)*, Vol. 61: 69–76.

Rao, G.N.V., 1985. Wind tunnel testing. Int. Course on Design of Civ. Engg. Structures for Wind load. Deptt. of Civ. Engg., University of Roorkee.

Russell, L.R., 1971. Probability distribution of hurricane effects. *J. Waterways, Harbours and Coastal Engg.*, Proc. ASCE, 97: 139–154.

Sachs, P., 1978. *Wind Forces in Engineering.* Oxford: Pergamon Press.

Simiu, E., 1974. Wind spectra and dynamic along-wind response. *J. Struct. Div., Proc. ASCE*, 100 (ST9): 1897–1910.

Simiu, E., J. Bietry and J.J. Filliben, 1978. Sampling errors in estimation of extreme winds. *J. Struct. Div., Proc. ASCE*, 104 (ST3): 491–501.

Simiu, E. and R.H. Scanlan, 1986. *Wind Effects on Structures.* John Wiley.

Thom, H.C.S., 1967. Towards a universal climatological extreme wind distribution. Proc. Int. Res. Seminar on Wind Effects on Buildings and Structures, Vol. I, Ottawa (Canada).

Thom, H.C.S., 1968. New distribution of extreme winds in United States. *J. Struct Div., Proc. ASCE*, 94: 1787–1801.

Tryggvason, B.V., D. Surry and A.G. Davenport, 1976. Predicting wind induced response in hurricane zones. *J. Struct. Div., Proc. ASCE*, 102: (ST12): 2333–2350.

Van Koten, H. 1968. Meting van Bervegingsnelheden aan de T.V. Nastte Wieringer Werf. T.N.O. Rapport N. B68–294/5/24.

Velozzi, J. and E. Cohen, 1968. Gust response factors. *J. Struct. Div. Proc. ASCE*, 94 (ST6): 1295–1313.

Vickery, B.J., 1971. On the reliability of gust loading factors. *Civ. Engg. Tran., The Institution of Engineers,* Australia.

Vickery, B.J., 1979. Wind load on chimneys. The Univ. of Western Ontario, Res. Report: BLWT—2—79.

Venkateswarlu, B., M. Armugam and S. Arunachalam, 1985. Risk analysis of cyclonic wind data. Proc. National Seminar on Tall Reinforced Concrete Chimneys. 25–27, April, 1985, New Delhi (India).

Yang, J.N., and Y.K. Lin, 1981. Along-wind motion of multistory building. *J. Engg. Mech. Div., Proc. ASCE,* 107 (EM2): 295–307.

9. Response of Offshore Structures to Wave Loading

9.1 Introduction

Offshore structures, both mobile and fixed, have existed since the beginning of the civilization. The oil crisis in early seventies, and the discovery of oil at increasing depth in offshore regions around the world has stimulated design and erection of a wide range of offshore platforms. The exponential growth in the number and coverage of papers presented at the annual offshore Technology Conference (OTC), and International Conference on Behaviour of Offshore Structures (BOSS) is a good indication of the explosive world wide growth of offshore engineering in less than two decades.

Offshore structures operate in a very complex and hostile environment. They are subjected to temporally and spatially varying random loads due to winds, waves, currents, earthquakes, ice and thermal gradient. The complexity of wind and earthquake loads is compounded by the wave environment. The wave environment itself is inherently nonlinear, and even if linearized, it becomes strongly nonlinear on transformation into loads due to hydrodynamic drag effects. The long-term behaviour of loads is nonstationary, and due to nonlinear functional dependence on wave environment, it is also non-Gaussian. The loads induced in the 'splash-zone' near the surface are intermittent and complex. The marine soil on which offshore structures are erected has a strong influence on the response of the structure. The behaviour of soil shows complex variation under cyclic loading adding to the complexity of structure-foundation system models. The compliant structures, such as tension leg platforms, undergo large excursions which result in structural nonlinearities. The highly corrosive ocean environment causes progressive deterioration of material properties, which must be incorporated in long-term engineering design procedures. It is clear, therefore, that an analytical treatment of offshore structures is a difficult engineering design problem.

We have already discussed various stochastic methods to determine the response of structures to earthquake and wind. With some modifications, these methods apply to offshore structures. In this chapter, we shall discuss the analysis and design of offshore structures for hydrodynamic loads induced by wave environment. The physical nature of wave environment and its stochastic models are discussed first. The stochastic properties of hydrodynamic loading due to wave-structure interaction are then derived. The nonlinear functional dependence of loads on Gaussian wave environment resulting in non-Gaussian distribution is discussed, and the validity of statistical linearization procedures is examined. The response of structures to short-term stationary models for a given sea state is determined, and convoluted with the joint distribution of sea states to derive long-term distributions.

An offshore structure must be designed for strength, fatigue and serviceability. Due to the random nature of the operating environment, and the nonlinearities inherent in the loading, and structural behaviour, design of offshore structures involves formidable analytical and computational effort. A sound engineering judgement is critical for the implementation of the design procedures. The present methods for the analysis and design based on random vibration analysis are discussed. A few examples are included to illustrate the application of the methods.

9.2 Nature of Sea Waves and Sea States

Sea is a large body of water bounded by irregular shore line, an undulating permeable bottom and a wavy free surface. The waves at the sea surface are generated primarily by the wind action and are one of the most complex and everchanging phenomena in nature. In a calm sea, the waves start building up slowly as the wind begins to blow, attain large heights as the wind persists, and gradually dissipate as the wind subsides. A strong correlation, therefore, exists in the nature of sea waves and the velocity, duration and direction of the wind. The waves exhibit distinct characteristics in 'deep-sea' ($d/L > 0.5$), where d is the depth of water, and L the wave length, and in 'near-shore' regions which include intermediate-depth and long waves at shallow-depth ($d/L < 0.05$). In deep sea, the surface waves are not influenced by the sea bottom. As the waves enter the intermediate and shallow depth regimes, the nature of sea waves begins to be affected by sea bottom topography, permeability, friction and the contours of the shore line. The waves are transformed by a complex interaction between phenomena, such as, shoaling, refraction, breaking, diffraction and reflection. The behaviour of the sea waves in the near-shore region is of great importance in coastal structures, such as, break waters, sea walls, and for offshore structures located in shallow waters. For a comprehensive treatment of these problems, the reader is referred to the text by Goda (1985). Since the offshore structures for oil exploration are moving increasingly to greater depth, we shall confine our discussion to wind and wave climate in deep sea.

The wind generated sea waves develop gradually through a complex mechanism of transfer of energy from air to the sea water (Kinsman, 1965). The process involves direct transfer of motion from air to sea water through shear forces and an alternating sequence of low and high pressure regions. As the wind starts blowing, high-frequency (short-wavelength) waves are initially generated, which break and transfer energy to lower-frequency components, which are further reinforced by the wind action. This transfer of energy down the frequency scale continues until a cut-off frequency is reached, at which the phase velocity of the wave is the same as the wind velocity. Below this frequency waves do not receive energy from the wind. The sea waves consist mostly of high-frequency components, if the wind duration is not sufficient for lower frequency waves to develop, and the sea is said to be 'duration limited.' Near the coastline, the fetch may limit the development of low-frequency components and at such locations sea is said to be 'fetch-limited.' A 'fully developed sea' condition may be attained at a location away from the coast, if the wind continues to blow for sufficiently long time. For a wind of certain duration and fetch, there exists a cut-off frequency below which there is little energy in the waves.

A disturbed sea surface forms a complex wave pattern made up of small and large wavelets moving in many directions. However, waves as a whole follow the direction of the wind. Figure 9.1 shows a contour map of the surface elevation of a disturbed sea obtained through computer simulation. The blank areas in between

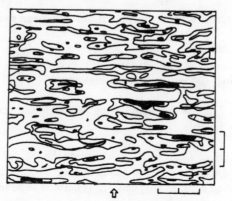

Fig. 9.1. Simulated surface elevation contours of random waves (Goda, 1976)

the contours represent water surface below the mean water level. Figure 9.2 shows a typical strip chart record of the surface elevation as a function of time obtained

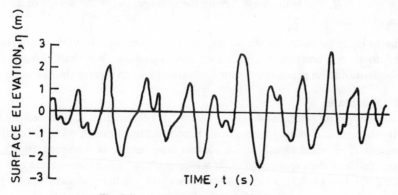

Fig. 9.2. A typical sample of wave record.

from a wave gauge at a fixed location in sea. The distinct feature of the wave patterns in Figs. 9.1 and 9.2 is the spatial and temporal irregularity in the nature of sea waves. Pierson (1952) was the first to recognize the irregularity of sea waves as a fundamental property and suggest steps to incorporate this fact in the engineering design procedures. We shall discuss this aspect in detail.

9.2.1 Some Definitions
We will now give definitions of some basic wave parameters with reference to Figs. 9.1 and 9.2.

ZERO LINE:represents the mean level of the water surface deduced from the record.

A WAVE: is the profile between two successive up (or down) crossings of the zero line.

WAVE PERIOD (T_z): is the time interval between the two successive zero-up-crossings.

WAVE LENGTH (APPARENT): is the horizontal distance between two successive zero-up-crossings.

WAVE HEIGHT (H): is the vertical distance between the highest and the lowest points between the two successive zero-up-crossings.

Highest Wave (H_{max}, T_{max}): is the wave having the height and period of the highest individual wave in a record.

HIGHEST $1/n$th WAVE $(H_{1/n}, T_{1/n})$: is the wave having the average height and period of the highest $1/n$ of the total number of waves in a record. An imaginary wave train having height and period equal to $H_{1/n}$ and $T_{1/n}$ is defined as he highest $1/n$th wave.

SIGNIFICANT WAVE (H_s, T_s) or highest one-third wave $(H_{1/3}, T_{1/3})$: is the highest $1/n$th wave with $n = 3$. An imaginary wave train having height and period equal to H_s and T_s is defined as the significant wave.

MEAN WAVE $(\bar{H}, \bar{T}_z)$: is the wave having the average height and period of all the waves in a record.

Due to irregular nature of the waves, the parameters defined above vary from record to record and are, therefore, random variables. We shall discuss the distributions of these parameters and the relations between them in the next section.

9.2.2 Sea States

For the design of offshore structures against wave loading, it is necessary to specify the sea states during the service life of structures through wave forecasting techniques. The 'wind roses', which contain information on the speed, direction and percentile of the expected winds at a location, can be transformed into a wave scatter diagram (Fig. 9.3), which defines the sea states in terms of significant wave heights and mean zero crossing periods (Darbyshire and Draper, 1963).

Scatter diagrams can also be constructed directly from wave measurements taken at a site for a period of one year or more. Two approaches are available to use the wave scatter diagram for design: (i) wave height exceedance curve; or (ii) wave spectrum for each sea state.

The wave height exceedance curve gives the number of waves in a given period of time against wave height. The curve is plotted on a semi-log paper and can be approximated by a straight line. The curve may be plotted for all waves, or waves from a particular direction. The maximum wave for a given return period can be deduced from the exceedance curve. Usually, a Rayleigh distribution fits the exceedance curve and provides the mathematical basis for prediction.

The wave spectrum approach is based on a psd of each sea state in the scatter diagram. The conditional probabilistic response of the structure can be obtained

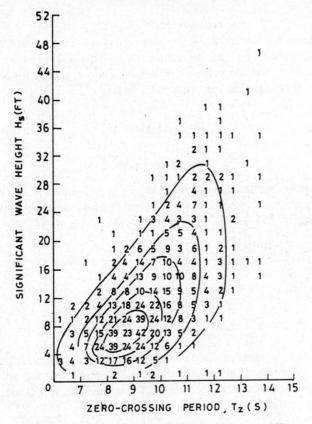

Fig. 9.3.   Wave scatter diagram (Draper and Square, 1967).

for each sea state through random vibration analysis. The long-term response is then obtained by convoluting this response with the distribution of all the sea states. The wave spectrum approach provides a consistent mathematical basis for the design of offshore structures during an extreme storm, or cumulative damage due to fatigue, and is recommended for the design of large structures.

9.3 Stochastic Description of Deep Sea Waves

We have seen that the wind generated deep sea waves consist of a large number of wavelets of varying heights and periods. While the waves as a whole follow the direction of the wind, individual wavelets move randomly in several directions. The sea surface, therefore, exhibits both spatial and temporal randomness, and can be modelled as a random function of the space coordinates $\bar{s}^T = (s_1, s_2)$, and time t. Such a function can be represented by a superposition of an infinite number of small amplitude waves with random phase, frequency and direction.

In Appendix 9A, we discuss the properties of small amplitude harmonic waves propagating freely on the surface of water of constant depth h in a particular direction. We show that the profile of the wave can be expressed as

$$\eta(\bar{s}, t) = a_0 \cos(ks_1 \cos\theta + ks_2 \sin\theta - \omega t + \varepsilon) \tag{9.1a}$$

$$= a_0 \cos(\bar{k} \cdot \bar{s} - \omega t + \varepsilon) \qquad (9.1b)$$

where η denotes the elevation of water surface above the mean sea level (MSL); a_0 the wave amplitude; $k = 2\pi/L$ the wave number; θ the angle between the s_1-axis and the direction of wave propagation; ω the wave frequency; and ε the phase angle. The wave number k, and the frequency ω are connected by the *dispersion relation* (9A.14)

$$\omega^2 = gk \tanh(kh) \qquad (9.2a)$$

or

$$L = gT^2/2\pi \tanh(2\pi h/L) \qquad (9.2b)$$

The random sea surface can be modelled by a superposition of infinite number of freely propagating individual waves (9.1) and expressed by the following series (Longuet-Higgins, 1957):

$$\eta(\bar{s}, t) = \sum_{n=1}^{\infty} \eta_n(\bar{s}, t) = \sum_{n=1}^{\infty} a_n(\omega_n, \bar{k}_n) \cos(k_n s_1 \cos \theta_n$$

$$+ k_n s_2 \sin \theta_n - \omega_n t + \varepsilon_n) \qquad (9.3a)$$

$$= \sum_{n=1}^{\infty} a_n(\omega_n, \bar{k}_n) \cos(\bar{k}_n \cdot \bar{s} - \omega_n t + \varepsilon_n) \qquad (9.3b)$$

where frequencies ω_n are densely distributed between zero and infinity; directions θ_n are densely distributed between $-\pi$ and π, and the phase angles ε_n are randomly and uniformly distributed between 0 and 2π. The infinitesimal amplitudes a_n satisfy the following relations:

$$\sum_{\omega}^{\omega+d\omega} \sum_{\theta}^{\theta+d\theta} \frac{a_n^2}{2} = G(\omega, \theta) \, d\omega \, d\theta \qquad (9.4)$$

where $G(\omega, \theta)$ is called the directional wave spectrum density function, or the *directional wave spectrum*. It represents the distribution of energy $(1/2 \, \rho \, ga^2)$ as a function of frequency ω and direction θ.

For a unidirectional wave distribution generated by a uniform wind in the positive s-direction, we have

$$\eta(s, t) = \sum_{n=1}^{\infty} a_n \cos(k_n s - \omega_n t + \varepsilon_n) \qquad (9.5)$$

and

$$\sum_{\omega}^{\omega+d\omega} \frac{a_n^2}{2} = G(\omega) \, d\omega \qquad (9.6)$$

where $G(\omega)$ is unidirectional one-sided psd. If the wind is assumed to blow for an infinite duration over an infinite sea, $G(\omega)$ represents the psd of a *fully developed sea*.

The surface elevation defined by (9.3) or (9.5) represents a stationary, Gaussian random process. The Gaussian property follows from the Central Limit Theorem

(Rice, 1954; Longuet-Higgins, 1952) and the statistical independence of component waves. The probability density function of η is, therefore, given by

$$p(\eta) = \frac{1}{(2\pi)^{1/2} \sigma} \exp\left(-\frac{\eta^2}{2\sigma^2}\right) \tag{9.7}$$

where $\sigma = (\lambda_0)^{1/2}$. We have assumed $E[\eta] = 0$, since η is measured from MSL.

Gross observations of the sea have generally confirmed a Gaussian distribution for the sea surface elevation. However, the surface elevation cannot be strictly Gaussian due to physical limit on the height of the waves imposed by breaking, and distortion of independent phase assumption due to nonlinear interactions between waves. Longuet-Higgins (1952) proposed a non-Gaussian distribution (a Gram Charlier distribution) to account for nonlinear wave interactions, which also shows discrepancies for extreme values (Johnson-Laird, 1970). If the ripples on the sea surface are taken into account, the Gaussian assumption breaks down. Since we are interested in the offshore structures, which are not affected by ripples, we shall assume that (9.7) applies.

The elevation of the sea surface is everchanging in space and time and is actually nonstationary. However, over a short period of time, say 20–60 min, the sea state may be expected to remain constant and the surface elevation at a point can be assumed to be a quasi-stationary random process represented by (9.3) or (9.5). Since the surface elevation has been assumed to be Gaussian, it is also ergodic over a short period of time, if it does not contain periodic (regular wave) components (Koopmans, 1974). This permits computation of the statistics of wave profile by taking time averages of a sample wave record.

9.3.1 Short-term Statistical Properties of Wave Parameters

The profile of random waves at a fixed point in the sea can be expressed as

$$\eta(t) = \sum_{n=1}^{\infty} a_n \cos(\omega_n t + \varepsilon_n) \tag{9.8}$$

The amplitude a_n and phase ε_n in (9.8) have different meanings from those in (9.3) and (9.5) where they represent amplitudes and phase angles of freely propagating independent waves. In (9.8), a_n and ε_n are the result of a mathematical representation of all waves propagating in different directions, but having the same frequencies. These waves are added together, and the result is expressed as a sum of harmonic functions. The component waves in (9.8) do not represent the physical reality by themselves. In a purely mathematical sense, (9.8) is the Fourier expansion of a random time-varying function (Goda, 1985).

The profiles of the sea waves, Fig. 9.2, have narrow-band characteristics, but contain a few crests and troughs, which are not of sufficient amplitude to cross the MSL. The probability density function of the height of crests (or troughs) of such random functions can be expressed in terms of spectral moments λ_i

$$p(\alpha) = \frac{1}{(2\pi)^{1/2}} \left[\frac{\varepsilon}{\sigma} \exp\left(-\frac{1}{2}\frac{\alpha^2}{(\varepsilon\sigma)^2}\right) + (1-\varepsilon^2)^{1/2} \frac{\alpha}{\sigma} \exp\left(-\frac{1}{2}\left(\frac{\alpha}{\sigma}\right)^2\right) \right.$$

$$\left. \times \int_0^{\alpha(1-\varepsilon^2)^{1/2}/(\varepsilon\sigma)} \exp\left(-\frac{u^2}{2}\right) du \right] \tag{9.9}$$

where α is the crest height measured from MSL, and

$$\varepsilon^2 = \frac{\lambda_0 \lambda_4 - \lambda_2^2}{\lambda_0 \lambda_4} \tag{9.10}$$

For narrow-band random processes, $\varepsilon \to 0$, and (9.9) reduces to

$$p(\alpha) = \frac{\alpha}{\sigma} \exp\left(-\frac{1}{2}\left(\frac{\alpha}{\sigma}\right)^2\right), \quad \alpha > 0 \tag{9.11}$$

the Rayleigh distribution.

For broad-band processes, $\varepsilon \to 1$, and (9.9) reduces to

$$p(\alpha) = \frac{1}{(2\pi)^{1/2} \sigma} \exp\left(-\frac{1}{2}\left(\frac{\alpha}{\sigma}\right)^2\right), \quad -\infty < \alpha < \infty \tag{9.12}$$

the Gaussian distribution.

The *spectral width parameter* ε can be estimated directly from a record by the following relation (Cartright and Longuet-Higgins, 1956)

$$\varepsilon = [1 - (1 - 2\gamma^2)]^{1/2} \tag{9.13}$$

where γ is the ratio of negative crests to the total number of crests in a record.

We have defined the wave height as the vertical distance between the highest and lowest points between two successive zero-upcrossings. For the nth wave in a record, it is given by

$$H_n = (\eta_{\max})_n - (\eta_{\min})_n \tag{9.14}$$

Since the profile of the surface elevation at a point in the sea is a narrow-band random process, $\eta(t)$ represents a harmonic wave with slowly varying phase and amplitude, and $H_n \simeq 2(\eta_{\max})_n = 2\alpha_n$. The distribution of the wave height, H, is therefore expected to be Rayleigh. The probability density function of H is

$$p_H(h) = \frac{h}{4\sigma^2} \exp\left(-\frac{h^2}{8\sigma^2}\right) \tag{9.15}$$

and the distribution function

$$F_H(h) = P(H \le h) = 1 - \exp\left(-\frac{h^2}{8\sigma^2}\right) \tag{9.16}$$

Longuet-Higgins (1952) demonstrated the applicability of Rayleigh distribution to wave heights in the sea, and since then this distribution has been extensively used. The validity of Longuet-Higgins assertion applies strictly to waves which have very small fluctuations in wave periods and whose heights exhibit beat-like characteristics. Real sea waves usually exhibit fairly large fluctuations in individual wave period and, therefore, depart slightly from Rayleigh distribution. Holmes and Howard (1971) have examined the validity of Rayleigh distribution by comparison with data obtained from laboratory wind-wave flume, and prototype records of waves in deep water. Forristal (1978) proposed the following probability distribution function to fit the real sea data

$$F_H(h) = 1 - \exp\left(- (h/\sigma)^{2.126}/8.42\right) \tag{9.17}$$

The difference between (9.16) and (9.17) is small and, in most cases, Rayleigh distribution provides a good approximation to the distribution of wave heights defined by zero-upcrossing method. This is true not only for wind waves and swell individually, but also for combined sea state of wind waves and swell (Goda, 1985).

For a Rayleigh distribution of wave heights (9.15), the wave parameters defined in section 9.2 can be derived and related to each other as follows:

$$\bar{H} = E[H] = (2\pi)^{1/2} \sigma \tag{9.18}$$

$$H_{rms} = 1.129 \, \bar{H} = 2(2)^{1/2} \sigma \tag{9.19}$$

$$H_{1/n} = H_{rms}\left[(\ln n)^{1/2} + \frac{n}{2}\sqrt{\pi}\,(1 - \mathrm{erf}\,((2\ln n)^{1/2}))\right] \tag{9.20}$$

$$H_s = H_{1/3} = 1.598 \, \bar{H} \tag{9.21}$$

$$H_{1/10} = 2.032 \, \bar{H} \tag{9.22}$$

and

$$E[H_{max}] = H_{rms}\left[(\ln N)^{1/2} + \frac{.5772}{2(\ln N)^{1/2}}\right] \tag{9.23}$$

where N is the total number of waves in the record.
For $N > 50$, (9.23) can be approximated by

$$E[H_{max}] \simeq H_{rms}\,(\ln N)^{1/2} \tag{9.24}$$

The Wave Period
The wave period T_z is defined as the time interval between two successive zero-upcrossings. Following the work of Rice (1944) and Woading (1955), Longuet-Higgins (1975) has derived the probability density function for wave periods of sea waves

$$p(\tau) = \frac{1}{2(1 + \tau^2)^{3/2}} \tag{9.25}$$

in which

$$\tau = \frac{T_z - \bar{T}_z}{v\,\bar{T}_z} \tag{9.26}$$

and

$$v = \left(\frac{\lambda_2}{\lambda_0}\right)^{1/2} \tag{9.27}$$

where $\bar{T}_z$ is the mean wave period.

The distribution (9.25) can simulate the characteristics of the observed waves, such that, short-period waves tend to have small heights. Goda (1985) has pointed out that (9.25) leads to a smaller value of significant period as compared to mean period, which is inconsistent with the observed characteristics of the sea waves.

The periods of individual waves in a wave record exhibit a distribution narrower than that of the wave heights. If the wind waves and swell coexist, distribution becomes broader and may become bi-modal with two peaks corresponding to mean periods of wind waves and swell. Hence, the wave period does not have a universal distribution, as in the case of wave height.

Joint Distribution of Wave Heights and Periods
Longuet-Higgins (1975) has proposed the following joint distribution of wave heights and periods

$$p(x, \tau) = \frac{x^2}{(2\pi)^{1/2}} \exp\left[-x^2 (1 + \tau^2)/2\right] \qquad (9.28)$$

where $x = H/2\sigma$.

The conditional distribution of wave period is given by

$$p(\tau \mid x) = \frac{p(x, \tau)}{p(x)} = \frac{x}{(2\pi)^{1/2}} \exp\left(-x^2 \tau^2/2\right) \qquad (9.29)$$

which is a normal distribution. From (9.29) and (9.26), conditional standard deviation of $T_z/\bar{T}_z$

$$\sigma\left(\frac{T_z}{\bar{T}_z} \mid H\right) = \left(\frac{2}{\pi}\right)^{1/2} \frac{\bar{H}}{H} \nu \qquad (9.30)$$

which indicates that the distribution of wave periods is inversely proportional to wave height, that is, waves of the smallest height exhibit wider distribution in wave period.

Assuming a Rayleigh distribution for the square of the wave period T_z, Battjes (1970) has proposed a bivariate Rayleigh distribution for wave heights and periods. Houmb and Overvik (1976) have proposed a two-parameter Weibull distribution to represent the conditional distribution (9.29).

Spectral Properties of the Sea Waves
We have seen that the short-term behaviour of sea waves can be modelled by a stationery random process. The distribution of the wave energy as a function of the frequency ω and wave direction θ is given by the directional wave spectrum $G(\omega, \theta)$ (9.4). For a unidirectional wave, $G(\omega)$ represents the unidirectional psd (9.6).

The characteristics of the psd of sea waves have been well established through analyses of a large number of wave records from different regions of the world. The spectra computed from the measured data at a station are called actual spectra. The spectral properties of the sea waves change with the sea state and are specific for a observation station. Several standard spectra have been derived from measured spectra by fitting expressions having free parameters (Bretschneider, 1959; Pierson and Moskowitz, 1964; Mitsuyasu, 1968; Hasselmann et al, 1973; Ochi and Hubble, 1976). Estimates of these parameters for a location allows the standard spectra to be used for different locations. The spectra due to Pierson and Moskowitz (P-M spectrum) has been empirically derived for a fully developed sea and has the special feature that the sea state is specified by the wind velocity as a principal parameter. The P-M spectrum can be expressed as

$$G(\omega) = \frac{\alpha g^2}{\omega^5} \exp\left[-\beta\left(\frac{g}{\omega V}\right)^4\right] \tag{9.31}$$

in which V is the wind velocity defining the sea state, g is acceleration due to gravity, and α and β are dimensionless constants given by

$$\alpha = 4\pi^3\left(\frac{H_s}{g\bar{T}_z^2}\right)^2 \tag{9.32}$$

$$\beta = 16\pi^3\left(\frac{V}{g\bar{T}_z}\right)^4 \tag{9.33}$$

The wind velocity is conventionally taken as 19.5 m/s and is related to H_s and $\bar{T}_z$ by

$$H_s = 2.12 \times 10^{-2} V^2$$

$$\bar{T}_z = 0.81\, 2\pi V/g$$

where V is measured in m/s. For the North Sea, α and β are taken as 0.0081 and 0.74 respectively. The spectral moments for the psd defined by (9.31) are

$$\lambda_0 = \frac{\alpha V^4}{4\beta} \tag{9.34}$$

$$\lambda_2 = \frac{\alpha V(\pi)^{1/2}}{4} \tag{9.35}$$

A spectrum called JONSWAP derived by Hasselman et. al. (1973), based on the results of joint waves observation programme for the fetch-limited North Sea, is similar to P-M spectrum, but includes a *peak enhancement factor*, which controls the sharpness of the spectral peak. The JONSWAP spectrum was developed to take into account the higher peaks in a storm situation. The functional form of JONSWAP spectrum is given by

$$G(\omega) = \frac{\alpha g^2}{\omega^5} \exp\left[-\frac{5}{4}\left(\frac{\omega_m}{\omega}\right)^4\right]\gamma^{\exp[-(\omega-\omega_m)^2/2\sigma^2\omega_m^2]} \tag{9.36}$$

where α is same as in P-M spectrum, ω_m is the frequency corresponding to maximum energy in P-M spectrum, that is

$$\omega_m = \left(\frac{4}{5}\beta\right)^{1/4}\frac{g}{V} \tag{9.37}$$

γ *the overshoot parameter*, is the ratio of the maximum spectral energy to the corresponding maximum of P-M spectrum, and

$$\sigma = \sigma_a, \text{ for } \omega \leq \omega_m$$
$$= \sigma_b, \text{ for } \omega > \omega_m \tag{9.38}$$

The average values of σ_a and σ_b are 0.07 and 0.09 respectively. The overshoot parameter $1 < \gamma < 7$, has a mean value of 3.3 for North Sea. It can be expressed as a function of H_s and T_p, the peak period (Fig. 9.4).

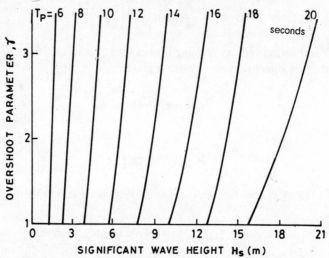

Fig. 9.4.   Overshoot parameter γ as a function of H_s and T_p (Chakrabarti and Snider, 1975).

Figure 9.5 shows a comparison of the P-M and JONSWAP spectra for $\gamma = 3.3$. It clearly shows the sharper peak of JONSWAP spectrum and its shift towards

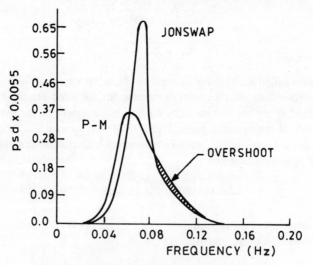

Fig. 9.5.   P-M and JONSWAP spectra for $\gamma = 3.3$ (Chakrabarti and Snider, 1975).

higher frequencies. Chakrabarti and Snider (1975) have analyzed the spectra of a North Sea storm at various stages and shown that JONSWAP spectrum represents the evolution of the storm more closely than P-M spectrum.

The actual wave spectra often exhibit significant deviations from the standard forms (9.31, 9.36). In situations where a combination of swell and wind waves exists, a bi-modal spectrum is obtained with a secondary peak corresponding to the representative frequency of swell or wind waves depending on their relative magnitudes. It has been suggested that JONSWAP spectrum may be used to model swell, as it propagates away from the wave generating area with $\gamma = 3 - 10$, depending on the travel distance (Goda, 1983).

It is clear from the above discussion that there is a large measure of uncertainty about the choice of a spectral shape, estimation of parameters for a particular location, and the evolution of spectral shape during a storm. The adaptation of a standard spectrum to a particular site improves with an increase in the number of free spectral parameters. The six parameter bi-modal spectrum proposed by Ochi and Hubble (1976) has the flexibility to fit different kinds of actual spectra. The Ochi-Hubble spectrum is also able to represent almost all stages of the sea condition associated with a storm.

Directional Wave Spectra

We have seen that the sea surface consists of a large number of short-crested waves propagating in various directions. The concept of *directional wave specrum*, $G(\omega, \theta)$, has been introduced to describe the state of the superimposed directional components. It represents the distribution of wave energy in the product space defined by frequency ω and direction θ, and can be expressed as

$$G(\omega, \theta) = G(\omega)\, S(\omega; \theta) \tag{9.39}$$

where $S(\omega; \theta)$ is the directional spreading function satisfying the normalization condition

$$\int_{-\pi}^{\pi} S(\omega; \theta)\, d\theta = 1 \tag{9.40}$$

$S(\omega; \theta)$ represents the relative magnitude of the directional distribution of wave energy in the direction θ; and $G(\omega)$ represents the absolute value of the wave energy density.

Due to difficulties associated with reliable field measurements, knowledge of directional distribution of the energy of sea waves is limited and a standard functional form for directional wave spectrum has not been established. On the basis of detailed field measurements with a special cloverleaf-type instrument and other data, Matsuyasu et al (1975) have proposed the following functional form for $S(\omega; \theta)$

$$S(\omega; \theta) = S_0 \cos^{2S}\left(\frac{\theta}{2}\right) \tag{9.41}$$

where θ is measured counter clockwise from the principal wave direction, and

$$S_0 = \frac{1}{\pi} 2^{2s-1} \frac{\Gamma(s+1)}{\Gamma(2s+4)} \tag{9.42}$$

to satisfy (9.40). In (9.42), Γ denotes Gamma function. The parameter s represents the degree of directional energy concentration. It is a function of the frequency and has a peak value around the frequency of the spectral peak ω_p. Its value decreases as the frequency moves away from ω_p towards both lower and higher frequencies. Goda and Suzuki (1975) have proposed the following relation for engineering applications:

$$s = s_{max} \left(\frac{\omega}{\omega_p} \right)^5 ; \omega \le \omega_p$$

$$= s_{max} \left(\frac{\omega}{\omega_p} \right)^{-2.5} ; \omega > \omega_p \tag{9.43}$$

The parameters s_{max} and ω_p may be estimated from the following relations:

$$s_{max} = 11.5 \left(\frac{\omega_p V}{g} \right)^{-2.5} \tag{9.44}$$

and

$$\omega_p = 1/(1.05\, T_s) \tag{9.45}$$

where V denotes the wind speed. Using Wilson's (1965) formula and (9.44), the mean value of s_{max} can be related to deep water wave steepness H_0/L_0. Goda and Suzuki (1975) have proposed the following mean values of s_{max} for engineering applications in deep water conditions:

Wind waves $\qquad\qquad\qquad\qquad s_{max} = 10$

Swell with short decay distance $\qquad s_{max} = 25$

Swell with long decay distance $\qquad s_{max} = 75$

Several other forms of $G(\omega; \theta)$ have been proposed. The first practical formula was based on a suggestion of Arthur (1949):

$$S(\omega; \theta) = \frac{2}{\pi} \cos^2 \theta ; \quad |\theta| \le \pi/2$$

$$= 0 \qquad\qquad |\theta| > \pi/2 \tag{9.46}$$

The above spreading function is independent of ω, and was employed in the P-N-J method for wave forecasting. It is employed in many applications due to its simple form and gives almost the same distribution of total energy as (9.41) for $s_{max} = 10$. Other forms of directional spreading functions are available in (Kinsman, 1965; Borgman, 1967 and Panicker, 1974).

9.3.2 Stochastic Model of Storms

A storm is defined as a sea state in which the significant wave height $H_s(t)$ (or visual wave height $H_v(t)$ remains above a predefined level H'_s (1-2 meters). The time interval between an upcrossing of the level H'_s and the subsequent downcrossing is defined as the storm duration S (Fig. 9.6).

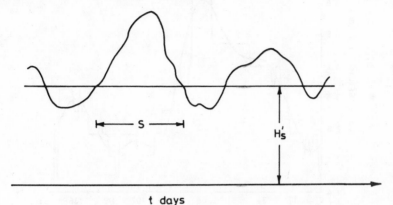

Fig. 9.6. Definition of a storm.

A storm may have a duration extending over several days. In the preceding section, we have seen that over time periods of the order of an hour, the surface elevation at a location may be treated as stationary. During a storm, the surface elevation is nonstationary with time-dependent mean $\bar{H}_s(t)$, superimposed with stationary random fluctuations $\Delta H_s(t)$ so that

$$H_s(t) = \bar{H}_s(t) + \Delta H_s(t) \tag{9.47}$$

A storm is, therefore, characterised by duration S, shape $\bar{H}_s(t)$, and inter-arrival time τ, each of which is random in nature. A stochastic model of the storm specifies the distributions of each of these parameters and makes it possible to generate synthetic storms from historical data. Shyam Sundar (1978) has proposed a 'best-fit' Fourier series expansion to model $\bar{H}_s(t)$ for a given storm duration S in the form

$$\bar{H}_s(t) = \frac{a_0}{2} + \sum_{n=1}^{M} [a_n \cos(\omega_n t) + b_n \sin(\omega_n t)] \tag{9.48}$$

where $\omega_n = 2\pi n/S$, and

$$a_n = \frac{2}{S} \int_0^s \bar{H}_s(t) \cos(\omega_n t)\, dt, n = 0, 1, 2 \ldots M \tag{9.49}$$

$$b_n = \frac{2}{S} \int_0^s \bar{H}_s(t) \sin(\omega_n t)\, dt, n = 1, 2 \ldots M \tag{9.50}$$

The Fourier coefficients in (9.48) are determined from historical records and the number of terms depend on the storm duration. Figure 9.7 shows best-fit of a historical storm.

On the basis of a careful and detailed analysis of historical data, Shyam Sundar (1978) has proposed following distributions for a stochastic model of the storms.

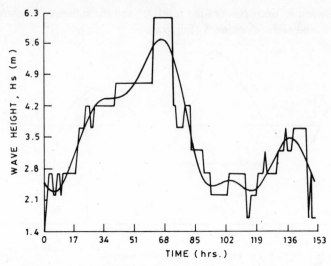

Fig. 9.7. Typical historical storm and the 'best-fit' (Angelide, 1978).

STORM DURATION S: A two parameter Weibull distribution

$$F_S(s) = \exp\left[\left(-\frac{s}{s_c}\right)^\beta\right] \tag{9.51}$$

where S is in hours and s_c and β are the parameters of the distribution.

COEFFICIENT a_0: The distribution of $a_0/2$, conditioned on S, is represented by a three parameter Weibull distribution

$$F_{a_0/2}\left(\frac{a_0}{2}\,|\,S\right) = \exp\left[-\left(\frac{a_0/2 - \alpha_0}{\alpha_c - \alpha_0}\right)\lambda\right] \tag{9.52}$$

To determine the parameters α_0, α_c and λ, the historical data is divided into storms of various durations.

COEFFICIENTS a_n, b_n: The distributions of a_n, b_n, conditioned on S and $a_0/2$, are represented by independent, normal distributions. In some cases, it is necessary to incorporate skewness in the distribution.

RANDOM FLUCTUATION $\Delta H_s(t)$: $\Delta H_s(t)$ is modelled as a stationary random process and its distribution conditioned on S and $a_0/2$, is assumed to be normal.

INTER-ARRIVAL TIMES BETWEEN STORMS, τ: The storm occurrence is assumed to be Poisson and, therefore, the inter-arrival time has an exponential distribution

$$F(\tau) = \exp\left(-\frac{\tau}{\tau_c}\right) \tag{9.53}$$

where τ_c is the mean of τ.

A stochastic model of the storms is fully described by the above distributions. Angelides (1978) has determined the parameters of these distributions, based on the visual data obtained by weather station 'M' in North Atlantic during the period 1961–75.

9.3.3 Long Term Statistical Properties of Sea Waves

For the design of offshore structures, it is necesary to predict the sea states during the service life of the structure. In the 'design wave' approach, it is necessary to specify the height and period of the design wave for a return period of 50 or 100 years. The wave data is commonly recorded for about 20 minutes after every three hours. The bivariate distribution of H_s and T_z is constructed from this data collected over a one year period. The long-term distribution of wave heights H is then obtained from

$$F_L(h) = P(H \le h) = \frac{1}{\overline{T_z^{-1}}} \int_0^\infty \int_0^\infty \frac{1}{t_z} F(h \mid h_s, t_z)\, p(h_s, t_z)\, dh_s\, dt_z \qquad (9.54)$$

where $p(h_s, t_z)$ is the bi-variate density function of H_s and T_z, and $\overline{T_z^{-1}}$ is the long term average number of waves per annum given by

$$\overline{T_z^{-1}} = \int_0^\infty \int_0^\infty \frac{1}{t_z} p(h_s, t_z)\, dh_s\, dt_z \qquad (9.55)$$

The long-term distribution $F_L(h)$ is usually one of the extreme value distributions. Log-normal (Darbyshire, 1956; Dattari, 1973), Weibull (Bretschneider, 1965; Battjes, 1970), and Gumbel (Saetre, 1974) distributions have been used by various investigators. Assuming Gumbel distribution, we have

$$F_L(h) = \exp\left[-\exp\left\{-a(h - \beta)\right\}\right] \qquad (9.56)$$

Design Wave Height

Let H_D be the 'design wave' height. For Poisson process model, the mean arrival rate of special waves $(H > H_D)$ is given by

$$v = [1 - F_L(H_D)]\, \overline{T_z^{-1}} \qquad (9.57)$$

The probability of atleast one arrival in N years ('failure') is given by

$$p = 1 - \exp\left(-vN\right) \qquad (9.58)$$

Substituting for v from (9.57) in (9.58), and rearranging terms

$$F_L(H_D) = 1 + \frac{\overline{T_z}}{N} \ln\left(1 - p\right) \qquad (9.59)$$

From (9.56) and (9.59), the 'design wave' height H_D corresponding to probability of exceedance p in N years is given by

$$H_{D;N,p} = \beta - \frac{1}{a} \ln\left[-\ln\left\{1 - \frac{\overline{T_z}}{N} \ln\left(1 - p\right)\right\}\right] \qquad (9.60)$$

If $\nu N \ll 1$, (9.59) may be expressed as

$$p \simeq \nu N$$

and (9.60) simplifies to

$$H_{D;N,p} \simeq \beta - \frac{1}{a} \ln \left[- \ln \left\{ \frac{N - p\bar{T}_z}{N} \right\} \right] \tag{9.61}$$

The 'return-period' for 'design-wave' is given by

$$T_{H_D} = \frac{1}{\nu} = \frac{\bar{T}_z}{1 - F_L(H_D)} \tag{9.62}$$

Hence,

$$F_L(H_D) = 1 - \frac{\bar{T}_z}{T_{H_D}} \tag{9.63}$$

and the 'design-wave' height for a return period T_{H_D} is given by

$$H_{D;T} = H_{D;N,p} \simeq \beta - \frac{1}{a} \ln \left[- \ln \left\{ \frac{T_{H_D} - \bar{T}_z}{T_{H_D}} \right\} \right] \tag{9.64}$$

Further, the exceedance probability

$$p = 1 - \exp \left(- \frac{N}{T_{H_D}} \right) \tag{9.65}$$

$$\simeq \frac{N}{T_{H_D}}, \text{ if } \frac{N}{T_{H_D}} \ll 1 \tag{9.66}$$

9.3.4 Nonlinearity of Sea Waves

We have modelled the random sea waves by linear superposition of an infinite number of infinitesimal amplitude component waves, defined by (9.3) or (9.5). The phase angle ε_n is assumed to be independent and uniformly distributed in the interval $[0, 2\pi]$. The model implies that the distribution of instantaneous surface elevation is normal. Analysis of actual sea waves, however, indicates deviation from the normal distribution due to presence of nonlinear wave components. Due to the presence of nonlinear components, some of the phase angles get related to each other and are, therefore, statistically dependent. The component waves in such cases do not satisfy the dispersion relation (9.2).

The deviation from the normal distribution is indicated by skewness and kurtosis parameters defined as

Skewness:

$$\sqrt{\beta_1} = \frac{1}{\eta_{rms}^3} \frac{1}{N} \sum_{i=1}^{N} (\eta_i - \bar{\eta})^3 \tag{9.67}$$

Kurtosis:

$$\beta_2 = \frac{1}{\eta_{rms}^4} \cdot \frac{1}{N} \sum_{i=1}^{N} (\eta_i - \bar{\eta})^4 \tag{9.68}$$

For normal distribution, $\sqrt{\beta_1} = 0$ and $\beta_2 = 3.0$. Records of sea waves, generally, have a positive skewness implying the observed fact that wave crest heights are greater than the wave trough depths. For wind generated waves in deep water, skewness is found to be proportional to wave steepness. Hwang and Long (1980) have proposed the following empirical relation

$$\sqrt{\beta_1} = 2\pi H / L \tag{9.69}$$

Goda (1983) has proposed the following parameter as a measure of wave nonlinearity:

$$\Pi = (H/L_A) \coth^3 k_A h \tag{9.70}$$

$$\omega^2 = gk_A \tanh k_A h, \text{ and } k_A = 2\pi/L_A \tag{9.71}$$

subscript A in (9.70) and (9.71) denotes small amplitude Airy wave theory. Figure 9.8 shows a plot of skewness against nonlinearity parameter Π. The solid line represents skewness of the theoretical profiles of regular, finite amplitude waves.

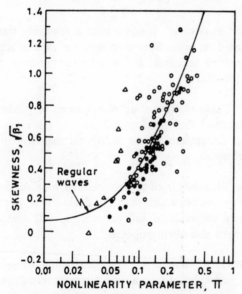

Fig. 9.8. Wave data of skewness of surface elevation against nonlinearity parameter Π. (Goda, 1983b) $\{h/L = 0.2\text{--}0.65\ (\Delta);\ 0.07 - 0.2\ (\text{o})\}$.

The kurtosis of actual wave records is on an average greater than 3.0. This implies that actual distribution has a peak higher than the corresponding normal distribution, and has long tails on both sides.

The nonlinear components of a sea wave may be analysed by the calculation of interaction among linear spectral components (Masuda et al, 1979). The primary (linear), secondary, and other higher order interactions tune in the frequency spectrum and can be identified through such an analysis (Tick, 1963). The nonlinear effects are reflected in auxiliary peaks of the spectrum at higher harmonics of the frequency corresponding to the primary peak. These cause an apparent increase in spectral moments and spectral width parameters, decrease in spectral peakedness parameter, and flattening of the slope of the spectrum in high frequency range (Goda, 1983).

9.4 Wave Forces on Offshore Structures

Offshore structures are subjected to dynamic loads due to wind, currents, waves, and if located in a seismically active region, to earthquakes. Of these, the loads due to waves are, usually, dominant and will be discussed in this section. In Chapters 7 and 8, we have discussed the application of random vibration theory to the probabilistic design of structures against earthquake and wind loading. With some minor modifications, it can be extended to the design of offshore structures, in combination with the wave loading.

The wave forces on an offshore structure depend on the characteristics of the wave environment, and the geometric and dynamic properties of the structure. The choice of the structural form of a fixed offshore structure is primarily governed by the soil characteristics and the depth of water. Commonly used structural forms are:

JACKET OR TEMPLATE PLATFORM: It consists of a relatively transparent and stiff space-frame supported on piles. Such platforms are used in regions with a large thickness of soft marine sediment, and are, by far the most common form for moderate depths   (30–300 m).

TOWER PLATFORM: These platforms are an extension of jacket platform to deep waters. The foundation of these platforms consists of pile clusters and skirt piles. The cross-sectional dimensions of the individual members of both jacket and tower platforms are small.

GRAVITY PLATFORM: It consists of a large cellular base supporting a few large diameter concrete towers on which rests the deck. These platforms are used in regions with heavily consolidated soil at, or near, mud line. Steel and hybrid gravity platforms have also been proposed.

COMPLIANT PLATFORM: It is a relatively flexible structure which moves to mitigate the effect of loads. Examples of such a structure are guyed towers and tension-leg platform in deep sea.

Several methods have been developed for the determination of wave forces on offshore structures. The applicability of these methods depends on the relative magnitude of the typical dimension, a, of the structural members with respect to the wave length L, and wave height H. The choice of the method for the determination of wave forces on offshore structures is dictated by the relative significance of viscous and diffraction effects. Garrison and Rao (1977) have classified the wave force theories in terms of two dimensionless ratios: the *wake parameter* $(H/2a)$, and the *scattering parameter* $(2\pi a/L)$, as shown in Fig. 9.9.

Hogben (1976) has delineated the regimes dominated by viscous, inertia and diffraction effects in terms of H and a. He has also indicated the effect of Reynolds

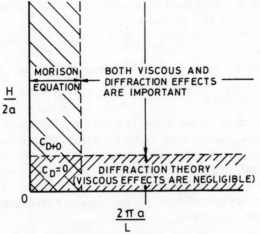

Fig. 9.9. Wave force theory classification (Garrison and Rao, 1971).

number. Basically, viscous effects are important for large amplitude and long period waves ($a/H < 0.2$); and the diffraction effects dominate when the presence of the body alters the wave field ($a/L > 0.2$). Hogben (1974) has proposed following criteria as a rough guide to indicate the importance of different effects:

$a/L > 1$: Conditions approximate to pure reflection.
$a/L > 0.2$: Diffraction increasingly important.
$a/w_0 > 0.2$: Inertia increasingly important.
$a/w_0 < 0.6$: Incipience of lift and drag.
$a/w_0 < 0.2$: Drag increasingly important.

where w_0 is the orbit width parameter of the water-particle. For deep water, $w_0 = H$.

The interaction between waves and structural member gives rise to following forces (Fig. 9.10):

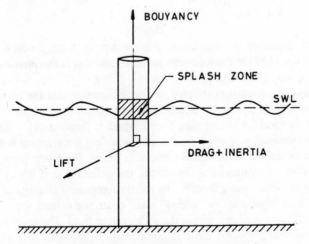

Fig. 9.10. Hydrodynamic forces on a cylinder.

   (i) fluctuating buoyancy force in the vertical direction due to variation in the immersed length of a vertical member;

  (ii) wave impact or 'slamming' in the horizontal members near the surface (splash zone);

 (iii) wave 'slapping' on the vertical members near the surface;

 (iv) in-line drag and inertia forces; and

  (v) transverse lift force.

Besides these forces, the oscillation of a body in a fluid due to wave action, or any other excitation, produces waves on the sea surface, which radiate the momentum away from the body. This gives rise to radiation damping. The effect is related to the size of the body and may be important in the diffraction regime.

We shall first discuss the in-line and lift forces on slender members for which diffraction effects are insignificant. We shall conclude our discussion with a brief review of the diffraction theory.

9.4.4   Wave Forces on Slender Members

In-line Force: Morison Equation

Based on experimental data, Morison et al (1950) proposed an empirical formula for the in-line force on a slender stationary vertical cylinder in presence of surface waves. The formula known as *Morison equation*, gives the in-line force per unit length.

$$F(s_3, t) = F_I + F_D$$

$$= M_I \dot{U} + C_H U \mid U \mid \tag{9.72}$$

in which

$$M_I = C_M \rho \frac{\pi D^2}{4} = (C_m + 1) \rho \frac{\pi D^2}{4} = M_a + M_w \tag{9.73}$$

$$C_H = C_D \frac{\rho D}{2} \tag{9.74}$$

where D is the diameter of the cylinder; ρ is the density of the water; U is the fluid particle velocity, $\dot{U}$ is the fluid particle acceleration; C_M is the inertia coefficient, and C_D is the drag coefficient.

The Morison equation is postulated on the assumption that the presence of the member does not modify the wave field. The force is made up of two components: (i) F_I, the inertia force (or virtual mass force), which is proportional to the horizontal fluid particle acceleration; and (ii) F_D, the nonlinear drag force, which is proportional to the square of the horizontal fluid particle velocity. The water particle velocity and acceleration are evaluated at the axis of the cylinder, as if the cylinder were absent, using the linear wave theory. The inertia component consists of two terms: one due to 'hydrodynamic' or 'added' mass contribution and the other due to variation of pressure gradient within the accelerating fluid. The inertia component is proportional to D^2, whereas the drag component is proportional to D, and hence inertia force is expected to dominate for large diameter members.

The estimates of C_M and C_D are sensitive to the wave environment, surface roughness, the experimental set up and show a large variability (Angelides, 1978). Morison et al (1950) determined the following average values of C_M and C_D at the instants of zero velocity and zero acceleration respectively:

$$C_M = 1.508 \pm 0.197 \tag{9.75}$$

$$C_D = 1.626 \pm 0.414 \tag{9.76}$$

The values show some scatter, but no trend as functions of h/L, H/L or Reynolds number N_R. We shall discuss the uncertainty in the estimates of these coefficients at the end of this section.

$F(s_3, t)$ in (9.72) is in-line horizontal force per unit length on a vertical tubular member. The force and moment on a finite length of the member can be obtained by integrating it over the appropriate length. The force F is a linear function of acceleration $\dot{U}$, but a nonlinear function of velocity U. For short-term behaviour, $U(t)$ and $\dot{U}(t)$ may be assumed to be stationary and, therefore, orthogonal random processes. Further, if they are assumed to be Gaussian, it follows from the theory of functions of random variables that the probability density function of $F(t)$ is given by

$$p_F(f) = \frac{1}{2\pi M_I \sigma_U \sigma_{\dot{U}}} \int_{-\infty}^{\infty} \exp\left[-\frac{1}{2}\left(\frac{u^2}{\sigma_U^2} + \frac{\dot{u}^2}{\sigma_{\dot{U}}^2}\right)\right] du \tag{9.77}$$

where

$$\dot{u} = (f - C_H u|u|)/M_I \tag{9.78}$$

It can be shown that the moments of F are given by (Pierson and Holmes, 1965)

$$E[F] = 0 \tag{9.79}$$

$$E[F^2] = M_I^2 \sigma_{\dot{U}}^2 + 3C_H^2 \sigma_U^4 \tag{9.80}$$

$$E[\dot{F}^2] = M_I^2 \sigma_{\ddot{U}}^2 + 4C_H^2 \sigma_U^2 \sigma_{\dot{U}}^2 \tag{9.81}$$

and

$$E[F^4] = 3M_I^4 \sigma_{\dot{U}}^4 + 18M_I^2 C_H^2 \sigma_U^2 \sigma_{\dot{U}}^4 + 105C_H^4 \sigma_U^8 \tag{9.82}$$

Level Crossing Rate and Peak Distribution of F(t)
The expected rate of level crossing and the distribution of the peaks of $F(t)$ can determined from the joint distributions of $F(t)$, $\dot{F}(t)$ and $\ddot{F}(t)$ by the well known relations (6.4*), (6.28*) and (6.30*)

$$N^+(\alpha) = \int_0^\infty \dot{f} p(\alpha, \dot{f}) d\dot{f} \tag{9.83}$$

$$M(\alpha) = \int_\alpha^\infty df \int_{-\infty}^0 |\ddot{f}| p(f, o, \ddot{f}) d\ddot{f} \tag{9.84}$$

and

$$F_p(\alpha) = 1 - \frac{M(\alpha)}{MT} \qquad (9.85a)$$

where $N^+(\alpha)$ is the expected rate of crossing the level α with positive slope; $M(\alpha)$ is the expected rate of occurrence of peaks; $MT = M(\alpha)$ is the total rate of occurrence of peaks; and $F_p(\alpha)$ is the distribution function of peaks.

If the force spectrum is assumed to be narrow-band, the distribution (9.85a) can be simplified and expressed in terms of joint distribution of $F(t)$ and $\dot{F}(t)$ in the form

$$F_p(\alpha) = 1 - \frac{\int_0^\infty \dot{f} p(\alpha, \dot{f})\, d\dot{f}}{\int_0^\infty \dot{f} p(0, \dot{f})\, d\dot{f}} \qquad (9.85b)$$

Further, if $F(t)$ and $\dot{F}(t)$ can be assumed to be independent (9.85b) simplifies to

$$F_p(\alpha) = 1 - \frac{p(\alpha)}{p(0)} \qquad (9.85c)$$

If the occurrence of peaks can be assumed to be independent, the distribution of extreme force is given by

$$F_E(\alpha) = [F_P(\alpha)]^{MT} \qquad (9.86)$$

The above results have been derived under the assumption that mean current is zero. Grigoriu (1984) has derived the distribution of the force and, its moments, in the presence of current.

Long-term Distribution
Equation (9.77) gives the short-term distribution of $F(t)$ for a specific sea state $(H_s, T_z, \bar{\theta})$. The long-term distribution of F can be derived in the same way as (1.1) and can be expressed as

$$p_L(f) = \int_{\bar{\theta}} \int_0^\infty \int_0^\infty p(f \mid h_s, t_z, \bar{\theta})\, p(h_s, t_z, \bar{\theta})\, dt_z\, dh_s\, d\bar{\theta} \qquad (9.87)$$

where $p(f \mid h_s, t_z, \bar{\theta})$ is given by (9.77).

Linearized Force
Since $F(t)$ is a nonlinear function of U and $\dot{U}$, the distribution (9.77) is non-Gaussian. A Gaussian approximation of F can be obtained by linearizing (9.72) through a minimization of the mean-square error (Borgman, 1967), yielding linearized Morison equation

$$F(s_3, t) = M_I U + C_H \sigma_U \left(\frac{8}{\pi}\right)^{1/2} U \qquad (9.88)$$

and the variance of F is given by

$$E[F^2] = \sigma_F^2 = M_I^2 \, \sigma_U^2 + \frac{8}{\pi} \, C_H^2 \, \sigma_U^4$$

The linearized force F is Gaussian and, therefore, rate of level crossing and distribution of peak and extreme value distributions can be expressed in a closed form.

Alibe (1986) has derived the rate of level crossings and distribution of peaks for nonlinear Morison Eq. (9.72), treating the total force as the sum of two independent forces representing the inertia and drag components. He has determined the distribution and the mean of the maximum value of the force in a given time interval, assuming independent Poisson crossings. Comparison of the mean value for Gaussian approximation based on linearized Morison equation, and the non-Gaussian distribution shows, that linearization yields non-conservative estimates of peak force, especially when drag terms dominate.

Spectral Properties of the Wave Forces
In appendix 9A, we show that the horizontal particle velocity $(U_1 = U)$ (9A.23)

$$U(s_3, t) = \omega \left[\frac{\cosh [k(s_3 + h)]}{\sinh (kh)} \right] \eta(t) = \omega f(s_3) \, \eta(t) \tag{9.89}$$

Hence, the psd of U and $\dot{U}$ are given by

$$\Phi_{UU}(\omega, s_3) = \omega^2 f^2(s_3) \, \Phi_{\eta\eta}(\omega)$$

$$\Phi_{\dot{U}\dot{U}}(\omega, s_3) = \omega^4 f^2(s_3) \, \Phi_{\eta\eta}(\omega)$$

The psd of the wave force defined by (9.72) can be expressed as (Borgman, 1967b; 1972)

$$\Phi_{FF}(\omega, s_3) = M_I^2 \, \Phi_{\dot{U}\dot{U}}(\omega) + C_H^2 \left[\frac{8}{\pi} \, \sigma_U^2 \, \Phi_{UU}(\omega) \right.$$

$$\left. + \frac{4}{3\pi} (\sigma_U)^{-2} \times \{\Phi_{UU}(\omega)\}^{*3} + \ldots \right] \tag{9.90}$$

where

$$\{\Phi_{UU}(\omega)\}^{*n} = \int_{-\infty}^{\infty} \{\Phi_{UU}(\omega')\}^{*(n-1)} \, \Phi_{UU}(\omega - \omega') \, d\omega'$$

For linear approximation (9.88),

$$\Phi_{FF}(\omega, s_3) = M_I^2 \, \Phi_{\dot{U}\dot{U}}(\omega) + \frac{8}{\pi} \, C_H^2 \, \sigma_U^2 \, \Phi_{UU}(\omega)$$

$$= \left[M_I^2 \, \omega^2 + \frac{8 C_H^2 \, \sigma_U^2}{\pi} \right] \omega^2 f^2(s_3) \, \Phi_{\eta\eta}(\omega) \tag{9.91}$$

Thus in the linearized approximation, the terms containing the convolutions of the velocity spectra are neglected. These terms may not always be negligible. The spectrum $\{\Phi_{UU}, (\omega)\}*^3$ has two peaks. The second peak occurring at the frequency which is three times the peak frequency of $\Phi_{\eta\eta}(\omega)$, and may coincide with one of the natural frequencies of the structure and, thus, make a significant contribution to the response.

Morison Equation for Flexible or Moving Members

The original Morison equation can be modified to include the effect of flexibility on the motion of a member, by replacing the absolute velocity by relative velocity and including an added mass term associated with the acceleration of the structure (Berge and Penzien, 1974). The modified equation is

$$F(s_3, t) = (C_M - 1)\frac{\rho\pi D^2}{4}\ddot{Y} + \rho\frac{\pi D^2}{4}\dot{U} + \frac{1}{2}\rho C_D D\,\dot{Y}|\dot{Y}| \qquad (9.92)$$

$$= M_a\ddot{Y} + M_w\dot{U} + C_H\dot{Y}|\dot{Y}|$$

where $\dot{Y} = U - \dot{X}$ is the relative velocity, and $X(t)$ is the displacement of the member in the direction of the wave propagation. In (9.92), U should be measured at the instantaneous deflected position of the member. However, for narrow-band systems, it may be measured at undeflected member coordinates without significant error (Malhotra and Penzien, 1970b). In the sequel, we shall make this assumption.

In (9.92), the first term is the hydrodynamic mass term and it depends on the motion of the member. The second term which represents the force due to flow, is independent of the motion of the member. It is called Froude Krylov term. The third term is the drag term. Equation (9.92) can be linearized in the same way as (9.88) and expressed as

$$F(s_3, t) = (C_M - 1)\frac{\rho\pi D^2}{4}\ddot{Y} + \frac{\rho\pi D^2}{4}\dot{U} + \left(\frac{1}{2}\rho D\right)C_D\left(\frac{8}{\pi}\right)^{1/2}\sigma_{\dot{Y}}\dot{Y}$$

$$= M_a\ddot{Y} + M_w\dot{U} + C_H\left(\frac{8}{\pi}\right)^{1/2}\sigma_{\dot{Y}}\dot{Y} \qquad (9.93)$$

It must be noted that to arrive at the above linearized equation, it is assumed that Y is Gaussian, which is an approximation.

When the structure is stationary, $X(t) = 0$ and (9.92) reduces to (9.72). If the water is at rest $U(t) = \dot{U}(t) = 0$, (9.92) reduces to

$$F(s_3, t) = -\rho\frac{\pi D^2}{4}(C_M - 1)\ddot{X} - \frac{1}{2}\rho DC_D\,\dot{X}|\dot{X}|$$

$$= -\rho\frac{\pi D^2}{4}C_m\ddot{X} - \frac{1}{2}\rho DC_D\,\dot{X}|\dot{X}|$$

$$= -M_a\ddot{X} - C_H\dot{X}|\dot{X}| \qquad (9.94)$$

which gives the force exerted by the fluid on the body due to its motion. In (9.94), the force should include an additional term due to radiation damping. This term is proportional to member velocity and is usually small for slender member (Taylor, 1974).

Morison Equation for Inclined Members

The Morison equation was originally postulated for vertical members. The horizontal force was expressed in terms of the horizontal component of the particle velocity and the effect of vertical component (tangential to the member axis) was neglected in the calculation. Many offshore structures, specially jacket platforms, contain bracing members which are inclined to the vertical. Several methods have been proposed to extend Morison equation to inclined members (Borgman, 1958, Wade and Dwyes, 1976; Chakrabarti et al, 1975). We shall follow the method due to Borgman (1958), which replaces horizontal velocity and acceleration in Morison equation with components of fluid-particle velocity and acceleration normal to the member axis. In the vector notation,

$$\bar{F}(s_3, t) = M_I \dot{\bar{U}}_n + C_H \bar{U}_n |\bar{U}_n| \tag{9.95}$$

where $\bar{U}_n$ and $\dot{\bar{U}}_n$ are normal water-particle velocity and acceleration vectors respectively. If $\bar{e}$ represents a unit vector along the axis of the member, (Chakrabarti et al, 1975)

$$\bar{U}_n = \bar{U} - (\bar{U} \cdot \bar{e})\,\bar{e} = (\mathbf{I} - \bar{e}\bar{e}^T)\bar{U} \tag{9.96}$$

where $\bar{U}$ is the water-particle velocity vector.

Wade and Dwyer (1976) have compared four approaches for calculation of wave forces on inclined members and have found a variation of about 22% for base shear and overturning moments between the four methods. The method due to Borgman (9.95) is non-conservative but is consistent with classical Morison equation.

Angelides (1978) has proposed a simple procedure to determine horizontal forces on inclined members. In deep water, the water particles move in near circular orbits and, hence, it is reasonable to assume that force per unit length remains the same for any orientation of a member lying in the vertical plane parallel to the direction of wave propagation. Segments of an inclined member can, therefore, be replaced by vertical members as shown in Fig. 9.11. The horizontal component of the wave force on an inclined member is then equal to the sum of the horizontal

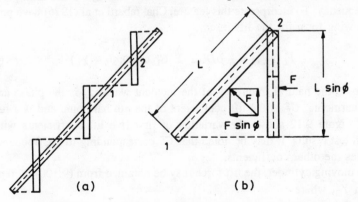

Fig. 9.11. Discretization of inclined members on vertical plane.

forces on discrete vertical segments. The horizontal members inclined in the direction of wave propagation may also be treated in the same way, although there is no physical basis for this procedure. The proposed discretization of inclined members also resolves, in an approximate manner, the question of variation of phase along an inclined member.

Lift Forces on Slender Members

The vortex shedding due to flow past a body gives rise to transverse lift force. For steady flow past a cylinder, the Strouhal number N_s, determines the frequency of vortex shedding. In the flow field due to wave motion, however, the situation is quite complex due to regular velocity fluctuations (Bidde, 1971, Stansby, 1978). To sustain vortex shedding, the orbital velocity of water particles should remain nearly constant for atleast three to four times the vortex shedding frequency. Hence, lift forces are likely to occur for long waves and small cylinders. Since the water particle velocity decreases with depth, the lift forces due to waves are confined to the surface.

Keulegan and Carpenter (1958) have studied the forces on cylinders in oscillating fluid and have shown that lift forces occur when Keulegan-Carpenter number

$$N_{KC} = \frac{U_m T}{D} \geq 15 \qquad (9.97)$$

in which U_m is the maximum horizontal water particle velocity; and T is the time period of the wave.

If the vortex shedding occurs due to wave action, the lift force per unit length can be expressed as

$$F_L(s_3, t) = K_L U |U| \qquad (9.98)$$

where

$$K_L = \frac{1}{2} \rho D C_L \qquad (9.99)$$

and C_L is the lift coefficient.

It is observed that frequency of the lift force is always some multiple of the wave frequency. To incorporate this feature, Chakrabarti et al (1976) have proposed the following form for the lift force

$$F_L(s_3, t) = \frac{1}{2} \rho D U_m^2 \sum_{n=1}^{N} C_L^n \cos(n\omega_0 + \chi_n) \qquad (9.100)$$

where ω_0 is the natural frequency of the incident wave; χ_n is the phase angle of the nth harmonic; C_L^n is the lift coefficient of the nth harmonic and is a function of N_{KC}. Figure 9.12 shows the variation of first five lift coefficients with N_{KC} based on test results. It may be noted that C_L^n corresponding to the frequency $2\omega_0$ dominates the other coefficients.

For a moving cylinder, the lift force may be obtained from (9.100) by replacing U_m with $\dot{Y}_m$, where

$$\dot{Y}_m = \max(|\bar{U} - \dot{\bar{X}}|) \qquad (9.101)$$

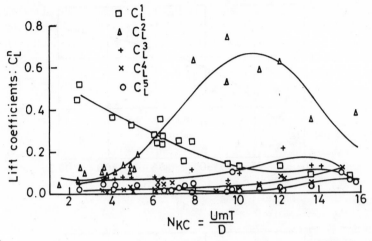

Fig. 9.12.   First five coefficients of lift force (Chakrabarti et al, 1976).

in which | | denotes vector modulus. In (9.101), we use vectors to determine the maximum relative velocity, since the motion of the member will not be in the same direction as the wave field.

Uncertainty in the Coefficients of Morison Equation

The estimates of inertia and drag coefficients C_M and C_D, in Morison equation depend on several critical parameters, such as, Reynolds number N_R, Keulegan-Carpenter number N_{KC}, roughness of the cylinder surface, and show significant variability. Several investigators have conducted and analyzed laboratory and field experiments to determine C_M and C_D and to relate them to critical parameters (Morison et al 1950; Keulegan and Carpenter, 1958; Wiegel et al, 1957; Thrasher and Aagaard, 1969; Dean and Aagaard, 1970; Kim and Hibbard, 1975; Chakrabarti et al, 1976, Sarpkaya et al, 1976a, b and 1977). The uncertainty in the estimates of C_M, C_D depends upon the reliability of the wave theory used in predicting the wave kinematics from measured surface variations. In case of field experiments, estimates of C_D are distorted, if ocean currents are present. In many studies, the estimates of these coefficients have not been related to critical flow parameters (N_R, N_{KC}), and the wide range of values, which these parameters admit have not been investigated. Hence, there may be considerable uncertainty in the choice of C_M, C_D, for a particular problem, if the environmental conditions differ from the available sources of information. We briefly review the results of some of these investigations.

Keulegan and Carpenter (1958) investigated, both theoretically and experimentally, the forces acting on cylinders in an oscillating flow field. The observed values of C_M and C_D varied over a wave cycle and were averaged. Figures 9.13(a), (b) show the variation of cycle averaged values against Keulegan-Carpenter number N_{KC}.

Keulegan and Carpenter did not observe any correlation with Reynolds number. Agreement between measured forces and forces calculated using linear theory were good except near $N_{KC} = 15$, where the differences were of the order of 20%. Two programs identified as Project I and II provided estimates of C_M, C_D based on various wave theories and measured sea surface data (Thrasher and Aagaard,

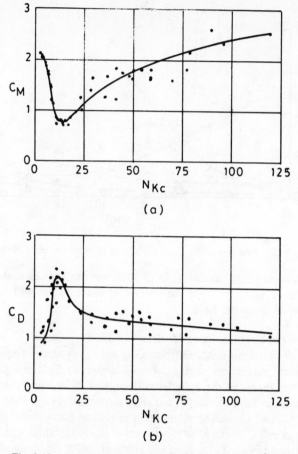

Fig. 9.13. Cycle averaged values of (a) C_M and (b) C_D against Keulegan Carpenter Number, N_{Kc} (Keulegan and Carpenter, 1958).

1969; Evans, 1969; Wheeler, 1969; Dran and Aagaard, 1970). The variation between calculated and measured local force varied upto 50%. Kim and Hibbard (1975) estimated C_M and C_D by fitting Morison equation to simultaneous measurements of wave-induced velocities and forces, and, thus, avoided the uncertainty due to choice of wave theory and ocean current. For $2 \times 10^5 < N_R < 8 \times 10^5$, they estimated

$$C_D = 0.61 \pm 24\%$$

$$C_M = 1.20 \pm 22\%$$

and reported good agreement between measured and calculated forces for $C_D = 0.61$ and $C_M = 1.20$.

In a series of careful and comprehensive experimental studies, Sarpkaya et al (1976 a, b and 1977) have obtained estimates of C_M and C_D for smooth and roughened circular cylinders for $N_R \leq 7 \times 10^5$, $N_{KC} \leq 200$, and roughness ratio k/D: 0.001–0.02. Measured values of wave kinematics and wave forces were used, thus, eliminating the uncertainties due to choice of a particular wave theory and steady current. Figures 9.14 and 9.15 show their estimates of C_D and C_M as functions of N_R, N_{KC} and relative roughness. The scatter in the data is generally small. Agreement between measured and calculated forces was found to be excellent, except for $10 < N_{KC} < 20$, and low values of N_R.

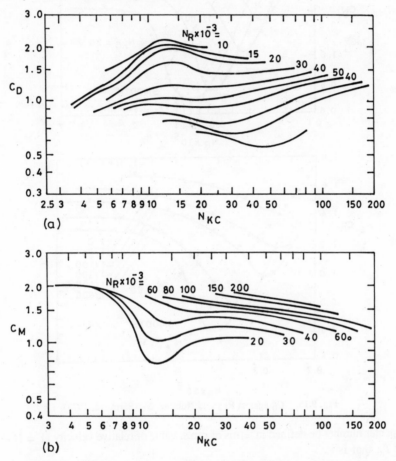

Fig. 9.14. (a) C_D, and (b) C_M, against Keulegan-Carpenter Number, N_{KC} (Sarpkaya, 1976b).

Among the available data on C_M and C_D, the experimental results of Sarpkaya are most reliable and complete. However, since Sarpkaya's experiments were conducted for a single harmonic rectilinear flow around fixed cylinders, three sources of uncertainties are introduced in the evaluation of wave force based on his data: (i) the fluid particles are not moving in a rectilinear manner; (ii) the sea state is random; and (iii) the cylinder is flexible.

As regards the first source of uncertainty, Sarpkaya (1976c) claims that his values of C_M and C_D are conservative. To deal with uncertainties due to second and third sources, Angelides (1978) has proposed that Reynolds and Keulegan-

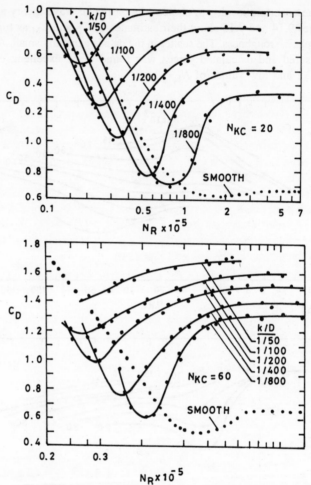

Fig. 9.15. C_D against Reynold Number, N_R (Sarpkáya, 1977).

Carpenter number be defined in terms of r.m.s. value of relative velocity ($\dot{Y} = U - \dot{X}$) and T_z, that is

$$N_R = \frac{D\sigma_{\dot{Y}}}{V} \tag{9.102}$$

$$N_{KC} = \frac{T_z\sigma_{\dot{Y}}}{D} \tag{9.103}$$

and C_M and C_D be estimated using N_R and N_{KC} defined above. Further research is needed to study the sources of uncertainty indicated above.

The estimation of wave forces is also influenced by the uncertainty due to marine growth on structures. The accumulation of marine growth, increases the diameter of the cylinder and, therefore, the wave forces. It also increases the drag coefficient C_D due to increase in relative roughness.

9.4.2   Wave Forces on Members in Diffraction Regime

A body, which is large enough to modify the incident wave field, is considered to be in the diffraction regime. Gravity platforms, which have been developed to withstand severe wave conditions in locations such as North Sea, fall in this category. The viscous effects are negligible in the diffraction regime and the linear diffraction theory based on potential flow is usually used to calculate forces and moments on bodies (Havelock, 1940; MacCamy and Fuchs, 1954). In the framework of linear theory, superposition is employed to decompose the wave environment into incident waves, diffracted waves, and if the body is assumed to move, the radiated waves. For a single circular cylinder, MacCamy and Fuchs (1954) developed the original analytical solution using Bessel functions. Analytical solutions for more than one circular cylinder have been developed by Spring and Monkmeyer (1974), and Chakrabarti (1978). Green's function (John, 1950; Black, 1975; Garrison and Stacey, 1977), and integral equation (Wehausen, 1971) methods have been used to calculate forces on arbitrarily shaped bodies. Finite element methods have also been used for diffraction problems (Yue et al, 1976; Brebbia and Walker, 1978; Zienkewicz et al, 1978).

We give below the principal results for the forces on a single cylinder in random short crested sea (Huntington and Thompson, 1976; Huntington, 1978). For a detailed derivation and treatment of more than one cylinder, submerged bodies, and application of finite element methods, the reader is referred to Brebbia and Walker (1979).

Consider a circular surface-piercing a cylinder shown in Fig. 9.16. We assume that the fluid is inviscid, the flow is irrotational and the incident waves are of

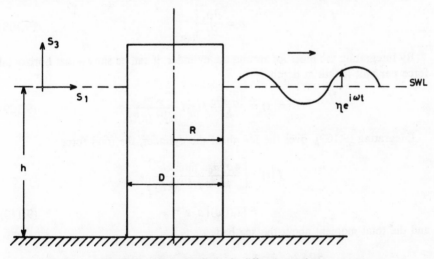

Fig. 9.16.   Vertical cylinder in diffraction regime.

small steepness so that linear wave theory is applicable. The potential function for a long-crested sine wave propagating in positive s_1-direction can be expressed as (9A.20)

$$\phi_i = \frac{iga_0}{\omega} f(s_3) \exp\left[-i(ks_1 - \omega t)\right] \qquad (9.104)$$

in which

$$f(s_3) = \frac{\cosh{[k(s_3 + d)]}}{\cosh{kd}} \qquad (9.105)$$

In polar coordinates, (9.104) transforms to

$$\phi_i = \frac{iga_0}{\omega} f(s_3) \left[\sum_{n=0}^{\infty} \alpha_n i^n J_n(kr) \cos n\theta \right] e^{i\omega t}$$

where J_n is a Bessel function of the first kind and order n; $\alpha_0 = 1$, $\alpha_n = 2$ for $n \geq 2$; and r, θ are polar coordinates.

Assuming that the waves diffracted by the cylinder can be described by a similar expansion, their potential function can be expressed in terms of Hankel functions H_n, and the total potential for the incident and diffracted waves can be expressed as:

$$\phi_t = \frac{iga_0}{\omega} f(s_3) \left[\sum_{n=0}^{\infty} \{ \alpha_n i^n J_n(kr) + a_n H_n^{(2)}(kr) \} \cos n\theta \right] e^{i\omega t} \qquad (9.106)$$

where a_n are constants obtained by setting water particle velocity to zero normal to the cylinder at $r = R$. Denoting differentiation with respect to r by $'$, we have

$$\phi_t = \frac{iga_0}{\omega} f(s_3) \left[\sum_{n=0}^{\infty} \left\{ \alpha_n i^n J_n(kr) + \frac{J_n'(kR)}{H_n^{(2)'}(kR)} H_n^{(2)}(kr) \right\} \cos n\theta \right] e^{i\omega t} \qquad (9.107)$$

The dynamic pressure at the surface of the cylinder is given by

$$p = \left. \frac{\partial \phi_t}{\partial t} \right|_{r=R} \qquad (9.108)$$

By integrating the pressure around the cylinder, it can be shown that horizontal force per unit length at depth s_3

$$F(s_3, t) = \frac{4\rho g a_0}{k} f(s_3) \frac{e^{i\omega t}}{H_1^{(2)'}(kR)} \qquad (9.109)$$

Integrating (9.109), over the length of the cylinder, the total force

$$F(t) = \left[\frac{4\rho g a_0}{k^2} \frac{\tanh kh}{H_1^{(2)'}(kR)} \right] e^{i\omega t}$$

$$= [T_F(\omega)] a_0 e^{i\omega t} \qquad (9.110)$$

and the total moment about the sea bed

$$M(t) = \frac{4\rho g a_0}{k^3} \frac{1}{H_1^{(1)'}(kR)} \{ kh \tanh kh - 1 + \sec kh \} e^{i\omega t}$$

$$= [T_m(\omega)] a_0 e^{i\omega t} \qquad (9.111)$$

If $\Phi_{\eta\eta}(\omega)$ is the psd of the unidirectional wave spectrum, then the psd of the horizontal force

$$\Phi_{FF}(\omega) = |T_F(\omega)|^2 \, \Phi_{\eta\eta}(\omega) \tag{9.112}$$

and psd of the base moment

$$\Phi_{MM}(\omega) = |T_M(\omega)|^2 \, \Phi_{\eta\eta}(\omega) \tag{9.113}$$

where transfer functions $T_F(\omega)$ and $T_M(\omega)$ are given by (9.110) and (9.111), respectively.

We now consider the short-crested random sea state as the linear sum of trains of long-crested uncorrelated, random waves from all directions. Since the wave trains from different directions are uncorrelated, the forces and moments are induced at right angles to the principal wave direction even for symmetric wave spectrum. In the $s_1 - s_2$ plane (horizontal plane), let θ be the angle of wave incidence measured from s_1. Let F_1 and F_2 be the components of the force in s_1 and s_2 directions. Then,

$$\Phi_{F_1 F_1}(\omega) = \int_{-\pi}^{\pi} \Phi_{\eta\eta}(\omega, \theta) |T_F(\omega)|^2 \cos^2 \theta \, d\theta \tag{9.114a}$$

and

$$\Phi_{F_2 F_2}(\omega) = \int_{-\pi}^{\pi} \Phi_{\eta\eta}(\omega, \theta) |T_F(\omega)|^2 \sin^2 \theta \, d\theta \tag{9.114b}$$

where $\Phi_{\eta\eta}(\omega, \theta)$ is the directional spectrum. Assuming that the angular distribution of energy is same in all directions, that is

$$\Phi_{\eta\eta}(\omega, \theta) = \Phi_{\eta\eta}(\omega) \, S(\theta) \tag{9.115}$$

where $S(\theta)$ is the directional spreading function satisfying the normalisation condition (9.40), the transfer functions relating the wave spectrum to the spectra of forces in directions s_1 and s_2 are given by

$$|T_{F_1}(\omega)| = \left[\frac{\Phi_{F_1 F_1}(\omega)}{\Phi_{\eta\eta}(\omega)} \right]^{1/2} = k_1 |T_F(\omega)| \tag{9.116a}$$

and

$$|T_{F_2}(\omega)| = \left[\frac{\Phi_{F_2 F_2}(\omega)}{\Phi_{\eta\eta}(\omega)} \right]^{1/2} = k_2 |T_F(\omega)| \tag{9.116b}$$

where

$$k_1 = \left[\int_{-\pi}^{\pi} S(\theta) \cos^2 \theta \, d\theta \right]^{1/2} \tag{9.117a}$$

$$k_2 = \left[\int_{-\pi}^{\pi} S(\theta) \sin^2 \theta \, d\theta \right]^{1/2} \tag{9.117b}$$

and

$$\Phi_{F_i F_i}(\omega) = k_i^2 |T_F(\omega)|^2 \, \Phi_{\eta\eta}(\omega), \, i = 1, 2 \tag{9.118}$$

For $S(\theta)$ defined by (9.56), $k_1 = 0.87$ and $k_2 = 0.5$. Thus, in short-crested sea, the force in line with the principal wave direction is 13% less than the force estimated using long-crested waves with the same total energy. The transverse force is 57% of the longitudinal force, and 50% of the force estimated using the long-crested waves. The resultant force on the cylinder is a vector sum of the two components. Similar expressions can be derived for the moments.

Huntington (1978) has reported results of measurements of transfer functions on a large cylinder in short-crested sea, and compared theoretical and experimental results. The agreement is good, justifying the use of linear theory. A comparison of force psd calculated from diffraction theory and Morisons' equation shows that, for large cylinders, Morisons' equation underestimates the spectra by as much as 6% below a critical frequency, and grossly overestimates the spectrum values above the critical frequency (Brebbia and Walker, 1979).

A gravity platform consists of a large diameter base supporting a group of cylinders. The base represents a large submerged body in a diffraction flow field. Besides a horizontal force, the base is also subjected to a vertical force on the top face which adds to the overturning moment. The forces and moments can be obtained from the linear diffraction theory with appropriate boundary conditions (Brebbia and Walker, 1979).

9.4.3   Wave Forces on Members in Presence of Currents

Ocean currents are caused by a complex interaction between several physical phenomena which include: tidal forces, coriolis forces, and atmospheric effects such as, barometric slope, wind slope and wind stress (Defant, 1961; MairWood, 1969). The drift currents, which are caused by wind stress, decay and change direction with depth according to Ekman Spiral. The decrease in the magnitude of the current with depth is proportional to $\exp(ks_3)$ (s_3 is negative), and, therefore, long wave lengths decay very slowly giving nearly uniform velocity distribution.

The presence of current in the wave environment modifies the forces on members due to: (i) change in water particle velocities; (ii) change in amplitude and steepness of surface waves; (iii) standing wave pattern on water surface produced by the member; and (iv) vortex shedding. Generally, the surface effects are more important for large diameter members, and the slender members are affected by change in drag forces and resonant reaction between members called 'galloping'.

For slender members, the effect of current can be included by writing the Morison equation in the following vector form:

$$\bar{F}(s_3, t) = M_I(\dot{\bar{U}} + \dot{\bar{U}}_c) + C_H(\bar{U} + \bar{U}_c)\,|\,\bar{U} + \bar{U}_c\,| \qquad (9.119)$$

where $\bar{U}_c$ is the current vector. For steady current, $\dot{\bar{U}}_c = 0$ and (9.119) reduces to

$$\bar{F}(s_3, t) = M_I\dot{\bar{U}} + C_H(\bar{U} + \bar{U}_c)\,|\,\bar{U} + \bar{U}_c\,| \qquad (9.120)$$

The steady flow past a member due to current may also cause vortex shedding and a harmonic lift force perpendicular to the direction of the current. The lift force per unit length can be expresed as

$$\bar{F}_L(s_3, t) = K_L\,|\,\bar{U}_c\,|\,\bar{U}_c \exp(i\pi f t) \qquad (9.121)$$

where $\bar{U}_c$ is the particle velocity vector due to current; f is the vortex shedding frequency given by

$$N_s = \frac{Df}{|\bar{U}_c|} \tag{9.122}$$

in which N_s is the Strouhl number which varies with the Reynolds number N_R (Fig. 9.17).

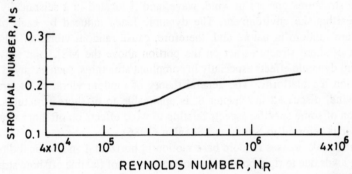

Fig. 9.17. Strouhl number for vortex shedding behind oscillating circular cylinder (Schmidt, 1965).

The coefficient K_L in (9.121) is given by

$$K_L = \frac{1}{2}\rho d C_L$$

where ρ is the density of water and C_L is the lift coefficient. The above analysis is applied to a rigid stationary member. If the member vibrates, the eddies may be triggered by the vibration at or near the resonant frequency of the member. In such cases, the lift and drag forces may be upto four or five times the forces on a rigid member (Laird, 1962).

In most offshore structures, some members lie in the wake of other members. This results in interactions due to change in flow directions, shedding of vortices and turbulent wake at high Reynolds number. The understanding of these interaction effects is very limited (Yamamoto and Nath, 1976).

For large-diameter members, the surface effects due to current are more important. Kotik and Morgan (1961) have derived the theoretical results for the force induced by standing wave pattern. Hogben and Standing (1975) have presented experimental results for determining the force on the body due to standing waves. Hogben (1974) has shown that this effect is not significant for offshore structures in North Sea. The presence of current modifies the amplitude and velocity of propagation of the waves (Longuet-Higgins and Steward, 1960). It can be shown (Brebbia and Walker, 1979) that the frequency corresponding to a wave of a particular wavelength is increased, if the current has a positive component in the wave direction. This will modify the wave height spectrum. The diffracted and reflected waves are also modified by the current. At present, a coherent theory of diffraction in presence of currents is not available. Work of Hogben (1976) in North Sea indicates that interaction between currents and scattered waves should not cause large increase in the loading. However, the effect may be significant for tow-outs.

Spanos and Chen (1981), Gudmestad and Connor (1983), Alibe (1986), have investigated the effect of current on the spectral characteristics of hydrodynamic forces, and on the response of offshore structures.

9.5 Response of Offshore Structures to Wave Loading

Offshore structures operate in wind, wave, and if located in a seismically active region, earthquake environment. The dynamic loads induced by each of these sources are random in nature and, therefore, cause random vibration. The wind loads on offshore structures act on the portion above the MSL, and may cause significant dynamic effects especially in compliant structures, such as guyed towers and tension leg platforms. The general theory of random vibration of structures due to wind, discussed in Chapter 8, is applicable to offshore structures. For a discussion of some specific aspects relating to wind effects on offshore structures, the reader is referred to Simiu and Scanlan (1986).

An earthquake causes random base motion in horizontal and vertical directions inducing loads due to rigid body and vibratory motion of flexible offshore structures. Due to fluid-structure interaction, the loads induced by the earthquake and surface waves get coupled (Venkataramana, et al, 1989). We shall briefly discuss the combined action of wave and earthquake loads at the end of this section. We shall first discuss the random vibration of offshore structures due to wave action.

9.5.1 Structure-Foundation Models

In sections 9.3 and 9.4, we have discussed the short-term and long-term stochastic models of sea waves, and the wave induced forces on offshore structures. To determine the dynamic response, it is necessary to construct idealized mathematical models of offshore structures. Discrete models consisting of a set of nodal points interconnected by linear elastic elements are commonly used in view of the complex geometry and other characteristics of structure-foundation systems. In offshore structures, the soil-structure interaction has a strong influence on the dynamic response and, therefore, it must be included in the dynamic analysis.

In section 9.4, we have described the basic features of different types of offshore platforms. Figures 9.18b and 9.19b show typical discrete models of jacket and gravity type offshore structures. It is convenient to use the substructure concept to separate the discrete model of an offshore structure into the structure and the foundation sub-systems, respectively (Gutirrez, 1976). The loads on the structure are transmitted to the foundation through the interface nodal points.

In the structure sub-system, the mass is lumped at the nodal points, and in the most general formulation, each node has six degrees-of-freedom. Let n denote the total number of degrees-of-freedom of a discrete model. Let n_b degrees-of-freedom be associated with the nodal points at the structure-foundation interface boundary, and $n_s = n - n_b$ represent the degrees-of-freedom of the remaining nodes. Correspondingly, the nodal displacement vector $\bar{X}$ can be partitioned as

$$\bar{X} = \left\{ \frac{\bar{X}_s}{\bar{X}_b} \right\} \tag{9.123}$$

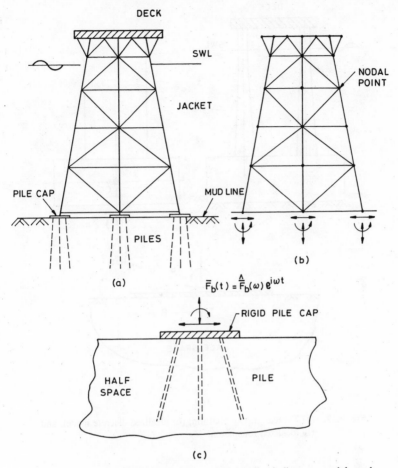

Fig. 9.18. (a) Typical jacket platform; (b) idealized discrete model; and (c) idealized foundation.

where $\bar{X}_s$ and $\bar{X}_b$ are n_s-dimensional and n_b-dimensional vectors, respectively.

The foundation sub-systems for typical jacket and gravity type offshore structures are shown in Figs. 9.18c and 9.19c, respectively. The influence of the soil-structure interaction on the dynamic response of a structure is incorporated in the analysis through the force-displacement relations associated with n_b degrees-of-freedom at the interface boundary of the foundation sub-system. These relations may be derived by a steady-state dynamic analysis of the foundation under harmonic excitation at the interface boundary. Let

$$\bar{F}_b(t) = \hat{\bar{F}}_b(\omega)\, e^{i\omega t}$$

and

$$\bar{X}_b(t) = \hat{\bar{X}}_b(\omega)\, e^{i\omega t} \tag{9.124}$$

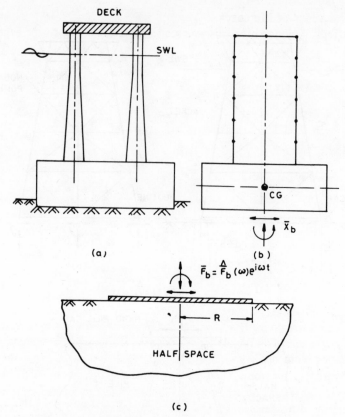

Fig. 9.19. (a) Typical gravity platform; (b) idealized discrete model; and (c) idealized foundation.

Then,

$$\bar{\hat{F}}_b(\omega) = [\mathbf{K}_{bb}^F(\omega) + i\omega\,\mathbf{C}_{bb}^F(\omega)]\,\bar{\hat{X}}_b(\omega) \qquad (9.125)$$

where $\mathbf{K}_{bb}^F(\omega)$ and $\mathbf{C}_{bb}^F(\omega)$ are frequency dependent foundation stiffness and damping matrices respectively.

For a gravity structure, foundation sub-system can be modelled by a massless disc supported on half space, as shown in Fig. 9.19c. The half-space may be assumed to be uniform or layered, and elastic or viscoelastic, and the elements of matrices $\mathbf{K}_{bb}^F$ and $\mathbf{C}_{bb}^F$ can be obtained analytically (Velestos and Wei, 1971; Luco and Westman; 1971; and Luco, 1979). These matrices can also be determined through a finite element analysis of disc-foundation medium.

The jacket structures are usually supported on a pile-group through a rigid pile cap. A realistic modelling of such a system is relatively difficult, and analytical solutions are available for simple cases only. Finite element analysis can be used to determine foundation impedance matrices in (9.125).

9.5.2 Equations of Motion

The equations of motion of the discrete model described above can be expressed as

$$\mathbf{M}\ddot{\bar{X}} + \mathbf{C}\dot{\bar{X}} + \mathbf{K}\bar{X} = \bar{F}(t) \tag{9.126}$$

where $\mathbf{M}$, $\mathbf{C}$ and $\mathbf{K}$ are mass, damping and stiffness matrices, which can be partitioned as

$$\mathbf{M} = \begin{bmatrix} \mathbf{M}_{ss} & \mathbf{M}_{sb} \\ \mathbf{M}_{sb}^T & \mathbf{M}_{bb} \end{bmatrix}$$

$$\mathbf{C} = \begin{bmatrix} \mathbf{C}_{ss} & \mathbf{C}_{sb} \\ \mathbf{C}_{sb}^T & \mathbf{C}_{bb} + \mathbf{C}_{bb}^F(\omega) \end{bmatrix}$$

and

$$\mathbf{C} = \begin{bmatrix} \mathbf{K}_{ss} & \mathbf{K}_{sb} \\ \mathbf{K}_{sb}^T & \mathbf{K}_{bb} + \mathbf{K}_{bb}^F(\omega) \end{bmatrix} \tag{9.127}$$

Note that, matrices $\mathbf{M}$, $\mathbf{C}$ and $\mathbf{K}$ have frequency dependent components due to soil-structure interaction. If soil-structure interaction effects are neglected, or if frequency dependence of foundation impedance matrices is ignored, these matrices have constant elements.

The hydrodynamic force vector $\bar{F}(t)$ represents the forces at the nodes. In appendix-9B, we discuss the determination of these forces under various assumptions. Two cases are distinguished: (i) Morison wave loading; and (ii) diffraction wave loading.

Morison Wave Loading

The hydrodynamic load vector $\bar{F}(t)$ is obtained by aggregating the hydrodynamic load on each member. It includes the effects of inertia, drag and fluid-structure interaction. In appendix 9B we show (9B.12), that for Morison wave loading assumption

$$\bar{F}(t) = \mathbf{M}_I \ddot{\bar{U}} - \mathbf{M}_a \ddot{\bar{X}} + \mathbf{C}_H \mid \bar{U} - \dot{\bar{X}} \mid (\bar{U} - \dot{\bar{X}}) \tag{9.128}$$

where $\mathbf{M}_I = \rho C_M \mathbf{V}$; $\mathbf{M}_a = \rho(C_M - 1)\mathbf{V}$; and $\mathbf{C}_H = \frac{1}{2}\rho C_D \mathbf{A}$

Substituting (9.128) in (9.126), and denoting $\dot{\bar{Y}} = \bar{U} - \dot{\bar{X}}$, it can be shown on rearranging the terms that

$$(\mathbf{M} + \mathbf{M}_a)\ddot{\bar{Y}} + \mathbf{C}\dot{\bar{Y}} + \mathbf{C}_H \mid \dot{\bar{Y}} \mid \dot{\bar{Y}} + \mathbf{K}\bar{Y} = (\mathbf{M} + \mathbf{M}_w)\ddot{\bar{U}} + \mathbf{C}\bar{U} + \mathbf{K}\int_0^t \bar{U}\,d\tau \tag{9.129}$$

where $\mathbf{M}_w = \mathbf{M}_a - \mathbf{M}_I = -\rho\mathbf{V}$.

The third term in the above equation is nonlinear. The equation may be solved numerically by Monte-Carlo simulation of the wave environment, (Borgman, 1972; Shinozuka et al, 1979; and Datta, 1983), or by using one of the techniques of nonlinear random vibration analysis.

In appendix 9B, we discuss a nonlinear approximation of hydrodynamic force vector under the assumption that the structural velocities are small as compared to the water particle velocities. Under this assumption, it is shown that (9B.14)

$$\bar{F}(t) = \mathbf{M}_I \dot{\bar{U}} - \mathbf{M}_a \ddot{\bar{X}} + C_H |\bar{U}| \bar{U} - \hat{C}_H \dot{X} \qquad (9.130)$$

where $\hat{C}_H = 2\rho \mathbf{C}_D \mathbf{A} \langle |\bar{U}| \rangle$.

Substituting (9.130) in (9.126) and rearranging terms

$$(\mathbf{M} + \mathbf{M}_a) \ddot{\bar{X}} + (\mathbf{C} + \hat{\mathbf{C}}_H) \dot{\bar{X}} + \mathbf{K}\bar{X} = \mathbf{M}_I \dot{\bar{U}} + C_H |\bar{U}| \bar{U} = \bar{P}_A(t) \quad (9.131)$$

The equation of motion is now linear, but the excitation is a nonlinear function of particle velocity vector, whose psd is given by (9.90). We have noted that some elements of the damping and stiffness matrices are frequency dependent and, therefore, the solution must be obtained in the frequency domain. A step-by-step procedure based on FFT can be used to determine the response (Penzien and Tseng, 1978).

For most engineering applications, it is necessary to consider only a few modes. For offshore structures, Malhotra and Penzien (1970a) have shown that for a typical structure the contribution of the third mode to deck displacement lies in the range 0.9–1.5 percent. It is, therefore, possible to reduce the size of the equations of motion by considering only a few modes. Penzien and Tseng (1978) have proposed a scheme for generating a system of reduced equations of motion using a few lower modes of the fixed-base structure.

Linearized Equations of Motion

The nonlinear equations of motion (9.129) can be linearized using the method of equivalent linearization (Malhotra and Penzien, 1970b; Nigam, 1983) and expressed as

$$(\mathbf{M} + \mathbf{M}_a) \ddot{\bar{Y}} + (\mathbf{C} + \tilde{\mathbf{C}}_H) \dot{\bar{Y}} + \mathbf{K}\bar{Y} = (\mathbf{M} + \mathbf{M}_w)\dot{\bar{U}} + \mathbf{C}\bar{U} + \mathbf{K}\int_0^t \bar{U} d\tau \qquad (9.132)$$

where

$$\tilde{C}_{Hij} = \frac{C_{Hjj} \langle \dot{Y}_j | \dot{Y}_j | \rangle}{\langle \dot{Y}_j^2 \rangle}; \qquad i = j$$

$$= 0; \qquad\qquad i \neq j \qquad (9.133)$$

Since the wave field is assumed to be zero mean, ergodic, Gaussian process, the forcing function in (9.132) has the same properties, and $Y_j(t)$ being the response of a linear system, is also a zero mean, ergodic, Gaussian process. Hence,

$$\sigma_{\dot{Y}_j}^2 = E[\dot{Y}_j^2] = \langle \dot{Y}_j^2 \rangle$$

and

$$\frac{\langle \dot{Y}_j | \dot{Y}_j | \rangle}{\langle \dot{Y}_j^2 \rangle} = \left(\frac{8}{\pi}\right)^{1/2} \sigma_{\dot{Y}_j} \qquad (9.134)$$

Equation (9.133) can now be expressed as

$$\tilde{C}_{Hij} = C_{Hjj}\left(\frac{8}{\pi}\right)^{1/2} \sigma_{\dot{Y}_j}, \quad i = j$$
$$= 0; \quad i \neq j \tag{9.135}$$

The linearized equation (9.132) can be rewritten as

$$(M + M_a)\ddot{\bar{X}} + (C + \tilde{C}_H)\dot{\bar{X}} + K\bar{X} = M_1\ddot{\bar{U}} + \tilde{C}_H\bar{U} = \bar{P}(t) \tag{9.136}$$

which represents a linear mdf system subjected to stationary, Gaussian excitation. Neglecting soil-structure interaction, or assuming the soil impedance matrices to be constant, the response of the system can be obtained by using the normal mode approach. Unfortunately, the modal matrix of the undamped system does not diagonalize the damping matrix. Two options are available: (i) (9.136) may be reduced to a system of 2n first-order equations and the response obtained using the normal mode method due to Foss (1956) (Nigam, 1983); or (ii) 9.136) may be uncoupled using a mean square error minimization procedure (Malhotra and Penzien, 1970b) described below.

Let ϕ be the modal column matrix of the undamped system such that, $\bar{X} = \phi\bar{Z}$ and

$$\phi^T(M + M_a)\phi = I$$

$$\phi^T K\phi = \text{diag}(\omega_1^2, \omega_2^2, \ldots \omega_n^2)$$

$$\phi^T(C + \tilde{C}_H)\phi = C_0 \text{ (coupled)} \tag{9.137}$$

The matrix C_0 is now replaced by a diagonal matrix C^*, such that the error vector

$$\bar{e} = (C_0 - C^*)\dot{Z} \tag{9.138}$$

is minimized in the mean square. It can be shown that

$$C_{jj}^* = C_{0jj} + \sum_{\substack{k=1 \\ k \neq j}}^{n} \frac{C_{0jk}\langle \dot{Z}_k \dot{Z}_j \rangle}{\langle \dot{Z}_j^2 \rangle} \tag{9.139}$$

Equation (9.136) can now be reduced to a system of n uncoupled equations

$$\ddot{Z}_j + 2\omega_j \zeta_j \dot{Z}_j + \omega_j^2 Z_j = P_j^*(t); j = 1, 2, \ldots n \tag{9.140}$$

where $\zeta_j = C_{jj}^*/2\omega_j$, and

$$P_j^*(t) = \bar{\phi}_j^T[M_1\dot{\bar{U}} + \tilde{C}_H\bar{U}] \tag{9.141}$$

The psd matrix of the displacement vector $\bar{X}$ can be expressed as

$$\Phi_{XX}(\omega) = \phi\,\Phi_{ZZ}(\omega)\,\phi^T \tag{9.142}$$

and

$$\Phi_{Z_jZ_k}(\omega) = H_j^*(\omega)\,H_k(\omega)\,\Phi_{P_j^*P_k^*}(\omega) \tag{9.143}$$

where $H_j(\omega)$ is the frequency response function of (9.140). From (9.141), it follows that (Malhotra and Penzien, 1970b)

$$\Phi_{P_j^* P_k^*}(\omega) = \sum_{r=1}^{n} \sum_{s=1}^{n} \phi_j^r \phi_k^s [M_{Ir} M_{Is} \, \Phi_{\dot{U}_r \dot{U}_s}(\omega) + \tilde{C}_{Hr} \tilde{C}_{Hs}$$

$$\times \Phi_{U_r U_s}(\omega) + i\omega \, (\tilde{C}_{Hr} M_{Is} - \tilde{C}_{Hr} M_{Is}) \, \Phi_{U_r U_s}(\omega)] \qquad (9.144)$$

and from (9A.23) a typical cross-psd is of the form

$$\Phi_{U_j U_i}(\omega) = \omega^2 \frac{\cosh k(s_{3j} + h) \cosh k(s_{3i} + h)}{\sinh^2 (kh)} \, \Phi_{\eta\eta}(\omega) \qquad (9.145)$$

where $\Phi_{\eta\eta}(\omega)$ is the wave height spectrum. Thus, the Gaussian response can be obtained in a closed form for an assumed wave height spectrum.

The damping coefficients in (9.136) and (9.140) contain the moment functions of the response (9.135, 9.139). An iterative procedure is, therefore, required to solve these equations. Malhotra and Penzien (1970b) have derived closed form expressions for these moments and shown that iterative procedure converges rapidly.

Frequency Domain Method

We have noted that, due to soil-structure interaction, some elements of the damping and stiffness matrices of structure-foundation system are frequency dependent. The frequency domain method, therefore, provides a direct procedure to determine the response of linear systems. Consider the linear Eq. (9.136). Let $\tilde{\bar{X}}(\omega)$ and $\tilde{\bar{P}}(\omega)$ represent the Fourier transforms of $\bar{X}(t)$ and $\bar{P}(t)$ respectively. Then,

$$\tilde{\bar{X}}(\omega) = \mathbf{H}(\omega) \, \tilde{\bar{P}}(\omega) \qquad (9.146)$$

where $\mathbf{H}(\omega)$ is the frequency response matrix of (9.136), and is given by

$$\mathbf{H}(\omega) = [\omega^2 (\mathbf{M} + \mathbf{M}_a) + i\omega (\mathbf{C} + \tilde{\mathbf{C}}_H) + \mathbf{K}]^{-1} \qquad (9.147)$$

The psd matrix of the displacement vector $\bar{X}(t)$ is given by

$$\Phi_{\bar{X}\bar{X}}(\omega) = \mathbf{H}(\omega) \, \Phi_{\bar{p}\bar{p}}(\omega) \, \mathbf{H}^*(\omega)^{\mathrm{T}} \qquad (9.148)$$

where

$$\Phi_{\bar{p}\bar{p}}(\omega) = [\mathbf{M}_I \, \Phi_{\ddot{\bar{U}}\ddot{\bar{U}}}(\omega) \, \mathbf{M}_I + \tilde{\mathbf{C}}_H \, \Phi_{\bar{U}\bar{U}}(\omega) \, \tilde{\mathbf{C}}_H$$

$$+ i\omega \, (\mathbf{M}_I \, \Phi_{\bar{U}\bar{U}}(\omega) \, \tilde{\mathbf{C}}_H - \tilde{\mathbf{C}}_H \, \Phi_{\bar{U}\bar{U}}(\omega) \, \mathbf{M}_I)] \qquad (9.149)$$

Diffraction Wave Loading

For typical gravity structures, the hydrodynamic forces on the base and columns supporting the deck, are dominated by diffraction effects. Since the viscous forces can be neglected, these forces are a linear function of the surface elevation. An analytical treatment of diffraction loading is, therefore, much simpler.

In section 9.4, we have derived the expression for the force per unit length of a fixed vertical circular cylinder in the diffraction regime (9.109). To determine the forces and moments at a node, this can be integrated over the appropriate length. The psd of each of these forces and moments in the directions s_1 and s_2 can be expressed by the equations of the same form as (9.114) in terms of the directional spectrum.

The equations of motion can be expressed as

$$(\mathbf{M} + \mathbf{M}_a)\,\ddot{\bar{X}} + (\mathbf{C} + \mathbf{C}_R)\,\dot{\bar{X}} + \mathbf{K}\bar{X} = \bar{F}(t) \tag{9.150}$$

where $\mathbf{M}_a$ is the added mass matrix, and $\mathbf{C}_R$ is the radiation damping matrix contributed by the outward radiation of energy in the form of waves due to motion of the body. The excitation vector $\bar{F}(t)$ is a stationary Gaussian random process and the response can be obtained by the classical methods of random vibration analysis.

Normally, gravity structures consist of a group of columns which interact with each other. The transfer functions required in (9.114) interact for such cases, and also the radiation damping is usually obtained by numerical procedures or through model tests (Hafokjold et al, 1973). For a detailed discussion of diffraction effects, the reader is referred to Brebbia and Walker (1979), Hogben and Standing (1974); Garrison and Chow (1972); and Venkataramana et al (1989).

Combined Earthquake and Wave Loading
During an earthquake, a structure is subjected to base motion, which can be represented by a three-dimensional random displacement vector $\bar{Z}(t)$. The total displacement vector of a discrete model of an offshore structure due to seismic base motion and vibration caused by combined hydrodynamic and earthquake loading can be expressed as

$$\bar{X}_t = \bar{X} + \bar{X}_g \tag{9.151}$$

in which $\bar{X}$ is the displacement relative to the fixed base, and

$$\bar{X}_g = \mathbf{B}\bar{Z} \tag{9.152}$$

where $\mathbf{B}$ is $n \times 3$ matrix (7.69).

The equation of motion of the structure can be expressed as

$$\mathbf{M}\ddot{\bar{X}}_t + \mathbf{C}\dot{\bar{X}} + \mathbf{K}\bar{X} = \bar{F}(t) \tag{9.153}$$

For linearized Morison wave loading, (9.153) can be expressed as

$$(\mathbf{M} + \mathbf{M}_a)\,\ddot{\bar{X}} + (\mathbf{C} + \tilde{\mathbf{C}}_H)\,\dot{\bar{X}} + \mathbf{K}\bar{X} = \mathbf{M}_I\dot{\bar{U}} + \tilde{\mathbf{C}}_H\,\bar{U} - (\mathbf{M} + \mathbf{M}_a)\,\mathbf{B}\ddot{\bar{Z}} - \tilde{\mathbf{C}}_H\mathbf{B}\dot{\bar{Z}}$$

$$= \bar{P}(t) \tag{9.154}$$

where the coefficient matrices are same as in (9.136) with $\dot{Y}_j = U_j - \dot{X}_{tj}$ in (9.133)–(9.135).

The forcing function in (9.154) is a function of two mutually independent, stationary and Gaussian random vectors $\bar{U}(t)$ and $\bar{Z}(t)$. The response can be obtained by the classical methods of random vibration theory (Nigam, 1983; Venkataramana, et al, 1989).

If the wave action is neglected, that is water is assumed to be still ($\tilde{U} = 0$) when earthquake occurs, (9.154)) reduces to

$$(\mathbf{M} + \mathbf{M}_a)\, \ddot{\bar{X}} + (\mathbf{C} + \tilde{\mathbf{C}}_H)\, \dot{\bar{X}} + \mathbf{K}\bar{X}$$

$$= -(\mathbf{M} + \mathbf{M}_a)\, \mathbf{B}\ddot{Z} - \tilde{\mathbf{C}}_H\, \mathbf{B}\dot{Z} = \bar{P}(t) \tag{9.155}$$

which is of the same form as (7.69) with 'added mass' and damping terms due to fluid-structure interaction.

Since simultaneous occurrence of a critical sea storm and a strong-motion earthquake is very unlikely, an offshore structure may be analysed independently for earthquake loading using (9.155). Malhotra and Penzien (1971) have found that a flexible structure may experience excessive deflection due to wave action, whereas a stiff structure may develop excessive transverse shear forces due to a major earthquake. In designing offshore structures for wave and earthquake forces, it is, therefore, desirable to seek optimum stiffness.

In the diffraction regime applicable to large diameter gravity structures, motion due to earthquake may make a significant contribution to the hydrodynamic force due to radiated waves. Numerical procedures are generally used to include this effect (Black et al, 1971).

Single-degree-of-freedom (sdf) System

We will now discuss the random vibration of a sdf model of offshore structures. We have noted that the first few modes make a major contribution to the dynamic response of offshore structures. A sdf model is, therefore, of considerable interest in obtaining a close approximation to the actual response, and may be used at the preliminary design stage.

Several approximate methods are available in the literature for the reduction of a complex dynamic system to a sdf system. We shall use the Galerkins' method based on weighted-residual principle (Nigam, 1983). We consider a simple offshore structure consisting of a cylindrical column supporting the deck. The mass of the deck (M) is lumped at the top as shown in Fig. 9.20.

Treating the column as an Euler beam, its equation of motion can be expressed as

$$\rho_c A_c \frac{\partial^2 X}{\partial t^2} + C \frac{\partial X}{\partial t} + \frac{\partial^2}{\partial s^2}\left(EI \frac{\partial^2}{\partial s^2}\right) X = F(s, t) \tag{9.156}$$

where $X(s, t)$ denotes the relative lateral displacement of the column; ρ_c is the mass of the column per unit length; A_c is the column cross section area; C is the damping coefficient; and EI is the bending stiffness.

The excitation $F(s, t)$ is the hydrodynamic load per unit length, which depends on the assumptions made regarding the loading regime. For example, if Morison loading regime is assumed, $F(s, t)$ is given by (9.92), and in the linearized form by (9.93). For diffraction loading, it is given by (9.109). Assuming, linearized Morison loading, (9.156) can be expressed as

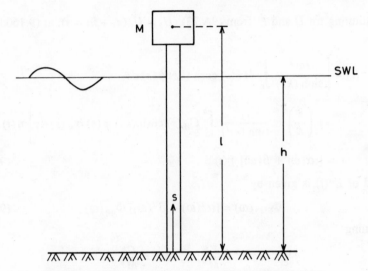

Fig. 9.20. Single-degree-of-freedom model of a offshore structure.

$$(\rho_c A_c + M_a) \frac{\partial^2 X}{\partial t^2} + (C + \tilde{C}_H) \frac{\partial X}{\partial t} + \frac{\partial^2}{\partial s^2} \left(EI \frac{\partial^2}{\partial s^2} \right) X = M_I \dot{U} + \tilde{C}_H U$$

$$= F(s, t) \qquad (9.157)$$

where

$$\tilde{C}_H = C_H \left(\frac{8}{\pi} \right)^{1/2} \sigma_{\dot{Y}} \qquad \text{and} \qquad \dot{Y} = U - \dot{X}$$

Let $\phi(s)$ be an admissible function satisfying atleast the rigid boundary conditions at $s = 0$, and let

$$X(s, t) = X_1(t)\, \phi(s) \qquad (9.158)$$

where $\phi(1) = 1$, so that X_1 represents the displacement of the deck. By Galerkins' method, the equation of motion of an 'equivalent' sdf system is given by

$$M^* \ddot{X}_1 + C^* \dot{X}_1 + K^* X_1 = F^*(t) \qquad (9.159)$$

where

$$M^* = \rho_c \int_0^1 A_c(s) [\phi(s)]^2 \, ds + \int_0^h M_a(s) [\phi(s)]^2 \, ds \qquad (9.160a)$$

$$C^* = \int_0^1 C(s) [\phi(s)]^2 \, ds + \left(\frac{8}{\pi} \right)^{1/2} \int_0^h C_H(s)\, \sigma_{\dot{Y}}(s) [\phi(s)]^2 \, ds \qquad (9.160b)$$

$$K^* = \int_0^1 EI(s) \left[\frac{\partial^2 \phi}{\partial s^2} \right]^2 \, ds \qquad (9.160c)$$

and

$$F^*(t) = \int_0^h M_I \dot{U} \phi(s) \, ds + \int_0^h \tilde{C}_H U \phi(s) \, ds \qquad (9.160d)$$

Substituting for U and $\dot{U}$ from (9A.23) ($U_1 = U$, $(s_3 + h) = s$), in (9.160d), we get

$$F^*(t) = \left[\frac{i\omega^2}{\sinh (kr)} \int_0^h MI(s) \cosh (ks) \phi(s) ds \right.$$

$$\left. + \left(\frac{8}{\pi} \right)^{1/2} \frac{\omega}{\sinh (kd)} \int_0^h C_H(s) \cosh (ks) \phi(s) \sigma_{\dot{Y}}(s) ds \right] \eta(t)$$

$$= [i\alpha(\omega) + \beta(\omega)] \, \eta(t) \qquad (9.161)$$

and psd of $F^*(t)$ is given by

$$\Phi_{F^*F^*}(\omega) = [\alpha^2(\omega) + \beta^2(\omega)] \, \Phi_{\eta\eta}(\omega) \qquad (9.162)$$

Defining

$$\omega^{*2} = \frac{K^*}{M^*}; \; 2\zeta^*\omega^* = \frac{C^*}{M^*}$$

(9.159) can be expressed as

$$\ddot{X}_1 + 2\zeta^*\omega^* \dot{X}_1 + \omega^{*2} X_1 = \frac{F^*(t)}{M^*} \qquad (9.163)$$

In (9.163), $F^*(t)$ is is a zero-mean stationary, Gaussian random process with psd given by (9.162). The psd of X_1 is given by

$$\Phi_{X_1X_1}(\omega) = |H(\omega)|^2 \, \Phi_{F^*F^*}(\omega) \qquad (9.164)$$

where

$$H(\omega) = [-\omega^2 + i2\omega\omega^*\zeta^* + \omega^{*2}]^{-1} . \qquad (9.165)$$

Since the system is linear, the response is also Gaussian with zero mean. The variance of X_1 is given by

$$\sigma_{X_1}^2 = \int_{-\infty}^{\infty} \Phi_{X_1X_1}(\omega) \, d\omega \qquad (9.166)$$

and, thus, its distribution is completely known. The distributions of peaks, maximum value and other properties of the response can now be derived in closed form.

It must be noted that the coefficient $\tilde{C}_H$ in (9.157) involves standard deviation of relative velocity, $\sigma_{\dot{Y}}$. An iterative procedure is, therefore, required to determine the response. However, if it is assumed that

$$\langle U \rangle \gg \langle \dot{X} \rangle$$

$\sigma_{\dot{Y}}$ may be replaced by σ_U, which is given by

$$\sigma_U^2 = \int_{-\infty}^{\infty} \Phi_{UU}(\omega) \, d\omega = \frac{\cosh^2 (ks)}{\sinh^2 (kh)} \int_{-\infty}^{\infty} \omega^2 \, \Phi_{\eta\eta}(\omega) \, d\omega \qquad (9.167)$$

and iteration is avoided.

Alibe (1986) has investigated and response of sdf systems using the Monte-Carlo simulation to determine the 'exact' response of nonlinear equations of motion, and also the response of linearized system using the frequency domain method. His results show that:

(i) the level crossing rate, and hence the peak forces, are grossly underestimated by the linearized system for high threshold levels;

(ii) for offshore structures with fundamental frequency in the range (0.4–3.0) rad/s, the mean and standard deviation determined from linearized system are in good agreement with the values obtained from simulation; and

(iii) for compliant structures, such as tension leg platforms with fundamental frequency of the order of 0.05, the variance of the response is also significantly underestimated.

The unsatisfactory performance of linearization method in predicting peak response is primarily due to the implicit assumption that response is Gaussian. This assumption is not valid for high values of threshold. It is recommended that linearization should not be used to estimate peak response. However, it can be used for fatigue studies involving relatively low thresholds.

In the above treatment, we have not included the effect of current. This can be done without any particular difficulty by assuming that particle velocity has the mean value $\bar{U}_c$, the velocity of the current (Spanos and Chen, 1981; Alibe, 1986).

Example 9.1: Response of an offshore platform to sea waves and earthquakes
          (Venkataramana et al, 1989)
Figure 9.21 shows the analytical model of an offshore structure-pile-soil system. The properties of the structure members, and the pile-soil foundation are given in

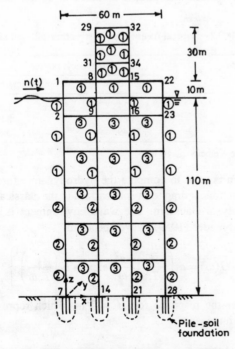

Fig. 9.21. Model of an offshore structure-pile-soil system.

Tables 9.1a and 9.1b. The structure-pile-soil system is analysed by the sub-structure method treating as an assemblage of two sub-systems; the structure sub-system

Table 9.1a   Details of structural members

Member No.	Diameter (m)	Thickness (mm)	c/s Area (m²)	Moment of Inertia (m⁴)
1	2.8	27	0.235	0.226
2	2.8	54	0.466	0.439
3	1.2	19	0.070	0.012

Table 9.1b   Details of the pile-soil foundation

Diameter of pile	0.6 m
Number of piles at each base node	10
Modulus of elasticity of pile	2.10^6 t/m²
Shear wave velocity in pile	$2.7 \cdot 10^7$ m/sec
Unit weight of soil	1.7 t/m³
Poisson's ratio for soil	0.4
Shear wave velocity in soil	100 m/sec

and the pile-foundation system. Each nodal point of the structure, except at the base, has three degrees of freedoms: horizontal (x-direction), vertical (z-direction) and rotation (about y-direction). The base nodes are restrained from vertical movement. The model has 98 degrees of freedom. The first three natural frequencies of the system model assuming rigid base and soil-structure interaction, are given in Table 9.2. The structural damping is assumed to be 2% for the first mode of vibration.

Table 9.2   Natural frequencies of structure-pile-soil system

Vibration Mode	Frequency: rad/sec	
	Rigid Base	Soil-structure Interaction
First	2.24	1.61
Second	11.88	8.87
Third	27.49	25.14

Shear Wave Velocity ... 10 m/s

Morison equation is used to compute the hydrodynamic forces with $C_M = 2.0$ and $C_D = 1.0$. The nonlinear drag term is linearized in the classical manner assuming that relative velocity is Gaussian. The wave environment is modelled by one-dimensional Bretschneider (1959), wave spectrum:

$$G(\omega) = a_1 \left(\frac{\bar{H}}{g \bar{T}_z^2} \right)^2 \frac{g^2}{\omega^2} \exp \left\{ -b_1 \left(\frac{1}{\bar{T}_z \omega} \right)^4 \right\}$$

We will consider the response for $\bar{H} = 7$ m, which represents the sea state during storms and hurricanes.

Response to Wave Action
We will first consider the response of offshore structure to wave action alone.

Specifically, we would study the effects of soil-structure interaction and leg spacing on the response.

Figure 9.22 shows the rms values of the displacement and the bending moment for rigid-base and soil-structure interaction models. In both cases, it is seen that

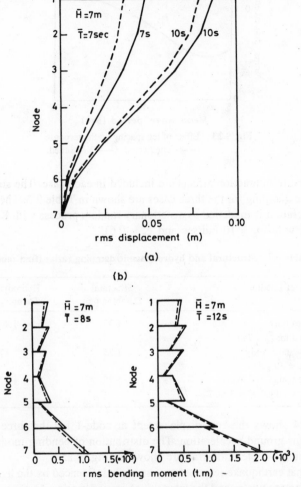

Fig. 9.22. rms—(a) displacements (m); (b) bending moments (t.m), (– – –) rigid base; (———) interaction.

soil-structure-interaction increases the response. Figure 9.23 shows the effect of leg spacing on the rms displacement response. It is seen that response values are overestimated when the effect of leg spacings are not included in the analysis.

Response to Earthquakes and Waves

We now consider the response of the offshore structure in Fig. 9.21, for the following three cases:

   (i) waves only;

   (ii) earthquakes only (in still water); and

   (iii) simultaneous action of both earthquakes and waves.

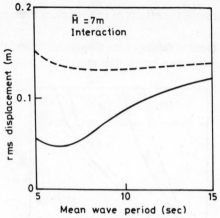

Fig. 9.23.   Effect of leg spacing on the response
(– – –) neglected; (———) included.

The soil-structure interaction effects are included in each case. The structural and hydrodynamic damping for the three cases are shown in Table 9.3. The earthquake ground acceleration is modelled as a stationary random process with Kanai-Tajimi spectra (7.2), with $\omega_g = 10$ rad/sec and $\zeta_g = 0.6$).

Table 9.3   Structural and hydrodynamic damping ratios (first mode)

Type of Loading	Structural Damping (%)	Hydrodynamic Damping (%)
Waves only ($\bar{H} = 3$ m; $\bar{T}_z = 7$ s)	1.35	1.40
Earthquakes only ($\sigma_{\ddot{u}g} = 30$ gal)	1.35	0.17
Earthquakes and Waves ($\sigma_{\ddot{u}g} = 30$ gal; $\bar{H} = 3$ m, $\bar{T}_z = 7$s)	1.35	1.07

Figure 9.24 shows the rms displacement at node-1 for the three cases as a function of rms ground acceleration. The distribution of bending moment for two values of ground acceleration rms are shown in Fig. 9.25.

It is seen that earthquake responses are strongly influenced by the hydrodynamic damping due to sea waves and the intensity of ground motion. In general, response due to simultaneous action of severe earthquake ground motion and strong sea waves are smaller than that due to earthquake alone. However, the probability of simultaneous occurrence of a strong earthquake and strong sea waves is very small. Occurrence of earthquakes in moderate sea states ($H_s \approx 3$ m) is more likely and may be used for design purposes.

Non-Gaussian Response of Offshore Structures
We have seen that, if the wave kinematics is treated in the framework of linear wave theory, and the hydrodynamic forces are 'linearized' for drag effects, the response of offshore structures is Gaussian. The second-order response statistics can be obtained by time, or frequency domain, methods of random vibration analysis, and closed form expressions can be derived for properties, such as, level crossings,

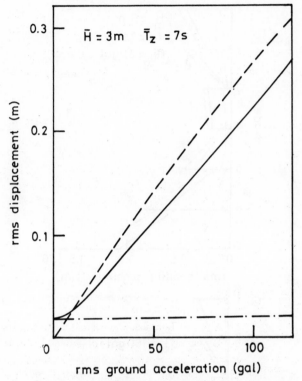

Fig. 9.24.   rms-displacement at node-1 (— · — · —) waves,
(– – –), earthquakes; and (——) waves and earthquakes.

peaks, fatigue damage. The design, including optimization, can be implemented within a reliability framework using well established procedures.

Field observations of real offshore platforms indicate that the response is non-Gaussian even for quasi-static or linear dynamic structural behaviour (Heavener et al, 1985). The non-Gaussian response may be attributed to the following sources of nonlinearities:

  (i) drag component of hydrodynamic force;
  (ii) wave kinematics; and
  (iii) intermittent forces in the 'sloshing zone' due to free surface fluctuations.

We have discussed the nonlinearity due to drag already. Kanegaonkar and Haldar (1987) have investigated the combined effect of nonlinear wave kinematics and free surface fluctuations on a jacket type offshore platform linearized for drag (9.136). They have determined the response moments upto fourth order under the following assumptions:

  (i) waves are stationary ergodic, narrow-band and Gaussian;
  (ii) wave kinematics is based on Stokes' second order theory for deep water, so that (Tung, 1984)

$$U = - (gk)^{1/2} \exp (ks_3) \eta; \tag{9.168}$$

  (iii) the effect of free-surface fluctuations is incorporated by defining the effective particle velocity field

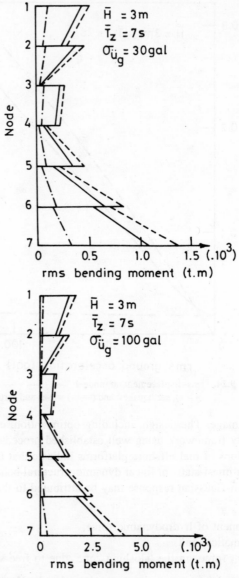

Fig. 9.25. rms-bending moment ($—\cdot—\cdot—$) waves,
($---$) earthquakes; and ($———$) waves and
earthquakes.

$$\tilde{U} = UH(\eta - s_3) \tag{9.169}$$

where $H(\cdot)$ is the Heaviside unit step function;

(iv) $\tilde{U}$ is assumed to be stationary with approximate psd (Tung, 1984)

$$\Phi_{\tilde{U}\tilde{U}}(\omega) = \left(\frac{\sigma_U}{\sigma_\eta}\right)^2 [\beta Z(\beta) + Y(\beta) \Phi_{\eta\eta}(\omega)] \tag{9.170}$$

where

$$\beta = s_3/\sigma_\eta$$

$$Z(\beta) = \frac{1}{(2\pi)^{1/2}} \exp\left(-\frac{\beta^2}{2}\right)$$

$$Y(\beta) = \int_\beta^\infty Z(\alpha)\, d\alpha$$

The particle velocity vector $\bar{U}$, in linearized Eq. (9.136) is now replaced by effective particle velocity vector $\tilde{U}$ defined above. Using the modal transformation, the cross-spectral density of the generalized forces P_j^* (9.141) is given by (9.144) with U replaced by $\tilde{U}$ and $\dot{U}$ replaced by $\dot{\tilde{U}}$. Kanegaonkar and Haldar (1987) have derived the expressions for the first four moments of the loads P_j^* in terms of the joint moments of $\tilde{U}$ and its derivative processes.

The response of the structure is resonance dominated for lower sea states, and the load can be considered to be strongly correlated over a period smaller than the natural period of the structure. For this case, Kanegaonkar and Haldar (1987) have shown that the rms value, coefficient of 'skewness' and 'excess' of the normal coordinates Z_j can be approximated as

$$\sigma_{Z_j} = \frac{1}{(\omega_j \Delta)^{1/2}} \frac{1}{2\zeta_j^{1/2}\omega_j} \sigma_{P_j^*} \tag{9.171}$$

and

$$\gamma_{z_j} = (\omega_j \Delta)^{1/2} \frac{16\zeta_j^{3/2}}{3(8\zeta_j^2 + 1)} \gamma_{P_j^*} \tag{9.172}$$

$$\xi_{z_j} = \frac{3\zeta_j}{2(12\zeta_j^2 + 1)} (\omega_j \Delta) \xi_{P_j^*} \tag{9.173}$$

in which γ and ξ are coefficients of skewness and excess, respectively, and Δ is the correlation distance beyond which the values of the load are weakly correlated and much smaller than the jth period of the system. In most cases, the resonant response corresponds to fundamental mode and γ and ξ may be associated with that mode.

For higher sea states, the load is no longer strongly correlated over a period much smaller than the natural period, and the response may be determined by a quasi-static analysis. The forcing function $\bar{P}(t)$ is given by (9.136) with $\bar{U}$ replaced by $\tilde{U}$, and, if $\mathbf{A} = [a_j^k]$ is the flexibility matrix of the structure

$$X_j(t) = \sum_{k=1}^n a_j^k P_k(t) \tag{9.174}$$

Comparing (9.174) with (9.141), it is clear that the cross-spectral density of $\bar{X}$ is given by (9.144) with ϕ_j^k replaced by a_j^k, and its moments can be derived in the same way as the moments of $\bar{P}^*$. Having determined the moments of upto fourth order, the non-Gaussian distribution function of the response may be assumed to be a mixture of weighted sum of a set of distributions, that is

$$F_X(x) = \sum_{j=1}^{n} C_j F_j \qquad (9.175)$$

with $C_j > 0$, and

$$\sum_{j=1}^{N} = C_j = 1 \qquad (9.176)$$

In (9.175) F_j are chosen such that their mean and variance are the same as that of $F_X(x)$, and the weight factors are chosen to minimize

$$\varepsilon = \sum_{j=3}^{4} (m_j - \hat{m}_j)^2 \qquad (9.177)$$

subject to (9.176). In (9.177), m_j, are the dimensionless central moments of the response of jth order; and $\hat{m}_j$, the dimensionless central moments of the linear combination (9.175) of the jth order.

Kanegaonkar and Haldar (1987) have analysed a typical jacket type offshore platform for seven sea states, and P-M spectrum. The distribution (9.175) is assumed to be a sum of the normal and shifted exponential distributions. Following general conclusions may be drawn from their results:

(i) For lower sea states, response is expected to be nearly Gaussian;

(ii) Skewness coefficient and coefficient of Kurtosis increase with increase in significant wave height, that is, higher sea states. This is indication of deviation from Gaussian distribution and is in agreement with field measurements; and

(iii) Free surface intermittent loading is partly responsible for non-Gaussian response. The deviation from Gausian distribution has a significant effect on fatigue life estimates.

Example 9.2 Non-Gaussian response of an offshore platform due to free surface fluctuations and nonlinear wave kinematics (Kanegaonkar and Haldar, 1987) Figure 9.26 shows a jacket type offshore platform fixed at the base. The structural data for the platform is given in Table 9.4.

The dynamic response of the platform is determined for unidirectional random waves characterized by a modified   P-M spectrum

$$G(\omega) = \left(\frac{H_s}{2}\right)^2 \frac{692.0}{\omega(1.09\bar{T}_z \omega^4)} \exp\left[-\frac{692.0}{(1.09\bar{T}_z \omega^4)}\right]$$

for seven sea states defined in Table 9.5. Stokes' deep-water second-order wave theory is used for estimating wave kinematics, and wave forces are calculated using Morison Eq. (9.72–9.74) with $C_D = 1.0$ and $C_M = 2.0$.

The dynamic response of the platform is determined with and without the effect of surface fluctuations represented by (9.164) considering both the linear and the second-order wave theories. The distribution in (9.175) is assumed to be a mixture of Gaussian and shifted exponential distribution. The response of the platform is resonance dominated for the lowest sea state (1). For other sea states, a quasi-steady analysis is in order, as the mean zero-crossing periods ($\bar{T}_z$) are much higher than the fundamental period of the platform. The coefficients of skewness and kurtosis are calculated for each sea state as measures of deviation from Gaussian behaviour.

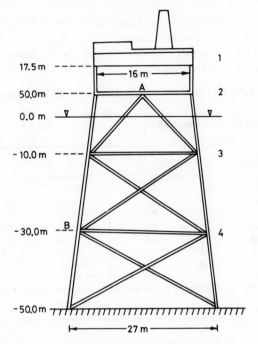

Fig. 9.26.   Offshore structure.

Table 9.4   Structural data-platform in Fig. 9.26

Deck legs	:	Diameter, D	2.00 m
	:	Thickness, t	50 mm
Jacket legs and diag.	:	D	1.20 m
	:	Vert. Plane, t	14 mm
Bracings and diag.	:	D	0.80 m
at level + 5 m	:	t	8 mm
Bracings and diag.			
at level – 10 m	:	D	1.20 m
	:	t	1.4 mm
Bracings at level – 30 m	:	D	1.20 m
	:	t	14 mm
Diagonals at level – 30 m	:	D	1.20 m
	:	t	16 mm
Young's modulus (steel)			205×10^8 kg/m^2
Mass density (steel)			7800 kg/m^3
Mass of deck			4800 tonnes
Fundamental frequency			3.38 rad/sec

Table 9.5   Sea States

Sea State	Significant Wave Height (m)	Mean Zero-crossing Period (sec)	% Ocurrence
1	0.76	4.0	19.27
2	2.28	7.0	49.00
3	3.86	8.3	24.00
4	5.33	9.5	6.00
5	6.86	10.0	2.65
6	8.38	11.5	1.35
7	9.90	12.3	0.73

Table 9.6 shows the rms value of the deck displacement without (Gaussian) and with (Non-Gaussian) fluctuations for the seven states. Figure 9.27 shows cumulative distribution function on normal probability paper for the seven sea states.

It is seen from the above results that:

Table 9.6   Dynamic deck displacement characteristics

Sea State	Deck Displacement Without Fluctuation (cm)			Deck Displacement with Fluctuations (cm)		
	rms (spectral)	C_1	C_2	rms (spectral)	C_1	C_2
1	0.056	1.0	0.0	0.056	1.0	0.0
2	0.141	1.0	0.0	0.141	0.9954	0.0465
3	0.211	1.0	0.0	0.210	0.9695	0.0305
4	0.260	1.0	0.0	0.245	0.8403	0.1597
5	0.326	1.0	0.0	0.296	0.7365	0.2635
6	0.352	1.0	0.0	0.316	0.7108	0.2892
7	0.396	1.0	0.0	0.354	0.6934	0.3066

Note: C_1, C_2—optimal coefficients of Gaussian and shifted exponential distribution in mixture distribution (9.175).

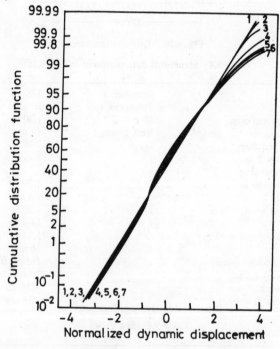

Fig. 9.27. Cumulative distribution of deck displacements for seven sea-states.

(i) For lower sea states, the response is nearly Gaussian. The deviation from Gaussian behaviour progressively increases with increase in wave height, that is, higher sea states.

(ii) The nonlinearity due to free surface intermittent loading is primarily responsible for non-Gaussian response. With additional nonlinearity due to drag, the response is expected to further deviate from the Gaussian distribution.

The above results are in general agreement with the measurements in the North Sea. In the next section, we show that the deviation from Gaussian distribution may be significant for fatigue life calculations.

9.6 Design of Offshore Structures

Design, construction and operation of offshore platforms is governed by the rules and codes of practice adopted by different countries from time to time (DOE, 1974; DnV, 1977; and API, 1981). The codes embody accumulated experience in the design, construction and operation of offshore platforms, and are a compendium of detailed procedures, data and charts consistent with the general approach adopted in the preparation of codes of practice. They represent a compromise between the needs of economy, safety, acceptable risk, standardization and practical considerations. Due to complex form of offshore platforms, and the environment in which they operate, the design steps and procedures involve considerable idealization at various stages.

The choice of a particular type of offshore structure is determined by its functions, such as, drilling, storage, material handling, pipeline booster, depth of sea, and geographic location. The structure must be designed to withstand dead loads, operational loads, environmental and temperature loads, besides loads during the construction phase, lifting, towing and final installation. It must meet the requirements of strength, fatigue and serviceability. We will confine our discussion to design based on random vibration analysis for strength and fatigue in a reliability framework. To implement such a design procedure, it is necessary to specify an acceptable system reliability based on value analysis, which includes both tangible and intangible costs. Consistent with the prescribed system reliability, design criteria are developed linking distributions of loads to the strength and fatigue (Marshall, 1969; and Bea, 1973 and 1975). In the preceding chapters, we have discussed the basic approach to reliability based design in random vibration environment. In this section, we shall cover special features connected with offshore structures.

An offshore structure (specially jacket type) contains a large number of members and joints and, therefore, failure may occur due to strength, or fatigue, at a large number of locations. From a practical design point of view, determination of stress and deformation time histories at each such location is a formidable task. Further, the design of tubular joints, which occur in large numbers in a typical jacket platform, involves special considerations, such as, punching shear, residual stresses and stress concentration. Simplified design procedures are, therefore, a must to reduce the enormous computational effort.

The forces induced in offshore structures are strongly time dependent and a dynamic analysis is, therefore, essential for a realistic determination of peak values and distribution of stresses and deformations. However, under conditions in which resonant effects are not significant, a simpler quasi-static analysis may be used. In such a case, the nonlinearities may be retained in the analysis yielding more accurate estimates of response statistics. Sometimes a hybrid procedure based on a combination of quasi-static and a simplified dynamic analysis may yield acceptable results (Cronin et al, 1978).

9.6.1 Design for Strength

Design for strength requires that a stress, or a combination of stresses, at every point of the structure remains within a specified bound during the service life of the structure. In the reliability framework, based on random vibration analysis, the design problem may be formulated as a fractional occupation time or a first-passage problem. This involves considerable computational effort.

The 'design-wave' method provides a relatively simple design procedure. In this method, the height of a design wave $H_{D;N,p}$ is determined from (9.60), (or approximately from (9.61)), for a specified service life (N), and exceedance probability (p). Design wave height for a specified return period, $H_{D;T}$, may be obtained from (9.64). The structure is designed on the basis of the design-wave height for the 'worst' direction. Note that, the worst direction will be different for different locations and modes of failure. To simplify the design procedure, the 'worst' direction is generally determined for the base shear (or moment) and used for the design of all members. Figure 9.28 shows a typical plot of failure probability against wave-height for a 100 yr service life.

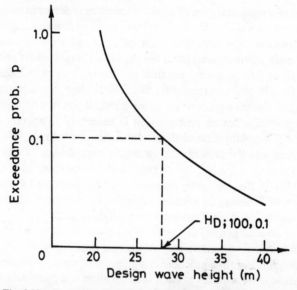

Fig. 9.28.   Probability of failure vs. design wave height ($N = 100$ yrs).

A design based on design-wave approach suffers from several limitations. A wave-height may be associated with a wide range of periods $(6.5H)^{1/2} < T < 20$, in which H is in meters, and T is in secs (DnV, 1977). Since the structural response depends not only on the height but also on the length of the design wave (hence frequency); and certain structures are particularly sensitive to variations in wave frequency, the stresses determined from design-wave approach may be in considerable error. A spectral analysis approach based on random vibration analysis, therefore, provides a realistic basis for design of offshore structures.

9.6.2 Design for Fatigue

Offshore structures are subjected to dynamic loads which induce time varying stresses and, therefore, cause fatigue. In view of the inaccessibility of and expense of inspection and repair, of offshore structures, a satisfactory fatigue performance is important. Due to complex and random behaviour of sea environment, possibility of resonant dynamic response, large size of offshore structures, complexities of welded details, and effect of corrosion on materials behaviour, the determination of fatigue life is an extremely time consuming task. The factors which significantly affect the fatigue life estimate include: "environmental conditions, hydrodynamic loading, structural modelling, soil-structure interaction, procedure for determining the stress response, stress concentration factors (SCF), and fatigue damage value." These factors may be grouped under different heads which constitute the elements of fatigue analysis. The major elements are: "environmental loading on the structure, local stress history at the joints and the fatigue damage model" (Gupta and Singh, 1986).

Three approaches are generally used to estimate the fatigue life:

  (i) discrete-wave analysis,
 (ii) spectral fatigue analysis, and
(iii) reduced spectrum analysis.

In Chapter 3, we have discussed the fatigue phenomenon, and the two commonly used methods for fatigue life estimation, namely, the linear damage theory based on Palmgren-Miner (P-M) law; and the theory of crack propagation. Despite its well known deficiencies, the linear damage theory in conjunction with S-N curve is extensively used in the design of offshore structures. We shall discuss the three approaches of fatigue analysis based on this theory. We must, however, note that one of the major uncertainties in the use of linear damage theory for offshore structures is the choice of S-N curve. There are many S-N curves which include: British Welding Institute BS-F) curve, American Welding Society (AWS-X) curve and the modified AWS-X curve as shown in Fig. 9.29.

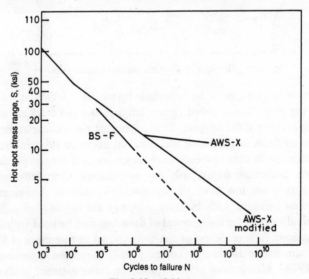

Fig. 9.29.   S-N curves.

The BS-F curve exists only for the range 10^5 to 2×10^6 cycles, and its application to high cycle fatigue damage requires extrapolation which may not be desirable.

The AWS-X modified curve is formed by the extension of AWS-X curve at the same slope beyond 2×10^6 cycles. It is clear that uncertainity is particularly significant in high cycle regions, where corrosion fatigue may severely limit the performance of welded tubular joints. It may be noted that BS-F : S-N curve is conservative as compared to AWS-X curves.

Let $Q(t)$ denote the random stress history at a point and let S denote the *stress range*. The determination of fatigue life based on P-M law requires estimates of two basic quantities:

(i) number of cycles $n(s)$, with range s, during the service life determined from the response analysis; and

(ii) number of cycles to failure $N(s)$, at stress range s, in a constant stress amplitude test, and represented by S-N curve.

In the preceding sections, we have discussed the methods for the determination of response statistics of offshore structures based on dynamic or quasi-static analysis. The quantity $n(s)$, can be determined through such an analysis. The determination of $N(s)$, for offshore structures, involves special considerations. In jacket platforms, welded joints are the weakest points, *hot-spots* (Fig. 9.30) for fatigue.

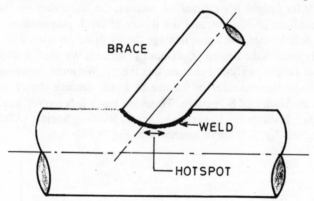

Fig. 9.30. Hot-spot in a typical welded tubular joint.

Two modes of failure can occur in tubular joints: (i) cracking at the toe of the weld joining the brace to the chord (brace fatigue) and (ii) cracking in the walls of the chord (punching shear fatigue). Stress concentrations occur at the joints due to nonuniform stiffness of the chord and material discontinuities in the area. It is very difficult to estimate stress concentration factors using simple empirical relations. It is, therefore, preferable to use S-N curves obtained from full-scale tests on joints. Such curves are produced by organisations, such as American Welding Society; Welding Institute, Lloyds. If such curves are not available, S-N curves based on nominal stresses for the material of the tubes may be used with appropriate stress concentration factors. Some of the well known semi-empirical formulae to evaluate SCF are due to Beale and Toprac (1967) (based on experimental data); and Visser (1974), Kuang et al (1975) (based on finite element analysis). These formulae incorporate effects of dimensions and geometrical parameters such as: thickness to diameter ratio of the chord, branch diameter to chord diameter ratio, branch thickness to chord thickness ratio, the angle between the branch and the chord; and diameter to length ratio of the chord for T joint etc. The SCF is also influenced by the detailing of the welded joint in addition to the joint configurations. Marshall (1974) has shown that rounded and grounded welds give lower SCF than normal welded joints.

Discrete Wave Analysis

The discrete wave method is based on a deterministic analysis of each sea state defined by scatter diagram (Fig. 9.31). The step-by-step procedure is shown in Fig. 9.31.

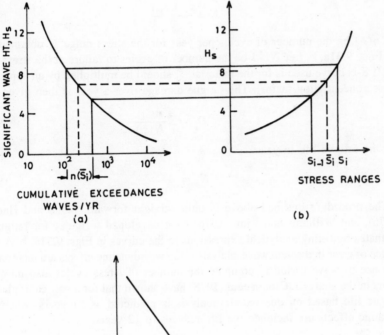

Fig. 9.31. Discrete wave analysis (a) wave height-cumulative exceedance diagram, (b) wave height-stress range diagram for each hot spot; and (c) SN-curve for the hot-spot.

The wave-height cumulative exceedance diagram (a), is obtained from the wave scatter diagram, usually for a year. The 'wave-height'—'stress-amplitude' diagram is then prepared for each hot-spot using the quasi-static analysis preferably based on a nonlinear wave theory (Stokes fifth order for deep water; cnoidal for shallow water). For this analysis, waves are applied from different directions to establish the stress-range. In structures with a large number of members, it is usual to assume that minimum and maximum total base shear (or overturning moment) also cause minimum and maximum stresses in all members. The dynamic amplification is incorporated, in this method, by a *dynamic load factor*, usually based on the response of a sdf system excited by a wave whose period is the same as the period of the discrete wave. The fatigue damage during a year is calculated by considering a discrete set of stress-ranges and summing up the damage over all discrete values as explained below.

Let s_i be a discrete value of the stress-range, and let

$$\bar{s}_i = \frac{s_i + s_{i+1}}{2} \tag{9.178}$$

Let $n(\bar{s}_i)$ be the number of cycles per year for the stress-range $\bar{s}_i$ obtained from the Figs. 9.31a, b. Let $N(\bar{s}_i)$ be the number of cycles to failure at the stress-range $\bar{s}_i$. (If S-N curve used is for the material, $\bar{s}_i$ should be multiplied by an appropriate stress concentration factor). The fatigue damage over a year is then given by

$$D_1 = \sum_i \frac{n(\bar{s}_i)}{N(\bar{s}_i)} \tag{9.179}$$

and the fatigue life in years

$$N_F = \frac{1}{D_1} \tag{9.180}$$

The procedure outlined above is quite straight forward. Nolte and Hansford (1976), and Williams and Rinn (1976) have developed formulae for fatigue life estimates by fitting analytical expressions to the curves in Figs. 9.31a, b. A major source of error in discrete wave analysis is the way dynamic effects are incorporated. The use of wave-period to compute the number of stress cycles also introduces errors in the analysis. Cronin et al (1978) have shown that for a particular platform, fatigue life based on quasi-static analysis is predicted as 51 years, whereas, if dynamic effects are included the life reduces to 12 years.

Spectral Fatigue Analysis

The spectral fatigue analysis is based on the psd of the stress history at a point. In Chapter 3, we have discussed the determination of fatigue life estimates based on the distribution of stress response. It was shown that for a narrow-band stationary response, the expected rate of fatigue damage

$$E[DR] = \frac{N^+(0)}{C} \int_0^\infty s^b p_s(s) \, ds \tag{9.181}$$

where C and b are material constants in the S-N curve; $N^+(0)$ is the expected rate of zero-crossing in secs, and $p_s(s)$ is the probability density of the stress-range S.

We have noted that for each sea state defined by $(H_s, T_z, \bar{\theta})$, surface elevation may be modelled as a Gaussian, narrow-band, stationary random process. The stress at a point may also be assumed to be stationary, and its distribution may be Gaussian, or non-Gaussian, depending upon the assumptions made in the analysis. For the diffraction regime, and for linearized Morison equation, the stress history is Gaussian. However, if the nonlinearities due to drag component and free surface effects are included in the analysis the stress response is non-Gaussian. For the Gaussian response, the distribution of stress-range is Rayleigh. For the non-Gaussian response, we have discussed the procedure for determining the distribution based on quasi-static, or approximate dynamic analysis (Kanegaonkar and Halder, 1987). In either case, the conditional expectation of the rate of fatigue damage for the sea state $(H_s, T_z, \bar{\theta})$ can be expressed as

$$E[DR|\,H_s, T_z, \bar{\theta}] = \frac{(N^+(0)\,|\,H_s, T_z, \bar{\theta})}{C} \int_0^\infty s^b\, p_s(s\,|\,H_s, T_z, \bar{\theta})\, ds \qquad (9.182)$$

and the unconditional (long-term) expected rate of fatigue damage per second is given by

$$E[DR] = \int_0^\infty \int_0^\infty \int_{-\pi}^\pi E[DR|\,H_s, T_z, \bar{\theta}]\, p(s\,|\,H_s, T_z, \bar{\theta})\, d\bar{\theta}, dT_z\, dH_s \qquad (9.183)$$

The expected fatigue life in years is given by

$$E[N_F] = \frac{\text{No. of seconds in a year}}{E[DR]} \qquad (9.184)$$

For the Gaussian assumption, the conditional expected rate of fatigue damage (9.182) can be expressed in the closed form (3.19)

$$E[DR|\,H_s, T_z, \bar{\theta}] = \frac{2\pi}{C} \frac{\sigma_{\dot{Q}}}{\sigma_Q} (\sqrt{2}\,\sigma_Q)^b\, \Gamma\!\left(1 + \frac{b}{2}\right) \qquad (9.185)$$

where σ_Q and $\sigma_{\dot{Q}}$ are functions of the sea state $(H_s, T_z, \bar{\theta})$.

Generally, the information regarding the frequency of occurrence of sea states is given in the form of scatter diagram for the discrete states (H_{s_i}, T_{z_j}). Let n_{ij} be the member of occurrence of the sea state (H_{s_i}, T_{z_j}) over a specified period of time. Then, the unconditional expected rate of fatigue damage per second can be expressed as

$$E[DR] = \sum_i \sum_j E[DR\,|\,H_{s_i}, T_{z_j}] \frac{n_{ij}\, T_{z_j}^{-1}}{\sum_i \sum_j n_{ij}\, T_{z_j}^{-1}} \qquad (9.186)$$

and the expected fatigue life is given by (9.184).

The spectral approach based on linearized random vibration analysis provides a consistent basis for fatigue life estimates. However, its disadvantage lies in the high cost, time and large computer capacity. It must be appreciated that economic constraints affect the application of spectral methods to fatigue analysis far more severely, than the strength analysis, because of the repetition required to construct the total response history over a long period of time covering all sea states and wave directions.

Reduced Spectral Method

It is possible to combine the discrete wave method and spectral fatigue analysis with a view to reduce the computational effort, and include the effects of nonlinearities (Cronin et al, 1978). In this method, the basic static wave load analysis package is used to construct the transfer function linking the sea surface spectrum to the spectrum of the stress history at a point through a quasi-static analysis. The dynamic amplification effect is incorporated by the transfer function for the first mode. The psd of the stress response is obtained as the product of the sea surface psd and the two transfer functions described above. Cronin et al (1978) have discussed the details of the methods and its limitations.

Fracture Mechanics Approach

The fatigue life of offshore structures is strongly influenced by the flaws and cracks caused by different manufacturing and installation processes. The micro-cracks introduced during welding and other inherent imperfections may significantly decrease the fatigue life of a structure as the cracks can grow rapidly with cyclic loading. The application of S-N curves provides fatigue life estimates of structures having no initial cracks. Since the crack initiation phase may consume a significant portion of total fatigue life, pre-existence of cracks in offshore structures introduce nonconservative errors into fatigue-life predictions based on S-N curves. The fracture mechanics approach, which accounts for the inherent initial crack size, is, therefore, better suited for fatigue life estimates of offshore structures.

In Chapter 3, we have discussed the fracture mechanics approach for estimates of fatigue life. The knowledge of initial crack size, the stress intensity factor (K) corresponding to the nature of loading and geometry of the joint, the crack growth rate as a function of the range of stress intensity factor (ΔK), and the fracture toughness enables one to estimate the fatigue life. Dover and Dharmavasan (1982) have expressed the parameter K for tubular joints in terms of characteristic stress field at the crack tip, the crack size and the thickness of the metal. The expression for the crack growth rate for a tubular welded connection subjected to cyclic load has been obtained experimentally by Dover et al (1981). Hibbard and Dover (1970) have modified the crack growth law proposed by Paris (1962), to account for the variations in the amplitude of the load in offshore structures. The range of stress intensity factor and characteristic stress around the crack tip have been weighted by the material property. These stresses are obtained by multiplying the nominal stresses by appropriate SCF's.

The fatigue-crack growth rate, based on weighted average stress range can be expressed as (Gupta and Singh, 1986)

$$\frac{da}{dN} = C[1.1\pi^{1/2}a^{1/2}\sigma_h)^m \tag{9.187}$$

where a is the crack size, σ_h is the weighted average stress range given by

$$\sigma_h = \{(\sum_{i=1}^{R} n_i(h_{\sigma_i})^m / \sum_{i=1}^{R} n_i\}^{1/m} \tag{9.188}$$

in which h_{σ_i} is the local stress range, n_i are the corresponding stress-cycles per year, and R is the total number of stress ranges.

Integration of (9.187) yields:

$$\tilde{N} = \frac{2}{(m-2)C(1.1\sigma_h)^m(\pi)^{m/2}}[a_i^{(2-m)/2} - a_f^{(2-m)/2}] \tag{9.189}$$

where $\tilde{N}$ is the total number of cycles to failure, a_i is the initial crack size, and a_f the final crack size. The fatigue life of the joint in years is given by

$$T = \tilde{N}/\sum n_i \tag{9.190}$$

Example 9.3:   Fatigue life estimates of an offshore platform
(Gupta and Singh, 1986)

We consider the fatigue life estimates of four joints (J_1, J_2, J_3 and J_4) of a doubly symmetric steel jacket offshore structure shown in Fig. 9.32. Gupta and Singh (1986) have investigated the fatigue life of the platform to uncertainities associated

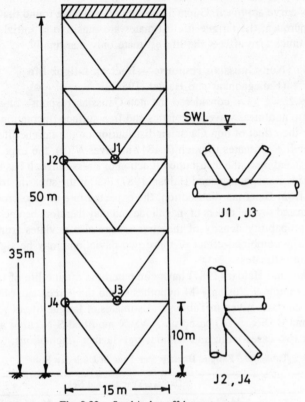

Fig. 9.32.   Steel jacket offshore structure.

with: long term sea statistics, hydrodynamic forces, current, structural modelling, soil-structure interaction, stress concentration factors and fatigue damage models. The long term sea model is represented by 15 sea states. We reproduce the results of the fatigue-life estimates of four joints based on different S-N curves and fracture mechanics approach in Table 9.7.

Table 9.7   Fatigue life estimates based on different S-N curves and fracture mechanics approach

Joint	Fatigue Life (yrs)			
	AWS-X	AWS-X Modified	BS-F	Fracture Mechanics
J1	229.0	166.0	27.3	10.2
J2	40.7	36.7	11.6	1.9
J3	13.9	13.2	6.7	1.1
J4	14.0	13.4	7.3	0.5

$a_i = 1.0$ mm; $a_f = 5.0$ mm; $C = 4.5 \times 10^{-12}$; $m = 3.3$

It is seen from the table that there is a wide variation in the fatigue life estimates for different S-N curves, specially for joint J_1 for which damage is dominated by low stress range. As expected, estimates based on BS-F are several orders of magnitude conservative as compared AWS-X curves. The fatigue life estimates based on fracture mechanics are much less than those based on S-N curves. This is to be expected as the welded joints have initial cracks which are not accounted for in the S-N curve approach. Gupta and Singh (1986) have found that in fracture mechanics approach, the fatigue life estimates are sensitive to initial crack size, but the final crack size affects the life estimate only marginally.

Example 9.4: Non-Gaussian response—effect on fatigue life
(Kanegaonkar and Haldar, 1987)

In Example 9.2, we have considered the non-Gaussian response of an offshore platform due to nonlinear wave kinematics and free-surface fluctuations. We shall now consider the effect of non-Gaussian distribution on the fatigue life estimates.

The fatigue life estimates through (9.181) involves $N^+(0)$, the expected rate of zero-crossings; and $p_s(s)$, the distribution function of stress-range. It has been shown (Grigoriu, 1984; Kanegaonkar and Haldar, 1987) that, if the stress history deviates significantly from Gaussian distribution, the expected rate of zero-crossings may be under-estimated several orders of magnitude and may therefore be unconservative. Further, the probability density of the stress range also deviates from Rayleigh distribution. The combined effect of these two deviations may be significant on the fatigue life estimates.

Kanegaonkar and Haldar (1987) have computed the fatigue life of the offshore platform in Example 9.2 using P-M hypothesis, and the percentage of occurrence of the seven sea states given in Table 9.5. Estimates of fatigue life of joints at two locations A and B (Fig. 9.26) based on AWS-X modified S-N curve are given in Table 9.8 for two cases with, and without free-surface fluctuations.

Table 9.8 Fatigue Damage Per Year in Each Sea State

Sea State	Damage/Year at Joint Near			
	M.S.L. A		Mud Line B	
	Without Fluctuations	With Fluctuations	Without Fluctuations	With Fluctuations
1	0.2492×10^{-5}	0.2492×10^{-3}	0.1363×10^{-4}	0.1363×10^{-4}
2	0.2629×10^{-3}	0.2764×10^{-3}	0.3796×10^{-2}	0.3680×10^{-2}
3	0.6091×10^{-3}	0.6881×10^{-3}	0.1106×10^{-1}	0.1115×10^{-1}
4	0.4137×10^{-3}	0.5215×10^{-3}	0.8754×10^{-2}	0.1012×10^{-1}
5	0.4642×10^{-3}	0.6173×10^{-3}	0.1023×10^{-1}	0.1280×10^{-1}
6	0.3059×10^{-3}	0.4137×10^{-3}	0.7667×10^{-2}	0.1018×10^{-1}
7	0.2651×10^{-3}	0.3706×10^{-3}	0.6955×10^{-2}	0.9498×10^{-2}
Total damage in one year	0.2323×10^{-2}	0.2886×10^{-2}	0.4848×10^{-1}	0.5723×10^{-1}
Life in years	430.5	346.4	20.63	17.47

It is seen from the above results that:

(i) the fatigue damage estimates based on Gaussian response assumption are unconservative at higher sea states; and

(ii) the ratio of fatigue damage, with nonlinear wave kinematics and free surface fluctuations, and without these factors, increases with an increase in significant wave height.

APPENDIX 9A: Hydrodynamics and Deterministic Water Wave

Sea is a large body of water with a free surface and an undulating bottom. The generation and propagation of waves on the sea surface has been an object of human curiosity throughout history. Interestingly, it was during the nineteenth century that fundamental theories of water waves were formulated providing a mathematical framework for study of wave related phenomena. The trochoidal wave theory for deep water waves was published by Gerstner in 1802. In 1844, Airy developed a small amplitude (linear) wave theory for full range of water depths. Significant contributions followed: Stokes (1847)—finite amplitude waves in deep water, later extended to intermediate depths; Russell (1844), Boussinesq (1871), and Rayleigh—solitary wave in shallow water; and Korteweg and de Vries (1895)— cnoidal wave theory.

Wave theories provide the mathematical basis for the study of waves and interpretation of experimental data. The actual sea waves are random in nature, but they can be modelled by superposition of regular waves of random amplitudes, periods and direction. For the study of wave forces in deep waters, which is our main concern in this chapter, linear wave theories are adequate. We reproduce the principal results of linear wave theory and the equations of hydrodynamics on which they are based. We make brief comment on Stokes higher order theories. For a detailed treatment of hydrodynamics and wave theories, the reader is referred to Lamb (1962), Stoker (1957), and Kinsman (1965).

9A.1 Equations of Hydrodynamics

We consider the motion of inviscid, incompressible homogeneous fluid of finite depth h, as shown in Fig. 9A.1. The origin of the coordinate system, fixed in space, is at still water level (SWL). Let $\bar{U} = (U_1, U_2, U_3)^T$ represent the Eulerian velocity vector at each point in the water. Under the assumptions stated above, the basic hydrodynamic equations and the boundary conditions governing the motion of the fluid-particles are:

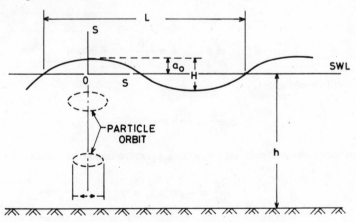

Fig. 9A.1. Propagation of linear wave.

Navier-Stokes Equations

$$\rho \frac{D\bar{U}}{Dt} = \rho \bar{F} - \bar{\nabla} p \tag{9A.1}$$

Continuity Equation

$$\overline{\nabla} \cdot \overline{U} = 0 \tag{9A.2}$$

where ρ is the fluid density, $\overline{F}$ is the body force vector; and p is the pressure, and

$$\frac{D}{Dt} = \frac{\partial}{\partial t} + \overline{U} \cdot \overline{\nabla}$$

Let $\phi(s_1, s_2, s_3)$ be the velocity potential function, such that

$$\overline{U} = \overline{\nabla} \phi = \left(\frac{\partial \phi}{\partial s_1}, \frac{\partial \phi}{\partial s_2}, \frac{\partial \phi}{\partial s_3} \right)^T$$

The continuity Eq. (9A.2) can be expressed as

$$\nabla^2 \phi = 0 \tag{9A.3}$$

where $\nabla^2 = \dfrac{\partial^2}{\partial s_1^2} + \dfrac{\partial^2}{\partial s_2^2} + \dfrac{\partial^2}{\partial s_3^2}$.

Neglecting the convective acceleration term in (9A.1),

$$\frac{p}{\rho g} + s_3 = -\frac{1}{g} \frac{\partial \phi}{\partial t} \tag{9A.4}$$

Boundary Conditions

FREE SURFACE ($s_3 = 0$)

$$U_3 = \frac{\partial \eta}{\partial t} + U_1 \frac{\partial \eta}{\partial s_1}, \text{ at } s_3 = \eta \tag{9A.5}$$

If U_i or $\partial \eta / \partial s_1$ is small,

$$U_3 = \frac{\partial \phi}{\partial s_3} = \frac{\partial \eta}{\partial t} \text{ at } s_3 = \eta \tag{9A.6}$$

But,

$$\phi \,|_{s_3 = \eta} = \phi \,|_{s_3 = 0} + \eta \frac{\partial \phi}{\partial s_3} \bigg|_{s_3 = 0} + 0(\eta^2)$$

For small disturbances,

$$\frac{\partial \phi}{\partial s_3} = \frac{\partial \eta}{\partial t} \text{ at } s_3 = 0 \tag{9A.7}$$

If surface-tension effects are neglected, pressure condition yields

$$\frac{\partial \phi}{\partial t} = -g\eta \text{ at } s_3 = \eta \tag{9A.8}$$

Again for small disturbances, we assume that (9A.8) holds at $s_3 = 0$. Combining (9A.7) and (9A.8), we have the linearized free surface boundary condition

$$\frac{\partial \phi}{\partial s_3} = -\frac{1}{g} \frac{\partial^2 \phi}{\partial t^2}, \text{ at } s_3 = 0 \tag{9A.9}$$

where g is the acceleration due to gravity.

BOTTOM ($s_3 = -h$)
Assuming impermeable bottom,

$$\frac{\partial \phi}{\partial s_3} = 0, \text{ at } s_3 = -h \tag{9A.10}$$

9A.2   General Solution
Assuming a separable solution of the form

$$\phi(s_1, s_2, s_3) = \bar{\phi}(s_1, s_2) f(s_3) \exp(i\omega t) \tag{9A.11}$$

the solution satisfying the continuity Eq. (9A.4) and the boundary condition (9A.9) and (9A.10) yields

$$\nabla^2 \bar{\phi} + k^2 \bar{\phi} = 0 \tag{9A.12}$$

and

$$f(s_3) = \frac{\cosh[k(s_3 + h)]}{\cosh(kh)} \tag{9A.13}$$

where k is the wave number satisfying the *dispersion relation*

$$\omega^2 = gk \tanh(kh) \tag{9A.14}$$

From (9A.8), we have the wave amplitude

$$\eta(s_1, s_2) = -i\frac{\omega}{g}\bar{\phi}(s_1, s_2) \tag{9A.15}$$

9A.3   Linear Wave Theory
We now consider a single harmonic wave of frequency ω propagating in the direction s_1. The free surface elevation is given by

$$\eta(s_1, t) = \bar{\eta}(s_1) \exp(i\omega t) \tag{9A.16}$$

and $\bar{\eta}(s_1)$ satisfies the one-dimensional Eq. (9A.12), that is

$$\frac{\partial^2 \bar{\eta}}{\partial s_1^2} + k^2 \bar{\eta} = 0 \tag{9A.17}$$

whose solution is

$$\bar{\eta}(s_1) = a_0 \exp(-iks_1) = \frac{H}{2}\exp\left(-i2\pi\frac{s_1}{L}\right) \tag{9A.18}$$

Hence,

$$\eta(s_1, t) = a_0 \exp[-i(ks_1 - \omega t)]$$

$$= \frac{H}{2}\exp\left[-i2\pi\left(\frac{s_1}{L} - \frac{t}{T}\right)\right] \tag{9A.19}$$

and the potential function is given by

$$\phi(s_1, s_3, t) = \frac{ig}{\omega} a_0 \exp\left[-i(ks_1 - \omega t)\right] \frac{\cosh\left[k(s_3 + h)\right]}{\cosh(kh)} \qquad (9A.20)$$

where $2a_0 = H$ is the wave height; L the wave length and T the time period. From (9A.3), the water-particle velocities are

$$U_1 = \frac{\partial \phi}{\partial s_1} = \frac{gk}{\omega} a_0 \exp\left[-i(ks_1 - \omega t)\right] \frac{\cosh\left[k(s_3 + h)\right]}{\cosh(kh)} \qquad (9A.21)$$

$$U_3 = \frac{\partial \phi}{\partial s_3} = \frac{-igk}{\omega} a_0 \exp\left[-i(ks_1 - \omega t)\right] \frac{\sinh\left[k(s_3 + h)\right]}{\cosh(kh)} \qquad (9A.22)$$

Using the dispersion relation (9A.14), above equations can be rewritten as

$$U_1 = \omega a_0 \exp\left[-i(ks_1 - \omega t)\right] \frac{\cosh\left[k(s_3 + h)\right]}{\sinh(kh)}$$

$$= \omega \frac{\cosh\left[k(s_3 + h)\right]}{\sinh(kh)} \eta(t) \qquad (9A.23)$$

$$U_3 = -i\omega a_0 \exp\left[-i(ks_1 - \omega t)\right] \frac{\sinh\left[k(s_3 + h)\right]}{\sinh(kh)}$$

$$= -i\omega \frac{\sinh\left[k(s_3 + h)\right]}{\sinh(kh)} \eta(t) \qquad (9A.24)$$

It follows from (9A.21) and (9A.22) that water particles describe an elliptical trajectory as shown in Fig. 9A.1.

The subsurface pressure distribution is given by (9A.4)

$$\frac{p}{\rho g} + s_3 = a_0 \left[\frac{\cosh\left[k(s_3 + h)\right]}{\cosh(kh)}\right] \exp\left[-i(ks_1 - \omega t)\right] \qquad (9A.25)$$

and the total energy per unit of the crest

$$E = \frac{1}{2} \rho g a_0^2 L \qquad (9A.26)$$

For deep waters, kh is large, $\tan kh = 1$, and the dispersion relation becomes

$$\omega^2 = gk \qquad (9A.27)$$

and the velocity potential can be expressed as

$$\phi(s_1, s_3, t) = i \frac{g}{\omega} a_0 \exp\left[-i(ks_1 - \omega t)\right] \exp(ks_3) \qquad (9A.28)$$

Note that, s_3 is negative in the water and, therefore, potential function decays with depth.

The ratio of the frequency ω and the wave number k is called the celerity or phase velocity

$$c = \frac{\omega}{k} = \frac{L}{T} \qquad (9A.29)$$

9A.4 Higher Order Wave Theories

The higher-order wave theories of Stokes' represent harmonic solution of continuity Eq. (9A.3) and nonlinear free surface boundary conditions (9A.6) and (9A.8). For a two-dimensional wave, the velocity potential is expressed by the approximate expansion

$$\phi(s_1, s_3, t) = \sum_{m=1}^{N} A_m \cosh(mks_3) \exp[-im(ks_1 - \omega t)] \qquad (9A.27)$$

where A_m are constants for specified values of depth, wave height and wave length, and are determined by satisfying the free-surface boundary conditions (Skjelbreia and Hendrickson, 1961). The number of terms in (9A.27) determines the order of the theory. For fifth order theory, $N = 5$. Practical applications of these theories involve extensive use of computers and programs for these theories are readily available.

Gerstner's trochoidal wave theory is an exact solution of finite amplitude waves in deep waters. The solution is obtained in a frame of reference moving with the wave and satisfies the nonlinear boundary conditions. The fluid motion is rotational and, therefore, velocity potential approach is not applicable making it difficult to use this theory for wave-structure interaction problems. Dean (1970, 1974) formulated the stream function theory in terms of stream function ψ. Since contours of ψ are are streamlines, it is possible to satisfy the kinematic boundary condition exactly by making the free surface a line of constant ψ. A stream function satisfying the continuity equation and the boundary condition on the sea bed can be expressed by the series:

$$\psi(s_1, s_3) = \sum_{m=1}^{N} A_m \sinh(mks_3) \cos(mks_1) \qquad (9A.28)$$

Besides coefficients A_m, the wave length L, and the wave of ψ on the surface are treated as unknowns; and for given H, T and h, these are determined numerically by minimizing the error in dynamic boundary condition at a large number of points on the free surface.

Comparisons of different wave theories are available in the works of Dean (1974), Le Me'haute (1969), Hogben and Standing (1975) and others. Regions for the validity of various wave theories have been identified in terms of independent parameters d/T^2 and H/T^2 by Dean (1974) and Le Me'haute (1969). Hogben and Standing (1975) have shown that for North Sea extreme weather conditions, Stokes fifth-order theory gives wavelength 7% longer than the linear theory. The total inertia force on a slender cylinder obtained from the two theories differ by about 7%, and the drag force by less than 13%. Linear theory is, therefore, considered adequate for calculation of wave forces, except for slender members near the surface, where drag component of the hydrodynamic force may be large.

APPENDIX 9B: Hydrodynamic Forces on Jacket-type Offshore Structures

A jacket type offshore structure is an assemblage of slender tubular members interconnected to form a three-dimensional tower which supports the deck. For dynamic analysis, the tower structures may be idealized by a discrete system consisting of a set of nodal points interconnected by linear elastic beam elements (Fig. 9.18b). The mass of the tower is lumped at the nodes, and in the most general treatment, each node may be assumed to have six degrees-of-freedom. Let the total number of degrees-of-freedom of the discrete model be n. The equation of motion of the offshore structure is given by (9.126), where $\bar{F}(t)$ is a n-dimensional hydrodynamic force vector, which includes the fluid-structure interaction effects, and is obtained by aggregating the hydrodynamic forces acting on each member.

Consider the mth member connecting nodes j and k as shown in Fig. 9B.1. The member is assumed to be a uniform, circular cylinder with diameter D, and length l. For primary members, l is the distance between the nodes, but secondary bracing members, l is the actual distance between the ends. The distance along the axis of the member is denoted by s, and $\bar{e}$ is a unit vector along the axis. Let $\bar{U}(s)$ and $\bar{X}(s)$ represent, respectively, the instantaneous water particle velocity and member displacement vectors at s. The normal components of these vectors in the plane defined by $\bar{e}$ are denoted by $\bar{U}_n(s)$ and $\bar{X}_n(s)$.

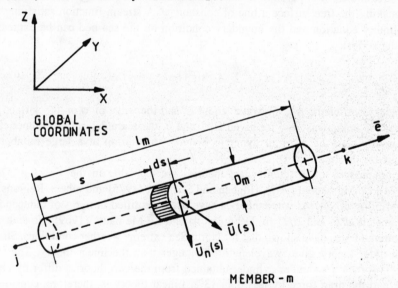

Fig. 9B.1. Typical member of the discrete model of a jacket type offshore structure.

The one-dimensional form of Morison equation (9.92 and 9.93) may be generalized to three-dimensional form to express the hydrodynamic force per unit length along member m at location s by the vector relation:

$$\bar{F}(s, t) = \rho A(C_M - 1) \ddot{\bar{Y}}_n(s) + \rho A \dot{\bar{U}}(s) + \frac{1}{2} \rho C_D D \, | \dot{\bar{Y}}_n(s) | \dot{\bar{Y}}_n(s) \qquad (9B.1)$$

where $\ddot{\bar{Y}}_n(s) = \dot{\bar{U}}_n(s) - \ddot{\bar{X}}_n(s)$.

$$\dot{\bar{Y}}_n(s) = \bar{U}_n(s) - \dot{\bar{X}}_n(s) \qquad (9B.2)$$

and

$$A = \pi D^2/4$$

Let the transformation matrix be defined as

$$\mathbf{N} = \mathbf{I} - \bar{\mathbf{e}}\,\bar{\mathbf{e}}^{\mathrm{T}}$$

Then, it can be shown that

$$\bar{U}_n(s) = \mathbf{N}\bar{U}(s);\ \dot{\bar{U}}_n(s) = \mathbf{N}\dot{\bar{U}}(s)$$

$$\dot{\bar{X}}_n(s) = \mathbf{N}\dot{\bar{X}}(s);\ \ddot{\bar{X}}_n(s) = \mathbf{N}\ddot{\bar{X}}(s) \qquad (9B.3)$$

and hence,

$$\dot{\bar{Y}}_n(s) = \mathbf{N}\dot{\bar{Y}}(s);\ \ddot{\bar{Y}}_n(s) = \mathbf{N}\ddot{\bar{Y}}(s) \qquad (9B.4)$$

Equation (9B.1) can now be expressed as

$$\bar{F}(s,t) = \rho A(C_{\mathrm{M}} - 1)\,\mathbf{N}\ddot{\bar{Y}}_n(s) + \rho A\mathbf{N}\ddot{\bar{U}}(s) + \frac{1}{2}\rho C_{\mathrm{D}} D\,|\,\mathbf{N}\dot{\bar{Y}}(s)\,|\,\mathbf{N}\dot{\bar{Y}}(s) \qquad (9B.5)$$

Let $\bar{F}_m(t)$ denote the 12-component force vector corresponding to 6 nodal displacements each at nodes j and k. Using appropriate interpolation functions, the vector function $\bar{F}(s,t)$ can be discretized, yielding (Penzien and Tseng, 1978)

$$\bar{F}_m(t) = \rho V_m(C_{\mathrm{M}m} - 1)\,\mathbf{L}_m\ddot{\bar{Y}}_m + \rho V_m\mathbf{L}_m\ddot{\bar{U}}_m$$

$$+ \frac{1}{2}\rho C_{\mathrm{D}m} A_m\,|\,L_m\dot{\bar{Y}}_m\,|\,L_m\dot{\bar{Y}}_m \qquad (9B.6)$$

where $\ddot{\bar{U}}_m,\ \dot{\bar{Y}}_m,\ \ddot{\bar{Y}}_m$ are 12-component vectors in the members' nodal coordinates; $V_m = A_m l_m$ is the effective volume; $A_m = D_m l_m$ is the effective drag area; and

$$\mathbf{L}_m \atop {(12\times12)} = \frac{1}{2}\begin{bmatrix} \mathbf{N} & 0 & 0 & 0 \\ 0 & 0 & 0 & 0 \\ 0 & 0 & \mathbf{N} & 0 \\ 0 & 0 & 0 & 0 \end{bmatrix} \qquad (9B.7)$$

Defining the effective volume and effective drag area matrices as $\mathbf{V}_m = V_m\mathbf{L}_m$; $\mathbf{A}_m = A_m\mathbf{L}_m$, respectively (9B.6) can be expressed as

$$\bar{F}_m(t) = \rho(C_{\mathrm{M}m} - 1)\,\mathbf{V}_m\ddot{\bar{Y}}_m + \rho\mathbf{V}_m\ddot{\bar{U}}_m + \frac{1}{2}\rho C_{\mathrm{D}m}\,|\,\mathbf{A}_m\dot{\bar{Y}}_m\,|\,\mathbf{A}_m\dot{\bar{Y}}_m \qquad (9B.8)$$

To determine the n-component hydrodynamic force vector $\bar{F}(t)$, it is necessary to aggregate the 12-component force vectors (9B.8) over all the N members constituting the system, that is

$$\bar{F}(t) = \sum_{m=1}^{N}\bar{F}_m(t) = \sum_{m=1}^{N}\left\{\rho(C_{\mathrm{M}m} - 1)\,\mathbf{V}_m\ddot{\bar{Y}}_m + \rho\mathbf{V}_m\ddot{\bar{U}}_m \right.$$

$$\left. + \frac{1}{2}\rho C_{\mathrm{D}m}\,|\,\mathbf{A}_m\dot{\bar{Y}}_m\,|\,\mathbf{A}_m\dot{\bar{Y}}_m\right\} \qquad (9B.9)$$

In vector notation, (9B.9) can be expressed as

$$\bar{F}(t) = \rho\,[\mathbf{C_M} - I]\,\mathbf{V}\,\ddot{\bar{Y}} + \rho\,\mathbf{V}\dot{\bar{U}} + \frac{1}{2}\,\rho\,\mathbf{C_D}\mathbf{A}\,|\,\dot{\bar{Y}}\,|\,\dot{\bar{Y}} \tag{9B.10}$$

$$= \rho\,[\mathbf{C_M} - I]\,\mathbf{V}(\dot{\bar{U}} - \ddot{\bar{X}}) + \rho\,\mathbf{V}\dot{\bar{U}} + \frac{1}{2}\,\rho\,\mathbf{C_D}\mathbf{A}\,|\,\bar{U} - \dot{\bar{X}}\,|\,(\bar{U} - \dot{\bar{X}}) \tag{9B.11}$$

where $\mathbf{C}_m$, $\mathbf{C_D}$ are matrices of inertia and drag coefficients, and $\mathbf{V}$ and $\mathbf{A}$ are matrices of projected volumes and areas, respectively. It is convenient to rewrite (9B.11) as

$$\bar{F}(t) = \rho\,\mathbf{C_M}\mathbf{V}\dot{\bar{U}} - \rho(\mathbf{C_M} - I)\,\mathbf{A}\ddot{\bar{X}} + \frac{1}{2}\,\rho\,\mathbf{C_D}\mathbf{A}\,|\,\bar{U} - \dot{\bar{X}}\,|\,(\bar{U} - \dot{\bar{X}})$$

$$= \mathbf{M_I}\dot{\bar{U}} - \mathbf{M_a}\ddot{\bar{X}} + \mathbf{C_H}\,|\,\bar{U} - \dot{\bar{X}}\,|\,(\bar{U} - \dot{\bar{X}}) \tag{9B.12}$$

where $\mathbf{M_I} = \rho\mathbf{C_M}\mathbf{V};\,\mathbf{M_a} = \rho(\mathbf{C_M} - I)\,\mathbf{V}$ and $\mathbf{C_H} = \frac{1}{2}\,\rho\,C_D\mathbf{A}$

A Nonlinear Approximation

If the water particle velocities are much greater than the velocity of the structure, that is, the temporal average

$$\langle\,|\,\bar{U}_n(s)\,|\,\rangle \gg \langle\,|\,\dot{\bar{X}}_n(s)\,|\,\rangle$$

the drag term in (9B.1) can be approximately expressed as

$$\frac{1}{2}\,\rho\,C_D D\,|\,\bar{U}_n(s) - \dot{\bar{X}}_n(s)\,|\,(U_n(s) - \dot{\bar{X}}_n(s))$$

$$\simeq \frac{1}{2}\,\rho\,C_D D\{|\,\bar{U}_n(s)\,|\,\bar{U}_n(s) - 2\,|\,\bar{U}_n(s)\,|\,\dot{\bar{X}}_n(s)\} \tag{9B.13}$$

and following the steps after (9B.1), (9B.12) can be expressed as (Penzien and Tseng, 1978)

$$\bar{F}(t) = \mathbf{M_I}\dot{\bar{U}} - \mathbf{M_a}\ddot{\bar{X}} + \mathbf{C_H}|\bar{U}|\,\bar{U} - \hat{\mathbf{C}}_\mathbf{H}\dot{\bar{X}} \tag{9B.14}$$

where $\hat{\mathbf{C}}_\mathbf{H} = 2\,\rho\,\mathbf{C_D}\mathbf{A}\,\langle\,|\,\bar{U}\,|\,\rangle$

Note that, (9B.14) retains the nonlinear drag term.

REFERENCES

Airy, G.B., 1975. *Tides and Waves. Encyc. Metrop.*

Alibe, B., 1986. Peak response of offshore structures to wave and current forces. Ph. D. Thesis, Cornell University.

American Petroleum Institute (API), 1989. Recommended practice for planning, designing and constructing fixed offshore platforms: API RP24.

Angelides, D.C., 1978. Stochastic response of fixed offshore structures in random sea. Ph. D. Thesis, Deptt. of Civil Engg., M.I.T.

Arthur, R.S., 1949. Variability in the direction of wave travel in ocean surface waves. *Ann. New York Acad. Sci.,* 52 (3): 511–522.

Battjes, J., 1970. Some properties of the bivariate Rayleigh density function. Internal Report, Univ. of Delft, Holland.

Bea, R.G., 1973. Selection of environmental criteria for fixed offshore platform design. OTC 1839, Vol. II: 185–196.

Bea, R.G., 1975. Development of safe environmental criteria for offshore structures. OI 75 : 7–15.

Beale, L.A. and A.A. Toprac, 1967. Analysis of inplane T and K welded tubular connections. Welding Research Council Bulletin No. 125.

Bidde, D.D., 1971. Laboratory study of lift forces on circular piles. *J. Waterways, Harbors and Coastal Eng. div. Proc. ASCE.*: 97 (WW4): 595–614.

Black, J.L., 1975. Wave forces on vertical axisymmetric bodies. *J. Fluid Mech.*, (67): 369–376.

Black, J.L., C.C. Mei and M.C.G. Bray, 1971. Radiation and scattering of waterwaves by rigid bodies. *J. Fluid Mech.*, (46): 151–164.

Borgman, L.E., 1958. Computation of ocean-wave force on inclined cylinders. *J. Geophys. Res.*, 39 (5): 585–888.

Borgman, L.E., 1967a. The estimation of parameters in a circular normal two dimensional wave spectrum. Tech. Rept. HEL. 1–9, Hydraulic Engineering Lab., Univ. Calif., Berkeley.

Borgman, L.E., 1967b. Spectral analysis of ocean wave force on piling. *J. Waterways and Harbors Div., proc. ASCE*, 93 (WW2): 124–156.

Borgman, L.E., 1972. Statistical models of ocean waves and wave forces. *Advances in Hydroscience* Vol. 8, Academic Press.

Boussinesq, J., 1981. The' orile de lintumescence liquide appele'e onde solitaire ou de translation se propageant dansun canal rectangulaire. *Iustitut de France, Acad' emie des Sciences, Comptes Rendus*: 755.

Brebbia, C.A. and S. Walker, 1979. *Dynamic Analysis of Offshore Structures*. London: Newnes-Butterworth.

Bretschneider, C.L., 1959. Wave variability and wave spectra for wind generated gravity waves. Tech, Memo. No. 118, Beach Erosion Board, Corps of Engineers.

Bretschneider, C.L., 1965. Generation of waves by wind-state of art. UNESCO Report SN-134-6.

Chakrabarti, S.K., 1973. Wave forces on pile including diffraction and viscous effects. *J. Hyd., Div., Proc, ASCE*, 99 (HYD): 1219–1233.

Chakrabarti, S.K., 1978. Wave forces on multiple vertical cylinders. *J. Waterways. Ports, Coastal and Ocean Div., Proc. ASCE*, 104 (WW2): 147–161.

Chakrabarti, S.K. and R.H. Sneider, 1975. Modelling of wind waves with JONSWAP spectra. Proc. Modelling: 120-139.

Chakrabarti, S.K. and W.A. Tam, 1975. Wave height distribution around vertical cylinder. *J. Waterways, Harbors, and Coastal Eng. Div. Proc. ASCE*, 101 (WW2): 225–230.

Chakrabarti, S.K., A.L. Tam and A.L. Wolbert, 1975. Wave forces on a randomly oriented tube. OTC paper No. 2190, Vol. I.: 433–448.

Chakrabarti, S.K., W.A. Wolbert and W.A. Tam, 1976. Wave forces on vertical circular cylinder. *J. Waterways, Harbors and Coastal Eng. Div., Proc. ASCE*, 102: (WW2): 203–221.

Cronin, D.J., P.S. Godfrey, P.M. Hook and T.A. Wyatt, 1978. Spectral fatigue analysis for offshore structures in *Numerical Methods in Offshore Engineering*, (ed.) O.C. Zienkewicz, New York: John Wiley.

Dao, B.V. and J. Penzien, 1982. Comparision of treatments of nonlinear drag forces acting on fixed offshore platforms. *App. Ocean Res.*, 4 (2): 66–72.

Darbyshire, J., 1955. An investigation of storm waves in North Atlantic Ocean. *Proc. Roy. Soc.*, London, A (230): 560–569.

Datta, T.K., 1983. Effect of nonlinear wave forces on the response history of offshore structure. OMAE Conference, Proc. paper No. IV: 227–234.

Dattari, L., 1973. Waves off Mangalore harbor-west coast of India. *J. Waterways, Harbors and Coastal Eng. Div., Proc. ASCE*, 99 (WW1): 39–58.

Dean. R.G., 1970. Relative validities of water wave theories. *J. Waterways, and Harbors Div., Proc. ASE*, 96 (WW1): 105–120.

Dean, R.G. 1974. Evaluation and development of water wave theories for engineering applications. Special Report No. 1, Vol. 1., 0.5. Army Coastal Eng. Res. Centre.

Dean, R.G. and D.M. Aagaard, 1970. Wave forces: data analysis and engineering calculation method. *J. Pet. Tech.*, 22: 368–375.

Defant, A., 1961. *Physical Oceanography* (2 vols). Pergamon Press.

Department of Energy (DOE), 1974. Guidance on design and construction of offshore installations. HMSO: London.

Dover, W.D. et al, 1981. Experimental and finite element comparisons of local stresses and compliance in tubular welded T-joints. Int. Conf. on steel in Marine Structures, Paris.

Dover, W.D. and S,. Dharmavasan, 1982. Fatigue fracture mechanics analysis of T and Y joints. Proc. OTC, OTC 4404.

Evans, D.J., 1969. Analysis of wave force data. Paper No. OTC 1005. Vol. I: 51–70.

Forristall, G.Z., 1978. On the statistical distribution of the heights in a storm. *J. Geophys. Res.*, 83 (15): 2353–2358.

Garrison, C.J. and P.Y. Chow, 1972. Wave forces on submerged bodies. *J. Waterways and Harbor Div. Proc. ASCE*, 98 (WW3): 375–392.

Garrison, C.J. and S. Rao, 1971. Interaction of wave and submerged objects. *J. Waterways, Harbors and Coastal Eng. Div. Proc. ASCE* (WW1): 259–277.

Garrison, C.J. and R. Stacey, 1977. Wave loads on North Sea gravity platforms: a comparison of theory and experiment. Paper No. OTC 2794: 513–524.

Goda, Y., 1976. A proposal of systematic calcualation procedures for the transformations and actions of irregular waves. Proc. Japan Soc. Civil Engrs., 253: 59–68.

Goda, Y., 1978. The observed joint distribution of periods and heights of sea waves. Proc. 16th Intl. Conf. Coastal Eng., Hamburg: 227–246.

Goda, Y., 1983 a. Analysis of wave grouping and spectra of long travelled swell. Rept. Port and Harbor Res. Instt., 22 (1): 3–41.

Goda, Y., 1983b. A unified nonlinearity parameter of water waves. Rept. Port and Harbor Res. Instt., 22 (3): 3–30.

Goda, Y., 1985. *Random Seas and Design of Maritime structures*. Tokyo: University of Tokyo Press.

Grigoriu, M., 1984. Extremes of wave forces. *J. Eng. Mech. Div., Proc. ASCE*, 110 (12): 1731–1742.

Grigoriu, M., 1985. Crossings of non-Gaussian translation process. *J. Engg. Mech. Div., Proc. ASCE*, 110 (4): 610–620.

Gudmestad, O.T. and J.J. Connor, 1983. Linearization methods and the influence of currents on the nonlinear hydrodynamic drag force. *J. App. Ocean Res.*, 5 (4): 184–194.

Gupta, A. and R.P. Singh, 1986. *Fatigue Behaviour of Offshore Structures*. New York: Springer Verlag.

Gutierrez, J.A., 1976. A substructure method for earthquake analysis of soil-structure interaction. Earthquake Engineering Research Centre, Univ. Of Calif., Berkeley, Rep. No. EERC 76–9.

Hafskjold, P.S., A. Törum and J. Eil, 1973. Submerged offshore concrete tanks. The Eighth International Navigation Congress of PIANC, Ottawa.

Hasselmann, K, et al, 1973. Measurement of wind-wave growth and swell decay during the Joint North Sea Wave Project (JONSWAP) *Deutche Hydr. Zeit*, Reehe, A (8): 1–95.

Havelock, T.M., 1940. The pressure of water waves on fixed obstacle, *Proc. Roy. Soc.*, A (175): 409–421.

Heavner, J.W., et al, 1985. Measured fatigue stress response of north sea jacket platforms. Proc. 4th. Int. Conf. on structural safety and Reliability, Vol. III, Kobe, Japan: 719–723.

Hogben, N., 1974. Wave resistance of surface piercing vertical cylinder in uniform currents. NPL Ship Div. Rept. 183. Hogben, N., 1976a. Wave load on structures. Proc. *Behaviour of Offshore Structures Conf.* Trondheim.

Hogben, N. and R.G. Standing, 1974. Wave loads on large bodies, Intl. Conf. on the Dynamics of Marine Vehicles and Structures in Waves, London.

Hogben, N. and R.G. Standing, 1975. Experience in computing wave loads on large bodies. Paper No. OTC 2189.

Hogben, N. 1977. Estimation of fluid loading on offshore structures. *Proc. Inst. of Civil Engrs*, Part 2: 515–512.

Holmes, P. and D. Howard, 1971. The statistical properties of random water waves. in *Dynamic Waves in Civil Engineering*, (ed.) D.A. Howells, Wiley Interscience: London.

Houmb, O.G. and T. Overvik, 1976. Parametrization of wave spectra and long-term joint distribution of wave height and period. *Proc. Behaviour of Offshore Structures*, 7 (1): 144–169.

Hsu, T.H., 1984. *Applied Offshore Structure Engineering*, Houston: Gulf.

Huntington, S.W., 1979. Wave loading on large cylinders in short crested sea. *Mechanics of Wave-Induced forces on Cylinders*. (ed.) T.L. Shaw, Pitman, London: 636–649.

Huntington, S.W. and O.M. Thompson, 1976. Forces on a large vertical cylinder in multi-directional random waves. Paper No. OTC 2539:

Huang, N.E. and S.R. Long, 1980. An experimental study of the surface elevation probability distribution and statistics of wind-generated waves. *J. Fluid Mech.*, (101): 179–200.

Jain, A.K. and T.K. Datta, 1986. Nonlinear dynamic analysis of offshore towers in frequency domain. *J. Engg. Mech. Div., Proc. ASCE.* 113 (4): 610–625.

John, F., 1950. On the motion of floating bodies II. *Commun. in Pure and App. Maths.*, 3 (45).

Johnson-Laird, P.J., 1970. Properties of surface elevation in random water waves. M. Eng. Thesis, Univ. of Liverpool.

Kanegaonkar, H.B. and A. Haldar, 1987. Non-Gaussian response of offshore platforms: dynamic. *J. Str. Div. Proc. ASCE*, 113 (a): 1982–1898.

Kanegaonkar, H.B. and A. Haldar, 1987. Non-Gaussian response of offshore platforms: fatigue. *J. Str. Div. Proc. ASCE*. 113 (a): 1899–1909.

Keulegan, G.H. and L.H. Carpenter, 1958. Force on cylinders and plates in an oscillating fluid. *J. of Research; National Bureau of Standards*, 60 (5): 423–440.

Kinsman, B. 1965. *Wind Waves, Their Generation and Propagation in the Ocean Surface*. Prentice Hall.

Kim, Y.K. and H.C. Hibbard, 1975. Analysis of simultaneous wave force and water particle measurements. OTC Paper No. 2192, Vol–I: 461–470.

Korteweg, D.J. and G. de Vries, 1895. On the change of form of long waves advancing in a rectangular canal and on the new type of long stationary waves. *Philosophical Mogazine*, 5 (39): 422.

Koopmans, L.H., 1970. *The Spectral Analysis of Time Series*. Academic Press.

Kotik, J. and R. Morgan, 1961. The uniqueness problem for wave resistance calculated from singularity distributions which are exact at zero Froude number *J. Ship Res.*, 13: 61–68.

Kuang, et al, 1975. Stress concentration in tubular joints. Proc. OTC, OTC–2205.

Laird, A.D.A., 1962. Water forces on flexible oscillating cylinder. *J. Waterway, Harbors Div. Proc. ASCE*, 88 (WW3): 125–138

Lamb, H., 1962. *Hydrodynamics* (6th ed.) Cambridge University Press.

Le Me' haute, B., 1969. *Shore Protection Manual*, U.S. Army Coatal Eng. Res. Centre.

Longuet-Higins, M.S., 1952. On the statistical distribution of height of sea waves. *J. Mar. Res.*, 11: 245–266.

Longuet-Higgins, M.S., 1957. The statistical analysis of random marine surface. *Phil Trans. Roy. Soc.* A (24a): 321–387.

Longuet-Higgins, M.S. 1975. On the joint distribution of the periods and amplitude of sea waves. *J. Geophys. Res.*: 80: 2688–2694.

Longuet-Higgins, M.S. and R.W. Stewart, 1960. Change in the form of short gravity waves on long waves and tidal currents. *J. Fluid Mech.* 8: 565–583.

MacCamy, R.C. and R.A. Fuchs, 1954. Wave forces on piles: a diffraction theory. U.S. Army Coastal Engineering Centre, Tech. Memo No. 69.

Malhotra, A.K. and J. Penzien 1970a. Response of offshore structures to random wave forces. *J. Str. Div. Proc. ASCE*. ST (10): 2155–2173.

Malhotra, A.K. and J. Penzien, 1970 b. Nondeterministic analysis of offshore structures. *J. Engg. Mech. Div.* Proc. ASCE EM (6): 985–1003.

Malhotra, A.K. and J. Penzien, 1971. Analysis of tall open structures subjected to stochastic excitation in *Dynamic waves in Civil engineering*. D.A. Howell (ed.), Wiley-Interscience: 505.

Marshall, P.W., 1969 Risk evaluation for offshore structures. *J Str. Div., Proc. ASCE*, 95 (ST12).

Marshall, P.W., 1974. General considerations for tubular joint design. *Welding in Offshore construction*, New castle.

Mitsuyasug. H. et al, 1975. Observation of the directional spectrum of ocean waves using clowerleaf busy. *J. Physical Oceanogr.* 5 (4): 750–760.

Morison, J.R., M.P.O.' Brien, J.W. Johnson and S.A. Schaaf, 1950. The force exerted by surface waves on piles, *Petroleum Transactions, AIME*, 189: 149–154.

MairWood, A.M., 1969. *Coastal Hydraulics*. London: Macmillan.

Nigam, N.C., 1983. *Introduction to Random Vibrations*. Cambridge, Mass.: The MIT, Press.

Nolte, K.G. and Hansford, J.E., 1976. Closed-form expression for determining the fatigue damage of structure to ocean wave. OTC 2606.

Ochi, M.K. and E.N. Hubble, 1976. On six parameter wave spectra. *Proc. 15th Int. Conf. Coastal Engg. Hawai*: 301–320.

Panicker, N.N., 1974. Review of techniques for directioal spectra *Proc. Intl. Symp. on Ocean Wave Measurement and analysis ASCE* (1).

Paris, P.C., 1962. The growth of fatigue cracks due to variations in load. Ph.D. Thesis, Lehigh University, Bethelehem, Pa.

Patel, M.H., 1989. *Dynamics of Offshore Structures*. London, Butterworth.

Penzien, J. and S. Tseng, 1978. Three dimensional dynamic analysis of offshore platforms. *Numerical Methods in Offshore Structures*. (eds) O.C. Zienkewicz, et. al., New York: John Wiley.

Pierson, W.J. and P. Holmes, 1965. Irregular wave force on a pile. *J. Waterways and Harbour Div, Proc ASCE* 91: (WW4): 1–10.

Pierson, W.J. and L. Moskowitz, 1964. A proposed spectral form for fully devloped wind sea based on the similarity theory of S.A. Kitaigorodiskii. *J. Geophys. Res.* 69: 5181–5190.

Pierson, W.J., Jr., G. Neumann and R.W. James, 1955. Pratical methods for observing and forecasting ocean waves by means of wave spectra and statistics. *U.S. Navy Hydrographic Office.* H.O. Pubs. No. 603.

Pierson, W.J., Jr. et. al. 1952. The theory of the refraction of a short crested Gaussian sea surface with application to Northern, North Jersey coast. *Proc. 3rd conf. Coastal Eng. Cambridge, Mass.*

Rice, S.O., 1944. Mathematical Analysis of Random Noise represented in *Selected Papers on Noise and Stochastic Processes,* N. Wax (ed.), 1954, Dover Pubs.

Russell, S., 1844. Report on Waves, British Association Reports. '

Saetre, H.J., 1974. On the high wave conditions in Northern North Sea. I.O.S. Wormley Report No. 3.

Sarpkaya, T., 1976a. In-line and transverse froces on cylinders in oscillatory flow at high Reynolds number. OTC–1976, Paper No. OTC 2533, Vol. II: 95–108.

Sarpkaya, T, 1976b. Vortex shedding and resistance in harmonic flow about smooth and rough circular cylinders. BOSS' 76. *Proc. First International Confrence,* Vol. I–220–235.

Sarpkaya, T., 1976c. Discussion by Hogben on Vortex shedding by Sarpkaya, Authors reply, Boss, Vol. II.: 312–315.

Sarpkaya, T., N.J. Collins and S.R. Evan, 1977. Wave forces on rough-walled cylinders at high Reynolds numbers. Paper No. OTC 2901, Vol III: 175–184.

Sarpkaya, T. and M. Isaacson, 1981. *Mechanics of Wave Forces on Offshore Structures.* New York: Van Norstand Reinhold.

Schmidt, L.V., 1965. Measurements of fluctuating air loads on a circular cylinder. *J. Aircraft, 2 (1):* 49–55.

Shinozuka, M., S-L.S. Frag and A. Nishitami, 1976. Time-domain structural response simulation in a short crested sea. *J. Energy Resources Technology* 101: 270–275.

Shyam Sunder, S., 1979. Stochastic modellig of ocean storms M.S. Thesis, Dept. of Civil Engg.; M.I.T.

Singh, R.P., 1980. Dynamic behaviour of steel jacket platforms,Ph. D,. Thesis, I.I.T. Delhi, India.

Skjelbria, L. and J. Hendrickson, 1961, Fifth order gravity wave theory. *Proc. Seventh Conf. On Coastal Engineering*

Spanos, P-T.D. and T.W. Chen, 1981. Random response to flow induced forces. *J. Engg. Mech. Div. Proc. ASCE* 107 (EM6): 1173–1190.

Spring, B. H. and P.L. Monkmeyer, 1974. Interaction of plane waves with vertical cylinders. *Proc. Conf. on Coastal Engg.* Vol. 2.

Stansby, P.K., 1979. Mathematical modellig of vortex shedding from circular cylinders in planar oscillatory flows, including the effects of harmonics. *Mechanics of Wave-Induced Forces on Cylinders,* (ed.) T.L. Shaw, Pitman, London: 450–460.

Stoker, J.J., 1957. *Waterwaves,* Interscience.

Stokes, G.G., 1847. On the theory of oscillatory waves. *Trans, Camb. Phil. Soc.,* 8, and supplementary scientific Papers–1.

Taylor, R.E. 1974. Structural dynamics of fixed and floating platforms in water. *The Dynamics of Marine Vehicles and Structures in Waves*: London.

Tick, L.J., 1963. Nonlinear probability models of ocean waves in *Ocean Wave Spectra,* Prentice Hall: 163–169.

Tickell. R.G., 1979. The probalistic approach to wave loading on maritime structures. *Mechanics of Wave Induced Forces on Cylinders.* T.L. Shaw, ed., San Francisco: Pitman Advanced publishing Program: 152–177.

Tickell, R.G. and M.H. Elway. A probabilistic description of forces on a member in a short-crested random sea. *Mechanics of Wave induced Forces on Cylinders.* T.L. Shaw, Led., San Francisco: Pitman Advanced Publishing Program: 561–576.

Thrasher, C.W. and P.M. Aagaard, 1969. Measured wave force data on offshore platforms. Paper No. OTC 1006, Vol. I: 71–82.

Tung, C.C., 1984. Statistical Properties of nonlinear waves. *Proc. 4th. Speciality conf. on Prob. Mech. and Structurals Reliability, ASCE,* New York: 111–114.

Venkataramana, K. et al, 1989. Stochastic response of offshore structures to sea wave and earthquake excitations with fluid—structure—soil interaction. Dept. of Civ. Engg., Kyoto University, Res,. Rep. No. 89—S.T.—01.

Wade, B.G. and M. Dwyer, 1976. On the application of Morison equation to fixed offshore platforms. Paper. No. OTC 2723.

Wehausen, J.V. 1971. The motion of floating bodies. Ann. Reviews of *Fluid Mech.* 3: 237–268.

Wheeler, J.D., 1969. Method for calculating forces by irregular waves. Paper No. OTC 1007, Vol. I: 83–94.

Wiegel, R.L., 1963. *Oceanographic Engineering.* New York: Prentice Hall.

Wiegel, R.L., K.E. Beebe and J. Moon, 1957. Ocean wave forces on circular cylindrical piles. *J. of Hyd. Div., Proc. ASCE*, 83 (HY2): 1–36.

Williams, A.K. and J.E. Rinne, 1976. Fatigue analysis of steel offshore structures. *Proc. Inst. Civil Engrs.* 60 (1): 635–654.

Woadiang, R.A., 1955. An approximate joint probability distribution for amplitude and frequency in random noise. *New Zealand J. of Sci. Tech.*, B (36): 537.

Yamamoto, T. and J.H. Nath, 1976. Hydrodynamic forces on a groups of cylinders. Paper No. OTC.-2499.

Yue, D.K.P., H.S. Chen and C.C. Mei, 1976. Water wave forces on three-dimensional bodies by a hybrid element method. *Tech, Rept. R.M. Parsons Lab.*, Deptt. of Civil Engg. M.I.T.

Zienkewicz. O.C. et al (eds.)., 1978. *Numerical Methods in Offshore Engineering.* New York: John Wiley.

10. Statistical Energy Analysis

10.1 Introduction

The classical normal mode method of vibration analysis of continuous structural systems is mainly applicable in the low frequency response regimes. The method and its variations to consider different forms of damping essentially decouple the equations of motion in the modal coordinates and the vibration response is obtained by superposing the modal responses corresponding to a finite number of modes by considering each mode as a sdf vibration system. Generally only a few lower order modes contribute most to the total response and consequent failures. Because of this the normal mode method has become one of the widely used methods of vibration analysis. The developments in the finite element analysis facilitating the ease of computation of natural frequencies and normal modes even for complex structural systems have also contributed to the wide applicability of the normal mode method. The extension of the normal mode method to random vibration problems is also direct and straightforward.

Although in principle the normal mode method may be valid at all frequencies, its application to vibration problems in the higher frequencies such as encountered in aerospace structural vibrations or coupled vibro-acoustic problems is fraught with a high degree of inaccuracy. The reasons for the doubtful validity of the normal mode method to deal with high frequency vibration problems are the following.

In the high frequency range, there are a large number of natural frequencies within a frequency band and hence the use of normal mode method will entail large computational effort to account for all the modes that contribute to the total response. Even with finite element techniques, the computation of natural frequencies and normal modes will be only of doubtful accuracy in the higher frequency ranges, since displacement functions associated with higher modes are extremely sensitive to the structural details and hence geometrical and material properties and boundary conditions have to be specified with utmost precision which is well neigh an impossible task. Moreover, many of the simplifying assumptions regarding structural analysis are not valid at high frequencies. For example, the mode shape of a simple supported beam corresponding to a higher mode will be so wrinkled with many half sine waves that the effects of the shear deformation and rotary inertia become important and have to be included in the analysis. Another reason for the impracticability of the normal mode method at higher frequencies is the fact that the response at a specific point in the structure is not of great relevance as the overall response averaged over the domain of the structure. The normal mode method generates a large mass of results corresponding to each point on the structure and it will be extremely difficult to handle and interpret meaningfully these results.

The need for an alternate method to understand and estimate vibrational response of complex multi-modal systems, simple in its approach and applicable in the high frequency regimes led to the development of the *statistical energy analysis (SEA)* by a group of scientists at the Bolt, Beranek, Newmen, Inc. Massachusetts, USA in the early sixties. Though many of the ideas leading to the development of the SEA method have been in use earlier in the area of room acoustics, it can be said that this group pioneered in systematically enunciating the various assumptions and approximations involved and the conditions of validity of the method. The first major applications of the method were in the area of predicting the random vibration characteristics of complex built up aerospace structures to broad band random excitation and the high frequency vibration of coupled acousto-structural systems.

Even though SEA has been primarily developed to deal with random vibration of coupled systems, the method is general in its approach and can be applied to the vibration analysis of coupled deterministic systems as well. The approach has come to be known as "statistical energy analysis" because the term "statistical" implies that averages are involved and that some measure of randomness is assumed in the excitation, the distribution of resonance frequencies and the matching conditions of various modes. The term "energy" refers to the fact that the average quantities used are closely related to the energy and power in the system. The approach considers the systems to be having parameters drawn from populations with random distributions. In the words of Lyon (1969), the main originator of the method "SEA has been described as a point of view of dealing with vibration of complex structures, and as such it employs a series of analytical and experimental methods most of which pre-date the identification of SEA. The view point is *statistical* because the systems under analysis are presumed drawn from populations with random parameters; *energy* is the independent dynamic variable chosen because, by using it distinctions between acoustical and mechanical systems disappear; and *analysis* emphasizes that SEA is an approach to problems rather than a set of techniques as such". SEA has thus developed into a branch of linear oscillatory dynamics to consider vibratory behaviour of complex mechanical and acoustical systems.

The method considers the energy flow into and out of a vibrating system and employs such concepts as modal energy and power flow. Vibrating systems whether they are mechanical, acoustical or fluid are considered to act as resonators and the total energy of the system and the energy dissipation are evaluated in terms of modal energies.

Lyon and Maidanik (1962) first derived the power flow relations between two independently and randomly excited linear oscillators with small linear coupling and showed that for conservative coupling the power flow was proportional to the difference of average energies between the oscillators. They also extended the results to the problem of power flow between two coupled multi-modal systems. This work formalized many of the important results of SEA. Smith (1962) also introduced many of the concepts and applied the method to the prediction of the response of complex structures to random acoustic excitation, where use has been made of Maidanik's (1962) work on radiation resistance of reverberant panels. Following these pioneering works a number of papers and reports have appeared in the nature of reviews and introduction of SEA procedures (Lyon and Maidanik, 1964; Smith and Lyon, 1965; Ungar, 1967; Fahy, 1974; Pope, 1974). Lyon's (1975) book is the most comprehensive treatise on the subject. Brief accounts of SEA can also be found in the books of Cremer, Heckl and Ungar (1973), and

Norton (1989). There have been a number of developments in the subject since the publication of Lyon's book and some of these are discussed in the works of Maidanik (1977), Woodhouse (1981a, b), Hodges and Woodhouse (1986) and Langley (1989).

In this chapter we present the SEA method starting from the basic derivation of the power flow calculations between two linearly coupled oscillators. The extension to power flow relations between a system of coupled oscillators is considered next. The concepts of modal density and coupling loss factors so essential in the application of SEA are then presented including analytical expressions for these parameters and descriptions of experimental methods to determine modal densities, internal and coupling loss factors. some applications of SEA in predicting the response of structures to random acoustic excitation and noise transmission problems are given. Finally some of the recent developments in SEA are briefly presented.

10.2   General Assumptions in SEA

In the SEA method, the complex structural or acoustic system is divided into a large number of coupled subsystems. In each subsystem one considers groups of multiple oscillators having resonant frequencies within certain frequency bands. The oscillators are assumed to be drawn from an ensemble with statistical description of relevant system parameters. Within each group of oscillators, energy is stored in the elastic and inertial elements and dissipated by damping. Energy is also transmitted to other subsystems via the coupling elements. The fundamental task in SEA applications is the calculation of energy flow relationship between subsystems in terms of system parameters, so that the energy balance relations can be derived. Average energies of the system can be obtained from energy balance relations and the vibration response which is dependent of the energies can consequently be obtained.

In deriving the energy balance relations some general assumptions are usually made. They are:

1. The coupling between the different subsystems is linear and conservative. (i.e., the coupling consists of only inertial, elastic and gyroscopic elements).
2. Only those groups of oscillators having resonance frequencies within the frequency band of interest have to be considered in the energy flow calculations.
3. The oscillators are subjected to broad-band random excitations which are statistically independent.
4. Energy is more or less equally shared between all resonant modes within a frequency band in a given subsystem.
5. The principle of reciprocity holds between the different subsystems.
6. The power flow between any two subsystems is proportional to the difference between the averaged coupled modal energies.

The significance of these assumptions will become clear when we present the power flow relationships.

10.3   Average Power Flow Between Two Coupled Oscillators

Before we consider the interaction between two or more interconnected elastic subsystems with each subsystem consisting of a number of modes, it is essential to consider the power flow relationship between two coupled sdf oscillators constituting the simplest coupled system.

10.3.1   Equations of motion with conservative coupling element

Consider two linear sdf oscillators idealized as mass-spring-dashpot systems coupled by a linear conservative element consisting of inertial, elastic and gyroscopic. coupling as shown in Fig. 10.1.

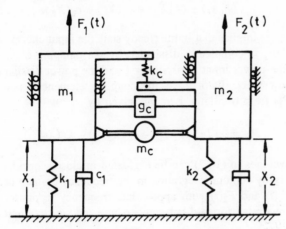

Fig. 10.1.   Gyroscopically coupled linear oscillators.

The equations of motion for the coupled system can be expressed as (Scharton and Lyon, 1968; Fahy, 1974):

$$(m_1 + m_c/4)\ddot{X}_1 + c_1\dot{X}_1 + (k_1 + k_c)X_1 + (m_c/4)\ddot{X}_2 - g_c\dot{X}_2 - k_cX_2 = F_1(t) \qquad (10.1)$$

$$(m_2 + m_c/4)\ddot{X}_2 + c_2\dot{X}_2 + (k_2 + k_c)X_2 + (m_c/4)\ddot{X}_1 + g_c\dot{X}_1 - k_cX_1 = F_2(t) \qquad (10.2)$$

where m_j, c_j and k_j, $j = 1, 2$ respectively are the mass, linear viscous damping coefficient and stiffness of the jth oscillator and m_c, k_c and g_c are the mass, stiffness and gyroscopic coupling coefficient of the coupling element. $F_1(t)$ and $F_2(t)$ are assumed to be stationary ergodic random excitations, statistically independent, and having uniform psds (ideally white noise excitations) acting on masses 1 and 2 respectively.

10.3.2   Average Power Flow Relations

The time averaged power supplied by the force F_1 is $\langle F_1\dot{X}_1\rangle$ which from (10.1) is given by

$$\langle F_1\dot{X}_1\rangle = (m_1 + m_c/4)\langle\ddot{X}_1\dot{X}_1\rangle + c_1\langle\dot{X}_1^2\rangle + (k_1 + k_c)\langle X_1\dot{X}_1\rangle$$
$$+ (m_c/4)\langle\ddot{X}_2\dot{X}_1\rangle - g_c\langle\dot{X}_2\dot{X}_1\rangle - k_c\langle X_2\dot{X}_1\rangle \qquad (10.3)$$

where the angular brackets $\langle.\rangle$ indicate time averages of the respective quantities contained within. In the steady-state when the response also becomes stationary the first and third terms in (10.3) vanish and hence

$$\langle F_1\dot{X}_1\rangle = c_1\langle\dot{X}_1^2\rangle + (m_c/4)\langle\ddot{X}_2\dot{X}_1\rangle - g_c\langle\dot{X}_2\dot{X}_1\rangle - k_c\langle X_2\dot{X}_1\rangle \qquad (10.4)$$

Similarly, the time averaged power supplied by the force $F_2(t)$ from (10.2) is

$$\langle F_2 \dot{X}_2 \rangle = c_2 \langle \dot{X}_2^2 \rangle + (\dot{m}_c/4) \langle \ddot{X}_1 \dot{X}_2 \rangle + g_c \langle \dot{X}_1 \dot{X}_2 \rangle - k_c \langle X_1 \dot{X}_2 \rangle \qquad (10.5)$$

Adding (10.4) and (10.5) and using the relations $\langle X_2 \dot{X}_1 \rangle = - \langle X_1 \dot{X}_2 \rangle$ and $\langle \ddot{X}_2 \dot{X}_1 \rangle = - \langle \ddot{X}_1 \dot{X}_2 \rangle$ due to stationarity we obtain

$$\langle F_1 \dot{X}_1 \rangle + \langle F_2 \dot{X}_2 \rangle \equiv c_1 \langle \dot{X}_1^2 \rangle + c_2 \langle \dot{X}_2^2 \rangle \qquad (10.6)$$

Equation (10.6) implies that in the steady state the input energy supplied by the forces F_1 and F_2 is completely dissipated by the damping elements. Since the difference in the power input to oscillator 1 and the power dissipated by it has to necessarily flow to oscillator 2, as the coupling element is nondissipative the power flow P_{12} from oscillator 1 to 2 is given by

$$P_{12} = (m_c/4) \langle \ddot{X}_2 \dot{X}_1 \rangle - g_c \langle \dot{X}_2 \dot{X}_1 \rangle - k_c \langle X_2 \dot{X}_1 \rangle \qquad (10.7)$$

The time averages in (10.7) can be evaluated in closed form by using random vibration theory by contour integration in the frequency domain of the product of the psds of $F_1(t)$ and $F_2(t)$ with appropriate frequency response functions. The integrations are straightforward but cumbersome and we state here only the final result (Lyon, 1975)

$$P_{12} = A \{\langle E_1' \rangle - \langle E_2' \rangle\} \qquad (10.8)$$

where $\langle E_j' \rangle$ is the time averaged blocked energy of oscillator j; the blocked energy of the oscillator is the sum of the kinetic and potential energies of the oscillator while it is coupled to the other oscillator in the system when the other oscillator is constrained not to have any motion. In (10.8) the blocked average energy $\langle E_j' \rangle$ is given by

$$\langle E_j' \rangle = \pi \Phi_j / \{\beta_j (m_j + m_c/4)\}, \quad j = 1, 2 \qquad (10.9)$$

where Φ_j is the constant two sided psd of the input force to the jth oscillator and

$$\beta_j = \eta_j \omega_j \quad \text{with} \quad \eta_j = c_j/(m_j + m_c/4) \qquad (10.10)$$

being the loss factor of the jth oscillator and

$$\omega_j^2 = (k_j + k_c)/(m_j + m_c/4) \qquad (10.11)$$

The proportionality constant A in (10.8) is given by

$$A = \frac{\beta_1 \beta_2}{(1 - \mu^2)d} \{\mu^2 [\beta_1 \omega_2^4 + \beta_2 \omega_1^4 + \beta_1 \beta_2 (\beta_1 \omega_2^2 + \beta_2 \omega_1^2)^2]$$

$$+ (\gamma^2 + 2\mu\kappa)[\beta_1 \omega_2^2 + \beta_2 \omega_1^2] + \kappa^2(\beta_1 + \beta_2)\} \qquad (10.12)$$

where

$$\mu = (m_c/4)/[(m_1 + m_c/4)\,(m_2 + m_c/4)]^{1/2}$$
$$\gamma = g_c/[(m_1 + m_c/4)\,(m_2 + m_c/4)]^{1/2}$$
$$\kappa = k_c/[(m_1 + m_c/4)\,(m_2 + m_c/4)]^{1/2}$$
$$d = \beta_1\beta_2[(\omega_1^2 - \omega_2^2)^2 + (\beta_1 + \beta_2)(\beta_1\omega_2^2 + \beta_2\omega_1^2)] + \mu^2(\beta_1\omega_2^2 + \beta_2\omega_1^2)^2$$
$$+ (\gamma^2 + 2\mu\kappa)(\beta_1 + \beta_2)(\beta_1\omega_2^2 + \beta_2\omega_1^2) + \kappa(\beta_1 + \beta_2)^2$$

It can also be shown that the blocked time averaged energy of oscillator j is also given by

$$\langle E_j' \rangle = (m_j + m_c/4) \langle \dot{X}_j^2 \rangle \tag{10.13}$$

which is twice the time averaged kinetic energy of the jth oscillator.

From (10.8) and the other relations following it, it can be concluded that the power flow from oscillator 1 to oscillator 2 is

(i) directly proportional to the difference between the time averaged blocked energies of the oscillators, with the constant of proportionality A dependent only on the system parameters.

(ii) as A is positive the average power flow is from the oscillator of higher energy to the oscillator of lower energy, and

(iii) since the constant of proportionality is symmetrical is the system parameters, the power flow relation is reciprocal in the sense that equal difference in the resonator energies in either direction results in equal power flow.

Lyon (1975) has also shown that the time averaged power flow from oscillator 1 to 2 is also given by

$$P_{12} = B\{\langle E_1 \rangle - \langle E_2 \rangle\} \tag{10.14}$$

where $\langle E_j \rangle$ $j = 1, 2$ are the actual time averaged energies of oscillators 1 and 2, without either of them being blocked and B is another proportionality constant given by

$$B = \{\mu^2[\beta_1\omega_2^4 + \beta_2\omega_1^4 + \beta_1\beta_2(\beta_1\omega_2^2 + \beta_2\omega_1^2)^2] + \gamma^2$$
$$+ 2\mu\kappa(\beta_1\omega_2^2 + \beta_2\omega_1^2) + \kappa^2(\beta_1 + \beta_2)\}/D \tag{10.15}$$

where

$$D = (1 - \mu^2)[(\omega_1^2 - \omega_2^2)^2 + (\beta_1 + \beta_2)\,(\beta_1\omega_2^2 + \beta_2\omega_1^2)] \tag{10.16}$$

Equations (10.8) and (10.14) represent the most fundamental results in SEA satisfying the assumption that the power flow is proportional to the difference between the averaged coupled modal energies.

10.3.3 Power Balance Relations

The average power dissipated by oscillator 1 is $c_1\langle \dot{X}_1^2 \rangle$ and that by oscillator 2 is $c_2\langle \dot{X}_2^2 \rangle$. They can be expressed in terms of the loss factors and the average energies of the oscillators in the form

$$c_j \langle \dot{X}_j^2 \rangle = (P_j)_{\text{diss}} = \eta_j \omega_j \langle E_j \rangle \quad j = 1, 2 \tag{10.17}$$

where $(P_j)_{\text{diss}}$ is the average power dissipated by the jth oscillator and η_j is the corresponding dissipation or internal loss factor.

A *coupling loss factor* between the two systems similar to the dissipation or internal loss factors η_js, can be analogously defined such that the power from one oscillator to the other to which it is coupled is related to the average energy of the first oscillator. Since the power flow depends on the average energy difference between the oscillators, the coupling loss factor is defined for zero average energy in the second oscillator. From (10.14) η_{12} is defined in the following way with $\langle E_2 \rangle = 0$.

$$P_{12}\big|_{\langle E_2 \rangle = 0} = B \langle E_1 \rangle = \eta_{12} \omega_1 \langle E_1 \rangle \tag{10.18}$$

Similarly,

$$P_{21}\big|_{\langle E_1 \rangle = 0} = B \langle E_2 \rangle = \eta_{21} \omega_2 \langle E_2 \rangle \tag{10.19}$$

Since the constant of proportionality B is symmetric in the system parameters, from (10.18) and (10.19) we obtain the relation $B = \eta_{12}\omega_1 = \eta_{21}\omega_2$ which when substituted in (10.14) yields

$$P_{12} = \eta_{12} \omega_1 \langle E_1 \rangle - \eta_{21} \omega_2 \langle E_2 \rangle \tag{10.20}$$

The power balance relations between the two oscillators can be established by noting the fact that the power input to one of the oscillators is equal to the sum of the power dissipated by it and the power flow from that oscillator to the other oscillator. Hence

$$P_1 = (P_1)_{\text{diss}} + P_{12} = \eta_1 \omega_1 \langle E_1 \rangle + \eta_{12} \omega_1 \{ \langle E_1 \rangle - \langle E_2 \rangle \} \tag{10.21}$$

$$P_2 = (P_2)_{\text{diss}} + P_{21} = \eta_2 \omega_2 \langle E_2 \rangle + \eta_{21} \omega_2 \{ \langle E_2 \rangle - \langle E_1 \rangle \} \tag{10.22}$$

where P_1 and P_2 respectively are the average power inputs to oscillators 1 and 2 due to external excitations.

When only one of the oscillators is externally excited and the other oscillator is excited only through the coupling element, we obtain

$$P_1 = \eta_1 \omega_1 \langle E_1 \rangle + \eta_{12} \omega_1 \{ \langle E_1 \rangle - \langle E_2 \rangle \} \tag{10.23}$$

$$0 = \eta_2 \omega_2 \langle E_2 \rangle + \eta_{21} \omega_2 \{ \langle E_2 \rangle - \langle E_1 \rangle \} \tag{10.24}$$

where it has been assumed that only oscillator 1 is subjected to external excitation $F_1(t)$ and $F_2(t) = 0$. From (10.24) the energy ratio is given by

$$\frac{\langle E_2 \rangle}{\langle E_1 \rangle} = \frac{\eta_{21}}{(\eta_2 + \eta_{21})} \tag{10.25}$$

Since all the power received by oscillator 2 will be dissipated by it, it is clear that

$$P_1 = \eta_1 \omega_1 \langle E_1 \rangle + \eta_2 \omega_2 \langle E_2 \rangle \tag{10.26}$$

The above result can also be obtained by substituting (10.25) in (10.23) and using the relation $\eta_{12}\omega_1 = \eta_{21}\omega_2$. From (10.25) we infer that the largest average energy of oscillator 2 will be at most equal to $\langle E_1 \rangle$, when the loss factor η_2 is very much less than the coupling loss factor $\eta_{21}(\eta_2 \ll \eta_{21})$. Thus equipartition of energy between the two oscillators occurs when the damping in the indirectly driven oscillator is very small and the coupling loss factor is large.

If in the energy balance relations (10.21) and (10.22) the internal loss factors η_1 and η_2 as well as the coupling loss factor η_{12} and η_{21} are known, from the average power inputs to the two oscillators, the average energies in the two oscillators can be obtained. Since the average energies of the oscillators are related to the oscillator mean square velocities by equation (10.17), the mean square velocities can be calculated and from them the other response quantities. This is the essence of the SEA method.

The assumption that $F_1(t)$ and $F_2(t)$ are white noise excitations in the derivation of the power flow relations is not overly restrictive. In estimating the time averaged quantities in (10.3) to (10.9), the psds of $F_1(t)$ and $F_2(t)$ being constants are taken out of the integrals and only the product of the appropriate frequency response functions is integrated between infinite limits by contour integration. If the interest is only in a finite frequency band, the infinite limits in the power flow calculations may be replaced by frequencies limiting the band. It is well known from the theory of vibration, that if damping is small, the frequency response functions are sharply peaked near the resonant frequencies, the greatest contribution to the integrals will be from frequency regions near the system resonances. Consequently little error will be entailed if the intervals of integration limits encompass system resonances. If the excitation psds are flat near the resonant frequencies, they can be taken outside the integrals. This is similar to the local white noise assumption frequently resorted to in random vibration analysis. Even if the excitation psds are not flat, the power flow relations are very good approximations provided the psds do not exhibit violent peaks near the system resonances.

10.3.4   Thermal Analogy

A simple thermal analogy can be had for the SEA power flow relations by considering the vibrational energy as analogous to thermal energy. Heat flows from a hotter source to a cooler sink at a rate proportional to the temperature difference, the constant of proportionality being a measure of the thermal conductivity. The flow of vibrational or acoustic energy in a structure or room behaves in the same way as the flow of heat. Provided that there are sufficient resonant or structural modes within a frequency band of interest, the average modal energy can be considered as equivalent to a measure of temperature. The number of modes per unit frequency within a frequency band (*modal density*) is analogous to the thermal capacity, the internal loss factors are analogous to radiation losses and the coupling loss factors which represent the strength of mechanical coupling between the subsystems are analogous to the thermal conductivity links between different elements in the thermal model.

10.4   Power Flow Relation between Two Sets of Coupled Oscillators

The power flow relations (10.8) and (10.14) between two coupled sdf oscillators cannot be directly extended to the general case of coupled groups of oscillators having arbitrarily strong coupling between them. Scharton and Lyon (1968) have shown that the power flow relations between two oscillators hold good in the special case of two sets of identical oscillators having identical mass and stiffness couplings. This result may be of interest in the application of SEA for periodic structural systems having identical elements.

Hodges and Woodhouse (1986) discuss the applicability of SEA assumption that the power flow is proportional to the difference in the average coupled modal energies in some detail and have derived power flow relations taking into account indirect coupling terms between blocked modes in the subsystems which are not directly coupled. But the application of such results is beset with practical problems, as much information is not available about blocked natural frequencies and the interaction forces between modes in a subsystem. The basic assumption of SEA that the power flow between coupled subsystems is proportional to the difference of average modal energies holds good for subsystems which are weakly coupled. Weak coupling means that the coupling forces are small compared to corresponding internal oscillator forces and implies that the power flow between a mode of one subsystem to a mode of another subsystem to which it is coupled can be evaluated without including the interaction of any other mode in the subsystems. One restriction however is that the generalized forces (modal) applied to the individual oscillators (modes) should be statistically independent. Thus it must be understood that SEA is a powerful engineering tool in the vibration analysis of complex coupled multimodal systems if applied appropriately and is very successful (i) when there is weak coupling between subsystems, (ii) when the external excitations are broadband and (iii) the modal densities of respective subsystems are high.

10.4.1 Modal Power Flow Relations

Consider two sets of subsystems (modes), one denoted by α and other denoted by β schematically shown in Fig. 10.2. The modes of each set are assumed not to be coupled to each other directly, but each mode of the α-set is assumed to be coupled to each mode of the β-set. The purpose now is to derive power flow relations

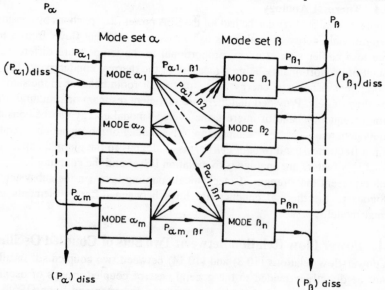

Fig. 10.2. Two coupled sets of modes.

between the modes of the α-set and the modes of the β-set under conditions, where each of the modes in the α and β sets may dissipate energy and may be supplied with energy from outside the system (external excitation). This exercise is very similar to the power flow relation derived between two oscillators.

Each mode of the subsystem α or β is considered as a sdf resonator and subjected to the generalized force corresponding to that mode. It is assumed that the psd of the generalized force is uniform over a finite band of frequency $\Delta\omega$ (local white noise assumption), within which there are $N_\alpha = n_\alpha\Delta\omega$ modes of the α set and $N_\beta = n_\beta\Delta\omega$ modes of the β set, where $n_\alpha(\omega)$ and $n_\beta(\omega)$ are the *modal densities* of subsystems α and β. In deriving the power flow relations the following further assumptions are made.

1. The natural frequency of each mode in the α and β sets lies within the frequency band $\Delta\omega$ with uniform probability. This means that each subsystem is drawn from similar populations having randomly distributed parameters. Apart from the randomness arising out of the external excitation, this assumption introduces additional randomness about the system parameters because of the uncertainties associated with them especially at higher frequencies.

2. There is equipartition of energy between all the modes of a subsystem in the frequency band $\Delta\omega$. This assumption requires that the mode groups should be properly selected for which this assumption is valid.

Considering mode i from the α set and mode j from the β set, the average power flow from the ith mode in the α set to the jth mode in the β set can be written analogous to (10.8) as

$$P_{\alpha_i\beta_j} = A_{\alpha_i\beta_j}\{\langle E'_{\alpha_i}\rangle - \langle E'_{\beta_j}\rangle\} \tag{10.27}$$

where $\langle E'_{\alpha_i}\rangle$ and $\langle E'_{\beta_j}\rangle$ are the average blocked energies of the ith mode of the α set and the jth mode of the β set respectively, given by

$$\langle E'_{\alpha_i}\rangle = \pi\Phi_{\alpha_i}(\omega_c)/(M_{\alpha_i}\eta_{\alpha_i}\omega_{\alpha_i}) \tag{10.28}$$

$$\langle E'_{\beta_j}\rangle = \pi\Phi_{\beta_j}(\omega_c)/(M_{\beta_j}\eta_{\beta_j}\omega_{\beta_j}) \tag{10.29}$$

where Φ_{α_i}, Φ_{β_j} are the two sided psds of the generalized forces, $M_{\alpha_i}, M_{\beta_j}$ are generalized masses $\eta_{\alpha_i}, \eta_{\beta_j}$ are the internal loss factors, $\omega_{\alpha_i}, \omega_{\beta_j}$ are uncoupled natural frequencies corresponding to the i, j modes and ω_c is a central frequency within the frequency band $\Delta\omega$.

10.4.2. Total power flow relations

The total average power flow from mode set α to mode set β can be expressed as a sum over all the modes in the frequency band $\Delta\omega$ given by

$$P_{\alpha\beta} = \sum_{i=1}^{N_\alpha}\sum_{j=1}^{N_\beta} A_{\alpha_i\beta_j}\{\langle E'_{\alpha_i}\rangle - \langle E'_{\beta_j}\rangle\} \tag{10.30}$$

The quantity $A_{\alpha_i\beta_j}$ depends on the frequency. Equation (10.30) can be also averaged over the frequency band $\Delta\omega$ to give

$$\langle P_{\alpha\beta}\rangle_{\Delta\omega} = \langle A_{\alpha\beta}\rangle_{\Delta\omega} \sum_{i=1}^{N_\alpha}\sum_{j=1}^{N_\beta} \{\langle E'_{\alpha_i}\rangle - \langle E'_{\beta_j}\rangle\} \tag{10.31}$$

where $\langle . \rangle_{\Delta\omega}$ indicates averaging over $\Delta\omega$. The time averaged power flow from mode set α to mode set β also averaged over the frequency band $\Delta\omega$ can also be expressed analogous to (10.14) in the form

$$\langle P_{\alpha\beta}\rangle_{\Delta\omega} = \langle B_{\alpha\beta}\rangle_{\Delta\omega} \sum_{i=1}^{N_\alpha} \sum_{j=1}^{N_\beta} \{\langle E_{\alpha_i}\rangle - \langle E_{\beta_j}\rangle\} \tag{10.32}$$

where E_{α_i} and E_{β_j} are the actual energies corresponding to the i, j modes and $\langle B_{\alpha\beta}\rangle_{\Delta\omega}$ is a frequency averaged proportionality constant which depends on the system parameters.

Since equipartition of energy has been assumed within each subsystem, (10.32) can be expressed as

$$\langle P_{\alpha\beta}\rangle_{\Delta\omega} = N_\alpha N_\beta \langle B_{\alpha\beta}\rangle_{\Delta\omega} \{\langle E_\alpha\rangle - \langle E_\beta\rangle\} \tag{10.33}$$

where $\langle E_\alpha\rangle$ is the average energy corresponding to each mode set α and $\langle E_\beta\rangle$ of the set β. If $E_{\alpha, \text{tot}}$ and $E_{\beta, \text{tot}}$ are the total energies corresponding to the α and β sets respectively, (10.33) can be alternatively written as

$$\langle P_{\alpha\beta}\rangle_{\Delta\omega} = N_\alpha N_\beta \langle B_{\alpha\beta}\rangle_{\Delta\omega} \left\{ \frac{\langle E_{\alpha, \text{tot}}\rangle}{N_\alpha} - \frac{\langle E_{\beta, \text{tot}}\rangle}{N_\beta} \right\} \tag{10.34}$$

$$= \omega_c \eta_{\alpha\beta} \left\{ \langle E_{\alpha, \text{tot}}\rangle - \frac{N_\alpha}{N_\beta} \langle E_{\beta, \text{tot}}\rangle \right\} \tag{10.35}$$

where $\eta_{\alpha\beta}$ is defined as $\eta_{\alpha\beta} = \langle B_{\alpha\beta}\rangle_{\Delta\omega} N_\beta / \omega_c$.

If we define $\eta_{\beta\alpha} = \langle B_{\alpha\beta}\rangle_{\Delta\omega} N_\alpha / \omega_c = N_\alpha \eta_{\alpha\beta} / N_\beta$ we obtain from (10.34)

$$\langle P_{\alpha\beta}\rangle_{\Delta\omega} = \omega_c \{\eta_{\alpha\beta}\langle E_{\alpha, \text{tot}}\rangle - \eta_{\beta\alpha}\langle E_{\beta, \text{tot}}\rangle\} \tag{10.36}$$

Equation (10.36) is analogous to (10.20), the power flow relation for two oscillators. $\eta_{\alpha\beta}$ and $\eta_{\beta\alpha}$ are called as the *coupling loss factors* in SEA terminology. Using the relations $N_\alpha = n_\alpha \Delta\omega$ and $N_\beta = n_\beta \Delta\omega$ in (10.34) we get

$$\langle P_{\alpha\beta}\rangle_{\Delta\omega} = \frac{N_\alpha N_\beta}{\Delta\omega} \langle B_{\alpha\beta}\rangle_{\Delta\omega} \left\{ \frac{\langle E_{\alpha, \text{tot}}\rangle}{n_\alpha} - \frac{\langle E_{\beta, \text{tot}}\rangle}{n_\beta} \right\} \tag{10.37}$$

$$= \omega_c \eta_{\alpha\beta} n_\alpha \left\{ \frac{\langle E_{\alpha, \text{tot}}\rangle}{n_\alpha} - \frac{\langle E_{\beta, \text{tot}}\rangle}{n_\beta} \right\} \tag{10.38}$$

where n_α and n_β are the modal densities of the mode sets α and β. It can be also shown that

$$n_\alpha \eta_{\alpha\beta} = n_\beta \eta_{\beta\alpha} \tag{10.39}$$

The above reciprocity relation will be very useful in power flow calculations.

10.4.3 Power balance equations

The power balance equations analogous to (10.21) and (10.22) can be written as

$$P_\alpha = (P_\alpha)_{\text{diss}} + P_{\alpha\beta} = \omega_c[\eta_\alpha\langle E_{\alpha, \text{tot}}\rangle + \eta_{\alpha\beta}\langle E_{\alpha, \text{tot}}\rangle - \eta_{\beta\alpha}\langle E_{\beta, \text{tot}}\rangle] \tag{10.40}$$

$$P_\beta = (P_\beta)_{\text{diss}} + P_{\beta\alpha} = \omega_c[\eta_\beta\langle E_{\beta, \text{tot}}\rangle + \eta_{\beta\alpha}\langle E_{\beta, \text{tot}}\rangle - \eta_{\alpha\beta}\langle E_{\alpha, \text{tot}}\rangle] \tag{10.41}$$

where P_α and P_β are the average input power to subsystems α and β averaged over time as well as the frequency band of interest $\Delta\omega$ and η_α and η_β are average internal loss factors of subsystems α and β again averaged over frequency $\Delta\omega$. Henceforth, the averaging operator $\langle . \rangle_{\Delta\omega}$ is suppressed for notational convenience.

If only the subsystem α is externally excited, while the set β is only indirectly excited through the coupling elements (Fig. 10.3), we have

$$P_\alpha = \omega_c[\eta_\alpha\langle E_{\alpha,\,\text{tot}}\rangle + \eta_{\alpha\beta}\langle E_{\alpha,\,\text{tot}}\rangle - \eta_{\beta\alpha}\langle E_{\beta,\,\text{tot}}\rangle] \tag{10.42}$$

$$0 = \omega_c[\eta_\beta\langle E_{\beta,\,\text{tot}}\rangle + \eta_{\beta\alpha}\langle E_{\beta,\,\text{tot}}\rangle - \eta_{\alpha\beta}\langle E_{\alpha,\,\text{tot}}\rangle] \tag{10.43}$$

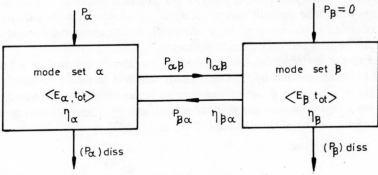

Fig. 10.3. Power flow between mode sets.

From (10.43) the energy ratio between the α and β sets is obtained as

$$\frac{\langle E_{\beta,\,\text{tot}}\rangle}{\langle E_{\alpha,\,\text{tot}}\rangle} = \frac{\eta_{\alpha\beta}}{\eta_\beta + \eta_{\beta\alpha}} = \frac{n_\beta}{n_\alpha}\frac{\eta_{\beta\alpha}}{\eta_\beta + \eta_{\beta\alpha}} \tag{10.44}$$

Equation (10.44) is the multi-modal equivalent of (10.25) which illustrates how energy ratios between coupled group of oscillators can be obtained from the internal and coupling loss factors. The energy flow relationships (10.42) and (10.43) illustrate the basic principles of SEA. With a knowledge of the modal densities, internal loss factors of two different subsystems and the coupling loss factors between the subsystems, the energy flow ratios can be estimated. Alternatively, from total energies of vibration and modal densities, information about the internal loss factors and coupling loss factors can be obtained.

Equation (10.44) can also be alternatively expressed as

$$\frac{\langle E_{\beta,\,\text{tot}}\rangle}{\langle E_{\alpha,\,\text{tot}}\rangle} = \frac{n_\beta\,\eta_{\alpha\beta}}{n_\beta\,\eta_\beta + n_\alpha\,\eta_{\alpha\beta}} \tag{10.45}$$

from which we obtain

$$\frac{\eta_{\alpha\beta}}{\eta_\beta} = \frac{n_\beta\langle E_{\beta,\,\text{tot}}\rangle}{n_\beta\langle E_{\alpha,\,\text{tot}}\rangle - n_\alpha\langle E_{\beta,\,\text{tot}}\rangle} \tag{10.46}$$

Equations (10.44)–(10.46) are valid only when the subsystem α is directly excited while the set β is indirectly excited. On the other hand if only subsystem β is excited while the set α is indirectly excited an analogous relation

$$\frac{\eta_{\alpha\beta}}{\eta_\alpha} = \frac{n_\beta\langle E_{\alpha,\,\text{tot}}\rangle}{n_\alpha\langle E_{\beta,\,\text{tot}}\rangle - n_\beta\langle E_{\alpha,\,\text{tot}}\rangle} \tag{10.47}$$

is obtained.

Using (10.46) and (10.47) the coupling loss factors of the subsystems can be experimentally determined if the modal densities and the internal loss factors of the individual subsystems are known. Considering two continuous structures coupled together by conservative elements, each structure may be taken to be constituting the mode sets α and β. Exciting the structure α at a number of points to satisfy the condition of statistical independence, the time and space averaged energies of both the structures can be obtained by measuring the response on both the structures at a number of points. Equation (10.45) can then be used to obtain an estimate of the coupling loss factor $\eta_{\alpha\beta}$. If the structure β is excited instead, $\eta_{\alpha\beta}$ can be obtained from (10.47). Modal densities for many of the structural elements and acoustical enclosures can be theoretically estimated, while for complex structures they have to be obtained experimentally. Very often information about the internal loss factors can be obtained only from experiments.

Sometimes, knowledge about the internal loss factors of subsystems α and β are difficult to obtain individually. The total steady state loss factor for the subsystem α coupled to subsystem β can be experimentally determined by measuring the power input into subsystem α and its vibrational energy. The relation to be used in such a case can be obtained from (10.42) and (10.43) as

$$(\eta_{\alpha,\,\text{tot}})_{\text{steady}} = \frac{P_\alpha}{\omega_c \langle E_{\alpha,\,\text{tot}} \rangle} = \eta_\alpha + \frac{\dfrac{n_\beta}{n_\alpha}\eta_\beta\,\eta_{\beta\alpha}}{\eta_\beta + \eta_{\beta\alpha}} \tag{10.48}$$

10.4.4 Power Balance Relations for Multiple Sets of Oscillators

The power balance equations (10.40) and (10.41) can be extended to multiple sets of oscillators with almost similar assumptions. For the general case of M groups of oscillators there are M simultaneous energy balance equations in the steady state which can be expressed in the matrix form as

$$\omega_c \begin{bmatrix} (\eta_1 + \sum\limits_{j=1}^{M} \eta_{1j})n_1 & -\eta_{12}n_1 & \cdots\cdots & -\eta_{1M}n_1 \\[2ex] -\eta_{12}n_2 & (\eta_2 + \sum\limits_{\substack{j=1 \\ j\neq 2}}^{M} \eta_{2j})n_2 & \cdots\cdots & -\eta_{2M}n_2 \\[2ex] \cdots\cdots & \cdots\cdots & \cdots\cdots & \cdots\cdots \\[1ex] -\eta_{M1}n_M & -\eta_{M2}n_M & \cdots\cdots & (\eta_M + \sum\limits_{j=1}^{M} \eta_{Mj})n_M \end{bmatrix}$$

$$\times \begin{bmatrix} \langle E_{1,\,\text{tot}} \rangle/n_1 \\[2ex] \langle E_{2,\,\text{tot}} \rangle/n_2 \\[1ex] \cdots\cdots \\[2ex] \langle E_{M1,\,\text{tot}} \rangle/n_M \end{bmatrix} = \begin{bmatrix} P_1 \\[2ex] P_2 \\[1ex] \cdots \\[2ex] P_M \end{bmatrix} \tag{10.49}$$

Equation (10.49) represents a set of matrix linear equations and provides a basis for a systematic analysis of complex groups of oscillators by the SEA method. The modal densities, internal loss factors of each subsystem and the coupling loss factors between the subsystems have to known for such an application. Therefore, defining the subsystems in a complex coupled systems of structures and other energy systems is of paramount importance in the proper application of SEA. Generally, the subsystems are so chosen that weak coupling exists between them in which case the coupling loss factors will be very small compared to internal loss factors.

10.5   Modal Density Expressions for Common Structural Elements

The modal density is an important parameter in the application of SEA procedures to estimate the vibration response of complex structural and acoustic systems and their combinations. It occurs in the power flow relations between coupled multi-modal systems. For many of the common structural elements like bars, beams, plates and regular acoustic volumes etc., the modal density expressions are available in the literature in closed form depending on the nature of vibration of modes like longitudinal, flexural or torsional. (Heckl, 1962; Bolotin, 1963; Hart and Shaw, 1971; Cremer, Heckl and Ungar, 1973; Lyon, 1975; Fahy, 1982; Ver and Holmer, 1971; Norton, 1989). Clarkson and Ranky (1983) and Ferguson and Clarkson (1986) have derived analytical expressions for the modal density of honeycomb plates and shells. In general, the modal density is a function of the frequency. Since in practical applications the frequency is expressed in Hz and since SEA procedures are used along with empirical results, it is customary to express frequency f in Hz rather than in terms of ω in rad/sec. The modal density in terms of the radian measure is related to the modal density in terms of Hz in a simple manner as $n(\omega) = n(f)/2\pi$. In this section we present modal density expressions for some of the common acoustic and structural elements. The natural frequency expression in terms of the mode count (modal number) is known for many structural elements. Expressing the mode number in terms of the frequency and differentiating the resulting expressions with respect to frequency the modal density expression is obtained.

10.5.1   Uniform Bars in Longitudinal Vibration

For example, the pth natural frequency expression for a uniform bar in longitudinal vibration with both of its ends fixed is given by

$$\omega_p = \frac{p\pi}{L}\left(\frac{EA}{\mu}\right)^{1/2}, \quad p = 1, 2 \tag{10.50}$$

where E is the Young's modulus, A is the area of cross-section, μ the mass per unit length and L the length of the rod. Expressing the modal count in terms of frequency in the form

$$p = \frac{L\omega_p}{\pi}\left(\frac{\mu}{EA}\right)^{1/2} \tag{10.51}$$

and differentiating with respect to ω_p we get the modal density expression as

$$n(\omega) = \frac{dp}{d\omega_p} = \frac{L}{\pi}\left(\frac{\mu}{EA}\right)^{1/2} = \frac{L}{\pi C_L} \tag{10.52}$$

where C_L is the longitudinal wave speed in the material of the bar given by $C_L = (E/\rho)^{1/2}$ with ρ being the density of the material. Expressing (10.52) in terms of f the frequency in Hz, we obtain

$$n(f) = 2\pi n(\omega) = 2L/C_L \qquad (10.53)$$

Although (10.53) is strictly valid for the bar to be fixed at both ends, in the higher frequency range as remarked earlier the effect of boundary conditions loses significance and the expression can be taken as the modal density of bars in longitudinal vibration. In this case the modal density is independent of frequency.

10.5.2   Uniform Bars in Flexure

For a simply supported uniform beam of length L the natural frequency expression is

$$\omega_p = \frac{p^2 \pi^2}{L^2} \left(\frac{EI}{\mu} \right)^{1/2} \qquad p = 1, 2 \ldots \qquad (10.54)$$

from which we obtain

$$p = \frac{L}{\pi} (\omega_p)^{1/2} \left(\frac{\mu}{EI} \right)^{1/4} \qquad (10.55)$$

where EI is the flexural rigidity of the beam and μ, its mass per unit length.

Differentiating (10.55) with respect to ω_p and dropping the subscript p in ω_p we obtain

$$n(\omega) = \frac{dp}{d\omega_p} = \frac{L}{2\pi} (\omega)^{-1/2} \left(\frac{\mu}{EI} \right)^{1/4} \qquad (10.56)$$

Since $n(f) = 2\pi n(\omega)$ we have

$$n(f) = \frac{L}{\sqrt{2\pi f}} \left(\frac{\rho A}{EI} \right)^{1/4} \qquad (10.57)$$

where ρ is the density of the material, and A the area of cross-section of the beam. Even though (10.57) is strictly valid for simply supported beams it can be taken as the general modal density expression for beams in flexure. In this case the modal density is a function of frequency.

10.5.3   Flat Plates in Flexural Vibration

The modal density expression for flat plates in flexural vibration is given by

$$n(f) = \frac{\sqrt{12} S}{2 C_L h} \qquad (10.58)$$

where $C_L = \{ E/\rho(1 - v^2) \}^{1/2}$ is the flexural wave speed in the plate, S is the surface area, h is the thickness of the plate and v is the Poisson's ratio and ρ is the density of the plate material.

The modal density for flat plates can be derived considering a rectangular plate with simply supported edges and from simple geometrical arguments. The natural frequency of a rectangular plate of length a and breadth b simply supported on all its edges is given by

$$\omega_{pr} = \sqrt{\frac{D}{\mu}} \, \pi^2 \left(\frac{p^2}{a^2} + \frac{r^2}{b^2} \right) \tag{10.59}$$

where $D = Eh^3/12(1 - v^2)$ is the flexural rigidity of the plate, $\mu = \rho h$ is the mass per unit area of the plate, ω_{pr} is the (p, r)th natural frequency corresponding to the modal numbers p and r in the two orthogonal directions along the length and. breadth of the plate. Considering a rectangular modal lattice consisting of the meshes of width $\pi(D/\mu)^{1/4}/a$ and $\pi(D/\mu)^{1/4}/b$ as shown in Fig. 10.4 the square of

the distance from the lattice point (p, r) to the origin is $(D/\mu)^{1/2}\pi^2 \left(\dfrac{p^2}{a^2} + \dfrac{r^2}{b^2} \right)$

which is the same as the (p, r)th natural frequency of the plate. The natural frequencies

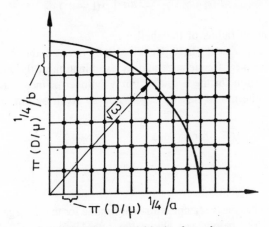

Fig. 10.4.   Modal lattice for a plate.

that fall below a given bounding value ω must therefore lie within the quarter circle of radius $\sqrt{\omega}$. Since each lattice point covers an area of magnitude $\pi^2 \sqrt{D/\mu}/ab$, leaving the points on the edge, the total number of natural frequencies below ω is obtained by dividing the area of the quarter circle by the area corresponding to a single natural frequency, that is

$$N = \frac{\pi}{4}\omega \Big/ \{\pi^2 (D/\mu)^{1/2}/ab\} = \frac{S\omega}{4\pi} \sqrt{\frac{\mu}{D}} \tag{10.60}$$

The modal density $n(\omega)$ is given by

$$n(\omega) = \frac{dN}{d\omega} = \frac{S}{4\pi} \sqrt{\frac{\mu}{D}} \tag{10.61}$$

which in terms of the cyclic frequency f is

$$n(f) = 2\pi n(\omega) = \frac{S}{2} \sqrt{\frac{\mu}{D}} \tag{10.62}$$

which is the same expression as in (10.58).

10.5.4   Thin Walled Cylindrical Shells

Modal densities of thin walled shells cannot be obtained in a simple way as for the previous cases. Hart and Shah (1971) discuss a number of theories which provide average values of modal densities. The ring frequency f_r of a cylindrical shell is an important parameter at which the cylinder vibrates uniformly in the breathing mode. Above the ring frequency, the structural wavelengths are such that a cylinder tends to behave like a flat plate. Below the ring frequency, the modal density varies because of grouping of structural modes of different circumferential mode orders. Many of the formulae discussed in Hart and Shah (1971) do not account for this grouping effect. It is obvious that the modal density expressions have to be different for frequencies below and above the ring frequency of the shell. The ring frequency of a cylindrical shell in Hz is given by

$$f_r = C_L/2\pi a = \frac{1}{2\pi a} \left\{ \frac{E}{\rho(1-v^2)} \right\}^{1/2} \tag{10.63}$$

where a is the mean radius of the shell.

For circular cylindrical shell, Heckl (1962) derived the modal density by determining the natural frequencies and by examination of the zeros of the point impedance expression for the cylinder. For a cylinder of mean radius a, length L and thickness h, the modal density expression as given by him are

$$n(f) = \frac{\sqrt{3}\,aL}{h} \sqrt{\rho/E} \quad \text{for } f > f_r$$

$$= \frac{\sqrt{3}\,aL}{h} \sqrt{\rho/E} \left\{ \frac{\pi}{2} + \sin^{-1} \left(\frac{fa}{\pi} \sqrt{\rho/E} - 1 \right) \right\} \quad \text{for } f < f_r \tag{10.64}$$

Szechenyi (1971) has derived approximate formulae for the average modal densities of cylindrical shells below and above the ring frequency. Clarkson and Pope (1981) have given semi-empirical relations for the modal density based on the approximations of Szechenyi. The modal density expressions are

$$n(f) = \frac{5S}{\pi C_L h} (f/f_r)^{1/2} \quad \text{for } f \le 0.48 f_r$$

$$= \frac{7.2S}{\pi C_L h} (f/f_r)^{1/2} \quad \text{for } 0.48 f_r < f \le 0.83 f_r$$

$$= \frac{2S}{\pi C_L h} \left[2 + \frac{0.596}{F - 1/F} \left\{ F \cos\left(\frac{1.745 f_r^2}{F^2 f^2} \right) - \frac{1}{F} \cos\left(\frac{1.745 f_r^2 F^2}{f^2} \right) \right\} \right]$$

$$\text{for } f > 0.83 f_r \tag{10.65}$$

where S is the surface area of the cylinder, and F is a bandwidth factor being the square root of the ratio of upper to lower cut off frequencies of the bandwidth. For octave bands $F = 1.414$ and for one third octave bands $F = 1.122$. Other formulae for the modal density of shells are given in Ferguson and Clarkson (1986).

Even these formulae do not account for the grouping of circumferential modes below the ring frequency. Keswick and Norton (1987) using the strain displacement relations of different shell theories have developed a computer prediction model for the determination of the modal densities taking into account the clustering of circumferential modes.

10.5.5 Honeycomb Plates

Honeycomb sandwich plates and shells are extensively used in aerospace applications and satellite structures because of their light weight and high stiffness. The broad band random excitation arising from the rocket motors is one of the major design considerations for satellite structures. The SEA procedure is used to advantage in estimating the response of these structures. The modal density of such honeycomb sandwich structures is therefore of much significance. Theoretical expressions for the modal density of flat honeycomb plates and honeycomb shells have been derived by Wilkinson (1968) and Erickson (1969) and Clarkson and Ranky (1983). Using the equation of motion for a honeycomb sandwich plate element as derived by Mead (1972), Clarkson and Ranky (1983) have derived an expression for the modal density of honeycomb sandwich plate, which is given by

$$n(f) = \frac{\pi abm}{g\beta} \, f \left[1 + \frac{\{m\omega^2 + 2g^2\beta(1 - v^2)\}}{\{m^2\omega^4 + 4m\omega^2 g^2\beta(1 - m^2)\}^{1/2}} \right] \qquad (10.66)$$

where a and b are panel dimensions, m is the total mass per unit area, f is the frequency in Hz, ω is the radian frequency, g is the core stiffness parameter. The core stiffness parameter g is defined as

$$g = \frac{G_x G_y}{h_2} \left(\frac{1}{E_1 h_1} + \frac{1}{E_3 h_3} \right) \qquad (10.67)$$

where h_1 and h_3 are the two face plate thicknesses, h_2 is the core thickness G_x and G_y are the shear moduli of the core material in the x and y directions respectively and E_1 and E_3 are moduli of elasticity of the face plate. The face plate longitudinal stiffness parameter β is given by

$$\beta = d^2 \frac{E_1 h_1 E_3 h_3}{E_1 h_1 + E_3 h_3} \qquad (10.68)$$

where $d = h_2 + (h_1 + h_3)/2$.

Formulae for the modal density of honeycomb cylindrical doubly curved and paraboloidal shells are given by Wilkinson (1968) and later by Erickson (1969) who included the effects of orthotropy, shear and rotary inertia in the derivation of the modal density. A summary of the various formulae for the modal density of honeycomb shells is given in Ferguson and Clarkson (1986).

10.5.6 Acoustic Volumes

The modal density expressions for acoustic volumes depending on whether they are single dimensional (long slender ducts), two dimensional (shallow cavities) or three dimensional enclosures are given by Lyon (1975). They are:

$$n(f) = 2L/c \quad \text{(one dimensional)} \qquad (10.69)$$

where L is the length of the duct and c is the speed of sound.

$$n(f) = \frac{\pi f A}{c^2} + \frac{P}{c} \quad \text{(two dimensional)} \qquad (10.70)$$

where A is the total surface area of the cavity and P is its perimeter.

$$n(f) = \frac{4\pi f^2 V}{c^3} + \frac{\pi f A}{c^2} + \frac{P}{c} \qquad \text{(three dimensional)} \quad (10.71)$$

where V is the volume of the enclosure, A is the total surface area and P is the total length of the edges comprising the enclosure.

10.6   Experimental Determination of Modal Density

If the subsystems modelled in the SEA method donot fall into any of the simple basic structural element type discussed in the previous section, it will be extremely difficult to establish the modal densities of such systems analytically. In such cases, the modal densities have to be determined experimentally.

The simplest and direct way of estimating the modal density experimentally is by means of the modal count technique. In this technique, the structure is excited sinusoidally at a point. The response at another point is measured and the frequency of the sinusoid is gradually varied. The number of peak responses within a frequency band is counted giving the modal count. Alternatively, the structure is excited at a point by an impact hammer and the response is obtained. The number of peaks in the frequency spectrum within a frequency band gives the modal count. The method is fraught with errors, because of modal overlap and the possibility of missing a number of natural frequencies depending on the locations of excitation and measurement. The errors can somewhat be reduced if a number of measurements are taken by changing the point of excitation and the point of measurement of the response. Even then it will be difficult to identify closely spaced natural frequencies.

Another method of experimentally determining the modal density of structures makes use of the theoretical relationship derived by Cremer, Heckl and Ungar (1973) between the frequency averaged value of the real part of the point mobility of a uniformly distributed multi-modal system and its modal density. Clarkson and Pope (1981, 1983), Brown (1984) and Keswick and Norton (1987) have used this relation in the experimental determination of the modal density of structures. They have shown that the modal density determined by this method is more reliable and accurate.

10.6.1   Point Mobility-modal Density Relation

Cremer, Heckl and Ungar (1973) have considered the power input to a flat plate due to a point load F acting at a location (x_0, y_0) and having a frequency response extending over a frequency band $\Delta\omega$. The power input due to the plate due to a distributed external harmonic pressure loading $p(x, y)$ can be expressed as

$$P = \frac{1}{2} \operatorname{Re} \left\{ \int_S p(x, y) \, v^*(x, y) \, dx \, dy \right\} \qquad (10.72)$$

where $v(x, y)$ is the velocity of the plate, the asterisk representing the complex conjugate and S is the total surface area of the plate. Re $(\cdot)$ denotes the real part. Upon using a normal mode expansion for $v(x, y)$ in (10.72) we obtain

$$P = \frac{1}{2} \operatorname{Re} \left\{ \sum_n v_n^* \int_S p(x, y) \, \varphi_n(x, y) \, dx \, dy \right\} \qquad (10.73)$$

where φ_n is the nth normal mode and v_n is the modal component of velocity. The velocity component v_n can be expressed as

$$v_n = \frac{i\omega \int_S p(x, y)\, \varphi_n(x, y)\, dx\, dy}{[\omega_n^2\,(1 + i\eta) - \omega^2]M_n} \tag{10.74}$$

where M_n is the generalized mass in the nth mode, ω_n in the nth natural frequency and η is the loss factor. ω is the frequency of the assumed harmonic variation in time of the external pressure loading $p(x, y)$, and i is the unit imaginary number.

Now, for a point force F acting at location (x_0, y_0), $p(x, y) = F\delta(x - x_0)\delta(y - y_0)$. The integrals in (10.73) and (10.74) reduce to $F\varphi_n(x_0, y_0)$, and the power input relation (10.73) reduces to

$$P = \frac{F^2\eta\omega}{2} \sum_n \frac{\omega_n^2\varphi_n^2\,(x_0, y_0)}{[(\omega_n^2 - \omega^2)^2 + \eta^2\omega_n^4]M_n} \tag{10.75}$$

In practice, exciting forces usually do not have a single frequency, but have a frequency content extending over a more or less broad frequency band. Similarly, they do not act at a single point in the structure, but distributed over the area of the plate. For weakly damped systems subject to excitation in a frequency band $\Delta\omega = \omega_2 - \omega_1$, which encompasses a number of resonance frequencies, the expression averaged over the frequency band $\Delta\omega$, as well as over the area of the plate gives

$$P = \frac{|F_\Delta|^2}{2} \frac{\pi}{2S\mu} \frac{\Delta N}{\Delta\omega} \tag{10.76}$$

where $|F_\Delta|^2$ is the mean square value of the force averaged over the frequency band $\Delta\omega$, S is the surface area, μ is the mass per unit area and $\Delta N/\Delta\omega$ is the modal density of the plate. Equation (10.76) illustrates that the natural modes of vibration can be considered as a set of independent simple single degree of freedom systems the basic assumption of SEA.

Equation (10.76) can also be expressed in terms of the mean square velocity of the plate as

$$P = \frac{1}{2} S\mu\,\eta\,\omega_c v_\Delta^2 \tag{10.77}$$

where ω_c is the center frequency of the band and v_Δ^2 is the mean square velocity of the plate averaged over the frequency band $\Delta\omega$. The power input to the plate can also be expressed in terms of the frequency averaged value of the point input admittance or mobility of the plate as

$$P = \frac{|F_\Delta|^2}{2}\, \mathrm{Re}\,\{Y(\omega)\} = \frac{|F_\Delta|^2}{2}\, \mathrm{Re}\left\{\frac{1}{Z(\omega)}\right\} \tag{10.78}$$

where $Y(\omega)$ is the average point input mobility and $Z(\omega)$ is the average point input impedance of the structure. Comparing (10.76) and (10.78) we get

$$\frac{\Delta N}{\Delta\omega} = n(\omega) = \frac{2S\mu}{\pi}\, \mathrm{Re}\,\{Y(\omega)\} \tag{10.79}$$

Equation (10.79) is of particular significance, because it is a relation between the admittances of finite and infinite systems. The response of a very large plate, beam etc., subjected to a point loading is least affected by the boundaries, because of the distance of the point of excitation from the boundaries. Therefore, the power input to a large finite structure is very nearly equal to that for a similar infinite structure, at least in the frequency average. Thus the driving point mobility $Y(\omega)$ must be identical to that of a corresponding infinite system. Therefore the modal density $n(\omega)$ can be obtained if the average point mobility $Y(\omega)$ is known or vice versa. It follows that generally the modal density is proportional to the area (or to the length or volume, for one dimensional or three dimensional structures, respectively) and independent of the boundary conditions, which may be seen to be true by considering the theoretical expressions for modal densities of structures and acoustic volumes presented in the previous section.

Thus the real part of point mobility $\mathrm{Re}\{Y(\omega)\}$ when averaged over the frequency band of interest and averaged over the domain of the structure is related to the modal density of the structure. The band averaged modal density within the frequency band $\Delta\omega$ is given by

$$n(\omega) = \frac{1}{\Delta\omega} \int_{\omega_1}^{\omega_2} \frac{2S\mu}{\pi} \, \mathrm{Re}\,\{Y(\omega)\}\, d\omega \qquad (10.80)$$

In terms of the cyclic frequency in Hz (10.80) can be written as

$$n(f_c) = \frac{1}{f_2 - f_1} \int_{f_1}^{f_2} 2S\mu \, \mathrm{Re}\,\{Y(f)\}\, df \qquad (10.81)$$

where f_c is the centre frequency of the band. In principle (10.81) is only applicable to structures with a uniform mass distribution. But experiments performed by Clarkson and Pope (1981, 1983), Brown (1984), Brown and Norton (1985) and Keswick and Norton (1987) have shown that the applicability of the expression even for structures with varying mass distribution provided the product $S\mu$ is replaced by the total mass of the structure.

10.6.2   Two-channel and Three-channel Measurement Techniques

As a first step in the experimental determination of the modal density by the use of point mobility-modal density relationship, one should experimentally obtain a reliable and accurate estimate of $\mathrm{Re}\{Y(f)\}$. Mobility is a complex frequency response function of an output velocity and input force. It is defined by

$$Y(f) = \frac{V(f)}{F(f)} \qquad (10.82)$$

where $V(f)$ and $F(f)$ are the Fourier transforms of the velocity and input force respectively. Point mobility is the ratio of the velocity to force at a specific point on the structure.

For an ideal system with a simple input-output model with no external noise or feedback as shown in Fig. 10.5, the point mobility is given by

Fig. 10.5.   Simple system without feed-back for point mobility.

$$Y(f) = \frac{G_{fv}(f)}{G_{ff}(f)} \tag{10.83}$$

where $G_{fv}(f)$ and $G_{ff}(f)$ are respectively the one sided cross spectral density of force and velocity and the auto spectral density of the force. In the experimental determination of $Y(f)$ the structure is subjected to broad band random excitation and the estimated power spectrum $G_{ff}(f)$ and the cross-spectrum $G_{fv}(f)$ are obtained yielding an estimate of the point mobility according to (10.83).

Alternatively transient methods of excitation have been used notably the rapid swept sine excitation (White 1971). In this case a transient rapid swept sine wave covering the frequency range of interest and of short duration is used. The transfer function can be determined and the point mobility estimated from relation (10.82). This method has been used by Clarkson and Pope (1981, 1983) in the determination of the modal density of a flat plate, a plain cylinder and a honeycomb panel for which analytical results are available. Modal density measurements have also been made for a corrugated plate, stiffened cylinder, corrugated cylinder and a doubly curved honeycomb panel for which analytical results are not available.

The experimentally obtained values of the modal densities are compared with the theoretical values and the agreement between the two has been reported to be reasonably good.

However, the expressions (10.82) or (10.83) obtained for the ideal system of Fig. 10.5 neglect the frequency response of the power amplifier and exciter system and the feedback due to structure-exciter interaction which may lead to bias errors that increase in the neighbourhood of a resonant frequency as $G_{ff}(f) \rightarrow 0$. This may result in negative values for the real part of the mobility representing a significant measurement error since the real part of the mobility must always be positive for a single point excitation of a structure. This problem was sought to be eliminated by the addition of a small damping tape around the edges of the plate. The characteristics of the impedance curve changed because of the increased damping as it was impossible to identify modes except at lower frequencies.

To overcome these errors, Brown (1984) has developed an improved three channel measurement technique for the point mobility which minimizes the feedback due to exciter structure interaction. This method is based on the one suggested by the work of Goyder (1981). A schematic of the method is shown in Fig. 10.6, which incorporates the signal $s(t)$ driving the power amplifier and the vibration exciter system in addition to the force velocity signals. In the figure, $H(f)$ represents the frequency response function of the power amplifier and the exciter system and

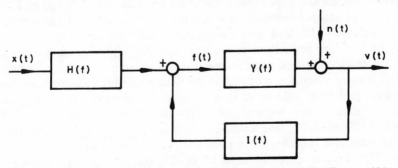

Fig. 10.6. Three channel frequency response function for point mobility (Brown, 1984).

$G(f)$ is the feedback frequency response function describing the electrodynamic shaker—structure interaction and $z(t)$ is some external noise at the output stage. The original signal $s(t)$ used to drive the power amplifier is usually a broad band random excitation and $f(t)$ and $v(t)$ are the measured force and velocity signals respectively. The point mobility is now estimated by

$$Y(f) = \frac{G_{sv}(f)}{G_{sf}(f)} \qquad (10.84)$$

where $G_{sv}(f)$ and $G_{sf}(f)$ are the cross spectral densities of the original test signal and the measured velocity signal respectively. This method gives well behaved estimates of real and imaginary parts of $Y(f)$ with no bias errors and leads to significantly improved results. The use of random excitation has added advantages in the calculation of other parameters like loss factors used in SEA. Once the point mobility is obtained, the modal density can be obtained from (10.81). A schematic of a typical experimental set up for the measurement of modal density through the point mobility is shown in Fig. 10.7.

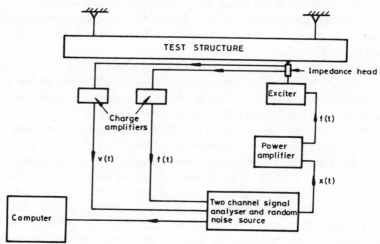

Fig. 10.7. Schematic of experimental setup for point mobility measurement (Norton, 1989.

Brown (1984) carried out a series of tests on flat plates and plain cylinders whose modal densities were known analytically. Both the two channel method used by Clarkson and Pope (1981), where the point mobility is given by (10.83) and the three channel technique in which the point mobility is given by (10.84) were employed. Also tests were carried out at three different locations on the plate, both without and with damping tape in order to provide comparison with previous results. The random excitation was filtered with a low pass filter with a cut off frequency equal to a quarter of the sample rate before being input. Anti-aliasing filters were used in all three channels at the same cut off frequency. Sufficient samples were acquired and a large sampling rate of the order of 10000 samples/sec was used giving a frequency resolution of 9.76 Hz. The material of the flat plate used in the experiment was aluminium with dimensions of 890 × 450 × 1.17 mm and with a total mass of 1.3 kg. The real parts of the mobility of the flat plate as obtained by the three channel and two channel techniques of an undamped plate based on a single point measurement are shown in Fig. 10.8a and b respectively. As observed from the figures not only are there no negative peaks

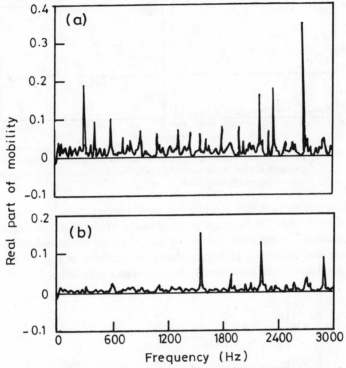

Fig. 10.8.   Real part of mobility of an undamped plate (a) by three channel technique, (b) by two channel technique (Brown, 1984).

in the three channel test, but the levels of the peaks are also much higher than the two channel test. The frequency averaged modal densities calculated from the point mobility plots are shown in Fig. 10.9. Even though both methods show a lot of scatter, it is evident that the three channel technique gives better estimates of modal densities closer to the theoretical values. Results of modal density measurements on an undamped and damped plate respectively again based on

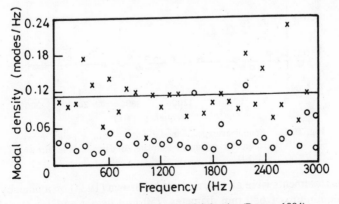

Fig. 10.9   Frequency averaged modal density (Brown, 1984)

single point measurement are shown in Fig. 10.10a and b. The addition of damping tape reduces the scatter around the theoretical results in both the cases especially

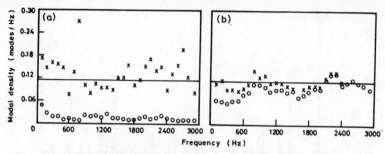

Fig. 10.10.   Modal density based on single point measurement. (a) undamped plate, (b) damped plate. ×: three channel technique, (O) two channel technique, (———) theory (Brown, 1984).

at high frequencies with the two channel results approaching the three channel results at those frequencies. Average values of modal density based on measurements on three locations in the plate and by the use of the three channel technique for the damped and undamped plates are shown respectively in Fig. 10.11 a and b showing good agreement between the theoretical and experimental values. Modal

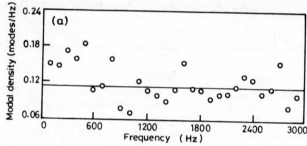

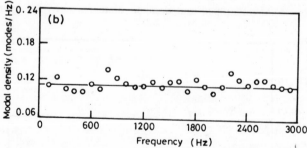

Fig. 10.11.   Spatial average of modal density with w point measurements: (a) undamped plate, (b) damped plate (Brown, 1984).

density measurements were also carried out by Brown (1984) on a plain cylindrical shell of aluminium of 650 mm diameter, 750 mm height and 1.5 mm thickness having a total mass of 6.05 kg and a natural frequency of 2530 Hz using the three channel technique. The results are shown in Fig. 10.12 showing good agreement between theory and experiment.

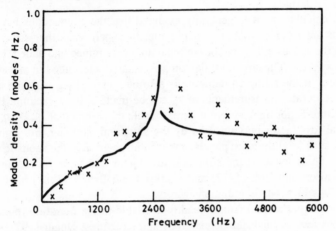

Fig. 10.12.   Modal density of undamped plain cylinder. (×): three channel technique, (——) theory (Brown, 1984).

10.7   Internal Loss Factors (Damping)

Another important parameter that enters the power flow relations in SEA is the internal loss factor (damping) which is a major source of uncertainty in the estimation of dynamic response of structures. Unlike the case of modal density no general analytical expressions are available for the estimation of internal loss factors and hence they have to be determined experimentally.

The damping mechanism in a large class of structural/mechanical and acoustic systems has a number of forms both linear and nonlinear. The most commonly accepted forms of linear damping are: (i) structural (hysteretic or viscoelastic) damping which is a function of the material properties of the structure and (ii) acoustic radiation damping associated with sound radiation from the surface of the structure into the surrounding fluid medium. Nonlinear damping mechanisms such as dry friction damping, squeeze film damping and damping due to gas pumping at structural joints are also additionally present especially at junctions of different structural elements.

The total internal loss factor may be separated in the form

$$\eta = \eta_s + \eta_{rad} + \eta_j \tag{10.85}$$

where η_s is the structural loss factor, η_{rad} is the radiation loss factor and η_j is the loss factor due to damping at the joints and connections. Each of the damping mechanisms is assumed to contribute independently to the total damping. If the boundaries between the different structural elements are rigidly connected, the joint damping η_j is usually small in which case the internal loss factor can be approximated by

$$\eta = \eta_s + \eta_{rad} \tag{10.86}$$

The internal loss factor of a subsystem in the SEA model is a measure of the rate of energy dissipation of the subsystem when it is considered as a separate system uncoupled to other subsystems in the SEA model. Therefore it is different from the total loss factor of the subsystem when it is coupled to other subsystems as defined in (10.48) even though the energy input is only to the subsystem whose loss factor is defined.

In SEA calculations it is generally assumed that the connections between the subsystems are rigid so that η_j is negligible. In experiments the structural loss factor and the radiation loss factor are separately determined and added to give the overall loss factor. The structural damping is usually determined in air and hence a component of the radiation damping is also measured along with the structural damping. Accurate measurements of η_s can be made only in vacuum. However, if the structure is not light weight it is reasonable to assume that $\eta_{rad} < \eta_s$ and that the internal loss factor is dominated by the structural damping. For light weight structures, such as aluminium panels, honeycomb shells etc., the acoustic radiation factor may be as much as the structural loss factor and some times can be even more. Rennison and Bull (1977) and Clarkson and Brown (1985) have reported measurements of acoustic radiation factors.

Not much experimental data is readily available in the literature regarding the internal loss factors of different materials and structures. Ungar (1971) provides a detailed discussion on the various damping mechanisms together with typical values of structural loss factors for a umber of structural materials. Ranky and Clarkson (1983) have given extensive experimental results regarding band averaged internal loss factors, including structural and acoustic radiation factors for aluminium plates and shells. Similarly Norton and Greenhalgh (1986) have presented a wide range of modal and band averaged internal loss factors of steel cylinders. Richards and Lenzi (1984) have obtained a large range of typical damping values of industrial machinery and have discussed different forms of linear and nonlinear damping mechanisms at structural joints.

When using experimentally measured internal loss factors for light weight structures obtained in a reverberant or in semi-reverberant room, it should be recognized that already the structure is coupled to an acoustic volume and hence the measured loss factor becomes the total loss factor of the structure-acoustic system and the radiation loss factor η_{rad} becomes a coupling loss factor between the structure and acoustic enclosure, that is $\eta_{rad} = \eta_{12}$ where the subscripts 1 and 2 refer respectively to the structure and the acoustic volume. Therefore care should be exercised in using appropriately the measured values of loss factors in SEA calculations.

The acoustic radiation loss factor of a structure can be expressed as (Cremer, Heckl and Ungar, 1973)

$$\eta_{rad} = \frac{\rho_0 c \sigma}{\omega_c \mu} \tag{10.87}$$

where σ is the acoustic radiation efficiency, ρ_0 is the density of the fluid medium into which sound is radiated by the structure, c is the speed of sound, ω_c is the center frequency and μ is the surface mass density of the structure. The radiation efficiency represents the ratio of the acoustic power radiated by the structure to that of the sound power radiated by an ideal piston source such as a rigid surface whose dimensions are greater than the acoustic wave length in the medium having the same surface area and root mean square velocity as the structure. The radiation efficiency is defined as

$$\sigma = \frac{P}{\rho_0 c \zeta \bar{v}^2} \tag{10.88}$$

where P is the acoustic power radiated by the structure, S is the total surface area of the structure and $\bar{v}^2$ denotes the average mean square velocity of the radiating surface. At low frequencies, σ is small and hence η_{rad} is also small. At high frequencies σ increases rapidly and may dominate the total damping in the system. Figure 10.13 shows some typical experimental results of radiation efficiencies.

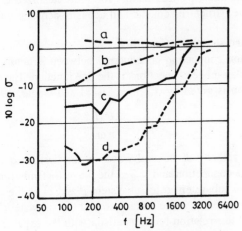

Fig. 10.13. Experimentally determined radiation efficiencies (Cremer, Heckl and Ungar, 1973). (a) 14 cm concrete ceiling, (b) engine blocks of various dimensions (c) and (d) steel pipe of 72 cm diameter and 1.3 mm wall thickness.

10.8 Coupling Loss Factors

The coupling loss factor $\eta_{\alpha\beta}$ is one of the most important parameters used in SEA calculations. It gives the degree of coupling between two coupled subsystems α and β. If $\eta_{\alpha\beta} < \eta_\alpha$ or η_β, then the coupling is weak between the subsystems. In SEA applications, it is convenient to select the subsystems so that they are weakly coupled in which case the SEA predictions will be more accurate. The different criteria that may be adopted in the selection of appropriate subsystems and their boundaries for a proper application of SEA have been described by Fahy (1974). Generally, when two systems of different material properties intersect, it is natural to choose the interface between them as a boundary. Points, edges, contours where freely propagating waves are substantially reflected due to impedance mismatches can be considered as boundaries dividing the subsystems.

Unlike the case of modal density for which theoretical expressions are available for some of the regular structural elements, no such general formulae are available for estimating the coupling loss factors between the subsystems. Like the internal loss factors, the coupling loss factors have also to be determined mainly from experiment. However, for some ideal connections between structural elements like junctions between plates, plate-cantilever beam junctions, beam-beam connections etc., and for couplings between structural elements and acoustic volumes, analytical expressions for the coupling loss factors are available (Lyon, 1975, Cremer, Heckl and Ungar, 1973).

Analytical coupling loss factor expressions are mostly obtained by wave transmission analysis, when the coupling loss factor $\eta_{\alpha\beta}$ is derived directly from the wave transmission coefficient $\tau_{\alpha\beta}$. The transmission coefficients can be evaluated in terms of wave impedances and mobilities. Cremer, Heckl and Ungar (1973) give a comprehensive coverage of wave transmission/attenuation coefficients for a wide range of structural discontinuities. In this section we summarize the analytical expressions of the coupling loss factors for some idealized connections between structure-structure, structure-acoustic and acoustic-acoustic couplings.

10.8.1   Structure-structure Coupling

The most common connection encountered between structural elements such as beam-plate and plate-plate connections is the line junction. The coupling loss factor for a line junction is given by (Cremer, Heckl and Ungar, 1973; Lyon, 1975).

$$\eta_{\alpha\beta} = \frac{2C_{\mathrm{B}}L\tau_{\alpha\beta}}{\pi\omega_{\mathrm{c}}S_{\alpha}} \tag{10.89}$$

where C_{B} is the velocity of flexural waves in the first structure (α-set), L is the length of the line connection and $\tau_{\alpha\beta}$ is the wave transmission coefficient of the line junction from subsystem α to subsystem β and ω_{c} is the centre frequency of the band of interest and S_{α} is the surface area of the subsystem α. Bies and Hamid (1980) have used this relation for comparison with the experimentally determined coupling loss factor of flat plates joined at right angles to each other. Wöhle et al (1981) have also used this relation for estimating the coupling loss factors for rectangular structural slab joints.

The normal incidence wave transmission coefficient for two coupled flat plates at right angles to each other is given by (Cremer, Heckl and Ungar, 1973).

$$\tau_{\alpha\beta}(0) = 2\{\psi^{1/2} + \psi^{-1/2}\}^{-2} \tag{10.90}$$

where

$$\psi = \frac{\rho_{\alpha}C_{\mathrm{L}\alpha}^{3/2}h_{\alpha}^{5/2}}{\rho_{\beta}C_{\mathrm{L}\beta}^{3/2}h_{\beta}^{5/2}} \tag{10.91}$$

and ρ is the density, C_{L} is the longitudinal wave speed and h is the thickness of the structure with the subscripts α and β referring to the subsystems.

Lyon (1975) has given the coupling loss factor for a cantilever beam-plate connection in terms of a junction moment impedance. The coupling loss factor for this case is given by

$$\eta_{\alpha\beta} = \frac{(2\rho_{\alpha}C_{\mathrm{L}\alpha}\kappa_{\alpha}A_{\alpha})^2}{\omega_{\mathrm{c}}M_{\alpha}} \, \mathrm{Re}\,[Z_{\beta}^{-1}] \, \frac{Z_{\beta}^2}{(Z_{\alpha} + Z_{\beta})} \tag{10.92}$$

where the subscript α refers to the cantilever beam and the subscript β refers to the plate, κ_{α} is the radius of gyration of the beam, M_{α} is the total mass, A_{α} is the cross-sectional area, Z_{α} the moment impedance of the beam, $C_{\mathrm{L}\alpha}$ is the longitudinal wave speed in the beam and Z_{β} the moment impedance of the plate. The moment impedances for a semi-infinite beam and an infinite plate are approximately given by (Cremer, Heckl, and Ungar, 1973)

$$Z_\alpha \simeq \frac{0.03 \rho_\alpha A_\alpha (C_{L\alpha} h_\alpha)^{3/2} (1 - i)}{(\omega_c/2\pi)^{1/2}} \qquad (10.93a)$$

$$Z_\beta \simeq \frac{C_{L\beta}^2 \rho_\beta h_\beta^3}{\left\{1 - \dfrac{4i \ln (0.9 ka)}{\pi}\right\}\left\{\dfrac{2.4 \omega_c}{\pi}\right\}} \qquad (10.93b)$$

where $a = h_\alpha/2$ is the moment arm of the applied force.

10.8.2 Structure-acoustic Volume Coupling
The coupling loss factor in this case is given by

$$\eta_{\alpha\beta} = \frac{\rho_\beta c \sigma}{\omega_c \mu_\alpha} \qquad (10.94a)$$

where the subsystems α and β respectively refer to the structure and acoustic volume. The quantities in (10.94a) have the same connotation as in (10.87), because the acoustic radiation loss factor for a structure becomes a coupling loss factor when the structure couples to an acoustic volume. From the reciprocity relation (10.39), the coupling loss factor between the acoustic volume and the structure is given by

$$\eta_{\beta\alpha} = \frac{\rho_\beta c \sigma n_\alpha}{\omega \mu_\alpha n_\beta} \qquad (10.94b)$$

where n_α is the modal density of the structure and n_β is the modal density of the acoustic volume. The coupling loss factor for non resonant, mass law sound transmission through a panel from a source room is given by

$$\eta_{rp} = \frac{cS}{4 \omega V_r} \tau_{rp} \qquad (10.95)$$

where as before, c is the speed of sound, S is the surface area of the panel, V_r is the volume of the source room and τ_{rp} is the sound intensity transmission coefficient from the source room through the panel, the subscripts r and p referring to respectively to room and panel.

10.8.3 Acoustic Volume-acoustic Volume Coupling
The coupling loss factor between two acoustic volumes is given by

$$\eta_{\alpha\beta} = \frac{cS}{4 \pi V_\alpha} \tau_{\alpha\beta} \qquad (10.96)$$

where S is the area of the coupling aperture, V_α is the volume of the acoustic cavity (α-set) and $\tau_{\alpha\beta}$ is the sound intensity transmission coefficient between the two volumes and c is the speed of sound.

10.9 Experimental Determination of Loss Factors and Coupling Loss Factors

10.9.1 Internal Loss Factor
The logarithmic decrement method and the half power bandwidth technique are

the common methods of measuring the modal internal loss factors. In SEA calculations, the modal internal loss factor is of only limited use. The loss factor averaged over the frequency band of interest is the parameter which is used in SEA calculations. Moreover, for structural systems with very low damping the measurement of loss factors by the logarithmic decay or the half power bandwidth method involves lot of errors and is not accurate. Hence, other methods are used in the experimental determination of the band averaged internal loss factors useful in SEA power flow calculations.

Bies and Hamid (1980) have proposed an experimental procedure for determining the band averaged internal loss factor and coupling loss factor based on steady state energy balance relations. The steady state energy flow technique has become a widely accepted procedure for the measurement of internal and coupling loss factors.

The method is based on the principle that in the steady state all of the power input to a subsystem is dissipated by the damping in the system. This results in the relation

$$\eta = \frac{P_{in}}{\omega E} \tag{10.97}$$

where η is the band averaged loss factor, P_{in} is the steady state power input to the structure, ω is the centre frequency of the frequency band and E is the space averaged total energy of vibration of the structure within the band.

Thus the determination of the internal loss factor by the stead -state energy flow technique requires an accurate estimation of the input power to the structure. The steady state power input into the structure is given by any of the following equivalent relations.

$$P_{in} = R_{fv}(0) = \int_{-\infty}^{\infty} \Phi_{fv}(\omega) \, d\omega \tag{10.98a}$$

$$P_{in} = \langle f(t)v(t) \rangle = \text{Re}\,[FV*]/2 \tag{10.98b}$$

$$P_{in} = \langle f^2(t) \rangle \, \text{Re}\,[Y] = |F|^2 \, \text{Re}\,[Y]/2 \tag{10.98c}$$

$$P_{in} = \langle v^2(t) \rangle \, \text{Re}\,[Z] = |V|^2 \, \text{Re}\,[Z]/2 \tag{10.98d}$$

In the above equations, R_{fv} is the zero lag cross correlation between the force $f(t)$ and the velocity $v(t)$, $\Phi_{fv}(\omega)$ is the cross spectral density function between $f(t)$ and $v(t)$, F and V respectively are the Fourier transforms of $f(t)$ and $v(t)$ and Re[Y] and Re[Z] respectively represent the real parts of the point mobility and point impedance. A schematic of instrumentation set up for the measurement of the band averaged internal loss factor by the steady-state energy flow technique is shown in Fig. 10.14.

If the excitation source in the experiment is a broad band random noise, measurements of internal loss factors based on the cross spectral density $\Phi_{fv}(\omega)$ provide more reliable estimates than those obtained by using the impedance relationships. This is so, because, near resonances, the real part of the impedance is very small involving differences between small numbers resulting in large errors in the estimation of the power input. This is especially so for systems having low damping. Brown and Clarkson (1984) have demonstrated that the real part of impedance can be used to estimate the internal loss factor more accurately, if a transient deterministic excitation is used instead of the broad band random excitation.

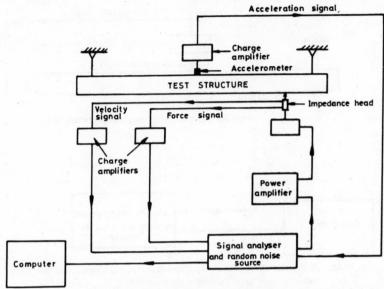

Fig. 10.14. Schematic of experimental set up for measuring internal loss factors by steady-state energy flow technique (Norton, 1989).

Norton and Greenhalgh (1986) have shown that properly controlled reverberation decay procedures produce reliable estimates of internal loss factors, which are generally more consistent than produced by the steady state power flow technique. In this method the structure is excited by a constant bandwidth random noise burst through a non contacting electro-magnetic exciter. The decay of the response immediately after the stoppage of the excitation is obtained, averaged and digitally filtered. The reverberation time T_{60} which is the time required for the response energy to reduce to one-millionth of the original value is obtained form the filtered response. This corresponds to a 60 dB drop in the response level. The band averaged internal loss factor is obtained from the relation

$$\eta = \frac{\ln 10^6}{\omega T_{60}} \tag{10.99}$$

where ω is the centre frequency of the frequency band. A schematic of the instrumentation set up to measure the internal loss factor by the reverberation decay random noise burst technique is shown in Fig. 10.15. This method allows for a rapid estimation of internal loss factors of structures and acoustic volumes.

A comparison of the broad band averaged internal loss factors obtained by the steady state energy flow technique and the reverberation decay random burst technique for cylinders made of mild steel is shown in Fig. 10.16. (Brown and Norton, 1985). The frequency range investigated corresponds to frequencies well below the ring frequencies of the cylinders. The parameters $\beta = h/(2\sqrt{3}a_m)$ and $\Delta = L/a_m$ respectively are the non dimensional thickness parameter and the ratio of the cylinder length L to the mean radius a_m and h is the cylinder wall thickness. It is observed from the figure that the internal loss factors measured by the steady state energy flow technique show much variation with respect to frequency as compared to the internal loss factors measured by the reverberation decay method. It can be also seen from the figure that this variation is much more when the input power

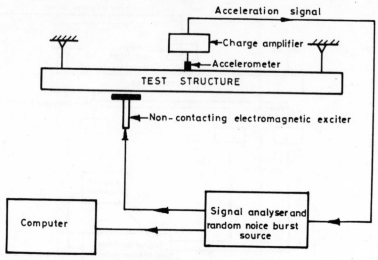

Fig. 10.15.   Schematic of experimental set up for measuring internal loss factors by random noise burst technique (Norton, 1989).

is measured based on the product of $|v^2(t)|$ and $\text{Re}[Z]$ than when the measurement is based on the cross spectral density between input force and velocity.

10.9.2   Coupling Loss Factors

The coupling loss factor between subsystems constituting the SEA method is an important parameter and is more often determined experimentally. Extreme care has to be exercised while determining the coupling loss factors by experiments because they are of an order of magnitude lower than the corresponding internal loss factors, especially when the subsystems are lightly coupled. Coupling loss factors can be either measured in laboratory under controlled experimental conditions or measured in situ.

In the laboratory the measurement of coupling loss factors is accomplished by limiting the number of subsystems to two. One of the subsystems, the α set is externally excited and the space and time averaged energies of both the subsystems α and β are measured. The coupling loss factor is obtained from the relation (10.46), that is

$$\eta_{\alpha\beta} = \frac{\eta_\beta n_\beta E_\beta}{n_\beta E_\alpha - n_\alpha E_\beta} \qquad (10.100)$$

For the evaluation of the coupling loss factor, the energies of the subsystems of the α and β sets E_α and E_β, the internal loss factor η_β and the modal densities n_α and n_β of the subsystems α and β have to be known. The modal densities can be either determined experimentally as specified in section 10.7 or can be estimated analytically. The internal loss factor can be obtained experimentally by the steady state energy flow technique or the reverberation decay random noise burst technique.

While performing the experiments for the measurement of coupling loss factor, sufficient care should be taken to minimize the energy losses in the joints connecting the two subsystems so that all of the energy flow is only between the coupled subsystems. For example, in the case of plates coupled to shells, the connection could be through fine wire point supports. Otherwise, the ends of the coupled

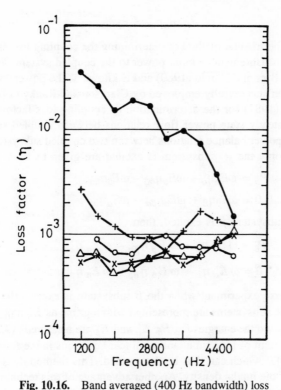

Fig. 10.16. Band averaged (400 Hz bandwidth) loss factors for steel pipes (Brown and Norton, 1985).

○ : by steady state technique, $\beta = 0.012$, $\Delta = 44.4$; $P_{in} = v^2(t)$. Re[Z];

+ : by steady state technique, $\beta = 0.012$, $\Delta = 44.4$; $P_{in} = F(t).v(t)$;

× : burst random noise technique, $\beta = 0.009$, $\Delta = 47.3$;

Δ : burst random noise technique, $\beta = 0.009$, $\Delta = 94.6$;

• : burst random noise technique, $\beta = 0.009$, $\Delta = 45.9$.

structure could be supported on foam rubber pads to simulate free-free end conditions. It is advisable to repeat the experiment by exciting the β subsystem and measuring the space and time averaged vibrational energies of both subsystems α and β. In this case relation (10.47) is used in estimating the coupling loss factor, namely

$$\eta_{\alpha\beta} = \frac{\eta_\alpha n_\beta E_\alpha}{n_\alpha E_\beta - n_\beta E_\alpha} \qquad (10.101)$$

The above relation assumes that the joint damping η_j is small. If η_j is not negligible the above procedure will entail errors because the uncoupled values of the internal loss factors obtained separately will not be equal to the coupled internal loss factor.

The coupling loss factor $\eta_{\beta\alpha}$ can be obtained from the reciprocity relationship

$$n_\alpha \eta_{\alpha\beta} = n_\beta \eta_{\beta\alpha} \tag{10.102}$$

Another experimental method of determining the coupling loss factors in situ is based on the measurement of input power to the coupled system. The method was developed by Bies and Hamid (1980) and is known as the power injection method which has been successfully employed by Clarkson and Ranky (1984) and Norton and Keswick (1987) for the determination of coupling loss factors. This method involves the steady state power flow relations between coupled subsystems. For example, the power balance relations between two coupled subsystems in the SEA model when only the α subsystem is excited are given by

$$P_\alpha = \omega E_\alpha \eta_\alpha + \omega E_\alpha \eta_{\alpha\beta} - \omega E_\beta \eta_{\beta\alpha} \tag{10.103}$$

$$0 = \omega E_\beta \eta_\beta + \omega E_\beta \eta_{\beta\alpha} - \omega E_\alpha \eta_{\alpha\beta} \tag{10.104}$$

Similarly if subsystem β is excited, then

$$P_\beta = \omega E'_\beta \eta_\beta + \omega E'_\beta \eta_{\beta\alpha} - \omega E'_\alpha \eta_{\alpha\beta} \tag{10.105}$$

$$0 = \omega E'_\alpha \eta_\alpha + \omega E'_\alpha \eta_{\alpha\beta} - \omega E'_\beta \eta_{\beta\alpha} \tag{10.106}$$

In the reverse experiment when the β subsystem is excited the band averaged energies of the subsystems are represented with a prime as E'_α and E'_β. If the input powers P_α, P_β and the energies E_α, E_β, E'_α and E'_β are measured (10.103) to (10.106) represent a system of four simultaneous linear equations in the four unknowns η_α, η_β, $\eta_{\alpha\beta}$ and $\eta_{\beta\alpha}$ which can thus be determined. This method does not require the knowledge of the modal density. Another advantage of this technique is that it can be extended to more than two coupled subsystems when the energy balance relations represent a set of simultaneous matrix equations for the solution of the coupling loss factors.

Accurate estimates of the coupling loss factors by this method depend largely on the accuracy with which the input power and energy levels of the subsystems can be measured. The loss factor estimated are very sensitive to the energy measurements. The effect of the experimental errors can be reduced by using an iterative procedure giving the best fit in a least square sense. Also the energy associated with the mass loading of the exciter on the structure, contact damping due to contact between exciter and structure, exciter-structure feedback interaction etc. have to be minimized as much as possible. This can be achieved by a proper choice of exciter arrangements like the spherical point contact, flexible wire rod and free-floating magnetic exciter arrangements. For details regarding these aspects readers are referred to the works of Norton (1989), Keswick and Norton (1987).

Typical coupling loss factor measurements carried out by Clarkson and Ranky (1984) for a plate-plate junction, with the planes of the plates perpendicular to each other are shown in Figs. 10.17 and 10.18. The theoretical results and the experimental results based on raw data as well as based on least-square iteration are shown in the figures. The results of Fig. 10.17 correspond to the case when no external damping was added to the plates, while the results of Fig. 10.18 correspond to the case when 8 mm width of self-adhesive damping was added to the periphery of the two plates. The effect of adding the damping tape was to increase the individual internal loss factors and to increase the modal overlap between successive modes of vibration. This facilitates ease of experimental measurements and provides a better fit to the theoretical results.

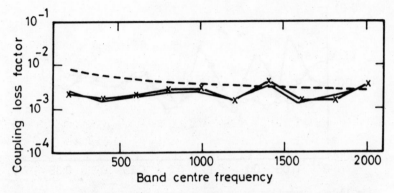

Fig. 10.17. Coupling loss factor for plate-plate junction without damping tape. (– – –) theory; (– × – × –) raw data; (——) iterated result (Clarkson and Ranky, 1984).

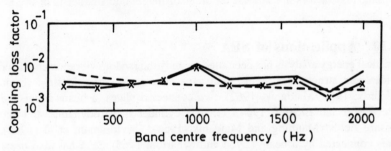

Fig. 10.18. Coupling loss factor for plate-plate junction with damping tape. (– – –) theory; (– × – × –) raw data; (——) iterated result, (Clarkson and Ranky, 1984).

Coupling loss factor measurements for coupled steel cylindrical shells obtained by the power injection technique (Norton and Keswick, 1987) are shown in Fig. 10.19.

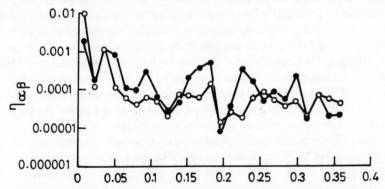

Fig. 10.19. Coupling loss factors for two cylindrical shells with a flanged joint. (— • —) air gap joint; (—∘—) rubber gasket joint (Norton and Keswick, 1987).

The coupling loss factors are for 65 mm diameter, 1 mm wall thickness shell each 1.5 m long and 109 mm diameter. The corresponding internal loss factors uncoupled and in situ are shown in Fig. 10.20. The differences between the two measurements are due to the coupling damping. It may be also seen from the figure that the

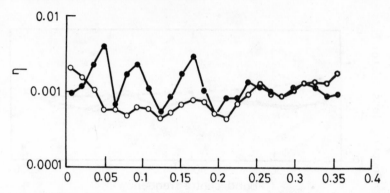

Fig. 10.20. Internal loss factors for a cylindrical shell. (—•—) *in situ,* i.e. coupled via an air gap joint; (—○—) uncoupled (Norton and Keswick, 1987).

coupling loss factors are at least an order of magnitude smaller than the internal loss factors.

10.10 Applications of SEA

Statistical energy analysis has been used to study the random vibration characteristics of idealized structural systems such as a plate in an acoustic field (Lyon and Maidanik, 1962; Maidanik, 1962), two connected plates, a beam connected to a plate (Lyon and Eichler, 1964), circular cylinder in aerodynamic and acoustic pressure fields (Manning and Maidanik, 1964); Chandramani et al (1966), two plates connected by a beam (Lyon and Scharton, 1965). SEA has also been used to estimate noise and vibration transmission in space vehicle structures and vibration of electronic components mounted inside (Manning, Lyon and Scharton, 1966; Scharton and Yang, 1967; Clarkson et al, 1981; Pocha, 1977).

Acoustic volumes by virtue if their large modal densities are ideally amenable to be modelled as subsystems in an overall SEA chain. Many of the applications of SEA are therefore concerned with the vibration and noise transmission characteristics of structural systems in a sound field or backed by acoustic cavities. The sound transmission characteristics of a panel dividing two reverberant rooms was studied using SEA by Crocker and Price (1969), Kamaraju (1977). Sawley (1969), Jensen (1977) and Kihlman and Plunt (1977) have likewise used SEA to estimate the structure borne sound transmission in ship structures. Miller and Faulkner (1983) have estimated aircraft interior noise through fuselage side walls using SEA. Richards (1981) has extended SEA procedures in the prediction of impact noise with application to machinery noise control. Stimpson et al (1986) have applied SEA in predicting the sound radiation from built up vibrating structures. The applications of SEA listed above are only samples and not exhaustive. In the sequel we present a few examples illustrating the applications of SEA.

10.10.1 Random Vibration of a Beam-plate System

This is one of the earliest application of SEA considered by Lyon and Eichler (1964). A connected structure consisting of a beam cantilevered to a flat plate as shown in Fig. 10.21 is considered. The plate and the beam form the two subsystems in the SEA model. It has been assumed that energy is transmitted mainly through flexural waves between the coupled beam-plate system. Further the two structural

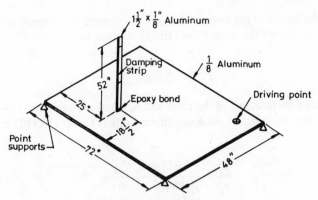

Fig. 10.21. Beam-plate structure.

elements are assumed to be strongly coupled in that the coupling loss factor between the plate and the beam is much larger than the internal loss factors of the beam and plate. The plate is subjected to external excitation while the beam is only indirectly excited.

From (10.44) we obtain

$$\frac{E_b}{E_p} = \frac{n_b}{n_p} \frac{\eta_{bp}}{\eta_b + \eta_{bp}} \tag{10.107}$$

where E represents the spatially averaged vibrational energy, n represents the modal density with the subscripts b and p referring to beam and plate, η_b the internal loss factor of the beam and η_{bp} the coupling loss factor between the beam and plate. The angular brackets in the energy expressions representing the averaging operation is omitted for notational convenience. Equation (10.107) can be alternatively expressed in terms of a velocity spectrum normalized to give the energy spectrum when multiplied by the total mass of the corresponding structure. Thus,

$$\frac{(\Phi_{vv})_b}{(\Phi_{vv})_p} = \frac{M_p}{M_b} \frac{n_b}{n_p} \frac{\eta_{bp}}{\eta_b + \eta_{bp}} \tag{10.108}$$

where M_p and M_b are the total masses and $(\Phi_{vv})_p$ and $(\Phi_{vv})_b$ are velocity psds of plate and beam respectively. Under the assumption of strong coupling ($\eta_{bp} \gg \eta_b$), there is equipartition of modal energy between beam and plate and (10.108) reduces to

$$\frac{M_b(\Phi_{vv})_b}{n_b} = \frac{M_p(\Phi_{vv})_p}{n_p} \tag{10.109}$$

which can be expressed in terms of the space averaged mean square velocities

$$\frac{v_b^2}{v_p^2} = \frac{M_p n_b}{M_b n_p} \tag{10.110}$$

If the beam is of uniform rectangular cross-section ($b \times h_b$) the modal density expression (10.57) of the beam reduces to

$$n_b(\omega) = \frac{L}{3.376 \, (C_{Lb} h_b \omega)^{1/2}} \tag{10.111}$$

where L is the length of the beam and C_{Lb} is the flexural wave speed in the beam. The modal density of the plate from (10.58) is given by

$$n_b(\omega) = \frac{S}{3.628\, C_{Lp} h_p} \tag{10.112}$$

where S is total surface area of the plate and h_p is the plate thickness. Noting that $M_p = \rho_p S h_p$ and $M_b = \rho_b L b h_b$ and substituting (10.111) and (10.112) in (10.110) we obtain

$$\frac{v_b^2}{v_p^2} = \frac{1.075\, \rho_b h_p^2\, C_{Lp}}{\rho_b b h_b (C_{Lb} h_b \omega)^{1/2}} \tag{10.113}$$

In (10.113) ρ_b and ρ_p are the densities of the beam and plate material. Equation (10.113) gives a simple formula to calculate the mean square velocity of the beam in terms of the mean square velocity of the plate.

Lyon and Eichler (1964) have conducted experiments on the set up shown in Fig. 10.21 to verify the formula. The plate was excited at the driving point marked by broad band noise in octave and 8% bandwidths. The plate vibration was measured with an accelerometer at several locations, which were within one or two decibels of each other signifying more or less uniform vibration levels over the surface of the plate. The space averaged acceleration level was therefore obtained by a simple arithmetic average. The beam vibration was measured at the free tip with a non-contacting displacement sensor. The space averaged mean square velocity of the beam is one fourth the mean square tip velocity which can be shown to be so by considering the cantilever beam eigenfunctions. The coupling loss factor η_{bp} for the beam-plate system is given by (10.92) along with the impedance formulae (10.93a) and (10.93b). If the beam and plate are made of the same material as is the case with the experiment it can be shown that the coupling loss factor between the beam and plate is given by (Lyon and Eichler 1964)

$$\eta_{bp} = b/(4L) \tag{10.114}$$

For the dimensions of the beam $\eta_{bp} = 1.5/204 = 7.35 \times 10^{-3}$ which was much larger than the internal loss factors of the beam and plate satisfying the assumption of strong coupling. In the experiment a damping strip was added to the beam by increasing the thickness and length of which the internal loss factor of the beam can be increased within certain limits.

The mean square velocity ratio (10.110) is presented in Fig. 10.22 by the solid line denoted by "theoretical broad band" along with experimental results. The agreement between the two is reasonably good even in the 100 Hz band where only two beam modes were responding showing the efficacy of the SEA procedure.

10.10.2 Beam-plate-room System

The results derived in the previous section can be extended to include a large reverberant room in the SEA model. Consider a system comprising a large flat structure as would obtain in the case of a large radiating surface of a machine radiating sound in a reverberant room. The plate is excited by a cantilever beam attached to it. The beam could be a directly driven machine element. SEA can be used to predict the sound pressure level inside the room induced by the vibration

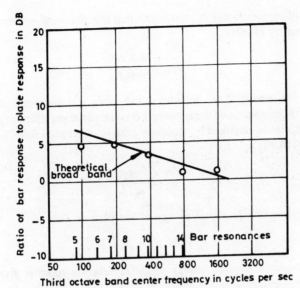

Fig. 10.22. Response ratio for an undamped beam canti-levered to a plate. (——) Theory; (o) octave band ratios (Lyon and Eichler, 1964).

of the beam and plate and radiation of sound from the plate. The three subsystems in the SEA model therefore are the beam, the plate and the acoustic volume of the reverberant room whose parameters are respectively referred by the subscripts b, p and a.

As in the previous example it is assumed that the vibrational energy is transmitted mainly through flexure. In this case, however, the beam element is the one which is directly excited instead of the plate. It is also assumed that there is no direct coupling between the beam and acoustic volume and that the structural elements, the beam and the plate are strongly coupled as before.

As the beam is the element which is excited the space averaged mean square velocity of the plate in terms of the space averaged mean square velocity of the beam ia given by the reciprocal of (10.113), that is

$$\frac{v_p^2}{v_b^2} = \frac{\rho_b b h_b (C_{Lb} h_b \omega)^{1/2}}{1.075 \, \rho_b h_p^2 C_{Lp}} \tag{10.115}$$

The ratio of the space averaged levels of the acoustic volume and the plate is given by (10.45), that is

$$\frac{E_a}{E_p} = \frac{\eta_{pa}}{\eta_a + \dfrac{n_p}{n_a} \, \eta_{pa}} \tag{10.116}$$

where η_{pa} is the coupling loss factor between plate and reverberant room, η_a is the internal loss factor and n_a the modal density of the acoustic volume.

The modal density of the acoustic volume is given by (10.70)

$$n_a(\omega) = \frac{\omega^2 V}{2 \pi^2 c^3} \tag{10.117}$$

and the coupling loss factor η_{pa} assuming that the plate radiates sound from both sides is given by (10.94) as

$$\eta_{pa} = \frac{2\rho_a c \sigma}{\omega \rho_p h_p} \qquad (10.118)$$

where ρ_a is the density if the medium in the room, c is the speed of sound in the room and V is the volume of the room and σ is the radiation efficiency. The internal loss factor η_a can be obtained by measuring the reverberation time in the room and is given by (10.99) as

$$\eta_a = \frac{13.816}{\omega T_{60}} \qquad (10.119)$$

The space averaged vibrational energy of the plate is given by

$$E_p = M_p v_p^2 = \rho_p h_p S v_p^2 \qquad (10.120)$$

The space averaged acoustic energy in the reverberant room is given by

$$E_a = \frac{\bar{p}^2 V}{\rho_a c^2} \qquad (10.121)$$

where $\bar{p}^2$ is the mean square sound pressure in the room. Substituting (10.117)-(10.121) into (10.116) we obtain

$$\bar{p}^2 = \frac{v_p^2 \rho_p h_p \rho_a S c^2}{V} \left\{ \frac{2\rho_a c \sigma/(\omega \rho_p h_p)}{\eta_a + 4\pi^2 \rho_a c^4 \sigma S/(3.628 \, C_{Lp} h_p^2 V \rho_p \omega^3)} \right\} \qquad (10.122)$$

10.10.3   Aircraft Interior Noise Prediction

In this section we present the example of interior noise prediction through an aircraft fuselage side wall using SEA as considered by Miller and Faulkner (1983). The fuselage structure is essentially a circular cylindrical shell supported by longitudinal stringers and circumferential frames in its interior. However, for the SEA model the fuselage structure is modelled as a series of curved, isotropic plates with the effect of the stiffeners considered by smearing them out over the area of the panels. Complicating effects such as internal pressurization, acoustic transmission through windows and aerodynamic excitation are excluded in their analysis. The schematic of the SEA power flow model is shown in Fig. 10.23. The three subsystems in the model are the external noise field, the fuselage shell and the aircraft interior space which are respectively denoted subsystems 1, 2 and 3. The important SEA parameters like the modal density, coupling loss factor and internal loss factor are discussed in the sequel.

Modal Density Expressions

The modal density expressions for a cylindrical shell are given by (10.64) and (10.65). Similar expressions as applicable to the fuselage structure have been developed by Scharton (1971) and Wilby and Scharton (1974) in three different frequency ranges and for acoustically fast (AF) and acoustically slow (AS) modes. They are,

Power input

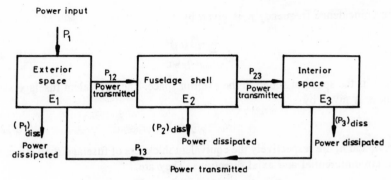

Fig. 10.23.   SEA power model for three element system.

$$n_{2AF} = \left(\frac{f}{f_r}\right)^2 \left(\frac{f_r}{f_c}\right) \frac{\sqrt{3}\,S_2}{h_e C_L}, \quad f < f_r$$

$$= 0, \qquad\qquad\qquad f_r < f < f_c$$

$$= \frac{\sqrt{3}\,S_2}{h_e C_L} \qquad\qquad f > f_c \qquad\qquad (10.123)$$

and

$$n_{2AS} = \left(\frac{f}{f_r}\right)^{0.67} \frac{\sqrt{3}\,S_2}{h_e C_L}, \qquad f < f_r$$

$$= \frac{\sqrt{3}\,S_2}{h_e C_L}, \qquad\qquad f_r < f < f_c$$

$$= 0 \qquad\qquad\qquad f > f_c \qquad\qquad (10.124)$$

Acoustically fast modes couple well with the acoustic field, effectively radiate sound from the whole surface of the structure and have bending wave speeds greater than the speed of sound in the exterior space. On the other hand acoustically slow modes do not couple well to the excitation and have bending wave speeds less than the speed of sound in the exterior space and effectively radiate into the far field. In (10.123) and (10.124) the subscript 2 generally refers to the subsystem 2, f_r is the ring frequency corresponding to a breathing mode in which the shell undergoes uniform expansion and contraction without flexure, f_c is the critical frequency corresponding to the coincidence of flexural and acoustical wavelengths, C_L is the longitudinal wave speed in the shell, S_2 is the total surface area of the fuselage side wall panel and h_e is the equivalent thickness of the fuselage shell structure accounting of the smearing effect of the stiffeners. The various quantities in (10.123) and (10.124) are defined in the following:

From (10.63),

$$f_r = \frac{C_L}{2\pi R_2} \quad \text{with } C_L = [E_2/\{\rho_2(1 - v_2^2)\}]^{1/2} \qquad (10.125)$$

where R_2 is the mean radius of the fuselage shell, E_2, ρ_2 and v_2 being respectively its Young's modulus, density and Poisson's ratio.

The coincidence frequency f_c is given by

$$f_c = \frac{\sqrt{3}\,(c_1)^2}{\pi\,h_e\,C_L} \tag{10.126}$$

where c_1 is the speed of sound in the exterior space. The equivalent fuselage shell thickness is given by

$$h_e = \sqrt{h_s h_x} \tag{10.127}$$

where h_s and h_x are respectively the equivalent thickness of fuselage shell structure in the circumferential and axial directions. They are,

$$h_s = h_2 \left\{ 1 + 12(1 - v_2^2)\left(\frac{E_s}{E_2}\right)\left(\frac{I_{os}}{d_s h_2^3}\right) \right\}^{0.33} \tag{10.128}$$

$$h_x = h_2 \left\{ 1 + 12(1 - v_2^2)\left(\frac{E_x}{E_2}\right)\frac{I_{ox}}{d_s h_2^3} \right\}^{0.33} \tag{10.129}$$

In (10.128) and (10.129) (E_s, E_x), (I_{os}, I_{ox}) and (d_s, d_x) respectively are the Young's moduli, area moments of inertia and spacing of the stringer and frame and h_2 is the baseline fuselage shell thickness.

From (10.123) and (10.124) we observe that below the ring frequency both the slow and fast modes exist, while above the coincidence frequency only the fast modes are present and between the ring and coincidence frequencies only the slow modes occur. The modal density of the fuselage shell is the sum of the modal densities corresponding to the acoustically fast and slow modes, that is

$$n_2(f) = n_{2AF}(f) + n_{2AS}(f) \tag{10.130}$$

The modal density of the interior space of the fuselage constituting the acoustic volume is given by (10.70) and repeated here for the subsystem 3 as

$$n_3(f) = \frac{4\pi V_3}{c_3^3} f^2 + \frac{\pi S_2}{2c_3^2} f + \frac{P_2}{8c_3} \tag{10.131}$$

where $V_3 = \pi R_2^2 L_{x2}$ is the enclosed volume inside fuselage, c_3 is the speed of sound in interior space, $S_2 = 2\pi R_2(L_{x2} + R_2)$ is the total surface area enclosing fuselage, L_{x2} is the fuselage shell axial length and $P_2 = 4(\pi R_2 + L_{x2})$ is the total edge length of surface area enclosing fuselage.

The exterior space constituting the subsystem 1 occupies a volume but has no definite shape. Equation (10.131) for the subsystem 1 reduces to

$$n_1(f) = \frac{4\pi V_1 f^2}{c_1^3} \tag{10.132}$$

where V_1 is the volume surrounding the fuselage shell, c_1 speed of sound in exterior space. Volume V_1 is arbitrarily chosen since the modal energy of the exterior space is independent of V_1. It is assumed that the volume V_1 is 500 times the enclosed volume of the fuselage.

Coupling Loss Factors

The coupling loss factors for the model of Fig. 10.23 are given by (10.94) as

$$\eta_{21} = \frac{\rho_1 c_1 \bar{\sigma}_2}{2\pi f h_e \rho_2} \quad \text{and} \quad \eta_{23} = \frac{\rho_3 c_3 \bar{\sigma}_{2'}}{2\pi f h_e \rho_2} \tag{10.133}$$

$$\eta_{12} = \frac{n_2(f)}{n_1(f)} \eta_{21} \quad \text{and} \quad \eta_{32} = \frac{n_2(f)}{n_3(f)} \eta_{23} \tag{10.134}$$

In Eqs. (10.133) $\bar{\sigma}_2$ and $\bar{\sigma}_2'$ are respectively the average radiation efficiencies of the fuselage shell for radiation to the exterior an interior spaces and are given by

$$\bar{\sigma}_2 = \frac{\sigma_{2AF} n_{2AF} + 0.667\, \sigma_{2AS} n_{2AS}}{n_2(f)} \tag{10.135}$$

and

$$\bar{\sigma}_2' = \frac{\sigma_{2AF} n_{2AF} + 2\,\sigma_{2AS} n_{2AS}}{n_2(f)} \tag{10.136}$$

where σ_{2AF} and σ_{2AS} are the radiation efficiencies corresponding to the acoustically fast and slow modes. For all practical purposes, σ_{2AF} can be taken to be unity in the entire frequency range for the AF modes. However, σ_{2AS} depends on frequency and is given by

$$
\begin{aligned}
\sigma_{2AS} &= \frac{4 c_1 P_r f \sqrt{2}}{\pi^4 S_2 f_c^2} & f < f_r \\[2mm]
&= \frac{c_1 \sqrt{2}}{S_2 f_c}\left(\frac{c_1}{f_c} g_1\left(\frac{f}{f_c}\right) + P_r g_2\left(\frac{f}{f_c}\right) \right) & f_r < f < f_c \\[2mm]
&= \left(1 - \frac{f_c}{f}\right)^{-0.5} & f > f_c
\end{aligned}
\tag{10.137}
$$

where

$$P_r = 2(L_{s2} + N_s L_{x2} + N_x L_{s2}) \tag{10.138}$$

with $L_{s2} = 2\pi R_2$ being the fuselage circumferential length, L_{x2} is the length of the fuselage shell, N_s the number of stringers and N_x the number or frames. The functions $g_2(f/f_c)$ and $g_1(f/f_c)$ are given by (Maidanik, 1962)

$$
\begin{aligned}
g_2(f/f_c) &= \frac{1}{4\pi^2(1 - f/f_c)^{1.5}}\left\{ \left(1 - \frac{f}{f_c}\right)\ln\left(\frac{(1 + f/f_c)}{(1 - f/f_c)}\right) + \frac{2f}{f_c} \right\} & f < f_c/2 \\[2mm]
&= 0 & \text{for } f \geq f_c/2
\end{aligned}
\tag{10.139}
$$

and

$$
\begin{aligned}
g_1(f/f_c) &= \frac{\dfrac{8}{\pi^4}\left(1 - \dfrac{2f}{f_c}\right)}{\left(\dfrac{f}{f_c}\right)^{0.5}\left(1 - \dfrac{f}{f_c}\right)^{0.5}} & f < f_c/2 \\[2mm]
&= 0 & \text{for } f \geq f_c/2
\end{aligned}
\tag{10.140}
$$

From (10.95) the coupling loss factor between the acoustic subsystems 1 and 3 is given by

$$\eta_{13} = \frac{\bar{\tau}_2 c_1 S_2}{8\pi f V_1} \tag{10.141}$$

where $\bar{\tau}_2$ is the average sound transmission coefficient

$$\bar{\tau}_2 = \left(\frac{\rho_1 c_1}{\pi f \rho_2 h_e}\right) \ln\left[1 + \left(\frac{\pi f \rho_2 h_e}{\rho_1 c_1}\right)^2\right] \tag{10.142}$$

From the reciprocity relation (10.39)

$$\eta_{31} = \frac{n_3(f)}{n_1(f)}\, \eta_{13} \tag{10.143}$$

Internal Loss Factors

From experimental measurements the internal loss factor of the fuselage structure was estimated to lie in the range of 0.01–0.02. The loss factor was assumed to be constant over the frequency spectrum. The internal loss factor of the fuselage structure was taken to be $\eta_2 = 0.02$. The internal loss factor for the interior acoustic space is taken to be

$$\eta_3 = \frac{c_3 S_2 \bar{\alpha}_3}{8\pi f V_3} \tag{10.144}$$

where $\bar{\alpha}_3$ is the average absorption coefficient for the interior space and is related to the reverberation time T_{60}.

Energy Balance Relations

The average energies for the exterior space and the structure within a frequency band Δf are given by

$$E_1(f) = \frac{\Phi_{p1}(f)\Delta f V_1}{\rho_1 c_1^2} \tag{10.145}$$

and

$$E_2(f) = \frac{\Phi_{a1}(f)\Delta f V_1}{4\pi^2 f^2} \tag{10.146}$$

where $\Phi_{p1}(f)$ and $\Phi_{a2}(f)$ are respectively the psds of sound pressure in the exterior field and the acceleration of the fuselage shell.

The power balance equations for the system of Fig. 10.23 from (10.49) can be alternatively written after making use of the reciprocity relation as

$$\begin{bmatrix} \eta_2 + \sum\limits_{\substack{j=1 \\ j\neq2}}^{3} \eta_{2j} & -\eta_{32} \\[2em] -\eta_{23} & \eta_2 + \sum\limits_{\substack{j=1 \\ j\neq2}}^{3} \eta_{2j} \end{bmatrix} \begin{Bmatrix} \dfrac{E_2}{E_1} \\[1.5em] \dfrac{E_3}{E_1} \end{Bmatrix} = \begin{Bmatrix} \eta_{12} \\[1.5em] \eta_{13} \end{Bmatrix} \tag{10.147}$$

From (10.147) E_3/E_1 can be obtained and the acoustic energy in the interior space can be predicted and consequently the interior sound pressure level. The noise reduction (NR) from the exterior to the interior of the structure can be obtained from the relation

$$NR = 10 \log \left(\frac{E_3}{E_1} \right) dB \qquad (10.148)$$

Experimental Validation

The above analysis was used to predict the interior noise level by Miller and Faulkner (1983) of the C15 airplane fuselage. The experiment was conducted on a full scale demonstration component (FSDC) of the fuselage section from station 367 to 871 (Fig. 10.24) which was 10.28 m long consisting of a nose section forward of section 439 and a cargo compartment section aft of station 439. Most part of the fuselage shell was cylindrical with a constant radius of 274.3 cm from station 516 back to station 871. The shell was circular in cross-section forward of station 516-867. No acoustic absorption material was present in the interior. The fuselage was made up of 19 separate modules that were conventionally riveted and each module consisted of an external skin bonded to internal reinforcing members.

The external noise source was a loudspeaker and a control microphone was located at the plane of the mouth of the loudspeaker. Sixteen 1/2″ microphones were used to measure the external and interior sound pressure levels whose locations are shown in Fig. 10.24. One accelerometer was also used to measure the vibration level of the skin at the location shown. The acoustic and vibration data were recorded in an instrumentation tape recorder and analyzed using a one third octave band frequency analyzer by playing back the data tapes in the laboratory.

The predicted noise reduction through the FSDC using (10.148) was compared with the measured noise reduction in Fig. 10.25. The measured data are from microphones 4 (exterior) and 13 (interior) and were corrected for receiving space effects by substracting the difference in dB measured between the interior microphones 13 and 14. The corrected noise reduction agrees reasonably well with those predicted by SEA except for the 250 Hz one-third octave band which is in the region of the fuselage ring frequency where the exterior and interior sound pressure levels were nearly the same with the fuselage wall transparent to the noise excitation. Assuming the internal loss factor of the structure to be 0.002 instead of 0.02 resulted in the predicted level being lowered by 22.5 to 6.5 dB depending on the frequency range, showing that the SEA calculations are sensitive to the choice of the internal loss factors. The noise reduction predicted by mass law (Franken and Kerwin, 1958) shown alongside in the figure does not account for the peaks and valleys as does the SEA prediction.

10.10.4   Structure-borne Noise Transmission in Ship Hull Structures

We present here another example of the application of SEA in the estimation of structure-borne noise transmission through a model ship section which 1:5 replica of cross-section through a machinery room, casing and deckhouse of a large bulk carrier (Jensen 1977). Figure 10.26 shows the model ship section with the length of the cross-section being 5 frame distances corresponding to 4 m in full scale. The model is built up of welded, periodically stiffened steel plates. For the SEA

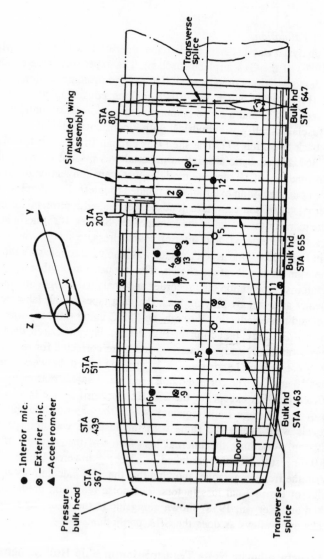

Fig. 10.24. Airplane fuselage and transducer locations (Miller and Faulkner, 1983).

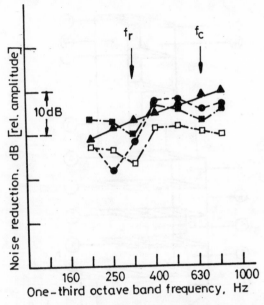

Fig. 10.25. Predicted and measured noise reduction for fuselage shell. (Miller and Faulkner, 1983). (▲—▲) mass law; (■—■) SEA with $\eta_2 = 0.02$; (□—□) : SEA with $\eta_2 = 0.02$; (●—●) measured (Miller and Faulkner, 1983).

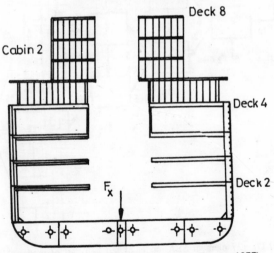

Fig. 10.26.   1 : 5 model of ship section (Jensen, 1977).

model, the section is considered to be subdivided into 27 sub-sections as shown in Fig. 10.27. The sub-sections consist of the double bottom section, deck-sections, hull-plate sections and deck-house sections, casing plates and transverse bulkheads. These 27 plate sections are assumed to be coupled in combinations of right-angle corners and T-junctions. The structure is assumed to be externally excited only at the sub-section No. 1 double bottom section as shown in the figure. The input power to the system arises out of the excitation and is termed as $P_{1, \text{in}}$. The internal

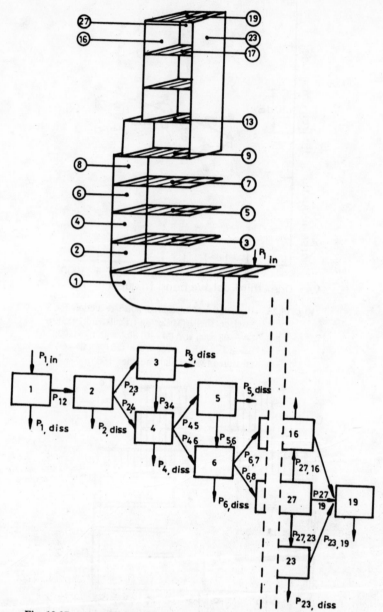

Fig. 10.27. 27 element SEA-model of left half-part of modal ship section (Jensen, 1977).

loss factors η_i corresponding to each sub-system were estimated from decay rate measurements before they were used in the final SEA model and ranged from 0.001 to 0.002.

The coupling loss factors are theoretically estimated according to (10.89) assuming the system to be consisting of only plates and vibrating only in the flexural mode namely

$$\eta_{ij} = \frac{C_{gi} L_{ij} \tau_{ij}}{\omega \pi S_i} \qquad (10.149)$$

where the indices α and β in (10.89) are interchanged with i and j and C_{gi} is the group velocity in the plate corresponding to subsystem i.

The plates are interconnected by three types of joints, namely stiff right angle corners, T-junctions and cross-junctions as shown in Fig. 10.28 which are very common in ship hull plate systems.

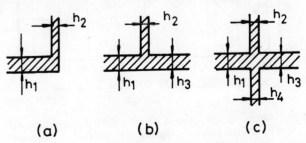

Fig. 10.28.   Three types of plate junctions: (a) right angle corner, (b) T-junction (c) cross-junction.

Corresponding to a right angle junction between two plate thicknesses h_1 and h_2 (Fig. 10.28a), the wave transmission coefficients for pure flexural waves impinging in a different direction normal to the junction is given by (Cremer, Heckl and Ungar 1973)

$$\tau_{12} = 2(r^{-5/4} + r^{5/4})^{-2} \qquad (10.150)$$

with $r = h_2/h_1$.

Likewise, the transmission coefficients for the T-junction are

$$\tau_{12} = \frac{\tau}{1 + \left(\dfrac{h_3}{h_2}\right)^{5/2}}, \qquad \tau_{13} = \frac{\tau}{1 + \left(\dfrac{h_2}{h_3}\right)^{5/2}} \qquad (10.151)$$

where $\tau = 2(\kappa^{-1/2} + \kappa^{1/2})^{-2}$ with $\kappa = (h_2/h_1)^{5/2} + (h_3/h_1)^{5/2}$

Similarly for the cross-junction

$$\tau_{12} = \frac{\tau}{1 + \left(\dfrac{h_3}{h_2}\right)^{5/2} + \left(\dfrac{h_4}{h_2}\right)^{5/2}}, \qquad \tau_{13} = \frac{\tau}{1 + \left(\dfrac{h_2}{h_3}\right)^{5/2} + \left(\dfrac{h_4}{h_3}\right)^{5/2}}$$

and

$$\tau_{14} = \frac{\tau}{1 + \left(\dfrac{h_2}{h_4}\right)^{5/2} + \left(\dfrac{h_3}{h_4}\right)^{5/2}} \qquad (10.152)$$

where $\tau = 2(\kappa^{-1/2} + \kappa^{1/2})^{-2}$ with $\kappa = (h_2/h_1)^{5/2} + (h_3/h_1)^{5/2} + (h_4/h_1)^{5/2}$.

Experimental Verification

The calculated differences ΔL_v between the mean vibration level at the double bottom section and the mean vibration level at deck 2, and deck 8 respectively are

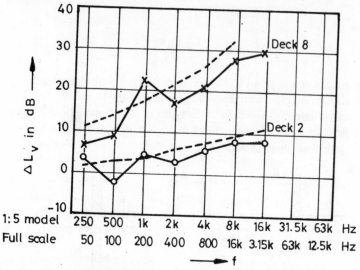

Fig. 10.29. Difference between mean vibration levels on the tank top and the different decks. (– – –) predicted by SEA. (———) measured (Jensen, 1977).

shown in Fig. 10.29 with the dotted lines. ΔL_v is defined as

$$\Delta L_v = (L_v)_{\text{bottom}} - (L_v)_{\text{deck i}} \tag{10.153}$$

The full lines in the figure show the corresponding measured values obtained from vibration measurements on the corresponding model section. For the experiments the model section was excited with an electrodynamic exciter driving the model at the machinery foundation at the tank top (at the arrow marked in Fig. 10.26). The exciter was driven by 1/1-octave band limited noise corresponding to a constant force excitation of 1 N (rms). The excited mean vibration level was measured with a small accelerometer on the double bottom, on the hull plates and on the various decks. A comparison of the corresponding measured and calculated values of the vibration level upwards through the model section shows a good agreement between SEA prediction and measurements except perhaps at low frequencies. The measured and calculated vibration levels were derived for rather small values of internal loss factors ($\eta_i \cong 0.001$–0.002). In order to investigate the accuracy of the SEA model for large values of loss factors, the ship model section was heavily damped at the hull plate between deck 1 and deck 2 by covering the outside of the hull plate by a constrained damping layer. Besides, the four frames and a heavy web-frame were also damped between the two decks by a covering of viscoelastic damping material. The measured and calculated insertion loss with the damping treatment for the hull plate between deck 1 and deck 2 (the section covered with damping layer) is shown in Fig. 10.30 which again shows good agreement between the SEA calculations and experiments.

10.11 Recent Developments in SEA

10.11.1 Transient Response

Application of SEA technique has been mainly limited to steady state vibration analysis and stationary random excitations. Manning and Lee (1968), Mercer,

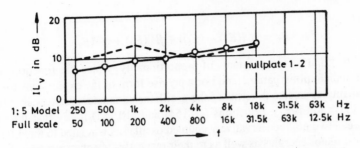

.Fig. 10.30.   Inertion loss at two of the hull plate sections with applied damping treatment. (– – –) predicted by SEA, (———) measured (Jensen, 1977).

Rees and Fahy (1971) have developed power flow relations between coupled oscillators under shock and transient excitations similar to the SEA power flow relations. More recently Powell and Quartararo (1987) used modified SEA relations to predict time varying vibration envelopes. They included as additional term to the power balance equations to account for the time varying vibrational energy and derived a set of first order differential equations and showed that broad band response envelopes could be predicted quite well. In their analysis they used the steady-state dissipation and coupling loss factors. But in transient vibration, the response energy and the power flow are time varying and it has been shown by Lai and Soom (1990a) that time varying coupling loss factors and internal loss factors are needed to accurately model time-varying vibration envelopes of complex vibrating systems. Lai and Soom (1990b) have derived a set of energy balance equations analogous to the conventional SEA power balance equations to model time-variant transient vibrations. It has also been shown by them that the conventional SEA loss and coupling loss factors are also applicable in the transient vibration analysis.

Starting from the energy balance relations between two coupled oscillators for transient excitations, where it has been shown that the time integrated energy flow between two conservatively coupled oscillators is related to the time integrated response energy of the oscillators, they have extended the energy flow and energy balance relations to coupled multimodal subsystems. They have verified their results by conducting experiments on a system consisting of two coupled aluminium plates. They have further shown that the results derived by them can be used to estimate experimentally conventional SEA parameters like internal and coupling loss factors using transient excitations on the system.

10.11.2 Non-conservative Coupling

The power flow relations of a vibrating system have been conventionally derived for conservatively coupled subsystems. In some practical structural vibration problems, the influence of coupling damping between substructures connected to each other on power flow have to be considered. These effects of coupling damping on power flow relations have been recently investigated by Sun et al (1987a), Fahy and Yao (1987) and Sun and Ming (1988). When coupling damping is included between the two oscillators it has been shown by them that the power flow from oscillator 1 to 2 and vice versa are given by

$$P_{12} = \alpha \left\{ \langle E_1 \rangle - \langle E_2 \rangle \right\} + \beta \langle E_1 \rangle + \gamma \langle E_2 \rangle \qquad (10.154)$$

and

$$P_{21} = \alpha \left\{ \langle E_2 \rangle - \langle E_1 \rangle \right\} + \beta' \langle E_2 \rangle + \gamma' \langle E_1 \rangle \qquad (10.155)$$

where α, β, γ and γ' are constants of proportionality, which are functions of the oscillator and coupling parameters. For a precise form of the expressions for these proportionality constants in terms of the oscillator parameters readers are referred to Fahy and Yao (1987). Equations (10.154) and (10.155) imply that the energy flow between two non-conservatively coupled oscillators is related to the difference between oscillator energies as well as to their respective absolute energies. Further, the energy flow in different directions is not equal. When the coupling damping is very small compared with other internal loss factor components the constants β, γ, β' and γ' are much less than α and one can use the power flow relations derived for conservative coupling. When the coupling damping is of the same order of magnitude as the structural and acoustic radiation damping it has the effect of increasing the internal loss factors of the coupled oscillators. Fahy and Yao (1987) have also shown in their analysis that coupling damping reduces the energy of the indirectly driven subsystem. For the case of two coupled oscillators, equipartition of energy does not form an upper limit on oscillator energy ratios as in the case of conservatively coupled oscillators.

Sun et al (1987b) conducted a series of experiments on a thick plate bolted to a thin plate in an L shape to verify the theoretical power flow relations with coupling damping. The ratio of energy stored in the plates was measured with and without different gaskets between the joint and the results were compared with those calculated from energy balance equations for non-conservatively coupled structures and the agreement between the two was found to be good.

10.11.3   Strongly Coupled Systems

One of the basic assumptions of SEA as originally described by Lyon and Maidanik (1962), that the sets of models which determine the power flow between various dynamical systems should be weakly coupled has often been overlooked and the SEA method is applied in many situations irrespective of the degree of coupling between subsystems. Fahy (1970) has addressed this problem and showed that the application of SEA without taking into account the coupling strength between the subsystems would result in some discrepancies. Newland (1966) has presented a perturbation analysis to consider the higher order effects due to strong coupling between the coupled subsystems. Davies (1972) has derived the power flow equations in the random vibration analysis of some idealized strongly coupled dynamic systems. Gulzia and Price (1977) have derived the power flow relations for a set of strongly coupled oscillators and considered the interaction between a mechanical structure and an acoustic field and showed that the SEA approach can be used even in the case of strongly coupled subsystems provided that it is applied to reverberant fields with high modal density. They devised an experiment consisting of a mechanically excited plate immersed in a tank of water representative of strong coupling and showed a good measure of agreement between the predicted value of radiation resistance by SEA calculations and the experimentally measured values. Chandiramani (1978) has derived expressions for the oscillator energies in a system of coupled oscillators excited at resonance by random sources and considered the problem of transition from weak to strong coupling between the oscillators. Smith (1979) has also addressed a similar problem and developed statistical methods to consider the effect of variable strength of coupling between the subsystems in SEA and the transition from weak to strong coupling. More recently Keane and Price (1987) have derived exact energy flow relations between strongly coupled subsystems with low modal density. They have also shown how the original SEA power flow relations could be recovered from their general results for the case of weak coupling and high modal density. Dimitriadis and Pierce (1988) have derived an exact analytical solution for the vibration transmission between two strongly coupled plates in a L-shaped configuration in terms of the uncoupled component modes of the individual plates and derived expressions for power flow and vibrational energies of the plates. Keane and Price (1989) have extended their formulation of the power flow equations inclusive of the strong coupling assumptions to the application of SEA for the vibration analysis of periodic structures. They have combined the theory of wave propagation in periodic structures and the SEA method and showed that the essential results of SEA could be retained. The combined SEA-wave propagation model takes into account the grouping of the natural frequencies in the free wave propagation zones or pass bands of the periodic element of the periodic structure. It has also been shown that the model can take care of the important case of structures with minor deviations from periodicity or of almost periodic structures. They have outlined the applications of their extended model to ship structures, off-shore platforms and satellite structures.

10.12 Mean Square Velocity-Stress Relations

We have seen that space averaged mean square vibration velocities can be estimated for coupled multimodal systems in the high frequency range by use of SEA procedures. But in the design of structures for strength and for service against fatigue the dynamic stress levels are of importance. Hence if the mean square vibration levels can be related to the dynamic stress levels then SEA will be of even greater utility. Hunt (1960) was one of the first to develop a relationship between kinetic energy, velocity and dynamic strain in beams and plates. Ungar (1962) has also developed similar relations. Stearn (1970, 1971) and Fahy (1971) have developed a theoretical analysis for establishing a simple relationship between mean square vibrational velocities and dynamic stresses for homogeneous structures like beams, plates and cylinders subjected to multimode frequency excitations. They used the concept of bending waves in a reverberant field in their analysis. Recently Norton and Fahy (1988) have conducted a number of experiments to establish the correlation of dynamic stress and strain with pipe wall vibrations for small diameter cylinders. Particular attention has been paid to the statistics of the distributions of the ratio of stress/strain into the cylinder wall velocity and comparisons have been made between predicted and actual rms stress/strain levels. They have also showed that the relationships can be used even in the regions of stress concentration.

It has been shown that for different forms of vibratory motions like flexure, torsion etc. that the mean square dynamic stress is directly related to the mean square velocity by the simple relation

$$\frac{\langle \sigma^2 \rangle}{\langle v^2 \rangle} = 1.61 \, k \, C_L^2 \rho^2 \tag{10.156}$$

where $\langle \sigma^2 \rangle$ and $\langle v^2 \rangle$ represent the space and time averaged mean square stress and velocity, and $C_L^2 = E/\rho$, with E and ρ being respectively the Young's modulus and density of the material and K is a proportionality constant, depending on the type of motion and system geometry. A Poisson's ratio of 0.3 has been assumed in the expression (10.156). A similar relation is obtained for the mean square strain as

$$\frac{\langle \varepsilon^2 \rangle}{\langle v^2 \rangle} = \frac{K}{C_L^2} \qquad (10.157)$$

where ε is the dynamic strain. The constant K varies over a small range near unity for normal stress situations. If stress concentration is present experimental results (Norton and Fahy, 1988) show that K could vary between 3 and 20.

REFERENCES

Bies, D.A. and S. Hamid, 1980. In-situ determination of loss and coupling loss factor by the power injection method. *J. Sound Vib.* 70(2): 187–204.

Bolotin, V.V., 1963. On the density of the distribution of natural frequencies of thin elastic shells. *J. Appl. Math. Mech.* 27(2): 538–543.

Brown, K.T., 1984. Measurment of modal density: An improved technique for use on lightly damped structures. *J. Sound and Vib.* 96(1): 127–132.

Brown, K.T. and B.L. Clarkson, 1984. Average loss factors for use in statistical energy analysis, *Vibration Damping Workshop, Wright-Patterson Air Force Base,* Ohio, USA.

Brown, K.T. and M.P. Norton, 1985. Some comments on the experimental determination of modal densities and loss factors for statistical energy applications. *J. Sound Vib.* 102(4): 588–594.

Chandiramani, K.L., 1978. Some simple models describing the transition from weak to strong coupling in statistical energy analysis. *J. Acoust. Soc. Am.* 63(4): 1081–1083.

Chandiramani, K.L., S.E. Widnall, R.H. Lyon and P.A. Franken, 1966. Structural response to inflight acoustic and aerodynamic environments. *BBN Report 1417.* Cambridge, Massachussets.

Clarkson, B.L. and K.T. Brown, 1985. Acoustic radiation damping. *J. Vib. Acous. Str. and Rel. Des.* 107: 357–360.

Clarkson, B.L. and R.J. Pope, 1981. Experimental determination of modal densities and loss factors of flat plates and cylinders. *J. Sound Vib.* 77(4): 535–549.

Clarkson, B.L. and R.J. Pope, 1983. Experimental determination of vibration parameters required in the statistical energy analysis method. *J. Vib. Acous. Str. Rel. Des.* 105: 337–344.

Clarkson, B.L. and M.F. Ranky, 1983. Modal density of honeycomb plates. *J. Sound. Vib.* 91(1): 103–118.

Clarkson, B.L. and M.F. Ranky, 1984. On the measurement of the coupling loss factor of structural connections. *J. Sound. Vib.* 94(2): 249–261.

Clarkson, B.L., R.J. Cummins, D.Eaton and J. Vessaz, 1981. Prediction of high frequency structural vibrations using statistical energy analysis (SEA). *European Space Agency Journal.* 5: 137–150.

Cremer, L., M. Heckl and E.E. Ungar, 1973. *Structure borne sound.* Springer Verlag. Heidelberg, Berlin.

Crocker, M.J. and A.J. Price, 1969. Sound transmission using statistical energy analysis. *J. Sound. Vib.* 9(3): 469–486.

Davies, H.G., 1973. Random vibration of distributed systems strongly coupled at discrete points. *J. Acous. Soc. Am.* 51: 507–515.

Dimitriadis, E.K. and A.D. Pierce, 1988. Analytical solution for the power exchange between strongly coupled plates under random excitation: A test of statistical energy analysis concepts. *J. Sound. Vib.* 121(3): 397–412.

Erickson, L.L., 1969. Modal densities of sandwich panels: Theory and experiments. *The Shock and Vib. Bull.* 39(3): 1–16.

Fahy, F.J., 1970. Response of a cylinder to random sound in the contained fluid. *J. Sound. Vib.* 13: 171–194.

Fahy, F.J., 1971. Statistics of acoustically induced vibration. *Proc. 7th Int. Cong. on Acous.* Budapest, 561–564.

Fahy, F.J., 1974. Statistical energy analysis—a critical review. *Shock. Vib. Digest.* 6: 14–33.

Fahy, F.J., 1982. Statistical energy analysis. *Noise and Vibration* Eds. White, R.G. and G.J. Walker. Chichester: Ellis Harwood Ltd.

Fahy, F.J. and D. Yao, 1987. Power flow between non-conservatively coupled oscillators. *J. Sound. Vib.* 114(1): 1–11.

Ferguson, N.S. and B.L. Clarkson, 1986. The modal density of honeycomb shells. *J. Vib. Acous. Str. Rel. Des.* 108: 399–404.

Franken, P.A. and E.M. Kerwin, 1958. Methods of flight vehicle noise predication. *WADC–TR–58–343.*

Goyder, H.G.D., 1981. Frequency response testing a nonlinear structure in a noisy environment with a distorting shaker. *Spring. Conf. Inst. of. Acous.* 37–40.

Gulzia, C. and A.J. Price, 1977. Power flow between strongly coupled oscillators. *J. Acous. Soc. Am.* 61: 1511–1515.

Hart, F.D. and K.C. Shaw, 1971. Compendium of modal densities for structures. *NASA* CR. 1773.

Heckl, M. 1962. Vibration of point-driven cylindrical shells. *J. Acous. Soc. Am.* 32(10): 1553–1557.

Hodges, C.H. and J. Woodhouse, 1986. Theories of noise and vibration transmission in complex structures. *Rep. Prog. Phy.* 49: 107–170.

Hunt, F.V., 1960. Stress and strain limits on the attainable velocity in mechanical vibrations. *J. Acous. Soc. Am.* 32(9): 1123–1128.

Jensen, J.O., 1977. Calculation of structure borne noise transmission in ships using statistical energy analysis approach. *Proc. Int. Symp. on Ship Board Acoustics.* Ed. J.H. Janssen. Elsevier Scientific Publ. Co. Amsterdam.

Kamaraju, P. 1977. Some studies on acoustic transmissibility of structures. Ph.D. Thesis. I.I.T. Madras.

Keane, A.J. and W.G. Price, 1987. Statistical energy analysis of strongly coupled system. *J. Sound. Vib.* 117(2): 363–386.

Keane, A.J. and W.G. Price, 1989. Statistical energy analysis of periodic structures, *Proc. Roy. Soc. Lond. A.* 423: 331–360.

Keswick, P.R. and M.P. Norton, 1987. A comparison of modal density measurement techniques. *Appl. Acous.* 20: 137–153.

Kihlman, T. and J. Plunt, 1977. Prediction of noise levels in ships. *Proc Int. Symp. on Ship Board Acoustics.* 299–317. Ed. J.H. Janssen. Elsevier Scientific Publ. Co. Amsterdam.

Lai, M.L. and A. Soom, 1990a. Prediction of transient vibration envelopes using statistical energy analysis techniques. *J. Vib. Acous.* 112: 127–137.

Lai, M.L. and A. Soom, 1990b. Statistical energy analysis for the time-integrated transient response of vibrating systems. *J. Vib. Acous.* 112: 206–213.

Langley, R.S., 1989. A general derivation of the statistical energy analysis equations for coupled dynamic systems. *J. Sound. Vib.* 135(3): 499–508.

Lyon, R.H., 1969. Statistical analysis of power injection and response in structures and rooms. *J. Acous. Soc. Am.* 45(3): 545–565.

Lyon, R.H., 1975. *Statistical Energy analysis of dynamical systems: Theory and application.* M.I.T. Press.

Lyon, R.H. and E. Eichler, 1964. Random vibration of connected structures. *J. Acous. Soc. Am.* 36: 1344–1354.

Lyon, R.H. and G. Maidanik, 1962. Power flow between linearly coupled oscillators. *J. Acous. Soc. Am.* 34: 623–639.

Lyon, R.H. and G. Maidanik, 1964. Statistical methods in vibration analysis. *AIAA J.* 2(6): 1015–1024.

Lyon, R.H. and T.D. Scharton, 1965. Vibrational energy transmission in a three element structure. *J. Acous. Soc. Am.* 38: 253–261.

Maidanik, G., 1962. Response of ribbed panels to reverberant acoustic fields. *J. Acous. Soc. Am.* 34: 809–862.

Maidanik, G., 1977. Some elements in statistical energy analysis. *J. Sound. Vib.* 52: 171–191.

Manning, J.E. and K. Lee, 1968. Prediction of mechanical shock transmission. *Shock and Vib. Bull.* 37(4): 65–70.

Manning, J.E., R.H. Lyon and T.D. Scharton, 1966. Transmission of sound and vibration to a shroud-enclosed spacecraft. NASA-CR-81688.

Manning, J.E. and G. Maidanik, 1964. Radiation properties of cylindrical shells. *J. Acous. Soc. Am.* 36: 1691–1698.

Mead, D.J., 1972. The damping properties of elastically supported sandwich plates. *J. Sound. Vib.* 24: 275–295.

Mercer, C.A., P.L. Rees and F.J. Fahy, 1971. Energy flow between two weakly coupled oscillators subject to transient excitation. *J. Sound. Vib.* 15(3): 373–379.

Miller, V.R. and L.L. Faulkner, 1983. Prediction of aircraft interior noise using the statistical energy analysis method. *J. Vib. Acous. Str. and Rel. Des.* 105: 512–518.

Newland, D.E., 1966. Calculation of power flow between coupled oscillators. *J. Sound. Vib.* 3: 262–276.

Norton, M.P., 1989. *Fundamentals of Noise and Vibration analysis for Engineers.* Cambridge University Press. London.

Norton, M.P. and F.J. Fahy, 1988. Experiments on the correlation of dynamic stress and strain with pipe wall vibrations for statistical energy applications. *Noise Control Engg.* 30(3): 107–111.

Norton, M.P. and R. Greenhalgh, 1986. On the estimation of loss factors in lightly damped pipeline systems. Some measurement techniques and their limitation. *J. Sound. Vib.* 105(3): 397–423.

Norton, M.P. and P.R. Keswick, 1987. Loss and coupling loss factors and coupling damping in non-conservatively coupled cylindrical shells. *Proc. Internoise*: pp. 651–654. Bejing. China.

Pocha, J.J., 1977. Acoustic excitation of structures analyzed by the statistical energy method. *AIAA J.* 15(2): 175–181.

Pope, L.D., 1974. Energies of resonant structures and reverberant acoustic fields. *Shock and Vib. Digest.* 6(7): 1–11.

Powell, R.E. and L.R. Quartararo, 1987. Statistical energy analysis of transient vibrations. ASME winter meeting on Statistical Energy Analysis. Boston: 535–547.

Ranky, M.F. and B.L. Clarkson, 1983. Frequency average loss factors of plates and shells. *J. Sound. Vib.* 89(3): 309–323.

Richards, E.J. 1981. On the prediction of impact noise, III: energy accountancy in industrial machines. *J. Sound. Vib.* 76(2): 187–232.

Richards, E.J. and A. Lenzi, 1984. On the prediction of impact noise IV: the structural damping of machinery. *J. Sound. Vib.* 97(4): 549–586.

Rennison, D.C. and M.K. Bull, 1977. On the modal density and damping of cylindrical pipes. *J. Sound. Vib.* 54(1): 39–53.

Sawley, R.J., 1969. The evaluation of a shipboard noise and vibration problem using statistical energy analysis. *ASME Symposium on Stochastic Processes in Dynamical Systems.* 63–73.

Scharton, T.D., 1971. Response of a ring–stringer stiffened cylinder to acoustical and mechanical excitation. NASA-CR-103172.

Scharton, T.D. and R.H. Lyon, 1968. Power flow and energy sharing in random vibration. *J. Acous. Soc. Am.* 43: 1332–1343.

Scharton, T.D. and T.M. Yang, 1967. Statistical energy analysis of vibration transmission into an instrument package. *SAE paper. no. 670876.*

Smith, P.W., 1962. Response and radiation of structural modes excited by sound. *J. Acous. Soc. Am.* 34: 623–639.

Smith, P.W., 1979. Statistical models of coupled dynamical systems and the transition from weak to strong coupling. *J. Acous. Soc. Am.* 65(3): 695–698.

Smith, P.W., and R.H. Lyon, 1965. *Sound and Structural Vibration.* NASA CR-160.

Stearn, S.M., 1970. Spatial variation of stress, strain and acceleration in structures subject to broad frequency band excitation. *J. Sound. Vib.* 12(1): 85–97.

Stearn, S.M., 1971. The concentration of dynamic stress in a plate at a sharp change of section. *J. Sound. Vib.* 15(3): 353–365.

Stimpson, G.J., J.C. Sun and E.J. Richards, 1986. Predicting sound power radiation from built-up structures using statistical energy analysis. *J. Sound. Vib.* 107(1): 107–120.

Sun, J.C., N. Lalor and E.J. Richards, 1987a. Power flow and energy balance of non-conservatively coupled structures I: Theory. *J. Sound. Vib.* 112(2): 321–330.

Sun, J.C., L.C. Chow, N. Lalor and E.J. Richards, 1987b. Power flow and energy balance of non-conservatively coupled structures II: Experimental verification of theory. *J. sound. Vib.* 112(2): 331–343.

Sun, J.C. and R.S. Ming, 1988. Distributive relationships of dissipated energy by coupling damping in non-conservatively coupled structures. *Proc. Internoise. Avignon, France* pp. 323–326.

Szechenyi, E., 1971. Modal densities and radiation efficiencies of unstiffened cylinders using statistical methods. *J. Sound. Vib.* 19(1): 65–81.

Ungar, E.E., 1962. Maximum stresses in beams and plates vibrating at resonance. *J. Engg. for Ind.* 84(1): 149–155.

Ungar, E.E., 1967. Statistical energy analysis of vibrating systems. *J. Engg. for Ind.* 89(4): 626–632.

Ungar, E.E., 1971. Damping of panels. In *Noise and Vibration control,* Ed. L.L. Beranek, McGraw-Hill.

Ver I.L. and C.I. Holmer, 1971. Interaction of sound waves with solid structures. In *Noise and Vibration control,* Ed. L.L. Beranek, McGraw Hill.

White, R.G., 1971. Evaluation of the dynamic characteristics of structures by transient testing. *J. Sound. Vib.* 15: 147–167.

Wilby, J.F. and T.D. Scharton, 1974. Acoustic transmission through a fuselage side wall. *NASA-CR* 132602.

Wilkinson, J.P.D., 1968. Modal densities of certain shallow structural elements. *J. Acous. Soc. Am.* 43: 245–251.

Woodhouse, J., 1981a. An introduction to statistical energy analysis of structural vibrations. *Appl. Acous.* 14: 455–469.

Woodhouse, J., 1981b. An approach to the theoretical background of statistical energy analysis applied to structural vibration. *J. Acous. Soc. Am.* 69: 1695–1709.

Wöhle, W., Th. Beckmann and H. Schreckenbach, 1981. Coupling loss factors for statistical energy analysis of sound transmission at rectangular slab joints, Parts I and II. *J. Sound. Vib.* 77(3): 323–344.

Author Index

Subject Index